EUROPEAN MINERALOGICAL UNION
NOTES IN MINERALOGY

Series editor: Giovanni Ferraris

Volume 10

ION PARTITIONING IN AMBIENT-TEMPERATURE AQUEOUS SYSTEMS

UNIVERSITY TEXTBOOK

Edited by
M. Prieto and H. Stoll

Mineralogical Society

Published by the European Mineralogical Union and the
Mineralogical Society of Great Britain & Ireland
London, 2010

The publication of this textbook and the organization of the related school were sponsored by:
European Mineralogical Union

On the front cover: Electron micrograph of a cross section through an experimentally grown crystal of composition $(Ba,Sr)SO_4$. The dark regions are Sr-rich while the paler regions are Ba-rich. The crystal shows both oscillatory and sector zoning of Ba and Sr. The crystal is ~150 mm long. (Image reproduced with permission of Nature Publishing Group from a paper by A. Putnis *et al.* (1992): Experimentally produced oscillatory zoning in the $(Ba,Sr)SO_4$ solid solution. *Nature*, **358**, 743–745.

ISSN: 1417 2917
ISBN: 978-0-903056-26-7

Published by the European Mineralogical Union and the Mineralogical Society of Great Britain & Ireland (12, Baylis Mews, Amyand Park Road, Twickenham TW1 3HQ, UK)
Printed by Burlington, Foxton, UK

The EMU Notes in Mineralogy Series

Published volumes

Volume	*Year*	*Editors*	*Title*
1	1997	S. Merlino	*Modular aspects of minerals*
2	2000	D.J. Vaughan R. Wogelius	*Environmental mineralogy*
3	2001	C.A. Geiger	*Solid solutions in silicate and oxide systems*
4	2002	C. Gramaccioli	*Energy modelling in minerals*
5	2003	D.A. Carswell R. Compagnoni	*Ultrahigh pressure metamorphism*
6	2004	A. Beran E. Libowitzky	*Spectroscopic methods in mineralogy*
7	2005	R. Miletich	*Mineral behaviour at extreme conditions*
8	2010	F.E. Brenker G. Jordan	*Nanoscopic approaches in Earth and Planetary Sciences*
10	2010	M. Prieto H. Stoll	*Ion-partitioning in ambient-temperature aqueous systems*

Forthcoming volumes

9	2010	G. Christidis	*Advances in the Characterization of Industrial Minerals*

Copies of the EMU Notes (volumes 1–7) are distributed in Europe by the larger member societies of the European Mineralogical Union:

Società Italiana di Mineralogia e Petrologia:
www.socminpet.it

Mineralogical Society of Great Britain & Ireland:
www.minersoc.org

Société Française de Minéralogie et de Cristallographie:
www.sfmc-fr.org

and by the Secretary of the EMU, Prof. Herta S. Effenberger:
www.univie.ac.at/Mineralogie/EMU

in America by the Mineralogical Society of America:
www.minsocam.org

Institutional orders as well as individual requests from outside Europe and America should be sent also be sent to the the Secretary of the EMU, Prof. Herta S. Effenberger:
www.univie.ac.at/Mineralogie/EMU

For volumes 8 onwards, the Mineralogical Society of Great Britain acts as co-publisher and copies may be ordered from www.minersoc.org

or from:
Mineralogical Society
12 Baylis Mews, Amyand Park Road
Twickenham TW1 3HQ
UK
E-mail: admin@minersoc.org
Tel. +44 (0)20 8891 6600
Fax: +44 (0)20 8891 6599

Contents

Preface: Ion partitioning in ambient-temperature aqueous systems

On the surface of the Earth, the intermingling of water and minerals gives rise to a diverse suite of reactions that determine the purity of water we drink, the fate of contaminants we emit, and the composition of minerals and biominerals that we use to interpret past environmental conditions from the sediment archive. Human societies have ubiquitous exposure to the outcome of these mineral–water reactions. Understanding in detail the ion partitioning in mineral–water interactions is of fundamental importance to geochemical studies and ultimately to society. The solid-solution properties of minerals are a significant part of the complexity, and also the importance, of these ion-partitioning reactions. Natural minerals always contain a certain proportion of trace elements in solid solution. These trace elements, precisely because of their rarity, often have a disproportionately large impact on living organisms as is the case for familiar toxic metals such as As and Cd. A clear understanding of ion partitioning behaviour is therefore essential for environmental objectives such as scavenging heavy metals from solution, remediating contamination in soils, or ensuring safe, long-term storage of anthropogenic CO_2 or radionuclides in geological reservoirs. Materials science has also taken a new look at the role of trace-element and ion partitioning in regulating biomineralization. Finally, the last several decades have seen a surge in interest in reconstructing past climate and environmental conditions from the sediment archive. An accurate interpretation of ion partitioning is essential to the correct interpretation of records from diverse systems like stalagmites, corals, or shells of marine foraminifera. Given this wide range of applications for ion partitioning, it is fortunate that theoretical and thermodynamic frameworks for modelling ion partitioning have advanced significantly in the last decade. We believe that it is an opportune time to convene experts on ion partitioning from a range of perspectives, from theoretical to applied, to exchange knowledge across these topics and through this exchange, maximize the advances that have been made in the discipline. We are pleased to be able to convene these experts in person, at the European Mineralogical School in Oviedo in June 2010 to share these advances with each other and with the next generation of geochemists. It is our hope that this book will serve most crucially as a bridge through which researchers in one aspect of ion partitioning will be able to productively venture into complementary systems and models to better solve their research goals and perhaps be inspired with new research questions.

The chapters in the book are arranged in thematic groups. The book begins with thermodynamic fundamentals and computer modelling of geochemical records. The first chapter by Manuel Prieto describes the equilibrium thermodynamics of ion partitioning for ionic substitutions with increasing degrees of complexity. Variations of the equilibrium distribution coefficients with changes in the background aqueous solution or in the composition of the solid as well as the methods are discussed in depth. Finally, the concept of supersaturation in systems involving solid solutions links this chapter with the following one, in which Andrew Putnis discusses the kinetic and mechanistic factors that influence the effective distribution of substituting ions under non-equilibrium conditions. Factors such as supersaturation, growth mechanisms, surface features, presence of organic molecules, *etc.* are illustrated with the help of a number of case studies. In the third chapter Dmitrii Kulik highlights some typical features of models of the equilibrium partitioning of inorganic ionic species between aqueous electrolyte and solid-solution phases. The emphasis is not so much on definitions and equations, but rather on providing some guidance about the methods and codes for geochemical thermodynamic modelling, and on showing how to use them in order to help understand ion partitioning in complex scenarios.

The subsequent section describes four examples of ion partitioning in low-temperature geochemical systems. Michael Bottcher and Martin Dietzel focus on selected binary carbonate and sulphate solid solution aqueous solution systems and give a presentation of relevant current topics related to low temperature metal ion partitioning during complex precipitation and dissolution behaviour. Christine Putnis and Lurdes Fenández-Díaz provide vivid examples of ion partitioning during low-temperature replacement reactions, underscoring the importance of the interaction not between the bulk aqueous solution but the solution in direct contact with the mineral surface. Christophe Monnin and Guilhem Hoareau give an account of the problems involved in characterizing chemical equilibria in the marine environment and focus on the development of an accurate model for seawater-type solutions, which is then applied to the analysis of the saturation state of seawaters and marine sediment pore fluids. Ian Fairchild and Adam Hartland describe ion partitioning between karst waters and host limestone and stalagmites, highlighting the links between ion partitioning and hydrology which can be exploited to reconstruct past changes in climate and hydrological balance. Speleothems also have potential to track broader biogeochemical cycles *via* substitution of ions such as sulphate or phosphate.

The application of ion partitioning to solve problems of environmental contamination is covered by three chapters. Athanasios Godelitsas and José M. Astilleros explain how toxic metals can be removed from waters and soils by sorption and precipitation of metal-bearing solid solutions on the surface of minerals and how these processes can control both the dispersion and accumulation of these pollutants and hence their geoavailability and bioavailabity in near-surface environments. Eric Oelkers and Sigurdur Gislason describe how carbonation of silicates may be effective for permanent sequestration of anthropogenic CO_2 and how carbonate mineral formation is favoured by the presence of reactive divalent-metal silicate minerals, in particular basalts and ultra-mafic rocks. Dirk Bosbach focuses on solid solutions that are relevant to long-term safety in nuclear-waste repositories. In particular, the structural incorporation of trivalent actinides and lanthanides into calcite and the incorporation of radium into barite are discussed in detail.

The final two chapters in the book describe application of ion partitioning in marine biogenic carbonates to infer past changes in climate and ocean conditions. Corals which grow in the sunlit ocean, have been a key target for reconstructing tropical ocean temperatures in the past. Foraminifera and coccolithophorids, the dominant carbonate producers in the open ocean, provide the longest archive of ocean conditions with seafloor sediments dating back to the Mesozoic. Anne Cohen and Glenn Gaetani compare element partitioning during coral biomineralization with abiogenic aragonite precipitation from highly supersaturated seawater. They illustrate how simple evolution of the aragonite saturation state and Rayleigh distillation of trace element composition in a closed seawater calcification reservoir is able to explain the variation in crystal habits and the evolution in element/Ca ratios in corals as a function in temperature without invoking any additional purely biological effects. From this a robust multi-element system can be used to calculate both growth temperature and degree of calcite production in corals. Heather Stoll reviews the main 'proxies' from trace element composition of foraminifera and coccoliths. In some cases ion partitioning can be assumed to be constant and temporal or spatial changes in the ratio of trace element/Ca in the ocean can be inferred. In other cases thermodynamic and/or biological effects on ion partitioning give rise to compositions dependent on ocean temperature, ocean carbonate saturation state or *pH*, or the physiology of the organism.

It is worth noting that the terminologies and parameters used to refer to ion partitioning vary from some chapters to others. Even though this inhomogeneity can be misleading, we have decided to keep the terms as initially proposed by the authors because, in each case, these terms are the most familiar for the corresponding scientific community. A commonly accepted terminology does not seem to exist, which is not necessarily a problem as far as there is a specification of the terms used in each chapter. Moreover, a general presentation of the different terminologies and definitions is given in chapter 1.

This volume would not have been possible without the cooperation and help of the authors, all of whom we thank. We thank Roberta Oberti and Fernando Gervilla for their enthusiastic encouragement and support. We are very grateful to Giovanni Ferraris for his extensive help in the review of the preliminary manuscripts. Finally, special thanks go to Kevin Murphy who made the main effort to complete this volume in record time.

Manuel Prieto and *Heather Stoll*
Oviedo, Spain
May, 2010

EMU Notes in Mineralogy, Vol. 10 (2010), Chapter 1, 1–42

Thermodynamics of ion partitioning in solid solution–aqueous solution systems

MANUEL PRIETO

Department of Geology, University of Oviedo, C/ Arias de Velasco, s/n, 33005 Oviedo, Spain
e-mail: mprieto@geol.uniovi.es

While the study of ion partitioning in solid solution–aqueous solution (SS–AS) systems has garnered scientific attention for decades, the terminologies used in ion-partitioning studies are often inconsistent. Moreover, the equilibrium thermodynamics underlying the partitioning process is often disregarded or misunderstood, although this background is indispensable in interpreting effective (non-equilibrium) distribution of substituting ions between minerals and their growth environments. Here, after an introduction to the different terminologies and definitions, the equilibrium thermodynamics of ion partitioning is reviewed for the simplest case of solid solutions with isostructural end-members and one mixing position per formula unit. Variations of the equilibrium distribution coefficients with changes in the background solution or in the composition of the solid as well as the methods to determine equilibrium values are discussed in depth. The simplifications when one of the substituting ions occurs in trace amounts (dilute solid solutions) and the effect of temperature are introduced in a later section, followed by a discussion of the thermodynamic treatment of more complex SS–AS systems: ion partitioning involving solids with more than one mixing position per formula unit, solid solutions with non-isostructural end-members, heterovalent substitutions, *etc.* Finally, the concept of supersaturation in SS–AS systems is introduced as a fundamental parameter to interpret non-equilibrium partitioning.

1. Introduction: terminology and definitions

While the distribution of substituting ions between minerals and their growth environments has received increased attention in the last few decades, the terminologies used in partitioning studies are often inconsistent. The terms 'partition coefficient', 'exchange coefficient', 'distribution coefficient', 'fractionation factor' together with the symbols D, K_D, and K_d are used indiscriminately to stand for various partitioning relations, and the definitions of these terms and symbols vary from one publication to another. In fact, more than a decade ago, a number of authors discussed in a joint paper the need for a common terminology (Beattie *et al.*, 1993). Unfortunately, such a commonly accepted terminology still does not seem to exist, requiring specification of the concepts that one is using.

The simplest expression that relates the distribution of an impurity ion between coexisting phases (*e.g.* α and β) is given by the ratio (C^{α}/C^{β}) between its concentrations in both phases. According to the classical Berthelot-Nernst distribution law (Berthelot,

DOI: 10.1180/EMU-notes.10.1

1872; Nernst, 1891), this ratio is expected to be a constant at equilibrium that depends on the temperature and pressure but not on the concentration of the impurity ion as long as it is very small. However, when dealing with varied solid-solution ranges, it is generally convenient to normalize this simple partition ratio by the value of the partition ratio for a complementary substituting ion, in agreement with the classical expression introduced by Henderson & Kracek (1927). In the geochemical literature, this normalized ratio is variably referred to as distribution coefficient, exchange coefficient, partition coefficient, or fractionation factor. In the case of a solid solution–aqueous solution (SS–AS) system in which B^{n+} and C^{n+} ions substitute for each other in the solid structure, such a normalized distribution coefficient is given by

$$D_{B/C} = \frac{C^s_{B^{n+}}}{C^s_{C^{n+}}} \bigg/ \frac{C^{aq}_{B^{n+}}}{C^{aq}_{C^{n+}}}, \tag{1}$$

where C^s and C^{aq} stand for the concentrations of the substituting ions in the solid (s) and aqueous (aq) phase, respectively. When $D_{B/C} > 1$, there is a greater B^{n+} concentration in the solid relative to its proportion in the aqueous phase; for $D_{B/C} < 1$, the relative B^{n+} concentration is smaller. When $D_{B/C} = 1$, the relative concentrations are similar in both phases.

Equation 1 does not presuppose that equilibrium between the solid and the aqueous solution has been reached. Because factors such as supersaturation and the solid-phase growth rate have an effect on the effective partitioning (*e.g.* Lorens, 1981; Kornicker *et al.*, 1991; Tesoriero & Pankow, 1996; Prieto *et al.*, 1997; Rimstidt *et al.*, 1998; Watson, 2004; Katsikopoulos *et al.*, 2008a), it is convenient to distinguish between equilibrium (*eq*) and measured or effective distribution coefficients. Although many studies (Beattie *et al.*, 1993) use distribution coefficients calculated based on a weight fraction, a molar basis provides the closest reflection of the thermodynamics controlling the ion partitioning in SS–AS systems and therefore will be applied in this work. Finally, the use of aqueous activities instead of aqueous concentrations in equation 1 is advantageous because the composition of natural waters can vary widely. As we will discuss in section 2.2, if the substituting ions have similar properties, the change in the background solution will affect the activity coefficients of both elements in the same way. In such a case, the distribution coefficient will remain fairly constant even with considerable changes in the aqueous solution composition (McIntire, 1963). However, this is not always the case, and a distribution coefficient based on aqueous activities is more practical for normalizing partitioning values for different aqueous solutions. Throughout this chapter, we will use the symbol $D_{\{B/C\}}$ to refer to a distribution coefficient calculated on an aqueous activity basis, and this type of coefficient will be used in most cases, unless otherwise stated. In their clarifying reviews on SS–AS systems, Glynn and co-workers mostly used $D_{\{B/C\}}$ (Glynn & Reardon, 1990; Glynn, 1991; an exception can be seen in Glynn *et al.*, 1990), which was also applied in the solid-solution tool of the 'popular' speciation program PHREEQC (Parkhurst & Appelo, 1999). In the next three sections, we consider binary solid solutions of isostructural end-members with a simple *BA*-type stoichiometry to simplify the expressions. The simplifications when one of the substituting ions occurs in trace amounts are introduced in section 5, followed by a discussion of the thermodynamic treatment of more complex SS–AS systems in section 6. Finally, the concept of supersaturation in SS–AS systems is introduced in section 7, which links with Chapter 2.

2. Equilibrium distribution coefficients in SS–AS systems

2.1. General relationships

In a hypothetical binary solid solution (*B*,*C*)*A*, ions B^{n+} and C^{n+} substitute randomly for each other in specific sites of a given crystal structure. Such a solid phase can be seen as consisting of two components, *BA* and *CA*, and the distribution of the substituting ions between the aqueous (*aq*) and solid (*s*) phases can be described by applying the law of mass action to the exchange reaction:

$$CA_{(s)} + B^{n+}_{(aq)} \longrightarrow BA_{(s)} + C^{n+}_{(aq)} \qquad (2)$$

for which the equilibrium condition can be written in the form:

$$K_{B/C} = \frac{a_{BA}\{C^{n+}\}}{a_{CA}\{B^{n+}\}} \qquad (3)$$

In equation 3, the quantities in curly brackets denote the activities of the substituting ions in the aqueous phase, while a_{BA} and a_{CA} represent the activity of components *BA* and *CA* in the solid phase. In this type of expression, the equilibrium distribution constant $K_{B/C}$ is defined in terms of the ratios of the activities and is independent of the phase composition. In natural systems, the aqueous phase will generally have a complex composition, but there is a number of models that relate concentration to activity in water for a wide range of ionic strengths (*e.g.* Appelo & Postma, 2005). Thermodynamic treatment requires consideration of a standard state for the aqueous species, which is generally chosen to be a hypothetical ideal 1 molal aqueous solution, so that:

$$\{B^{n+}\} = \gamma_{B^{n+}} \frac{m_{B^{n+}}}{m^o_{B^{n+}}} = \gamma_{B^{n+}} m_{B^{n+}} \qquad (4)$$

where $m^o_{B^{n+}}$ (=1 mol/kg water) is the standard state concentration, $m_{B^{n+}}$ is the molal concentration of the aqueous free B^{n+} ions, and $\gamma_{B^{n+}}$ is the corresponding aqueous activity coefficient. Thus, the activity of an aqueous species is equal to the dimensionless value of its molal concentration corrected by a dimensionless activity coefficient that stands for non-ideality. An analogous expression defines $\{C^{n+}\}$ as a function of $\gamma_{C^{n+}}$ and $m_{C^{n+}}$.

For solids, the standard state is defined as the pure solid. Thus, the activities of the components in the solid solution are related to their mole fractions by the expression:

$$a_{BA} = \gamma_{BA}\frac{X_{BA}}{X_{BA}^{\circ}} = \gamma_{BA}X_{BA}, \tag{5}$$

where X_{BA} is the mole fraction of the component BA in the solid phase and $X_{BA}^{\circ} = 1$ is the mole fraction in the standard state. In the same way as for aqueous species, a dimensionless activity coefficient γ_{BA} is used to correct for non-ideal behaviour. An analogous expression defines a_{CA} in terms of X_{CA} and γ_{CA}. In the case of a binary $(B,C)A$ solid solution, $X_{CA} = 1 - X_{BA}$ and both mole fractions stand for the same solid composition.

Equation 3 describes the equilibrium partitioning between the solid and aqueous phases. However, it does not completely describe equilibrium because it does not provide a measure of the solubility of the solid solution. Information about the aqueous activity of the non-substituting ion A^{n-} and thus about the solubility is lost. A complete description of the equilibrium solubility of a binary solid solution requires consideration of two reactions, one for each component of the solid solution, namely:

$$BA_{(s)} \longleftrightarrow B^{n+}_{(aq)} + A^{n-}_{(aq)} \tag{6}$$

and

$$CA_{(s)} \longleftrightarrow C^{n+}_{(aq)} + A^{n-}_{(aq)} \tag{7}$$

The equilibrium conditions for the reactions of equations 6 and 7 are given by:

$$K_{BA} = \frac{\{B^{n+}\}\{A^{n-}\}}{a_{BA}} \tag{8}$$

and

$$K_{CA} = \frac{\{C^{n+}\}\{A^{n-}\}}{a_{CA}} \tag{9}$$

where K_{BA} and K_{CA} are the thermodynamic solubility products of pure BA and CA solids, respectively. Because equations 8 and 9 describe thermodynamic equilibrium, so does any combination of these two equations. Thus, by dividing them, one obtains:

$$\frac{K_{CA}}{K_{BA}} = \frac{a_{BA}\{C^{n+}\}}{a_{CA}\{B^{n+}\}} = K_{B/C} \tag{10}$$

Therefore, the equilibrium constant $K_{B/C}$ in equation 3 is equal to the quotient between the solubility products of the pure end-members of the solid solution.

While $K_{B/C}$ is a thermodynamic constant that does not depend on the compositions of the solid or aqueous solutions, the distribution coefficient $D_{B/C}$ as expressed in equation 1 depends on both parameters. Using a molar basis, equation 1 becomes:

$$D_{B/C} = \frac{X_{BA}}{X_{CA}} \bigg/ \frac{m_{B^{n+}}}{m_{C^{n+}}} \tag{11}$$

and considering the definitions in equations 4 and 5, equation 11 can be rewritten as:

$$D_{B/C} = \frac{a_{BA}\,\gamma_{CA}}{a_{CA}\,\gamma_{BA}} \bigg/ \frac{\{B^{n+}\}\,\gamma_{C^{n+}}}{\{C^{n+}\}\,\gamma_{B^{n+}}}. \tag{12}$$

At thermodynamic equilibrium, the combination of equation 12 with the two equilibrium conditions in equations 8 and 9 gives an expression for $D_{B/C(eq)}$ in terms of the end-member solubility products:

$$D_{B/C(eq)} = \frac{K_{CA}\,\gamma_{CA}\,\gamma_{B^{n+}}}{K_{BA}\,\gamma_{BA}\,\gamma_{C^{n+}}} = K_{B/C} \cdot \frac{\gamma_{CA}}{\gamma_{BA}} \cdot \frac{\gamma_{B^{n+}}}{\gamma_{C^{n+}}}. \tag{13}$$

Using aqueous activities instead of molal concentrations, equation 11 becomes:

$$D_{\{B/C\}} = \frac{X_{BA}}{X_{CA}} \bigg/ \frac{\{B^{n+}\}}{\{C^{n+}\}}. \tag{14}$$

By combining equation 14 with the two equilibrium conditions in equations 8 and 9, the equilibrium expression for $D_{\{B/C\}}$ is obtained in terms of the end-member solubility products:

$$D_{\{B/C\}(eq)} = \frac{K_{CA}\,\gamma_{CA}}{K_{BA}\,\gamma_{BA}} = K_{B/C} \cdot \frac{\gamma_{CA}}{\gamma_{BA}}. \tag{15}$$

Notably, because both γ_{BA} and γ_{CA} depend on the solid solution composition, the equilibrium distribution coefficient $D_{\{B/C\}(eq)}$ as defined in equation 15 is not a constant, but rather a function that can change significantly with composition. Therefore, $D_{\{B/C\}(\mathrm{eq})}$ is independent of the solid-phase composition only in the case of ideal solid solutions. In such a case, $\gamma_{BA} = \gamma_{CA} = 1$ and equation 15 can now be written as:

$$D_{\{B/C\}(eq,id)} = \frac{K_{CA}}{K_{BA}} \tag{16}$$

and $D_{\{B/C\}(eq)}$ becomes a constant equivalent to $K_{B/C}$.

When the distribution coefficients are expressed in terms of molal concentrations, their values also depend on the aqueous solution composition. Relating equations 11 and 14 yields:

$$D_{B/C} = D_{\{B/C\}} \frac{\gamma_{B^{n+}}}{\gamma_{C^{n+}}} \tag{17}$$

which indicates that $D_{B/C(eq)}$ is not a constant even in the case of an ideal solid solution. From equation 13, the distribution coefficient $D_{B/C(eq)}$ can be seen as a product of three terms (McIntire, 1963). The first one ($K_{B/C}$) is a thermodynamic constant, and thus an 'intrinsic' factor independent of composition. The second term (γ_{CA}/γ_{BA}) is a 'solid interaction' factor that accounts for the non-ideality of the solid solution and depends on the solid composition. Finally, the third term ($\gamma_{B^{n+}}/\gamma_{C^{n+}}$) is an 'aqueous interaction' factor that represents the effect of the aqueous solution composition on the distribution coefficient.

The dependence of the equilibrium distribution coefficients on system composition is sometimes overlooked in the literature. However, although distribution coefficients can be assumed to be constant under certain conditions (*e.g.* trace impurities in virtually pure solids and diluted aqueous solutions), such an approach is not always valid. In general, both the aqueous phase and the solid solution depart from ideal behaviour and the activity coefficients (γ_{BA}, γ_{CA}, $\gamma_{B^{n+}}$, and $\gamma_{C^{n+}}$) are sensitive to major compositional changes.

2.2. Variation of $D_{B/C(eq)}$ with the aqueous phase composition

When the substituting ions have similar properties and the same charge, changes in the solution ionic strength will affect the activity coefficients of both ions in the same way. Thus, even though the activity coefficient of each ion changes rapidly with the addition of solute, the ratio $\gamma_{B^{n+}}/\gamma_{C^{n+}}$ in equation 17 will approach unity and $D_{B/C(eq)}$ will change slowly. As an example, Figure 1 shows the evolution of $\gamma_{Cd^{2+}}/\gamma_{Ca^{2+}}$ as a function of the ionic strength (I). The substitution of Cd^{2+} for Ca^{2+} in the calcite structure is interesting from an environmental point of view, and there are numerous papers

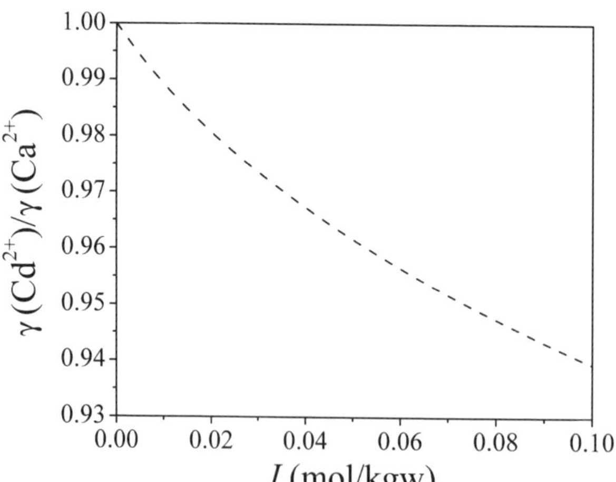

Fig. 1. Variation of the $\gamma_{Cd^{2+}}/\gamma_{Ca^{2+}}$ ratio as a function of the ionic strength. The ion-size parameters ($å_i$) used in equation 18 were 5 and 6 Å for Cd^{2+} and Ca^{2+}, respectively.

devoted to this solid solution (*e.g.* Davis *et al.*, 1987; Stipp *et al.*, 1992; Tesoriero & Pankow, 1996; Chiarello *et al.*, 1997; Prieto *et al.*, 2003). The calculation in Figure 1 was carried out using the Debye-Hückel (D-H) equation (see *e.g.* Appelo & Postma, 2005) for dilute electrolyte solutions ($I < 0.1$ mol/kg water):

$$\log \gamma_i = -\frac{A z_i^2 \sqrt{I}}{1 + B \mathring{a}_i \sqrt{I}} \tag{18}$$

where z_i is the charge number of ion i, $A = 0.5092$ and $B = 0.3283$ (at 25°C). The empirical ion-size parameter $\mathring{a}_i$ was taken to be 5 and 6 Å for Cd^{2+} and Ca^{2+}, respectively (*e.g.* see table 4.1 in Langmuir, 1997). As shown in Figure 1, the $\gamma_{Cd^{2+}}/\gamma_{Ca^{2+}}$ ratio decreases from 1 to ~0.94 with increasing (from 0 to 0.1 mol/kgw) ionic strength, which will affect the distribution coefficient when expressed on a molal concentration basis, according to equation 17. Moreover, ions in aqueous solution may form ion pairs or aqueous complexes, which reduces the concentration of 'free' ions in water and therefore lowers their availability for partitioning into the solid phase. Both complex formation and activity coefficients reduce the activity of the substituting ions, and their combined effect must be considered to evaluate the real significance of distribution coefficients calculated using an 'analytical' concentration. The effect is particularly significant in the presence of solutes that strongly interact with one of the two substituting ions, such as with Cd^{2+} and Ca^{2+} in the presence of chlorine. At moderate and high chlorine concentrations, dissolved cadmium is mostly present as cadmium chloride ($CdCl^+$, $CdCl_2^0$, and $CdCl_3^-$) aqueous complexes. As a consequence, the ratio between the concentrations of free Cd^{2+} and Ca^{2+} ions differs significantly from the analytical ratio.

To illustrate these effects, Figure 2 shows the equilibrium distribution of cadmium between NaCl-containing aqueous solutions and an otavite–calcite (Cd,Ca)CO_3 solid solution of composition $X_{Cd} = 0.01$. The calculations were carried out by equilibrating the solid solution with aqueous solutions with different concentrations of NaCl using the speciation program PHREEQC (Parkhurst & Appelo, 1999) and the PHREEQC.DAT database. Königsberger *et al.* (1991) suggested that otavite and calcite form a virtually ideal solid solution, a hypothesis also supported by the similar ionic radii of Cd^{2+} and Ca^{2+} and the linear shift of the lattice parameters with composition (Reeder, 1983). Therefore, we considered an ideal solid solution model in the calculations. Using the aqueous activities (dashed line), the equilibrium distribution coefficient does not depend on the aqueous solution composition. In agreement with equation 16 and when considering an ideal solid solution model, $D_{\{Cd/Ca\}(eq,id)} = K_{cal}/K_{otv} = 4{,}169$ at 25°C. Due to the much lower solubility of otavite ($K_{otv} = 10^{-12.1}$; Stipp *et al.*, 1993) compared with that of calcite ($K_{cal} = 10^{-8.48}$; Plummer & Busenberg, 1982), there is a strong tendency of the less soluble $CdCO_3$ component to partition into the solid phase, which explains the effectiveness of carbonate minerals in removing cadmium from the environment.

The equilibrium distribution coefficient is also very high when calculated based on the molal concentration (solid line with open circles in Fig. 2). However, it is no longer a

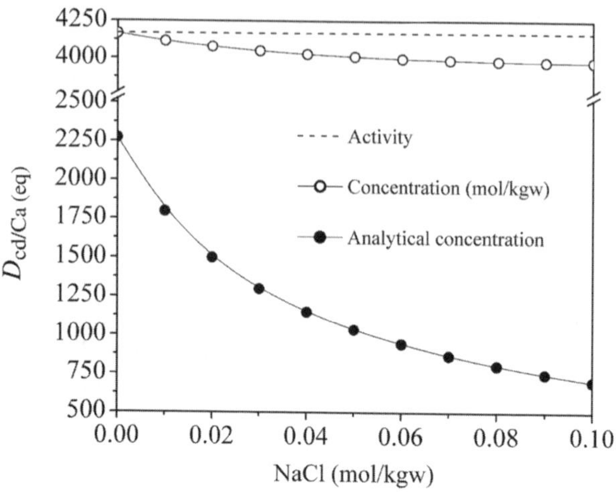

Fig. 2. Equilibrium distribution of cadmium between NaCl-containing aqueous solutions and an ideal (Cd,Ca)CO_3 solid solution of composition $X_{Cd} = 0.01$. The calculations were carried out using the speciation program PHREEQC (Parkhurst & Appelo, 1999) and the PHREEQC.DAT database. The upper curve (dashed line) represents the equilibrium distribution coefficient computed using aqueous activities. The solid line with open circles represents the distribution coefficient computed on a molal concentration basis. The third curve (solid line with filled circles) was calculated using total (analytical) concentrations of dissolved Ca and Cd.

constant but decreases with increasing NaCl concentration in the aqueous solution. Thus, for 0.1 mol/kgw NaCl, the coefficient $D_{Cd/Ca(eq,id)}$ has a value of ~3,975. This smaller value is due to the decrease in the $\gamma_{Cd^{2+}}/\gamma_{Ca^{2+}}$ ratio (and the corresponding increase in $m_{Cd^{2+}}/m_{Ca^{2+}}$) as the ionic strength increases (note that the activity ratio remains constant).

The third curve in Figure 2 (solid line with filled circles) was calculated using the total (analytical) concentrations of dissolved calcium and cadmium. The distribution coefficients calculated in this way are significantly smaller and decrease from ~2,272 to ~698 as the concentration of NaCl increases. These values are underestimated because the relative concentration of Cd^{2+} ions available for partitioning into the solid phase is overestimated. As previously discussed, a significant proportion of dissolved cadmium occurs in the form of aqueous complexes ($CdOH^+$, $CdCl^+$, $CdCl_2^o$, *etc.*) and should not be accounted for when calculating the distribution coefficient. In general, distribution coefficients should be determined using aqueous activities after suitable speciation of the aqueous solution. Values calculated using the total concentrations can only be correlated when they correspond to aqueous solutions with comparable ionic strengths and concentrations of the major elements. Unfortunately, many correlations reported in the scientific literature do not adhere to this principle. For example, in the case of cadmium, the rapid precipitation of biogenic carbonates (dominated by biochemical and kinetic processes) is invoked to account for the small $D_{Cd/Ca}$ values (<10) observed in foraminifera-seawater systems (Boyle, 1988; Delaney, 1989;

Mashiotta *et al.*, 1997). However, the distribution coefficients determined by these authors were expressed not using aqueous activities as in equation 14 but using total concentrations of Ca^{2+} and Cd^{2+} in seawater. In the case of seawater, this is equivalent to dividing the values obtained using equation 14 by a factor of ~25 (Carneiro *et al.*, 2010). Such a factor is due to the activity coefficients of Ca^{2+} and Cd^{2+} and to the complexing of cadmium with chlorine in seawater, which strongly decreases the concentration of free Cd^{2+} ions. The factor could be even larger if we consider the presence of organic ligands, which have been found to be a determining factor for cadmium behaviour, particularly in surface seawater (*e.g.* Bruland, 1992; Morel & Price, 2003). Thus, the reported $D_{Cd/Ca}$ values of ~2 in the foraminifera studies are equivalent to ~50 when using equation 14, a similar value to those measured in inorganic experiments at high precipitation rates (Lorens, 1981; Fernández-González *et al.*, 1999; Katsikopoulos *et al.*, 2008a).

In addition, the $\gamma_{Cd^{2+}}/\gamma_{Ca^{2+}}$ ratio depends on the expression used for calculating the activity coefficients, which can change from one database to another. For major ions, PHREEQC.DAT uses the Truesdell and Jones (T-J) equation (see Appelo & Postma, 2005), which is a modified version of the D-H formula with two specific fit parameters, a_i and b_i, for each ion:

$$\log \gamma_i = -\frac{Az_i^2\sqrt{I}}{1 + Ba_i\sqrt{I}} + b_i I. \tag{19}$$

The T-J equation has been found to be reliable up to $I = 0.7$ mol/kgw (Parkhurst, 1990). For Ca^{2+} ions, the values of a_i and b_i are 5.50 and 0.165, respectively. However, for minor ions such as Cd^{2+}, PHREEQC.DAT uses the Davies formula (Appelo & Postma, 2005), which is not ion-specific and gives the same value for all aqueous species with the same charge. Other databases use different expressions and values for the ion-specific parameters. For instance, while LLNL.DAT (Wolery, 1992) uses the D-H formula in equation 18, MINTEQ.DAT (Allison *et al.*, 1991) uses the T-J formula for major ions but with $a_i = 6.0$ instead of 5.50 for Ca^{2+}. In a similar way, the number of aqueous complexes and the value of the corresponding critical stability constants vary from one database to another, thus affecting the 'output' concentration of free ions. The disparity in the compiled thermodynamic data of numerous substances and aqueous species is well known (Oelkers *et al.*, 2009), arising from variability in experimental measurements, from derivations and extrapolations made from the original measurements, or from differences in the whole aqueous speciation. As a consequence, both the equilibrium and effective values of $D_{B/C}$ depend on the database used to make the speciation for determining $\gamma_{B^{n+}}$, $\gamma_{C^{n+}}$, $m_{B^{n+}}$, and $m_{C^{n+}}$. To compare equilibrium or effective values of distribution coefficients obtained in different environments, the speciation models must be consistent.

2.3. Non-ideal solid solutions

As previously discussed, γ_{BA} and γ_{CA} are equal to unity for ideal solid solutions, which means that $a_{BA} = X_{BA}$ and $a_{CA} = X_{CA}$ for the entire range of compositions. In such cases, $D_{\{B/C\}(eq)}$ is only controlled by the quotient between the solubility products of the end-members so that the less soluble component tends to preferentially incorporate into the solid phase. However, in non-ideal solid solutions, both γ_{BA} and γ_{CA} vary with composition and their determination is crucial in describing equilibrium partitioning in SS–AS systems. Many detailed descriptions of solid-solution thermodynamics are available in the literature (*e.g.* Saxena, 1973; Navrotsky, 1987; Putnis, 1992; Prieto, 2009); in this chapter, we only present the concepts relevant for expressing the solid-phase activity coefficients.

From a thermodynamic point of view, the degree of non-ideality of a solid solution is implicit in the chemical potential (μ) of each component (*BA* and *CA*) in the solid solution:

$$\mu_{BA} = \mu^{\circ}_{BA} + \mathrm{R}T \ln a_{BA} \tag{20}$$

and

$$\mu_{CA} = \mu^{\circ}_{CA} + \mathrm{R}T \ln a_{CA} \tag{21}$$

where T is the absolute temperature and R is the gas constant. Because the standard state (at which a_{BA} or $a_{CA} = 1$) for solids is the pure solid, the standard-state chemical potentials (μ°_{BA} and μ°_{CA}) of *BA* and *CA* are given by the molar Gibbs free energies of the pure end-members, G_{BA} and G_{CA}, respectively. Using chemical potentials, the total free energy (G_{SS}) of a solid solution of composition $X_{BA} = 1 - X_{CA}$ is given by:

$$\begin{aligned} G_{SS} &= X_{BA}\mu_{BA} + X_{CA}\mu_{CA} \\ &= X_{BA}(G_{BA} + \mathrm{R}T \ln a_{BA}) + X_{CA}(G_{CA} + \mathrm{R}T \ln a_{CA}) \end{aligned} \tag{22}$$

The tendency for solid-solution formation is related to the so-called free energy of mixing, which expresses the difference between the free energy of the solid solution and that of a compositionally equivalent mechanical mixture of the two end-members. In contrast to (*B*,*C*)*A*, which is a single phase with substitutional disorder and non-stoichiometric composition, a mechanical mixture is an aggregate of two phases. Therefore, the free energy (G_{MM}) of a mechanical mixture of pure *BA* and *CA* with composition X_{BA} is a linear combination of G_{BA} and G_{CA}:

$$G_{MM} = X_{BA}G_{BA} + X_{CA}G_{CA}. \tag{23}$$

From equations 22 and 23, the free energy of mixing (ΔG_M) is:

$$\Delta G_M = G_{SS} - G_{MM} = \mathrm{R}T(X_{BA} \ln a_{BA} + X_{CA} \ln a_{CA}) \tag{24}$$

Finally, the degree of non-ideality can be evaluated in terms of the excess free energy of mixing (ΔG_E), which is given by the difference between the free energy of mixing of the given solid solution and the free energy of mixing ($\Delta G_{M,id}$) of an equivalent ideal solid solution, namely:

$$\Delta G_E = RT(X_{BA} \ln a_{BA} + X_{CA} \ln a_{CA}) - RT(X_{BA} \ln X_{BA} + X_{CA} \ln X_{CA}) \quad (25)$$

The second term on the right side of equation 25 represents the free energy of mixing of the equivalent ideal solid solution, since $a_{BA} = X_{BA}$ and $a_{CA} = X_{CA}$ in an ideal solid solution. Considering the relationship between activity, activity coefficient and mole fraction shown in equation 5, equation 25 can be reduced to:

$$\Delta G_E = RT(X_{BA} \ln \gamma_{BA} + X_{CA} \ln \gamma_{CA}). \quad (26)$$

This last equation allows for the derivation of the activity coefficients from ΔG_E, according to:

$$\ln \gamma_{BA} = \left(\Delta G_E + X_{CA} \frac{\partial \Delta G_E}{\partial X_{BA}} \right) \Big/ RT \quad (27)$$

and

$$\ln \gamma_{CA} = \left(\Delta G_E - X_{BA} \frac{\partial \Delta G_E}{\partial X_{BA}} \right) \Big/ RT \quad (28)$$

By substituting these expressions for γ_{BA} and γ_{CA} into equation 15, the equilibrium distribution coefficient becomes:

$$D_{\{B/C\}(eq)} = \frac{K_{CA} \gamma_{CA}}{K_{BA} \gamma_{BA}} = \frac{K_{CA}}{K_{BA}} \cdot \exp\left(-\frac{\partial \Delta G_E}{\partial X_{BA}} \Big/ RT \right). \quad (29)$$

The exponential term in equation 29 accounts for the nature of the solid solution and its effect on the distribution coefficient. In the case of an ideal solid solution ($\Delta G_E = 0$), the exponential term becomes equal to unity and equation 29 turns into equation 16.

2.4. Variation of $D_{\{B/C\}(eq)}$ with the solid-phase composition

According to equation 29, to calculate equilibrium distribution coefficients, the derivative of the excess free energy of mixing with respect to X_{BA} is needed. In practice, when the value of ΔG_E is experimentally known for a number of compositions (see section 4), the excess free energy of mixing can be expressed as a function using a suitable fitting model. Guggenheim (1937) suggested that ΔG_E can be modelled by a fitting function that satisfies the required condition that $\Delta G_E = 0$ at the end-member compositions:

$$\Delta G_E = X_{BA} X_{CA} RT[a_0 + a_1(X_{BA} - X_{CA}) + a_2(X_{BA} - X_{CA})^2 + \cdots] \quad (30)$$

where a_0, a_1, a_2, ... are dimensionless fitting parameters independent of composition, but that may change with temperature and pressure. Once the Guggenheim parameters are known, equation 30 can be differentiated with respect to composition (note that $X_{BA} = 1 - X_{CA}$) to calculate the exponential term in equation 29. In most cases, two or three parameters are enough to quantify the degree of non-ideality (Glynn & Reardon, 1990). For ideal solid solutions, $a_0 = a_1 = a_2 = \cdots = 0$ so that $\Delta G_E = 0$.

When the two end-members of the solid solution are isostructural, both ΔG_E and ΔG_M can be represented by single curves across the whole composition range. In the simplest case of non-ideal behaviour, all but the first parameter in equation 30 are zero. As a consequence, the mixing properties become symmetric functions of composition, and the solid solution is termed 'regular'. For example, Figure 3a shows the excess free energy of mixing (dashed line) function calculated for the smithsonite-calcite solid solution (at 50°C) using $a_0 = 2.3$ (Glynn & Reardon, 1990; Crocket & Winchester, 1966). As can be observed, ΔG_E is a symmetric function about the mid-point ($X_{BA} = 0.5$) composition. This type of behaviour occurs in solid solutions in which

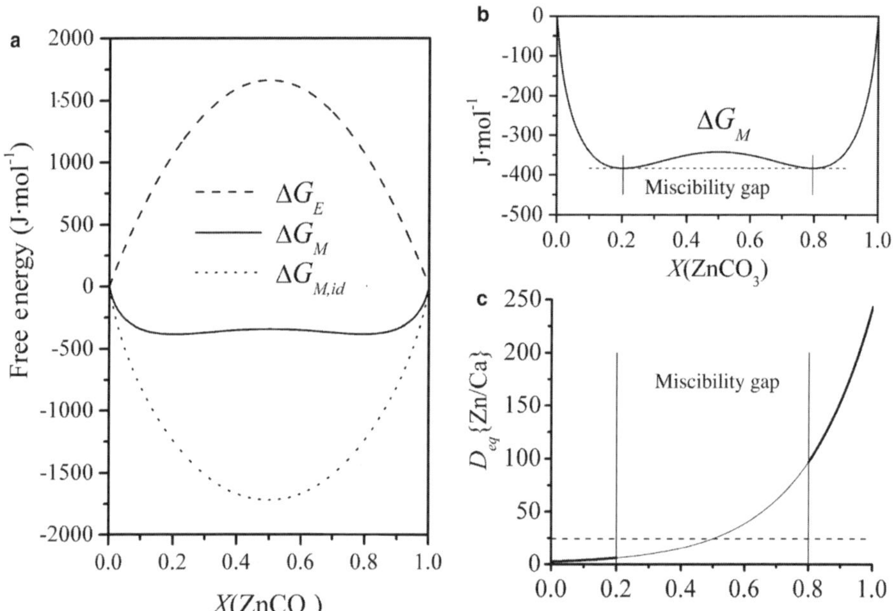

Fig. 3. (*a*) Free energy-composition curves for a smithsonite-calcite $(Zn,Ca)CO_3$ solid solution at 50°C calculated using a regular solid solution model with $a_0 = 2.3$ (Glynn & Reardon, 1990). In (*b*) the ΔG_M curve is plotted at a larger scale. The dashed line represents the miscibility gap determined by the common tangent points of the ΔG_M curve. The two minima occur at $X_{ZnCO_3} = 0.204$ and 0.796. (*c*) Dependence of $D_{\{Zn/Ca\}(eq)}$ on composition calculated using equation 29 with end-member pK values of 8.663 for calcite (Plummer & Busenberg, 1982) and 10.05 for smithsonite (Högfeldt, 1982). The thin segment of the curve corresponds to compositions within the miscibility gap and is meaningless from the point of view of equilibrium partitioning. The horizontal, dashed line represents the distribution coefficient of an equivalent ideal solid solution.

substituting *B* for *C* into the *CA* end-member has a similar energetic cost as substituting *C* for *B* into *BA*, which frequently occurs when there is a small difference in the size of the substituting atoms (Urusov, 1974). The fact that ΔG_E is positive for the whole range of compositions implies that the solid solution tends towards unmixing. Figure 3a also displays ΔG_M (solid line) and $\Delta G_{M,id}$ (dotted line) and Figure 3b shows the ΔG_M curve at a larger scale. The ΔG_M function is negative across the whole range, which means that the solid solution has a smaller free energy than an equivalent mechanical mixture of the two end-members for all compositions. However, the common tangent to the two minima delimits a miscibility gap from $X_{ZnCO_3} = 0.204$ to 0.796. A solid solution with a composition within these limits has a greater free energy than a compositionally equivalent mechanical mixture of two solid phases with compositions of 0.204 and 0.796. Therefore, exceeding these limits results in unstable or metastable solid solution members with respect to which an aqueous solution can never reach thermodynamic equilibrium. To reach equilibrium, such metastable members must separate into two solid phases with the compositions given by the gap limits.

The smithsonite-calcite $(Zn,Ca)CO_3$ solid solution illustrates the extent to which the equilibrium distribution coefficient depends on the solid-phase composition for non-ideal solid solutions. The effect is shown in Figure 3c, where $D_{\{Zn/Ca\}(eq)}$ is represented as a function of the $ZnCO_3$ mole fraction. The thin segment of the curve corresponds to compositions within the miscibility gap and is meaningless from the point of view of equilibrium partitioning. The distribution coefficient increases dramatically in the Zn region to reach values of ~240. In the Ca-rich region, $D_{\{Zn/Ca\}(eq)}$ increases with the Zn content at a smaller rate, but still rises from ~2 to ~6. These last values are significantly smaller than that (~24) of an equivalent ideal solid solution, which is represented by a horizontal dashed line in the figure. For an ideal solid solution, the distribution coefficient depends only on the solubility products of the end-members and the preferential partitioning of the $ZnCO_3$ component toward the solid phase is determined by the lower solubility of smithsonite.

While the regular model predicts symmetric miscibility gaps, the actual gaps are frequently asymmetric. Such is the case for the strontianite-aragonite $(Sr,Ca)CO_3$ solid solution, which (at 25°C) exhibits a thermodynamic miscibility gap from $X_{SrCO_3} \approx 0.005$ to ~0.86, as can be determined using a sub-regular solid solution model with $a_0 = 3.43$ and $a_1 = -1.82$ (Glynn & Reardon, 1990; Plummer & Busenberg, 1987). Figure 4a shows the corresponding ΔG_M curve, in which the miscibility gap is delimited by the common tangent to the two minima. The asymmetry of ΔG_M indicates that there is a greater tolerance for the incorporation of the smaller calcium atoms into strontianite than *vice versa*. Figure 4b shows the variation of $D_{\{Sr,Ca\}(eq)}$ with composition, with $D_{\{Sr/Ca\}(eq)} > 1$ in the Sr-rich miscibility region. However, $D_{\{Sr/Ca\}(eq)}$ remains roughly constant ($0.045 \leq D_{\{Sr/Ca\}(eq)} \leq 0.049$) and smaller than unity in the Sr-poor miscibility range (see Fig. 4c). Thus, while Sr^{2+} is incorporated preferentially over Ca^{2+} into strontianite-rich crystals, the opposite happens in aragonite despite the lower solubility of strontianite. The previous calculations indicate that the dependence of $D_{\{B/C\}(eq)}$ on composition cannot be overlooked, *e.g.* in the assessment of metal-ion behaviour in the environment. The distribution coefficients can be

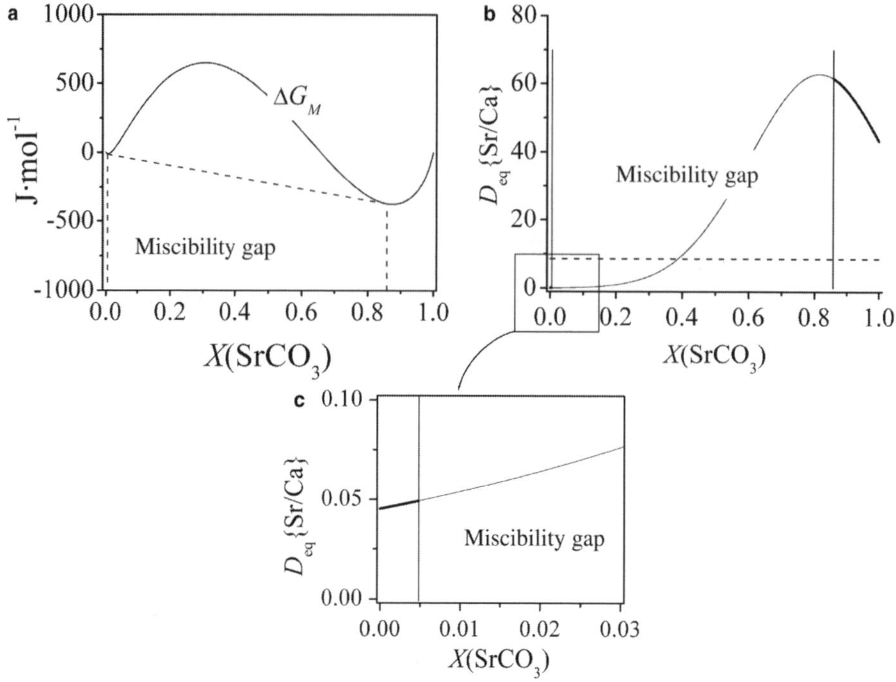

Fig. 4. (***a***) ΔG_M curve corresponding to the strontianite-aragonite $(Sr,Ca)CO_{3(orth)}$ solid solution at 25°C. Subregular model with $a_0 = 3.43$ and $a_1 = -1.82$ (Glynn & Reardon, 1990; Plummer & Busenberg, 1987). The dashed line represents the common tangent to the two minima and determines the limits (0.0048 and 0.858) of an asymmetric miscibility gap. Note that as the curve is asymmetric, the common tangent points do not coincide with the minima (*cf.* Plummer & Busenberg, 1987). Dependence of $D_{\{Sr/Ca\}(eq)}$ on composition calculated using equation 29 with end-member pK values of 8.336 for aragonite (Plummer & Busenberg, 1982) and 9.271 for strontianite (Busenberg *et al.*, 1984). The thin segment of the curve corresponds to compositions within the miscibility gap and is meaningless from the point of view of equilibrium partitioning. The horizontal, dashed line represents the distribution coefficient of an equivalent (8.61) ideal solid solution. (***c***) More detailed view of the Ca-rich miscibility region.

considered to be approximately constant only in the case of very dilute solid solutions. However, dilute solid solutions are special cases deserving separate discussion and will be considered in section 5.1.

Notably, it is not possible to consider changes in the composition of a solid solution in equilibrium with an aqueous solution without simultaneously considering variations in the composition of the aqueous phase. For this reason, in contrast to $D_{\{B/C\}(eq)}$, the composition of the SS–AS system as a whole determines the value of the distribution coefficient $D_{B/C(eq)}$ when expressed based on aqueous concentrations.

2.5. Excess chemical potential

The distribution coefficients can also be expressed in terms of the so-called excess chemical potentials (*e.g.* Holst, 1978). The excess chemical potential of the *BA* component in the (*B*,*C*)*A* solid solution is given by the difference between the actual chemical potential μ_{BA} and the chemical potential $\mu_{BA,id}$ of *BA* in an equivalent ideal solid solution:

$$\Delta\mu_{BA}^{E} = \mu_{BA} - \mu_{BA,id} \tag{31}$$

An analogous expression defines $\Delta\mu_{CA}^{E}$. Considering the definitions of μ_{BA} and μ_{CA} in equations 20 and 21, the relationship between activity and concentration in equation 5, and that the activity coefficients are equal to unity in the case of ideal solid solutions, we reach:

$$\Delta\mu_{BA}^{E} = \mathrm{R}T \ln \gamma_{BA} \tag{32}$$

and

$$\Delta\mu_{CA}^{E} = \mathrm{R}T \ln \gamma_{CA} \tag{33}$$

Solving these equations for γ_{BA} and γ_{CA} and substituting the results in equation 15, we obtain:

$$D_{\{B/C\}(eq)} = \frac{K_{CA}}{K_{BA}} \cdot \exp\left(\frac{\Delta\mu_{CA}^{E} - \Delta\mu_{BA}^{E}}{\mathrm{R}T}\right) \tag{34}$$

an expression equivalent to equation 29. In fact, from equations 26, 32 and 33,

$$\Delta G_E = X_{BA}\Delta\mu_{BA}^{E} + X_{CA}\Delta\mu_{CA}^{E} \tag{35}$$

and therefore

$$-\frac{\partial \Delta G_E}{\partial X_{BA}} = \Delta\mu_{CA}^{E} - \Delta\mu_{BA}^{E} \tag{36}$$

Although equation 34 does not exhibit any particular advantage with respect to equation 29, the concept of excess chemical potential is very useful in dealing with dilute solid solutions, as will be discussed further in section 5.1.

3. Equilibrium partitioning and solubility of solid solutions

3.1. Lippmann and Roozeboom diagrams

The parameters $K_{B/C}$, $D_{B/C(eq)}$, and $D_{\{B/C\}(eq)}$ in equations 10, 13 and 15 describe the equilibrium distribution of substituting ions but do not completely describe the equilibrium in SS–AS systems because they do not contain information about the non-substituting ion A^{n-}. As previously shown, the complete description of the equilibrium solubility of a binary solid solution requires the simultaneous consideration of two equilibrium conditions (equations 8 and 9). We can rewrite these equations in the form:

$$\{B^{n+}\}\{A^{n-}\} = K_{BA}a_{BA} = K_{BA}\gamma_{BA}X_{BA} \tag{37}$$

and

$$\{C^{n+}\}\{A^{n-}\} = K_{CA}a_{CA} = K_{CA}\gamma_{CA}X_{CA} \tag{38}$$

Because equations 37 and 38 describe a thermodynamic equilibrium, any combination of them also represents an equilibrium condition. As such, Lippmann (1977, 1980) defined a "total activity product" variable given by

$$\Sigma\Pi = (\{B^{n+}\} + \{C^{n+}\})\{A^{n-}\} \tag{39}$$

and the addition of equations 37 and 38 yields the so-called "solidus" relationship:

$$\Sigma\Pi_{eq} = K_{BA}\gamma_{BA}X_{BA} + K_{CA}\gamma_{CA}X_{CA} \tag{40}$$

where the "total solubility product" $\Sigma\Pi_{eq}$ is the value of the $\Sigma\Pi$ variable at thermodynamic equilibrium. While necessary, the solidus relationship is not a sufficient condition for equilibrium. For a complete description of thermodynamic equilibrium, a second relationship must be derived from equations 37 and 38. This second relationship expresses $\Sigma\Pi_{eq}$ as a function of the aqueous phase composition and was derived and termed "solutus" by Lippmann (1977, 1980):

$$\Sigma\Pi_{eq} = \frac{1}{\dfrac{X_{B,aq}}{K_{BA}\gamma_{BA}} + \dfrac{X_{C,aq}}{K_{CA}\gamma_{CA}}} \tag{41}$$

where $X_{B,aq}$ and $X_{C,aq}$ are the aqueous activity fractions of the aqueous B^{n+} and C^{n+} ions and are given by:

$$X_{B,aq} = \frac{\{B^{n+}\}}{\{B^{n+}\} + \{C^{n+}\}} \tag{42}$$

and

$$X_{C,aq} = \frac{\{C^{n+}\}}{\{B^{n+}\} + \{C^{n+}\}} \tag{43}$$

Like the solid-phase mole fractions, the aqueous activity fractions in a binary system are not independent, but both represent the same compositional variable, with $0 \leq X_{B,aq} \leq 1$ and $X_{C,aq} = 1 - X_{B,aq}$. The graphical representation of solidus and solutus yields a phase diagram that depicts the aqueous solubility of binary solid solutions. After the publication of clarifying papers by Glynn and co-workers (Glynn & Reardon, 1990; Glynn *et al.*, 1990), the Lippmann diagrams have become one of the most popular tools for the description of equilibrium as well as for the interpretation of crystallization processes in SS–AS systems (Königsberger *et al.*, 1991; Prieto *et al.*, 1993; Prieto *et al.*, 1997; Böttcher, 2000; Gamsjäger *et al.*, 2000; Glynn, 2000).

In constructing a Lippmann diagram, solidus and solutus must be represented on the ordinate against two superimposed scales, X_{BA} and $X_{B,aq}$, on the abscissa. Whereas solidus is a function of X_{BA}, solutus depends on $X_{B,aq}$. A Lippmann diagram can be used as a conventional phase diagram. Horizontal tie-lines can be drawn between the solutus and solidus curves thereby giving aqueous- and solid-phase compositions at thermodynamic equilibrium. Moreover, these equilibrium $X_{B,aq}$-X_{BA} pairs can be used to construct a Roozeboom-type plot (*e.g.* Prieto *et al.*, 2000), which is an alternative representation of ion partitioning in SS–AS systems. A recent review on SS–AS thermodynamics, including a detailed derivation of the most significant relationships and a number of examples of Lippmann and Roozeboom diagrams, was published by Prieto (2009).

As an illustration, Figure 5 shows the Lippmann and Roozeboom diagrams for the (Cd,Ca)CO_3-H_2O system at 25°C, considering an ideal solid solution model (see section 2.2). In Figures 5a and 5b, the dashed lines connect an aqueous ($X_{Cd,aq} = 0.0077$) and a solid phase ($X_{CdCO_3} = 0.97$) that coexist in thermodynamic equilibrium. In this particular case, the end-member solubility products differ by more than three orders of magnitude and the solidus and solutus curves plot far each

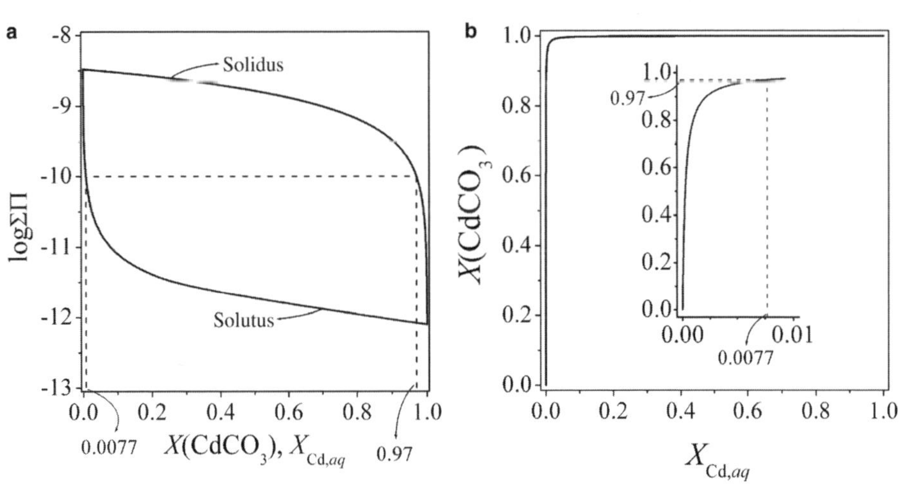

Fig. 5. Lippmann diagram (***a***) and Roozeboom $X_{Cd,aq}$-X_{CdCO3} plot (***b***) computed for the ideal (Cd,Ca)CO_3 solid solution. Due to the extremely low solubility of otavite (pK = 12.1; Stipp *et al.*, 1993) in comparison to that of calcite (pK = 8.48; Plummer & Busenberg, 1982), the solidus and solutus are plotted far from each other and the Roozeboom curve approximates to two perpendicular straight lines. The inset in (***b***) shows the vertical branch of the Roozeboom curve at a larger scale.

other. Due to the extremely low solubility of otavite compared with that of calcite, the Roozeboom curve approximately forms two perpendicular straight lines. The horizontal branch is virtually parallel to the abscissa axis from $X_{Cd,aq} \approx 0.1$ to 1, indicating that aqueous solutions in equilibrium with Ca-rich solids are essentially free of Cd^{2+}. Only a very narrow range of aqueous compositions can coexist in equilibrium with intermediate solid solution members. The solids tend to be either Ca-rich ($X_{CdCO_3} < 0.1$) or Cd-rich ($X_{CdCO_3} > 0.9$) over a very small range of fluid compositions ($0.000027 < X_{Cd,aq} < 0.00215$), which imposes a bimodal trend in crystallizing otavite–calcite solid solutions (Prieto *et al.*, 1997; Fernández-González *et al.*, 1999). All of these features result from the strong tendency of cadmium to partition into calcite, which is in agreement with the high value obtained for $D_{\{Cd/Ca\}(eq)}$ in section 2.2.

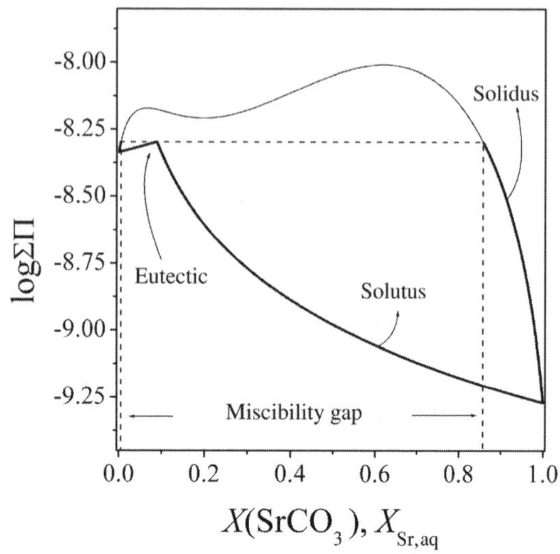

Fig. 6. Lippmann diagram for the $(Sr,Ca)CO_{3(orth)}$ solid solution at 25°C. Sub-regular model with $a_0 = 3.43$ and $a_1 = -1.82$ (Glynn & Reardon, 1990; Plummer & Busenberg, 1987) and end-member pK values of 8.336 for aragonite (Plummer & Busenberg, 1982) and 9.271 for strontianite (Busenberg *et al.*, 1984). The 'eutectic' point on the solutus is at equilibrium with the two extremes of the miscibility gap on the solidus. The thin segment of the solidus curve corresponds to unstable solid compositions and is meaningless.

Whereas the diagram in Figure 5a is quite simple (note that the plot is symmetric with respect to a 2-fold axis located at its centre), Lippmann's diagrams become asymmetric and the solutus curves frequently show peritectic, eutectic, or alyotropic points in the case of non-ideal solid solutions. For example, Figure 6 shows the Lippmann diagram for the $(Sr,Ca)CO_{3(orth)}$-H_2O system at 25°C using a sub-regular solid solution model (see section 2.4). The solutus curve shows a 'eutectic' point at $X_{Sr,aq} \approx 0.089$, $\log\Sigma\Pi \approx -8.29$, which corresponds to an aqueous solution that is simultaneously in equilibrium with the two solid compositions (~0.005 and ~0.86) that limit the miscibility ranges of this incomplete solid solution (see also Figure 4).

3.2. Stoichiometric solubility

When in contact with aqueous phases, solid solutions usually start to dissolve in a congruent way until an initial metastable equilibrium is reached. Then, to reach true equilibrium, the SS–AS system should undergo a dissolution-recrystallization process. However, the kinetics of this process are often very slow, and the system remains at "stoichiometric" saturation for a while (Thorstenson & Plummer, 1977).

As an illustration, we will consider the congruent dissolution reaction:

$$B_xC_{(1-x)}A_{(s)} + pB^{n+}_{(aq)} + qC^{n+}_{(aq)} + rA^{n-}_{(aq)} \longrightarrow (p+x)B^{n+}_{(aq)} + (q+1-x)C^{n+}_{(aq)} + (r+1)A^{n-}_{(aq)} \quad (44)$$

in which the transfer of ions between the solid and the aqueous phases occurs in a stoichiometric proportion equal to that of the solid. During congruent dissolution, the solid solution behaves as a stoichiometric solid with a fixed formula $B_xC_{(1-x)}A$, thus as a pure single-component solid with an activity equal to one. Therefore, by applying the mass action law to equation 44, we obtain:

$$K_{st} = \{B^{n+}\}^x\{C^{n+}\}^{(1-x)}\{A^{n-}\} \quad (45)$$

where the 'equilibrium' constant (K_{st}) is the stoichiometric solubility product. Taking into account that $x = X_{BA}$ and $1 - x = X_{CA}$ and applying equations 37 and 38 (see Prieto, 2009 for details) we arrive at:

$$K_{st} = (K_{BA}\gamma_{BA}X_{BA})^{X_{BA}}(K_{CA}\gamma_{CA}X_{CA})^{X_{CA}} \quad (46)$$

which expresses K_{st} in terms of the solubility products of the end-members. Finally, from equations 27 and 28, we can express γ_{BA} and γ_{CA} in terms of ΔG_E:

$$K_{st} = (K_{BA}X_{BA})^{X_{BA}}(K_{CA}X_{CA})^{X_{CA}} \exp(\Delta G_E/RT) \quad (47)$$

As we will discuss in section 4.1, this expression was used by Plummer & Busenberg (1987) as a tool for estimating non-ideality parameters from stoichiometric solubility measurements (see also Glynn & Reardon, 1990).

4. Estimation of equilibrium distribution coefficients

Equation 29 illustrates the relationship between $D_{\{B/C\}(eq)}$ and the composition derivative of the excess free energy of mixing. According to that equation, when the thermodynamic solubility products of the end-members are known, the $D_{\{B/C\}(eq)}$ values can be indirectly estimated if we are able to express ΔG_E as a function of composition. Conversely, if we have 'secure' equilibrium $D_{\{B/C\}(eq)}$ values for various solid compositions, the $\Delta G_E(X_{BA})$ function can be determined using Guggenheim's or similar fitting functions available in the literature (see Glynn & Reardon, 1990 and references therein). Although there are several ways for determining thermodynamic mixing parameters (Glynn, 2000), detailed descriptions are beyond the scope of the present review, which focuses only on some of the most usual methods. For any method chosen to estimate the mixing parameters, the first requisite is to ensure that homogenous solid solution samples, rather than mechanical mixtures or solids with compositional and/or structural heterogeneities (clustering, compositional zoning, *etc.*), are being used. Thus, prior to any measurement, a detailed structural characterization of the samples is necessary. For instance, the presence of anomalous peak broadening in powder X-ray diffraction patterns of the samples may be an index of heterogeneity (Andara *et al.*, 2005; Fernández-González *et al.*, 2007; Katsikopoulos *et al.*, 2009).

4.1. Stoichiometric solubility measurements

Stoichiometric saturation measurements are useful for estimating the excess free energy of mixing at low temperatures, particularly in the case of relatively insoluble minerals that do not undergo rapid dissolution-recrystallization (Glynn, 2000). Carefully controlled experiments are needed to ensure that stoichiometric saturation is reached and that the system has not undergone a later ripening process towards equilibrium (Böttcher, 1997). Once a K_{st} value is obtained, the corresponding excess free energy of mixing can be obtained by solving equation 47 for ΔG_E:

$$\Delta G_E = RT[\ln K_{st} - X_{BA} \ln (K_{BA}X_{BA}) - X_{CA} \ln (K_{CA}X_{CA})]. \tag{48}$$

Thus, equation 48 allows a Guggenheim function to be fitted from a set of experimental K_{st} values. Figure 7a shows the free energy of mixing obtained for the $(Sr,Ca)CO_{3(orth)}$ solid solution. The data points were calculated by applying equation 48 to experimental K_{st} values determined by Plummer & Busenberg (1987) and are fitted (continuous line) to a Guggenheim function of two terms ($a_0 = 3.24$ and $a_1 = -1.81$). The fitting parameters differ slightly from those computed by Plummer & Busenberg (1987) because no weighting of the data was applied prior to the fitting procedure. Figure 7b shows the original K_{st} data and illustrates the effect

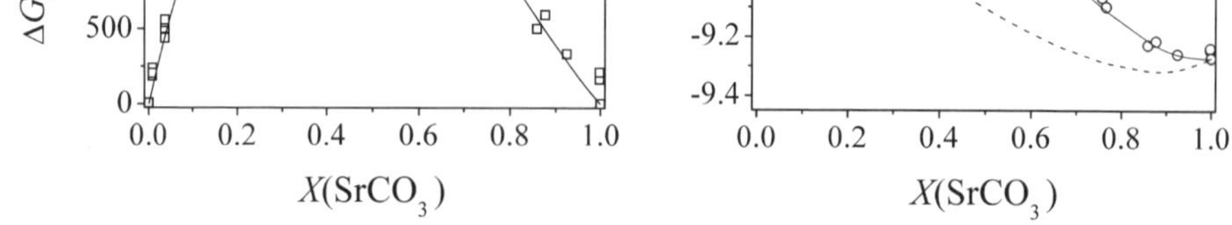

Fig. 7. (***a***) Excess energy of mixing determined for the $(Sr,Ca)CO_{3(orth)}$ solid solution using congruent dissolution data from Plummer & Busenberg (1987). The data points were fitted to a Guggenheim function of two terms ($a_0 = 3.24$ and $a_1 = -1.81$). (***b***) Comparison between the stoichiometric solubility of the actual solid solution (experimental data-points and continuous fitting curve) and that of an equivalent ideal solid solution (dashed line).

of non-ideality on the stoichiometric solubility. The fitting continuous line was calculated using equation 47 with $a_0 = 3.24$ and $a_1 = -1.81$, and the dashed lower curve corresponds to the equivalent ideal solid solution. The actual solid solution is more soluble than the equivalent ideal solid solution, which is typical of solid solutions with a positive excess free energy of mixing (Prieto, 2009).

4.2. Miscibility gap data

When a solid solution shows a symmetric miscibility gap, the $\Delta G_M(X_{BA})$ curve has two minima that coincide with the gap boundaries. In such a case, a regular solution model can be used and the single Guggenheim parameter a_0 can be determined by considering that the composition derivative of ΔG_M must be zero at both minima. In the literature, there are numerous examples of solid solutions with symmetric miscibility gaps that have been modelled in this way. For example, a $Ca(SeO_4,SO_4) \cdot 2H_2O$ solid solution, in which selenate substitutes for sulphate in the gypsum structure, presents a symmetric gap between 0.23 and 0.77. Considering a regular ($\Delta G_E = a_0 X_{BA} X_{CA} RT$) solid solution model and taking into account equations 24 and 25 yields:

$$\left(\frac{\partial \Delta G_M}{\partial X_{BA}}\right)_{0.23} = RT(a_0 - 2a_0 \cdot 0.23 + \ln 0.23 - \ln 0.77) = 0 \tag{49}$$

from which a value of $a_0 = 2.24$ can be determined (Fernández-González *et al.*, 2006). In the more general case of a sub-regular solid solution with an asymmetric miscibility gap, the common tangent points on the $\Delta G_M(X_{BA})$ curve do not coincide with the minima. Therefore, to determine a_0 and a_1 from the gap-limit compositions, we must solve the set of equations (Saxena, 1973):

$$\Delta G_{MBA}(X_{g1}) = \Delta G_{MBA}(X_{g2}) \tag{50}$$

and

$$\Delta G_{MCA}(X_{g1}) = \Delta G_{MCA}(X_{g2}) \tag{51}$$

where X_{g_1} and X_{g_2} are the common tangent points (the gap limits) and the functions ΔG_{MBA} and ΔG_{MCA} are given by:

$$\Delta G_{MBA} = \Delta G_M + X_{CA} \frac{\partial \Delta G_M}{\partial X_{BA}} \tag{52}$$

and

$$\Delta G_{MCA} = \Delta G_M - X_{BA} \frac{\partial \Delta G_M}{\partial X_{BA}} \tag{53}$$

Although miscibility gaps determined from laboratory experiments are a simple tool for determining the Guggenheim parameters, the method is limited because it does not guarantee that all of the obtained compositions are stable. Due to kinetic effects, the precipitation of metastable (out the miscibility range) compositions is always possible, even when crystallization occurs under moderate supersaturation conditions (Katsikopoulos *et al.*, 2009); even "intrinsically unstable" compositions lying between the spinodal points of the $\Delta G_M(X_{BA})$ curve can form in precipitation experiments. The initial precipitates usually have large surface energies and a large proportion of defects, which may favour the formation of extensive solid solution series. Such metastable solid phases will tend to undergo a recrystallization process to reach thermodynamic equilibrium. Despite these problems, Glynn (2000) recommends using miscibility gaps, as determined from mineral compositions observed in the field, to estimate mixing parameters in the absence of more accurate results. Even though there is a variety of factors that can influence the extent of a miscibility gap inferred from natural occurrences, the approach is still an improvement over the assumption of an ideal solid solution model. Geochemical occurrences have also been used by Kulik *et al.* (2000) to estimate non-ideality parameters.

4.3. Direct measurement of distribution coefficients

Distribution coefficients observed in natural systems or in the laboratory are often used to determine the excess free energy of mixing according to equation 29. For instance, Glynn & Reardon (1990), using Crocket & Winchester's (1966) experimental coefficients, proposed a regular model with $a_0 = 2.3$ for a $(Zn,Ca)CO_3$ solid solution and determined a symmetric miscibility gap matching the gap observed in natural occurrences. Unfortunately, the main problem of this method is to guarantee that the measured values correspond to true equilibrium. Factors such as supersaturation and solid-phase growth rate (Lorens, 1981; Mucci & Morse, 1983; Kornicker *et al.*, 1991; Tesoriero & Pankow 1996; Prieto *et al.*, 1997; Pina & Putnis, 2002; Katsikopoulos *et al.*, 2008a) significantly influence the effective distribution coefficients. The wide data dispersion in measured values (Rimstidt *et al.*, 1998) reflects the difficulty in achieving thermodynamic equilibrium. The influence of kinetic and mechanistic factors on ion partitioning is beyond the scope of this review and will be discussed in a specific chapter in this volume (Putnis, 2010). The determination of reliable equilibrium coefficients is necessary for interpreting non-equilibrium values and direct measurements are probably not the best method for that purpose.

4.4. Other experimental and computational methods

At first sight, measurements of solid-solution solubilities at true thermodynamic equilibrium are considered to be the best method for determining ΔG_E and estimating $D_{\{B/C\}(eq)}$ values. In practice, true equilibrium states are impossible to achieve on the laboratory time scale, except in the case of highly soluble solid solutions capable of rapid recrystallization (Glynn, 2000). In general, the experimental determination of mixing properties is very time-consuming and accompanied by many sources of uncertainty. As previously shown, stoichiometric solubility measurements may be inaccurate due to difficulties in achieving stoichiometric saturation. The formation of metastable compositions occurs frequently when using other methods, thus complicating the accurate determination of distribution coefficients and miscibility gaps. These difficulties explain the lack of low-temperature thermodynamic data for non-ideal solid solutions. Fortunately, the application of molecular simulation methods to solid solutions has received increasing attention over the last few years (*e.g.* Becker *et al.*, 2000; Vinograd *et al.*, 2006, 2008; Ruiz-Hernández *et al.*, 2010; Kulik *et al.*, 2010). Experiments such as calorimetric measurements of enthalpies of mixing, X-ray studies of compositional evolution of lattice parameters, determination of critical temperatures and critical mole fractions of mixing (*e.g.* Casey *et al.*, 1996; McBeath *et al.*, 1998; Glynn, 2000; Katsikopoulos *et al.*, 2009) can be used to investigate the mixing properties and to check and improve the calculation methods.

5. Trace-element partitioning in SS–AS systems

5.1. Thermodynamics of dilute solid solutions

The entry of trace amounts of substituting ions into the solid structure simplifies the thermodynamics relationships, providing an extremely dilute solid solution in which the host solid phase can be considered as a pure solid. To illustrate this approach, Figure 8 displays a typical example of the variation of the excess chemical potentials (see section 2.5) with solid composition. The calculation corresponds to the subregular $(Sr,Ca)CO_{3(orth)}$ solid solution ($a_0 = 3.43$ and $a_1 = -1.82$). The excess chemical potentials of both ($SrCO_3$ and $CaCO_3$) components approach zero as the

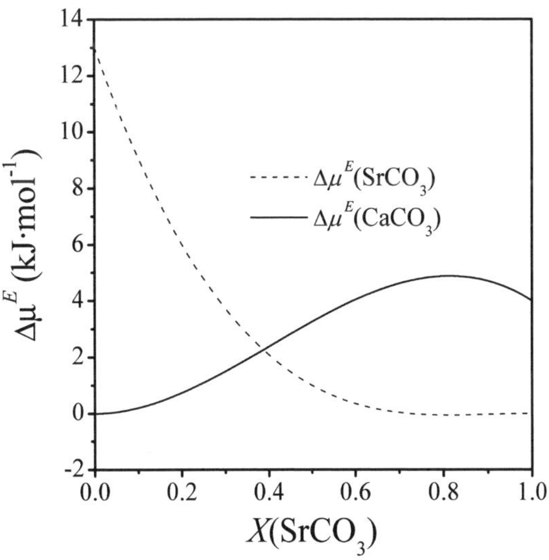

Fig. 8. Excess chemical potentials of the components of the $(Sr,Ca)CO_{3(orth)}$ solid solution as a function of composition ($a_0 = 3.43$ and $a_1 = -1.82$).

corresponding molar fractions draw near one. At trace concentrations of strontium, the activity coefficient of the aragonite component in the solid solution is virtually equivalent to one and *vice versa*. Therefore, in the general case of a binary $(B,C)A$ solid solution with trace concentrations of the BA component, $\Delta\mu^E_{CA} = \mathrm{R}T \ln \gamma_{CA} \approx 0$ and equation 34 simplifies to:

$$D_{\{B/C\}(eq)} = \frac{K_{CA}}{K_{BA}} \cdot \exp\left(\frac{-\Delta\mu^E_{BA}}{\mathrm{R}T}\right) \tag{54}$$

or

$$D_{B/C(eq)} = \frac{K_{CA}}{K_{BA}} \cdot \frac{\gamma_{B^{n+}}}{\gamma_{C^{n+}}} \exp\left(\frac{-\Delta\mu^E_{BA}}{\mathrm{R}T}\right) \tag{55}$$

when the distribution coefficient is expressed based on molal concentrations. Equation 55 is the classical expression for trace element partitioning (McIntire, 1963) and has been used widely in geochemical literature. For instance, Rimstidt *et al.* (1998) made an excellent compilation of experimental (effective) carbonate distribution coefficients and correlated them using this equation. Following these authors, equation 55 can be linearized by taking the logarithm of both sides:

$$\log D_{B/C} = \log \frac{K_{CA}}{K_{BA}} + \log \frac{\gamma_{B^{n+}}}{\gamma_{C^{n+}}} + \frac{-\Delta\mu^E_{BA}}{2.303\mathrm{R}T} \tag{56}$$

where 2.303 is the conversion factor between common and natural logarithms. Based on the assumption that the ratio $\gamma_{B^{n+}}/\gamma_{C^{n+}}$ approaches unity when the substituting ions are of the same charge, Rimstidt *et al.* (1998) speculated that a graph of $\log D_{B/C}$ against $\log(K_{CA}/K_{BA})$ should be a straight line with a slope of one and an intercept of $-\Delta\mu^E_{BA}/2.303\mathrm{R}T$. These authors fit a dataset of trace distribution coefficients in calcite for cations that form rhombohedral carbonates (Mg^{2+}, Co^{2+}, Fe^{2+}, Mn^{2+}, Cu^{2+}, Zn^{2+}, and Cd^{2+}) and another for cations that form orthorhombic carbonates (Sr^{2+}, Ba^{2+}, Pb^{2+}, and Ra^{2+}). Thus, Rimstidt *et al.* (1998) obtained two linear regressions in which the fact that the slopes differ from unity was attributed to non-equilibrium partitioning. Because the resulting empirical expressions have been widely cited in the geochemical literature, this method deserves some comment. A correlation between $\log D_{B/C}$ and $\log(K_{CA}/K_{BA})$ is expected since the difference between the solubility of the end-members is a key parameter in determining the tendency of a substituting ion to incorporate into a given mineral. However, a correlation using equation 56 implies consideration of the excess chemical potential $\Delta\mu^E_{BA}$ as a constant, which can be problematic and lead to misunderstandings. By definition, the excess chemical potential is specific for each solid solution system (see section 2.5), although for a given system it does not change significantly with composition at trace amounts. As an illustration, Figure 9 shows the excess chemical potential corresponding to trace amounts of different cations (Co^{2+}, Fe^{2+}, Mn^{2+}, Mg^{2+}, Zn^{2+}) into

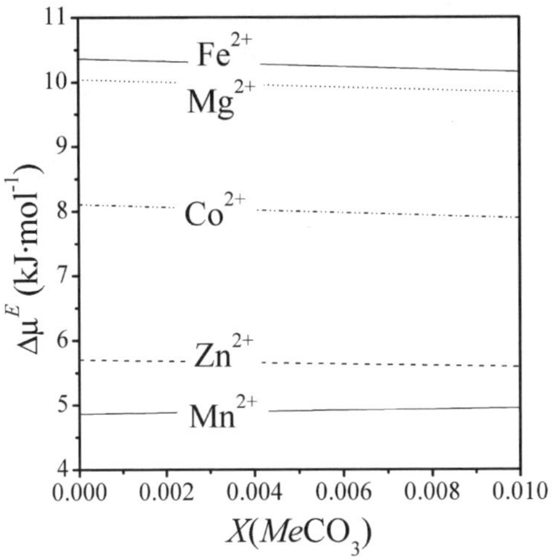

Fig. 9. Excess chemical potential corresponding to trace amounts of different divalent metals ($Me^{2+} = Co^{2+}$, Mn^{2+}, Fe^{2+}, Zn^{2+}, Mg^{2+}, and Fe^{2+}) in calcite. Calculation performed using Guggenheim parameters taken from different sources (Katsikopulos *et al.*, 2008b, 2009; Glynn & Reardon, 1990; Glynn, 2000).

calcite. The values differ significantly between them as well as from the meaningless fitting value of -1.14 kJ/mol obtained by Rimstidt *et al.* (1998). Notably, $\Delta\mu^E_{BA}$ is not the difference between the chemical potential of the *BA* component in a pure *BA* crystal and its chemical potential in the solid solution, which is $-RT\ln a_{CA}$ as deduced from equation 20 (*cf.* Rimstidt *et al.*, 1998). Instead, $\Delta\mu^E_{BA}$ is an excess parameter given by $RT\ln\gamma_{CA}$ (see section 2.5). Moreover, although Sr^{2+}, Ba^{2+}, Pb^{2+}, and Ra^{2+} can enter into calcite, they do not form calcite-type carbonates. As an alternative, Rimstidt *et al.* (1998) used the solubility products of the corresponding aragonite-type carbonates in equation 56, but this approach involves an error that is difficult to evaluate.

There are other thermodynamically based semi-empirical correlations available in geochemical literature, including a noteworthy method by Wang & Xu (2001). Based on the Sverjensky (1984) linear free-energy correlation expressions, these authors developed a model to predict trace-metal partitioning between minerals and aqueous solutions. We will recall this semi-empirical model in sections 6.2 and 6.3 as a useful tool for estimating certain thermodynamic unknowns in complicated SS–AS systems.

5.2. Effect of temperature

The influence of temperature on experimental distribution coefficients has been studied widely in the geochemical literature, particularly for carbonate minerals (Holland *et al.*, 1964; Kinsman & Holland, 1969; Mucci, 1987; Malone & Baker, 1999; Dietzel *et al.*, 2004). In fact, trace-element distribution coefficients measured in both biogenic and abiogenic natural carbonates are used frequently as palaeotemperature records in palaeoceanographic and palaeoenvironmental studies (Gagan *et al.*, 1998; Elderfield &

Ganssen, 2000; Fairchild & Treble, 2009; Huang & Fairchild, 2001; Stoll *et al.*, 2002; Gaetani & Cohen, 2006). In this section, we discuss the influence of temperature on the different terms that contribute to the equilibrium values expressed in equations 54 and 55. Reviews of the applications of measured distribution coefficients as temperature proxies are presented in various chapters in this volume (Böttcher & Dietzel, 2010; Fairchild & Hartland, 2010; Stoll, 2010).

The main effect of temperature on the equilibrium-distribution coefficients arises from the solubility product ratio term in equations 54 and 55. Thermodynamic solubility products change with temperature according to the integrated Van't Hoff equation (Langmuir, 1997), which is given by:

$$-\log K^T = -\log K^0 + \frac{\Delta H_r^0}{2.303 \cdot \mathrm{R}}\left(\frac{1}{T} - \frac{1}{298.15}\right) \quad (57)$$

where K^T is the solubility product at temperature T (in Kelvin), K^0 is the solubility product at the reference temperature (298.15 K = 25°C) and ΔH_r^0 is the standard enthalpy of the dissolution reaction (at 25°C). Equation 57 assumes that ΔH_r^0 is constant, which is reasonable as long as T is within ± 10 to 20°C of the reference temperature (Langmuir, 1997). Alternatively, when the thermodynamic solubility product has been measured experimentally for a number of temperatures, its value can be determined from an analytical expression:

$$\log K = A_1 + A_2 T + A_3 T^{-1} + A_4 \log T + A_5 T^{-2} \quad (58)$$

where $A_1, A_2, \ldots, A_5$, are empirical fitting parameters. Although this second method is more precise, the use of equation 57 is a suitable approximation for variations within a limited range of temperatures. PHREEQC.DAT and other databases incorporate ΔH_r^0 data or empirical A_i parameters to automatically correct the solubility products for the effect of temperature. As an example, Figure 10 displays the variation of the ratio $(K_{\mathrm{arag}}/K_{\mathrm{str}})$ between the solubility products of aragonite and strontianite as a function of temperature. The calculation was carried out using equation 58 with the empirical fitting parameters compiled in PHREEQC.DAT. The ratio decreases as the temperature increases, indicating a decreased strontium distribution coefficient (see equations 54 and 55) with increasing temperature. Such a negative correlation between the incorporation of strontium in aragonite and temperature is in agreement with experimental results reported by several authors (Kinsman & Holland, 1969; Dietzel *et al.*, 2004; Gaetani & Cohen, 2006).

The exponential term in equations 54 and 55 is also expected to change with temperature, except for so-called 'strictly regular' solid solutions. Strictly regular solid solutions are a type of regular solid solutions (only $a_0 \neq 0$ in equation 30) in which the product $\mathrm{R}T \cdot a_0$ is independent of temperature and pressure (Navrotsky, 1987). In such cases, ΔG^E, γ_{BA}, and the exponential term in equations 54 and 55 do not depend on temperature. In general, the products $\mathrm{R}T \cdot a_i$ change with temperature although they may be considered relatively constant over a reasonable temperature interval

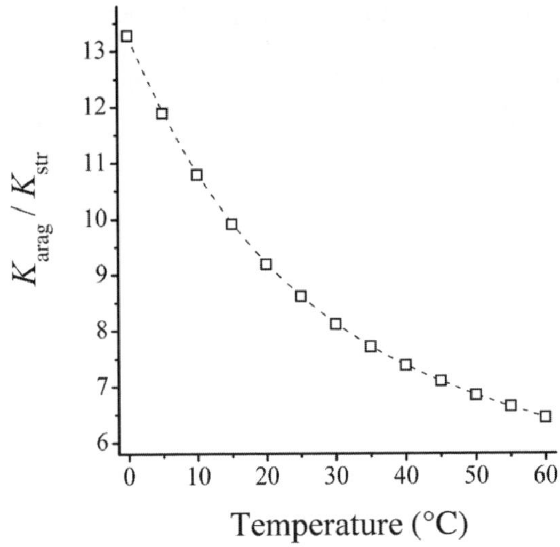

Fig. 10. Variation of the ratio (K_{arag}/K_{str}) between the solubility products of aragonite and strontianite as a function of temperature. Values of A_1, $A_2 \ldots, A_5$ (see equation 58) for aragonite and strontianite were taken from PHREEQC.DAT (Parkhurst & Appelo, 1999).

around the reference temperature. Finally, although the empirical parameters A and B in equations 18 and 19 depend on temperature, the quotient $\gamma_{B^{n+}}/\gamma_{C^{n+}}$ in equation 55 appears to be fairly 'insensitive' to changes in temperature. As a result, all of the terms in equation 55 except for the solubility product ratio remain constant over a moderate temperature range.

6. Complications

Whereas the previous considerations form the fundamental background of ion-partition thermodynamics, there are several complicating factors which need to be addressed for particular cases. (1) In the previous sections, we considered solid solutions with simple *BA*-type end-members, which involve only one mixing site per formula unit. When the stoichiometry implies several sites per formula unit, the thermodynamics remains simple, but the distribution coefficient expressions become more complex. (2) A second complication arises when the end-members are not isostructural, such as when trace amounts of large divalent cations such as Sr^{2+} or Ba^{2+} are incorporated into calcite. These large ions never form end-member carbonate minerals with a calcite-type structure; however, to apply equations 54 or 55, the solubility products of these 'fictitious' mineral phases are needed. (3) Ionic substitution may also be complicated by the need for charge balance when the minor element has a different valence than the carrier element for which it substitutes in the crystal structure. In such a case, neutrality is maintained by coupled substitution or by generation of extrinsic vacancies, which must be considered in formulating the corresponding distribution coefficients. (4) Partial equilibrium (Helgeson, 1968) commonly occurs in SS-AS systems where the initial reacting solids can become isolated from the aqueous phase by a layer of an overgrowing solid solution (Glynn *et al.*, 1990; Prieto, 2009).

In this section, we merely introduce complex solid-solution systems; however, these and other complications are considered in greater detail in a number of textbooks on the thermodynamics of natural systems, metamorphic petrology, *etc.* (*e.g.* Saxena, 1973; Spear, 1999; Anderson, 2005).

6.1. End-members with complex stoichiometry

In many solid solutions, there are two or more sites per formula unit for the substituting ions. For instance, when dealing with components like B_mA_n and C_mA_n, there are m mixing sites per formula unit of solid solution. Therefore, the exchange reaction in equation 2 becomes:

$$C_mA_{n(s)} + mB^{n+}_{(aq)} \longrightarrow B_mA_{n(s)} + mC^{n+}_{(aq)} \tag{59}$$

and the equilibrium condition in equation 3 is now given by

$$K_{B/C} = \frac{a_{B_mA_n}\{C^{n+}\}^m}{a_{C_mA_n}\{B^{n+}\}^m} \tag{60}$$

In the same way, the solubility product of the B_mA_n end-member becomes:

$$K_{B_mA_n} = \frac{\{B^{n+}\}^m\{A^{m-}\}^n}{a_{B_mA_n}} \tag{61}$$

and an analogous expression defines $K_{C_mA_n}$. Assuming ideal mixing of xB_mA_n and $(1-x)C_mA_n$ moles, the activities are given by $a_{B_mA_n} = x^m$ and $a_{C_mA_n} = (1-x)^m$, with the exponent m arising because the entropy of mixing is an extensive parameter and m moles of cations per mole of solid solution are being mixed (*e.g.* Navrotsky, 1987). To simplify the relationship between the activities of the components and the molar fractions of the substituting ions, it may be desirable to consider the chemical formulae on a 'one cation' basis, *i.e.* we can redefine the formula units as $BA_{n/m}$, $CA_{n/m}$ and $(B,C)A_{n/m}$. Thus, in the case of ideal mixing of $xBA_{n/m}$ and $(1-x)CA_{n/m}$, the activities $a_{BA_{n/m}}$ and $a_{BA_{n/m}}$ are equal to the cation mole fractions. This is a convenient formulation for application to element-distribution equilibria. With this approach, equation 60 becomes:

$$K_{B/C} = \frac{a_{BA_{n/m}}\{C^{n+}\}}{a_{CA_{n/m}}\{B^{n+}\}} \tag{62}$$

and the equilibrium distribution coefficient remains as in equation 34 (or as in equation 54 for trace elements), with the solubility products given by:

$$K_{BA_{n/m}} = (K_{B_mA_n})^{\frac{1}{m}} \tag{63}$$

and

$$K_{CA_{n/m}} = (K_{C_mA_n})^{\frac{1}{m}} \tag{64}$$

Thus, using the conventional formula, the equilibrium distribution coefficient becomes:

$$D_{\{B/C\}(eq)} = \left(\frac{K_{C_mA_n}}{K_{B_mA_n}}\right)^{\frac{1}{m}} \cdot \exp\left(\frac{\Delta\mu^E_{C_mA_n} - \Delta\mu^E_{B_mA_n}}{m\mathrm{R}T}\right) \tag{65}$$

or

$$D_{B/C(eq)} = \left(\frac{K_{C_mA_n}}{K_{B_mA_n}}\right)^{\frac{1}{m}} \cdot \frac{\gamma_{B^{n+}}}{\gamma_{C^{n+}}}\exp\left(\frac{-\Delta\mu^E_{B_mA_n}}{m\mathrm{R}T}\right) \tag{66}$$

when considering trace amounts of B^{n+} with D based on molal concentrations. The factor m in the exponential term (*cf.* McIntire, 1963) is due to $\gamma_{BA_{n/m}} = (\gamma_{B_mA_n})^{\frac{1}{m}}$ and $\Delta\mu^E_{B_mA_n}$ refers to the difference between the chemical potential of the component B_mA_n in the current solution and its chemical potential in an equivalent ideal solid solution.

The previous treatment can be extended to solid solutions that involve simultaneous mixing of ions on several sets of crystallographically non-equivalent sites. If each substituting ion can be unambiguously assigned to a given sub-lattice and there are no altervalent substitutions, such an extension should not be problematic.

6.2. Solid solutions with non-isostructural end-members

Solid solutions involving end-members (BA and CA) with different structures (α and β, respectively) can exist over a limited range of compositions. However, equations 20 and 21 are not sufficient for characterizing these types of solid solutions. In such cases, we must consider four different chemical potentials: (2) the chemical potentials μ^α_{BA} and μ^α_{CA} of components BA and CA in a solid solution with structure α; and (2) the chemical potentials μ^β_{BA} and μ^β_{CA} in a solid solution with structure β. Although the whole treatment becomes more complicated, equation 54 remains essentially the same for trace amounts of BA except for some minor specifications, such that:

$$D_{\{B/C\}(eq)} = \frac{K^\beta_{CA}}{K^\beta_{BA}} \cdot \exp\left(\frac{-\Delta\mu^E_{BA}}{\mathrm{R}T}\right) \tag{67}$$

where K^β_{BA} and K^β_{CA} are the solubility products of the end-members with structure β. Similarly, the excess chemical potential of BA becomes:

$$\Delta\mu^E_{BA} = \mathrm{R}T\ln\gamma^\beta_{BA} \tag{68}$$

where γ_{BA}^{β} is the activity coefficient of the trace component *BA* in the structure β of the carrier component *CA*. Pure *BA* with structure β can be a real solid phase at some temperatures and pressures, but in general, β-*BA* will be a physically unrealizable "crypto-modification" (Navrotsky, 1987) or fictitious phase. Such a situation occurs with the incorporation of Sr^{2+}, Ba^{2+}, Pb^{2+}, or Ra^{2+} in calcite because these large cations do not form calcite-type carbonates (see section 5.1). However, in order to apply equation 67, we need to know K_{BA}^{β}, the solubility product of a fictitious pure *BA* solid with structure β.

A practical way to solve this problem is by using Sverjensky's linear free-energy correlation expressions (Sverjensky, 1984; Sverjensky & Molling, 1992; Wang & Xu, 2001). This approach correlates the standard free energies of formation (ΔG_f^o) of different isostructural minerals (with a common anion) with the properties of the corresponding cations according to the expression:

$$\Delta G_{f,CA}^o = a_\beta \Delta G_{ns,C^{n+}}^o + b_\beta + c_\beta r_{C^{n+}} \tag{69}$$

where a_β, b_β, and c_β are fitting parameters determined for the specific isostructural family (β) to which mineral *CA* belongs and $r_{C^{n+}}$ represents the ionic radius of C^{n+}. The term ΔG_{ns}^o stands for the so-called standard non-solvation energy, which is given by the difference between the standard free energies of formation and solvation of the corresponding cation.

Once the fitting parameters are determined for the isostructural family β, the standard free energy of formation $\Delta G_{f,BA}^o$ of an unknown or fictitious β-*BA* solid can be calculated using equation 69. The solubility product of this fictitious phase can then be obtained from the expression:

$$\log K_{BA}^{\beta} = (\Delta G_{f,BA}^o - \Delta G_{f,B^{n+}}^o - \Delta G_{f,A^{n-}}^o)/2.303RT \tag{70}$$

where $\Delta G_{f,A^{n-}}^o$ and $\Delta G_{f,B^{n+}}^o$ are the standard free energies of formation of ions A^{n-} and B^{n+}, respectively. The parentheses in equation 70 contain the standard free energy of dissolution of the hypothetical β-*BA* solid. This correlation has been applied successfully to various mineral families (see Wang & Xu, 2001 and references therein) and allowed for the estimation of *pK* values of 7.61, 5.32, 10.33, and 4.83 for fictitious rhombohedral carbonates of Sr, Ba, Pb, and Ra, respectively. Figure 11 shows a linear free-energy correlation of the solubility products of some rhombohedral carbonates calculated using data compiled by Wang & Xu (2001). The data points estimated for some fictitious carbonates (open triangles) are also shown.

6.3. Non-isovalent substitutions

In the previous sections, we considered that the substituting ions are of the same valence. However, this is not always the case, particularly when one of the substituting ions occurs in trace amounts. Unfortunately, differences in valence complicate the previously discussed relationships because the maintenance of electrostatic

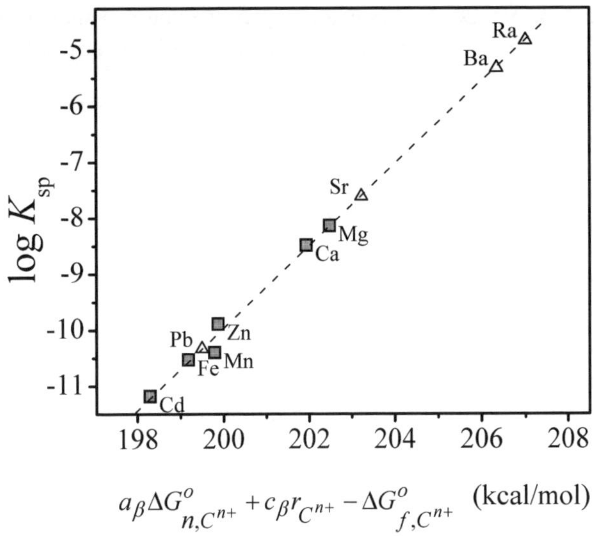

Fig. 11. Linear free energy correlation ($a_\beta = 0.969$; $c_\beta = 80.31$ kcal/mol) of the solubility products of some rhombohedral carbonates (modified from Wang & Xu, 2001). Data points (open triangles) estimated for some fictitious carbonates are also shown.

neutrality in the crystal structure requires a charge-compensation mechanism, either (1) the development of extrinsic vacancies or (2) a coupled substitution by a second trace element. In the first case, a typical exchange reaction might be written as:

$$CA_{(s)} + \frac{n}{m} B^{m+}_{(aq)} \longleftrightarrow B_{n/m} \square_{(1-n/m)} A + C^{n+}_{(aq)} \tag{71}$$

where $\square$ represents a vacancy.

When the charge is compensated by a second substitution, the exchange reaction can be written as:

$$2CA_{(s)} + B^{m+}_{(aq)} + D^{\nu+}_{(aq)} \longleftrightarrow BDA_{2(s)} + 2C^{n+}_{(aq)} \tag{72}$$

where the charge $\nu+$ of the balancing cation $D^{\nu+}$ is equal to $|m-n|$, and the equilibrium condition is given by:

$$K_{B/C} = \frac{a_{BDA_2} \{C^{n+}\}^2}{a^2_{CA} \{B^{m+}\} \{D^{\nu+}\}} \tag{73}$$

In both cases, determining equilibrium distribution expressions is difficult as we are working with components ($B_{n/m} \square_{(1-n/m)} A$ and BDA_2) the activities of which in the solid solution are unknown. Practical treatment for these non-isovalent mechanisms of substitution can be performed using Sverjensky's linear free-energy correlation expressions (Wang & Xu, 2001). For example, for the coupled substitution of equation 72, we can express the equilibrium condition in equation 73 in terms of the standard free energies of formation:

$$- RT \ln \frac{X_{BDA_2} \gamma_{BDA_2} \gamma^2_{C^{n+}} m^2_{C^{n+}}}{X^2_{CA} \gamma^2_{CA} \gamma_{B^{m+}} m_{B^{m+}} \gamma_{D^{\nu+}} m_{D^{\nu+}}} = \Delta G^o_{f,BDA_2} + 2\Delta G^o_{f,C^{n+}} - 2\Delta G^o_{f,CA} - \Delta G^o_{f,B^{m+}} - \Delta G^o_{f,D^{\nu+}} \quad (74)$$

Considering trace amounts of B^{m+} (and $D^{\nu+}$), the activity of the carrier component becomes $a_{CA} = \gamma_{CA} X_{CA} \approx 1$, whereas the unknowns $\Delta G^o_{f,BDA_2}$ and γ_{BDA_2} can be obtained by applying linear free energy correlations, such as in equation 69. A detailed description of this approach can be seen in Wang & Xu (2001), who applied the method to the coupled substitution of Ca^{2+} cations by REE^{3+} and Na^+ in the calcite structure.

6.4. The logarithmic distribution law

Due to the preferential partitioning effect, when a solid solution crystal grows from an aqueous phase, the substituting ions do not incorporate into the solid phase in the same proportion as in the aqueous phase. In practice, if the system is finite, the composition of both phases will change during the process, even when growth occurs near equilibrium. As a result, the solid composition will vary continuously from the centre of the crystal outward. There will be a difference in chemical potentials between the interior of the crystal and the aqueous phase and thus a tendency for migration of the components to equalize this difference. However, solid-state diffusion is an extremely sluggish process at ambient temperature, and the layers growing on the surface block the inner regions. We can consider this process as a sequence of 'partial equilibrium' states, in which each incremental crystal layer forms in equilibrium with the aqueous solution of a particular composition at that time. If m_{iA} and $m_{i,aq}$ are the molar amounts of cation *i* in the solid and aqueous solution, the typical differential equation for a Rayleigh process (Appelo & Postma, 2005) is:

$$\frac{dm_{BA}}{dm_{CA}} = \frac{-dm_{B,aq}}{-dm_{C,aq}} = \lambda_0 \frac{m_{B,aq}}{m_{C,aq}} \quad (75)$$

where λ_0 is a distribution coefficient. Integration of this equation from the initial (*ini*) to final (*fin*) composition gives:

$$\ln\left(\frac{m_{B,aq}(ini)}{m_{B,aq}(fin)}\right) = \lambda_0 \ln\left(\frac{m_{C,aq}(ini)}{m_{C,aq}(fin)}\right) \quad (76)$$

This expression is known as the logarithmic distribution law (Doerner & Hoskins, 1925) and λ_0 is usually called the logarithmic distribution coefficient. For $\lambda_0 > 1$, the component *BA* concentrates in the interior of the crystal and decreases outward. For

$\lambda_0 < 1$, the sequence is reversed. True Doerner-Hoskins behaviour implies that equilibrium between each infinitesimal incremental layer and the coexisting solution at its time of formation is given by equations 13 or 15. However, such behaviour does not represent true equilibrium but a sequence of partial equilibrium conditions. Over time, Doerner-Hoskins-type zoned crystals will dissolve and recrystallize as homogeneous crystals at true equilibrium with the aqueous phase. True Doerner-Hoskins behaviour is expected to occur only when growth rates are slow enough to permit equilibration between the solid-phase increments and solution and fast enough to prevent recrystallization of the growing crystals (Glynn *et al.*, 1990).

7. From equilibrium to effective distribution coefficients

The equilibrium relationships presented throughout the previous sections constitute an indispensable background for interpreting distribution coefficients measured in natural systems, but they are insufficient to account for many of the observed values. Kinetic and mechanistic factors significantly influence the effective distribution coefficients and will be considered separately in Chapter 2 (Putnis, 2010). Whatever geochemical scenario is considered, the first link between thermodynamics and crystallization behaviour is supersaturation. Supersaturation is a measure of the departure from equilibrium of a given system, which, in the case of stoichiometric minerals, is usually expressed by the quotient between the ionic activity product and the solubility product ($\Omega = IAP/K_{sp}$). Unfortunately, the concept of supersaturation in SS–AS systems is not straightforward and different expressions have been proposed (Prieto *et al.*, 1993; Astilleros *et al.*, 2003; Monnin & Cividini, 2006; Shtukenberg *et al.*, 2006; Prieto *et al.*, 2007). The interrelation and thermodynamic meaning of these expressions are discussed in greater detail by Prieto (2009), and here, we merely introduce the most common expressions.

7.1. Supersaturation in SS–AS systems

The simplest expression for supersaturation in SS–AS systems is given by (Monnin & Cividini, 2006; Shtukenberg *et al.*, 2006):

$$\xi = \frac{\Sigma\Pi}{\Sigma\Pi_{eq}} \tag{77}$$

where $\Sigma\Pi$ is the total activity product of the problem solution (see section 3.1) and $\Sigma\Pi_{eq}$ is the total solubility product corresponding to an aqueous solution on the solutus with the same aqueous activity fraction as the problem solution. Figure 12 illustrates the meaning of this concept for a hypothetical system. The problem solution plots above the solutus curve on the Lippmann diagram (point P in Fig. 12b) and has the same aqueous activity fraction as solution Q, which plots on the solutus. In turn, aqueous solution Q is in equilibrium with solid solution R (on the solidus) and the total solubility product ($\Sigma\Pi_{eq} = 10^{-8.5}$ in the example) is the same for both Q

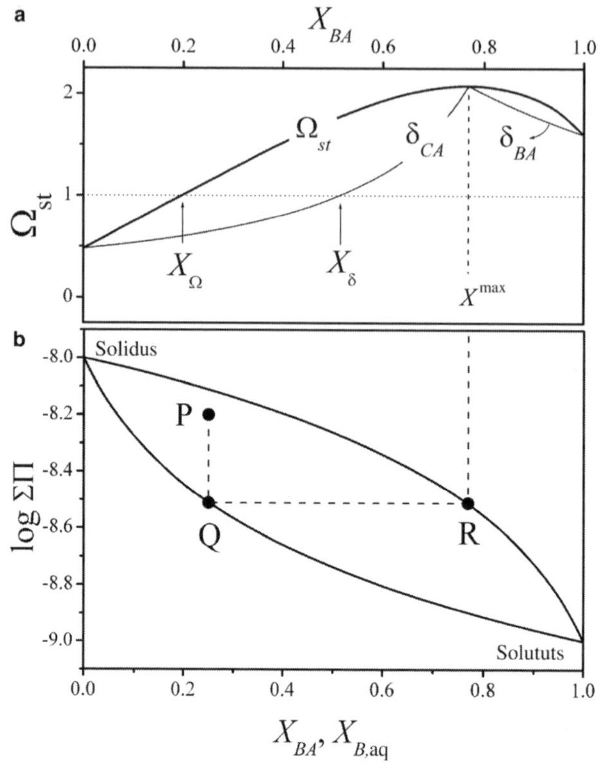

Fig. 12. Supersaturation functions on a Lippmann diagram. The calculation was carried out for a hypothetical solid solution. (***a***) Stoichiometric supersaturation Ω_{st} and δ functions. Note the three common points at $\delta_{CA}(0) = \Omega_{st}(0)$, $\delta_{BA}(1) = \Omega_{st}(1)$, and $\delta_{BA}(X^{max}) = \delta_{CA}(X^{max}) = \Omega_{st}(X^{max})$. The figure also shows the ranges, X_δ–1 and X_Ω–1, of compositions for which the aqueous solution is supersaturated. (***b***) The problem solution (point P) has the same aqueous activity fraction as solution Q. This last solution is at equilibrium with a solid solution (point R) of composition X^{max}.

and R. Therefore, the single value in equation 77 indicates the supersaturation of the problem solution P with respect to the solid solution R. However, aqueous solutions that plot above the solutus curve are supersaturated with respect to a range of solid solution compositions (Glynn & Reardon, 1990) and all members within this range are predisposed to precipitation. Therefore, in dealing with SS–AS systems, the supersaturation of a given solution cannot be expressed by a single value. To make predictions, the range of solid-solution compositions for which the problem solution is supersaturated and the degree of supersaturation (or undersaturation) with respect to all solid compositions should be known. The supersaturation must be expressed as a function of the solid composition, X_{BA}, and at least two different expressions for supersaturation are found in the literature. The first is the stoichiometric supersaturation function (Prieto *et al.*, 1993), which is based on the stoichiometric saturation concept (see section 3.2.) and is given by:

$$\Omega_{st}(X_{BA}) = \frac{\{B^{n+}\}^{X_{BA}}\{C^{n+}\}^{X_{CA}}\{A^{n-}\}}{(K_{BA}\gamma_{BA}X_{BA})^{X_{BA}}(K_{CA}\gamma_{CA}X_{CA})^{X_{CA}}} \quad (78)$$

Since stoichiometric saturation is not a true equilibrium state, but rather a metastable equilibrium condition, Ω_{st} does not represent the departure of the system from stability. For a given aqueous solution, *i.e.* for specific values of $\{B^{n+}\}$, $\{C^{n+}\}$, and $\{A^{n-}\}$, the function in equation 78 provides a departure from stoichiometric saturation with respect to all compositions of the solid solution. Figure 12a displays the $\Omega_{st}(X_{BA})$ function corresponding to solution P in Figure 12b. The curve has a maximum at $X^{\max}$, which represents the solid composition with respect to which the aqueous solution is most supersaturated. This maximum $\Omega_{st}(X^{\max})$ coincides with the value of ξ and occurs for the same solid composition (that of point R in Figure 12b). Therefore, $\Omega_{st}(X^{\max})$ is the only value of the $\Omega_{st}(X_{BA})$ function that represents a departure from a true equilibrium condition. The main advantage of using the $\Omega_{st}(X_{BA})$ function is that it allows not only for an immediate evaluation of $\xi = \Omega_{st}(X^{\max})$, but also for an estimation of the compositional range for which the aqueous solution is supersaturated. As shown in Figure 12a, solution P is stoichiometrically supersaturated ($\Omega_{st} > 1$) with respect to pure BA and to all solid compositions in the range of $X_{\Omega}-1$. If a solid with a composition within this range grows congruently, the free energy of the system will decrease. On the contrary, for compositions in the range of $0-X_{\Omega}$, the solid can dissolve congruently.

However, whereas congruent dissolution is a real dissolution mechanism, congruent growth is not a common process in either nature or the laboratory (Lafon, 1978). Therefore, the function in equation 78 is a good estimator for the driving force of dissolution processes, but should not be used in modelling crystal growth or ripening processes. Despite this, the supersaturated range with $\Omega_{st}(X_{BA}) > 1$ provides a useful qualitative tool for interpreting compositions of previously-produced nuclei (Andara *et al.*, 2005) and growth mechanisms (Pina *et al.*, 2000) in crystallization experiments.

Similar to equilibrium in binary SS–AS systems (equations 37 and 38), disequilibrium also requires that two equations be defined. Inspection of the equilibrium relationships in equations 37 and 38 allows for the definition of two 'partial' supersaturation functions:

$$\delta_{BA}(X_{BA}) = \frac{\{B^{n+}\}\{A^{n-}\}}{K_{BA}\gamma_{BA}X_{BA}} \quad (79)$$

and

$$\delta_{CA}(X_{CA}) = \frac{\{C^{n+}\}\{A^{n-}\}}{K_{CA}\gamma_{CA}X_{CA}} \quad (80)$$

In an attempt to find a more 'all-purpose' expression for supersaturation, Astilleros *et al.* (2003) explored these two functions, which meet at $X^{\max}$ (see Fig. 12a). Based on a 'geometrical' criterion, these authors proposed using equation 79 as an estimator of supersaturation for all solid compositions with $X_{BA} \geq X^{\max}$ and equation 80 for those with $X_{BA} \leq X^{\max}$. As shown in Figure 12a, the stoichiometric supersaturation and δ-functions have three points in common: $\delta_{CA}(0) = \Omega_{st}(0)$, $\delta_{BA}(1) = \Omega_{st}(1)$, and $\delta_{BA}(X^{\max}) = \delta_{CA}(X^{\max}) = \Omega_{st}(X^{\max}) = \xi$. For all other compositions, the functions diverge, with δ always being smaller than Ω_{st}. Moreover, the composition range of $X_{\delta}-1$ for which solution P is δ-supersaturated is considerably narrower.

The δ-functions simply define the range of solid compositions for which a given solution is 'strictly' supersaturated. In contact with solution P, a solid solution with a composition in the range of $X_{\delta}-1$ will grow. During that growth, both components will incorporate into the solid phase to decrease their chemical potentials. Net growth (increased total amount of solid) can also occur when solution P is in contact with solid solutions in the range of $X_{\Omega}-X_{\delta}$ (see Fig. 12a), but the solid will tend to become richer in *CA* and poorer in *BA*. A detailed discussion about the thermodynamic meaning of the supersaturated ranges is found in Prieto (2009).

7.2. Supersaturation and distribution coefficients

It is worth noting that there is a connection between maximum supersaturation and equilibrium partitioning. Such a relationship is obvious after inspection of Figure 12, since solution P has the same aqueous activity fraction as solution Q, which in turn is in equilibrium with a solid solution of composition $X^{\max}$. When a solid of composition $X^{\max}$ crystallizes, the effective distribution coefficient equals the corresponding equilibrium value $D_{\{B/C\}(eq)}$. One could think that the substituting ions will always tend to distribute according to the equilibrium values, since $X^{\max}$ is the composition for which the aqueous solution is most supersaturated and thus the most likely to precipitate. However, precipitate compositions are often not determined by maximum supersaturation values (Prieto *et al.*, 1997; Pina & Putnis, 2002; Putnis *et al.*, 2003). In many cases, the precipitates tend to be enriched in the more soluble of the two components, differing in this way from the composition expected from equilibrium partitioning considerations. The reasons why 'more-soluble' solid-solution compositions are kinetically favoured are discussed in Chapter 2 (Putnis, 2010). Delineating the range of solid compositions that are thermodynamically available to precipitate is fundamental to designing suitable kinetic models, which can be done using the supersaturation functions defined in the previous sections.

8. Concluding remarks

The study of ion partitioning in SS–AS systems has garnered scientific attention for decades. Distribution coefficients have been demonstrated to be a valuable record to interpret dissolution-crystallization processes in natural systems. Factors such as supersaturation, temperature, solution chemistry, overall growth rates, *etc.* have a significant influence on ion partitioning and, therefore, observed values can provide information about the original geochemical scenarios. To disclose this information, geochemists frequently use correlations or semi-empirical models. Whereas empirical correlations are useful tools for practical purposes, the thermodynamic relationships presented throughout the previous sections constitute an indispensable background for developing rigorous models.

Acknowledgements

Many of the concepts presented in this chapter have matured and progressed with the help of numerous researchers and collaborators. I would like to thank all these colleagues for discussions concerning solid solutions and related subjects, particularly Udo Becker, Dirk Bosbach, Michael Böttcher, Lurdes Fernández, Ángeles Fernández, Mario Gonçalves, Athanasios Godelitsas, Amalia Jiménez, Stephan Köhler, Dmitrii Kulik, Eric Oelkers, Carlos Pina, Andrew Putnis, Christine Putnis, Susan Stipp, and Heather Stoll. I would also like to express my gratitude to the many students past and present for their insightful contributions.

References

Allison, J.D., Brown, D.S. & Novo-Gradac, K.J. (1991) *MINTEQA2, a geochemical assessment model for environmetal systems.* Report EPA/600/3-91/0-21, U.S. Environmental Protection Agency, Athens, Georgia, USA, 107 pp.

Andara, A., Heasman, D., Fernández-González, A. & Prieto, M. (2005) Characterisation and crystallisation of the $Ba(SeO_4,SO_4)$ solid solution. *Crystal Growth and Design*, **5**, 1371–1378.

Anderson, G.M. (2005) *Thermodynamics of Natural Systems.* Cambridge University Press, Cambridge, UK, 648 pp.

Appelo, C.A.J. & Postma, D. (2005) *Geochemistry, Groundwater and Pollution.* Balkema Publishers, Leiden, The Netherlands, 649 pp.

Astilleros, J.M., Pina, C.M., Fernández-Díaz, L. & Putnis, A (2003) Supersaturation functions in binary solid solution-aqueous solution systems. *Geochimica et Cosmochimica Acta*, **67**, 1601–1608.

Beattie, P., Drake, M., Jones, J., Leeman, W., Longhi, J., McKay, G., Nielsen, R., Palme, H., Shaw, D., Takahashi, E. & Watson, B. (1993) Terminology for trace-element partitioning. *Geochimica et Comsochimica Acta*, **57**, 1605–1606.

Becker, U., Fernández-González, A., Prieto, M., Harrison, R. & Putnis, A. (2000) Direct calculation of thermodynamic properties of the barite-celestite solid solution from molecular principles. *Physics and Chemistry of Minerals*, **27**, 291–300.

Berthelot, M. (1872) One law which governs the distribution of a substance between two solvents. *Annales de Chimie et de Physique (4th series)*, **26**, 408–417.

Böttcher, M.E. (1997) Experimental dissolution of $CaCO_3$-$MnCO_3$ solid solutions in CO_2-H_2O solutions at 20°C, I. Synthetic low-temperature carbonates. *Solid State Ionics*, **101–103**, 1263–1266.

Böttcher, M.E. (2000) Editorial honoring Friedrich Lippmann. *Aquatic Geochemistry*, **6**, 115–117.

Böttcher, M.E. & Dietzel, M. (2010) Metal-ion partitioning during low-temperature precipitation and dissolution of anhydrous carbonates and sulphates. In: Ion partitioning in ambient temperature aqueous systems (M. Prieto and H. Stoll, editors). EMU-Notes in Mineralogy, **10**, 139–187. European Mineralogical Union and the Mineralogical Society, Twickenham, UK.

Boyle, E.A. (1988) Cadmium: Chemical tracer of deepwater paleoceanography. *Paleoceanography*, **3**, 471–489.

Bruland, K.W. (1992) Complexation of cadmium by natural organic ligands in the central North Pacific. *Limnology and Oceanography*, **37**, 1008–1017.

Busenberg, E., Plummer, L.N. & Parker, V.B. (1984) The solubility of strontianite ($SrCO_3$) in CO_2-H_2O solutions between 2 and 91°C, the association constants of $SrHCO_3^+$(aq) and $SrCO_3^0$(aq) between 5 and 80°C, and an evaluation of the thermodynamic properties of Sr^{2+}(aq) and $SrCO_3$(cr) at 25°C and 1 atm total pressure. *Geochimica et Cosmochimica Acta*, **48**, 2021–2035.

Carneiro, J., Prieto, M. & Stoll, H. (2010) The saturation state of seawater with respect to (Cd,Ca)CO_3, (Co,Ca)CO_3, (Ni,Ca)CO_3, and (Zn,Ca)CO_3 solid solutions (in prep.).

Casey, W.H., Chai, L., Navrostsky, A. & Rock, P.A. (1996) Thermochemistry of mixing strontianite [$SrCO_3$(s)] and aragonite [$CaCO_3$(s)] to form $Ca_{1-x}Sr_xCO_3$(s) solid solutions. *Geochimica et Cosmochimica Acta*, **60**, 933–940.

Chiarello, R.P., Sturchio, N.C., Grace, J.D., Geissbuhler, P., Sorensen, L.B., Cheng, L. & Xu, S. (1997) Otavite-calcite solid-solution formation at the calcite-water interface studied in situ by synchrotron X-ray scattering. *Geochimica et Cosmochimica Acta*, **61**, 1467–1474.

Crocket, J.H. & Winchester, J.W. (1966) Coprecipitation of zinc with calcium carbonate. *Geochimica et Cosmochimica Acta*, **30**, 1093–1109.

Davis, J.A., Fuller, C.C. & Cook, A.D. (1987) Mechanisms of trace metal sorption by calcite: Adsorption of Cd^{2+} and subsequent solid solution formation. *Geochimica et Cosmochimica Acta*, **51**, 1477–1490.

Delaney, M.L. (1989) Uptake of cadmium into calcite shells by planktonic foraminifera. *Chemical Geology*, **78**, 159–165.

Dietzel, M., Gussone, N. & Eisenhauer, A. (2004) Co-precipitation of Sr^{2+} and Ba^{2+} with aragonite by membrane diffusion of CO_2 between 10 and 50°C. *Chemical Geology*, **203**, 139–151.

Doerner, H.A. & Hoskins, W.M. (1925) Coprecipitation of radium and barium sulphates. *Journal of the American Chemical Society*, **46**, 662–675.

Elderfield, H. & Ganssen, G. (2000) Past temperature and $\delta^{18}O$ of surface ocean waters inferred from foraminiferal Mg/Ca ratios. *Nature*, **405**, 442–445.

Fairchild, I.J. & Treble, P.C. (2009) Trace elements in speleothems as recorders of environmental change. *Quaternary Science Reviews*, 28, 449–468.

Fairchild, I.J. & Hartland, A. (2010) Trace element variations in stalagmites: controls by climate and by karst system processes. In: Ion partitioning in ambient temperature aqueous systems (M. Prieto and H. Stoll, editors). EMU-Notes in Mineralogy, **10**, 259–287. European Mineralogical Union and the Mineralogical Society, Twickenham, UK.

Fernández-González, A., Prieto, M., Putnis, A. & López-Andrés, S. (1999) Concentric zoning patterns in crystallizing (Cd,Ca)CO_3 solid solutions from aqueous solutions. *Mineralogical Magazine*, **63**, 331–343.

Fernández-González, A., Andara, A., Alía, J.M. & Prieto M. (2006) Miscibility in the $CaSO_4 \cdot 2H_2O$–$CaSeO_4 \cdot 2H_2O$ system: Implications for the crystallisation and dehydration behaviour. *Chemical Geology*, **225**, 256–265.

Fernández-González, A., Andara, A. & Prieto, M. (2007) Mixing properties and crystallization behaviour of the scheelite-powellite solid solution. *Crystal Growth and Design*, **7**, 542–552.

Gaetani, G.A. & Cohen, A.L. (2006) Element partitioning during precipitation of aragonite from seawater: A framework for understanding paleoproxies. *Geochimica et Cosmochimica Acta*, **70**, 4617–4634.

Gagan, M.K., Ayliffe, L.K., Hoplry, D., Cali, J.A., Mortimer, G.E., Chappell, J., McCulloch, M.T. & Head, M.J. (1998) Temperature and surface ocean water balance of the mid-Holocene tropical western Pacific. *Science*, **279**, 1014–1018.

Gamsjäger, H., Königsberger, E. & Preis, W. (2000) Lippmann diagrams: Theory and application to carbonate systems. *Aquatic Geochemistry*, **6**, 119–132.

Glynn, P.D. (1991) MBSSAS: A code for the computation of Margules parameters and equilibrium relations in binary solid-solution aqueous-solution systems. *Computers and Geosciences*, **17**, 907–966.

Glynn, P.D. (2000) Solid-solution solubilities and thermodynamics: Sulfates, carbonates and halides, In: *Sulfate Minerals: Crystallography, Geochemistry and Environmental Significance* (C.N. Alpers,

J.L. Jambor & D.K. Nordstrom, editors). Reviews in Mineralogy and Geochemistry, **40**, 481–511. Mineralogical Society of America, Washington, D.C.

Glynn, P.D. & Reardon, E.J. (1990) Solid-solution aqueous solution equilibria: thermodynamic theory and representation. *American Journal of Science*, **290**, 164–201.

Glynn, P.D., Reardon, E.J., Plummer, L.N. & Busenberg, E. (1990) Reaction paths and equilibrium end-points in solid-solution aqueous solution systems. *Geochimica et Cosmochimica Acta*, **54**, 267–282.

Guggenheim, E.A. (1937) Theoretical basis of Raoult's law. *Transactions of the Faraday Society*, **33**, 151–159.

Helgeson, H.C. (1968) Evaluation of irreversible reactions in geochemical processes involving minerals and aqueous solutions: I. Thermodynamic relations. *Geochimica et Cosmochimica Acta*, **32**, 853–877.

Henderson, L.M. & Kraceck, F.C. (1927) The fractional precipitation of barium and radium chromates. *Journal of the American Chemical Society*, **49**, 739–749.

Högfeldt, E. (1982) *Stability Constants of Metal-Ion Complexes, Part A, Inorganic Ligands*. IUPAC Chemical Data Series N° 21. Pergamon Press, New York, 310 pp.

Holland H.D., Holland, H.J. & Munoz, J.L. (1964) The coprecipitation of cations with $CaCO_3$-II. The coprecipitation of Sr^{+2} with calcite between 90° and 100°C. *Geochimica et Cosmochimica Acta*, **28**, 1287–1301.

Holst, N.B. (1978) The use of thermodynamic excess functions in the Nernst distribution law. *American Mineralogist*, **63**, 83–86.

Huang, Y. & Fairchild, I.J. (2001) Partitioning of Sr^{2+} and Mg^{2+} into calcite under karst-analogue experimental conditions. *Geochimica et Cosmochimica Acta*, **65**, 47–62.

Katsikopoulos, D., Fernández-González, A. & Prieto, M. (2008a) Crystallization of the $(Cd,Ca)CO_3$ solid solution in double diffusion systems: The partitioning behaviour of Cd^{2+} in calcite at different supersaturation rates. *Mineralogical Magazine*, **72**, 433–436.

Katsikopoulos, D., Fernández-González, A., Prieto, A.C. & Prieto, M. (2008b) Co-crystallization of Co(II) with calcite: implications for the mobility of cobalt in aqueous environments. *Chemical Geology*, **254**, 87–100.

Katsikopoulos, D., Fernández-González, A. & Prieto, M. (2009) Precipitation and mixing properties of the "disordered" $(Mn,Ca)CO_3$ solid solution. *Geochimica et Cosmochimica Acta*, **73**, 6147–6161.

Kinsman, D.J.J. & Holland, H.D. (1969) The coprecipitation of cations with $CaCO_3$-IV. The coprecipitation of Sr^{2+} with aragonite between 16° and 96°. *Geochimica et Cosmochimica Acta*, **33**, 1–17.

Königsberger, E., Hausner, R. & Gamsjäger, H. (1991) Solid-solute equilibria in aqueous solution: V. The system $CdCO_3$-$CaCO_3$-CO_2-H_2O. *Geochimica et Cosmochimica Acta*, **55**, 3505–3514.

Kornicker, W.A., Presta P.A., Paige, C.R., Johnson, D.M., Hileman, O.E. & Snodgrass, W.J. (1991) The aqueous dissolution kinetics of the barium/lead sulfate solid solution series at 25 and 60°C. *Geochimica et Cosmochimica Acta*, **55**, 3531–3541.

Kulik, D.A., Kersten, M., Heiser, U. & Neumann, T. (2000) Application of Gibbs energy minimization to model early-diagenetic solid-solution aqueous-solution equilibria involving authigenic rhodochrosites in anoxic Baltic Sea sediments. *Aquatic Geochemistry*, **6**, 147–199.

Kulik, D.A., Vinograd, V.L., Paulsen, N. & Winkler, B. (2010) $(Ca,Sr)CO_3$ aqueous-solid solution systems: From atomistic simulations to thermodynamic modeling. *Physics and Chemistry of the Earth*, doi:10.1016/j.pce.2010.04.011.

Lafon, G.M. (1978) Discussion of "Equilibrium criteria for two-component solids reacting with fixed composition in aqueous phase. Example: the magnesian calcites". *American Journal of Science*, **274**, 1455–1468.

Langmuir, D. (1997) *Aqueous Environmental Geochemistry*. Prentice Hall, New Jersey, USA, 600 pp.

Lippmann, F. (1977) The solubility product of complex minerals, mixed crystals and three-layer clay minerals. *Neues Jahrbuch für Mineralogie Abhandlungen*, **130**, 243–263.

Lippmann, F. (1980) Phase diagrams depicting the aqueous solubility of binary mineral systems. *Neues Jahrbuch für Mineralogie Abhandlungen*, **139**, 1–25.

Lorens, R.B. (1981) Strontium, cadmium, manganese, and cobalt distribution coefficients in calcite as a function of calcite precipitation rate. *Geochimica et Cosmochimica Acta*, **45**, 553–561.

Malone, M.J. & Baker, P.A. (1999) Temperature dependence of the strontium distribution coefficient in calcite: An experimental study from 40° to 200°C and application to natural diagenetic calcites. *Journal of Sedimentary Research*, **69**, 216–223.

Mashiotta, T.A., Lea, D.W. & Spero, H.J. (1997) Experimental determination of cadmium uptake in shells of the planktonic foraminifera Orbulina universa and Globigerina bulloides: Implications for surface water paleoreconstructions. *Geochimica et Cosmochimica Acta*, **61**, 4053–4065.

McBeath, M.K., Rock, P.A., Casey, W.H. & Mandell, G.K. (1998) Gibbs energies of formation of metal-carbonate solid solutions: Part 3. The $Ca_xMn_{1-x}CO_3$ system at 298 K and 1 bar. *Geochimica et Cosmochimica Acta*, **62**, 2799–2808.

McIntire, W.L. (1963) Trace Element partition coefficients – a review of theory and applications to geology. *Geochimica et Cosmochimica Acta*, **27**, 1209–1264.

Monnin, C. & Cividini, D. (2006) The saturation state of the world's ocean with respect to $(Ba,Sr)SO_4$ solid solutions. *Geochimica et Cosmochimica Acta*, **70**, 3290–3298.

Morel, F.M.M. & Price, N.M. (2003) The biogeochemical cycles of trace metals in the oceans. *Science*, **300**, 944–947.

Mucci, A. (1987) Influence of temperature on the composition of magnesian calcite overgrowths precipitated from seawater. *Geochimica et Cosmochimica Acta*, **51**, 1977–1984.

Mucci, A. & Morse, J.W. (1983) The incorporation of Mg^{2+} and Sr^{2+} into calcite overgrowths: influences of growth rate and solution composition. *Geochimica et Cosmochimica Acta*, **47**, 217–233.

Navrotsky, A. (1987) Models of crystalline solutions. In: *Thermodynamic Modeling of Geological Materials – Minerals, Fluids and Melts* (I.S.E. Carmichael & H.P. Eugster, editors). Reviews in Mineralogy and Geochemistry, **17**, 35–69. Mineralogical Society of America, Washington, D.C.

Nernst, W. (1891) Distribution of a substance between two solvents and between solvent and vapor. *Zeitschrift für Physikalische Chemie*, **8**, 110–139.

Oelkers, E.H., Bénézeth, P. & Pokrovski, G.S. (2009) Thermodynamic databases for water-rock interaction. In: *Thermodynamics and Kinetics of Water-rock Interaction* (E.H. Oelkers & J. Schott, editors). Reviews in Mineralogy and Geochemistry, **70**, 1–46. Mineralogical Society of America, Washington, D.C.

Parkhurst, D.L. (1990) Ion-Association models and mean activity coefficients of various salts. In: *Chemical Modeling of Aqueous Systems II* (D.C. Melchior & R.L. Basset, editors). ACS Symposium Series, **416**, 30–43. American Chemical Society, Washington, D.C.

Parkhurst, D.L. & Appelo, C.A.J. (1999) *User's guide to PHREEQC (Version 2) – A computer program for speciation, batch-reaction, one-dimensional transport and inverse geochemical calculations.* U.S. Geological Survey Water Resources Investigations Report 99–4259, U.S. Geological Survey, Denver, 312 pp.

Pina, C.M. & Putnis, A. (2002) The kinetics of nucleation of solid solutions from aqueous solutions: a new model for calculating non-equilibrium distribution coefficients. *Geochimica et Cosmochimica Acta*, **66**, 185–192.

Pina, C.M., Enders, M. & Putnis, A. (2000) The composition of solid solutions crystallising from aqueous solutions: The influence of supersaturation and growth mechanisms. *Chemical Geology*, **168**, 195–210.

Plummer, L.N. & Busenberg, E. (1982) The solubilities of calcite, aragonite and vaterite in CO_2-H_2O solutions between 0 and 90°C, and an evaluation of the aqueous model for the system $CaCO_3$-CO_2-H_2O. *Geochimica et Cosmochimica Acta*, **46**, 1011–1040.

Plummer, L.N. & Busenberg, E. (1987) Thermodynamics of the aragonite-strontianite solid solutions: Results from stoichiometric solubility at 25 and 76°C. *Geochimica et Cosmochimica Acta*, **51**, 1393–1411.

Prieto, M. (2009) Thermodynamics of solid solution-aqueous solution systems. In: *Thermodynamics and Kinetics of Water-rock Interaction* (E.H. Oelkers & J. Schott, editors). Reviews in Mineralogy and Geochemistry, **70**, 47–85. Mineralogical Society of America, Washington, D.C.

Prieto, M., Putnis, A. & Fernández-Díaz, L. (1993) Crystallization of solid solutions from aqueous solutions in a porous medium: Zoning in $(Ba,Sr)SO_4$. *Geological Magazine*, **130**, 289–299.

Prieto, M., Fernández-González, A., Putnis, A. & Fernández-Díaz, L. (1997) Nucleation, growth, and zoning phenomena in crystallizing (Ba,Sr)CO_3, Ba(SO_4,CrO_4), (Ba,Sr)SO_4, and (Cd,Ca)CO_3 solid solutions from aqueous solutions. *Geochimica et Cosmochimica Acta*, **61**, 3383–3397.

Prieto, M., Fernández-González, A., Becker, U. & Putnis, A. (2000) Computing Lippmann diagrams from direct calculation of mixing properties of solid solutions: Application to the barite-celestite system. *Aquatic Geochemistry*, **6**, 133–146.

Prieto, M., Cubillas, P. & Fernández-González, A. (2003) Uptake of dissolved Cd by biogenic and abiogenic aragonite: a comparison with sorption onto calcite. *Geochimica et Cosmochimica Acta*, **67**, 3859–3869.

Prieto, M., Astilleros, J.M., Pina, C.M., Fernández-Díaz, L. & Putnis, A. (2007) Comment: Supersaturation in binary solid solution-aqueous solution systems. *American Journal of Science*, **307**, 1034–1045.

Putnis, A. (1992) *Introduction to Mineral Sciences*. Cambridge University Press, Cambridge, UK, 457 pp.

Putnis, A. (2010) Effects of kinetics and mechanisms of crystal growth on ion-partitioning in solid solution – aqueous solution (SS–AS) systems. In: Ion partitioning in ambient temperature aqueous systems (M. Prieto and H. Stoll, editors). EMU-Notes in Mineralogy, **10**, 43–64. European Mineralogical Union and the Mineralogical Society, Twickenham, UK.

Putnis, A., Pina, C.M., Astilleros, J.M., Fernández-Díaz, L. & Prieto, M. (2003) Nucleation of solid solutions crystallizing from aqueous solutions. *Philosophical Transactions of the Royal Society of London*, **361**, 615–632.

Reeder, R.J. (1983) Crystal chemistry of the rhombohedral carbonates. In: *Carbonates: Mineralogy and Chemistry* (R.J. Reeder, editor). Reviews in Mineralogy, **11**, 1–47. Mineralogical Society of America, Washington, D.C.

Rimstidt, J.D., Balog, A. & Webb, J. (1998) Distribution of trace elements between carbonate minerals and aqueous solutions. *Geochimica et Cosmochimica Acta*, **62**, 1851–1863.

Ruiz-Hernández, S.E., Grau-Crespo, R., Ruiz-Salvador, A.R. & de Leeuw, N.H. (2010) Thermochemistry of strontium incorporation in aragonite from atomistic simulations. *Geochimica et Cosmochimica Acta*, **74**, 1320–1328.

Saxena, S.K. (1973) *Thermodynamics of Rock-forming Crystalline Solutions*. Springer-Verlag, Berlin, 188 pp.

Shtukenberg, A.G., Punin, Y.O. & Azimov, P. (2006) Crystallization kinetics in binary solid solution-aqueous solutions systems. *American Journal of Science*, **290**, 164–201.

Spear, F.S. (1999) *Metamorphic Phase Equilibria and Pressure-Temperature-Time Paths*. Mineralogical Society of America, Washington, D.C., 799 pp.

Stipp, S.L., Hochella, M.F., Parks, G.A. & Leckie, J.O. (1992) Cd^{2+} uptake by calcite, solid-state diffusion, and the formation of solid-solution: Interface processes observed with near-surface sensitive techniques (XPS, LEED, and AES). *Geochimica et Cosmochimica Acta*, **56**, 1941–1954.

Stipp, S.L., Parks, G.A., Nordstrom, D.K. & Leckie, J.O. (1993) Solubility-product constant and thermodynamic properties for synthetic otavite, $CdCO_{3(s)}$, and aqueous association constants for the Cd(II)-CO_2-H_2O system. *Geochimica et Cosmochimica Acta*, **57**, 2699–2713.

Stoll, H.M. (2010) Ion partitioning and trace-element proxies in foraminifera and coccoliths. In: Ion partitioning in ambient temperature aqueous systems (M. Prieto and H. Stoll, editors). EMU-Notes in Mineralogy, **10**, 399–415. European Mineralogical Union and the Mineralogical Society, Twickenham, UK.

Stoll, H.M., Rosenthal, Y. & Falkowski, P. (2002) Climate proxies from Sr/Ca of coccolith calcite: Calibrations from continuous culture of Emiliania huxleyi. *Geochimica et Cosmochimica Acta*, **66**, 927–936.

Sverjensky, D.A. (1984) Prediction of Gibbs free energies of calcite-type carbonates and the equilibrium distribution of trace elements between carbonates and aqueous solutions. *Geochimica et Cosmochimica Acta*, **48**, 1127–1134.

Sverjensky, D.A. & Molling, P.A. (1992) A linear free energy relationship for crystalline solids and aqueous ions. *Nature*, **356**, 231–234.

Tesoriero, A.J. & Pankow, J.F. (1996) Solid solution partitioning of Sr^{2+}, Ba^{2+}, and Cd^{2+} to calcite. *Geochimica et Cosmochimica Acta*, **60**, 1053–1063.

Thorstenson, D.C. & Plummer, L.N. (1977) Equilibrium criteria for two-component solids reacting with fixed composition in an aqueous-phase. Example: The magnesian calcites. *American Journal of Science*, **277**, 1203–1223.

Urusov, V.S. (1974) Energetic criteria for solid solution miscibility gap calculations. *Bulletin de la Société Française de Minéralogie et de Cristallographie*, **91**, 217–222.

Vinograd, V.L., Winkler, B., Putnis, A., Gale, J.D. & Sluiter, M.H.F. (2006) Static lattice energy calculations of mixing and ordering enthalpies in binary carbonate solid solutions. *Chemical Geology*, **225**, 304–313.

Vinograd, V.L., Bosbach, D., Winkler, B. & Gale, J.D. (2008) Subsolidus phase relations in $Ca_2Mo_2O_8$–$NaEuMo_2O_8$-powellite solid solution predicted from static lattice energy calculations and Monte Carlo simulations. *Physical Chemistry Chemical Physics*, **10**, 3509–3518.

Wang, Y. & Xu, H. (2001) Prediction of metal partitioning between minerals and aqueous solutions: A liner free energy correlation approach. *Geochimica et Cosmochimica Acta*, **65**, 1529–1543.

Watson, E.B. (2004) A conceptual model for near-surface kinetic controls on the trace-element and stable isotope composition of abiogenic calcite crystals. *Geochimica et Cosmochimica Acta*, **68**, 1473–1488.

Wolery, T.J. (1992) *EQ3/6, a software package for geochemical modeling of aqueous systems*. Lawrence Livermore Nat. Lab. Report UCRL-MA-110662 Pt I-III, Livermore, CA, USA, 246 pp.

EMU Notes in Mineralogy, Vol. 10 (2010), Chapter 2, 43–64

Effects of kinetics and mechanisms of crystal growth on ion-partitioning in solid solution–aqueous solution (SS–AS) systems

ANDREW PUTNIS

Institute of Mineralogy, University of Münster, Corrensstrasse 24, 48149 Münster, Germany
e-mail: putnis@uni-muenster.de

Ion-partitioning between aqueous solutions and growing crystals is controlled by both the equilibrium thermodynamics of the solid solution–aqueous solution system and departures from equilibrium. In this chapter the various factors which influence ion-distribution coefficients are reviewed with specific emphasis on the role of supersaturation on crystal-growth mechanisms and growth rate, the role of the surface chemistry and structure of the crystal in controlling ion incorporation and finally on the role of organic molecules and background electrolytes in modifying the dynamics of water exchange around the ions which are potentially incorporated into a growing crystal. Examples to illustrate the general principles involved are taken from room-temperature experimental studies of crystal growth in solid solution–aqueous solution systems.

1. Introduction

Understanding the interaction of fluids with minerals is of fundamental importance to understanding how the Earth works. The rock cycle from weathering, sedimentation and diagenesis, to burial and metamorphism, and ultimately exhumation from depth involves aqueous fluid as the main agent driving chemical reactions, element transport and ultimately crustal geodynamics. At the fluid-mineral interface the mechanism involves dissolution and precipitation, processes which are coupled over a range of spatial scales. To understand these mechanisms in more detail we need to have information on the nature of fluids at various depths in the Earth.

One of the main problems which has limited our understanding of the extent of fluid-mineral interactions in the earth is that, while we can recover the rocks and in certain circumstances can observe the effect of fluid mineral interaction on the rock compositions, mineral assemblages and textures, direct evidence for the distinctive chemistry of the fluids is mostly lost. Fluid inclusions provide one avenue of investigation, but mostly we have to rely on indirect methods. These rely on interpreting the chemical and textural information stored in the minerals and 'reading backwards' to infer the mechanisms of fluid-mineral interaction and also the chemical signature.

DOI: 10.1180/EMU-notes.10.2

When minerals grow from an aqueous solution, their composition is invariably different from that of the solution. As we have seen from the previous chapter, the equilibrium between a solid solution and an aqueous solution, illustrated by a Lippmann diagram, defines the composition of a solid crystallizing from the aqueous solution, and hence an equilibrium partition coefficient for the major ions present. The compatibility of an element, or its tendency to be partitioned into the solid can be related to the solubility of the end members of the solid solution. In detail, the thermodynamic properties of the fluid and of the solid solution can be used to predict the equilibrium distribution coefficients. In theory therefore, the composition of the solid solution can be used to infer the composition of the fluid from which it crystallized. The same principles apply to trace element partitioning. The use of trace element contents of minerals is widely used to infer the trace element content of aqueous solutions in the Earth, again assuming that the equilibrium partitioning coefficient are known. However, this makes the assumption that crystallization took place near equilibrium, and therefore at a rate slow enough for equilibrium partitioning to be maintained.

In this chapter we will explore some of the problems with this approach, and the kinetic and mechanistic factors which can affect element partitioning. A detailed treatment would involve a review of the whole subject of the mechanisms of crystal growth and kinetics, a topic worthy of a whole textbook. Here we will merely outline some of the kinetic factors which should be considered and give examples, with references to further reading.

2. The effect of departures from equilibrium on element partitioning

The crystallization of a simple binary solid solution $(Ca,Cd)CO_3$ from an aqueous solution containing Ca^{2+}, Cd^{2+} and CO_3^{2-} can be used to illustrate some basic features of partitioning under equilibrium conditions, and the effect of departures from equilibrium. Figure 1a shows the Lippmann equilibrium diagram for this system. The very large difference in solubility between the end members results in an extremely strong partitioning of Cd into the solid phase, such that the carbonate phase crystallizing from a solution with only a small mole fraction of Cd^{2+} will have a composition of almost end-member $CdCO_3$. Figure 1b shows the equivalent data in a Roozeboom diagram where the solid line plots the equilibrium compositions of coexisting aqueous solution and solid-solution phases.

The Roozeboom diagram is a visual representation of the distribution coefficient. For a reaction such as $CaCO_3 + Cd^{2+} \Longleftrightarrow CdCO_3 + Ca^{2+}$ the distribution coefficient for Cd is written

$$D = \frac{X_{CdCO_3}}{X_{CaCO_3}} \Bigg/ \frac{\{Cd^{2+}\}}{\{Ca^{2+}\}} \qquad (1)$$

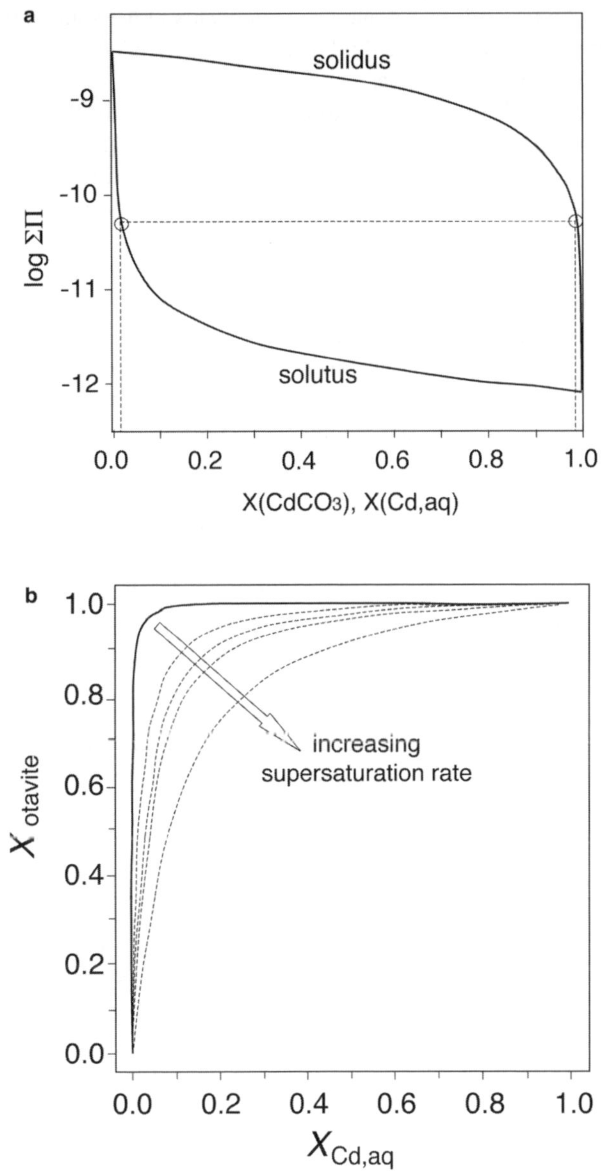

Fig. 1. (***a***) Lippmann diagram for the solid solution–aqueous solution system $CaCO_3$–$CdCO_3$–H_2O. The dashed line shows the composition of a solid solution (given by $X(CdCO_3)$ on the solidus line) which is in equilibrium with an aqueous solution (given by X(Cd,aq) on the solutus line). (***b***) A Roozeboom diagram showing the co-existing compositions of the $CdCO_3$ (otavite) solid solution and the aqueous solution. The solid line represents the coexisting compositions at equilibrium, while the dashed lines show the actual distribution of Cd between the solid and aqueous solution phases with increasing departure from equilibrium (modified from Katsikopoulos *et al.*, 2008).

which compares the ratio of the Cd and Ca components in the solid carbonate phase to the activity ratios of the ions in the fluid. Clearly a large value of D indicates a strong partitioning of Cd into the solid phase.

One consequence of the strong partitioning of Cd into the solid is that the aqueous solution may quickly become depleted in Cd^{2+} such that the equilibrium shifts and subsequent crystallization becomes Ca-rich as predicted from the Lippmann diagram. This results in a crystal with compositional zoning (Cd-rich core and Ca-rich rim).

Figure 1b also shows the effect of increasing departures from equilibrium on the partitioning of Cd into the solid solution (Katsikopoulos *et al.*, 2008). Departures from equilibrium can be induced in a number of ways, but at this point we just relate these to crystallization at increasingly high supersaturation rates. (The 'supersaturation rate' refers to the rate at which the supersaturation changes, for example, by increasing the rate at which the cations and anions mix during the crystallization process. The supersaturation rate can also be related to cooling rate, in cases where supersaturation is achieved by cooling an aqueous solution). Increasing departure from equilibrium results in a decrease in the distribution coefficient for Cd, in other words the amount of Cd incorporated in the solid is less than at equilibrium. Conversely, the amount of Ca incorporation increases. Thus when considering the relative partitioning of ion pairs, increasing supersaturation decreases the distribution coefficient of the ion most compatible in the solid, and increases the distribution coefficient of the less compatible ion.

The same principles apply to the partitioning of minor and trace elements into a crystal growing from aqueous solution. The partitioning of minor and trace elements into minerals is of considerable interest in geochemistry as it is used as a proxy for some aspect of the chemical or physical environment during crystallization (see chapter 12, Stoll, 2010). If such proxies are to be used with a degree of confidence then one factor among many which we need to understand is the effect of departures from equilibrium on the distribution coefficients.

One of the earliest studies of the effect of kinetics on minor-element distribution coefficients was by Lorens (1981) who determined the distribution coefficients for Sr^{2+}, Co^{2+}, Mn^{2+} and Cd^{2+} in calcite as a function of supersaturation, which increased the calcite precipitation rate. Sr^{2+} distribution coefficients increased with increasing precipitation rate, while Co^{2+}, Mn^{2+} and Cd^{2+} distribution coefficients decreased with increasing precipitation rate. This can be understood in terms of the relative compatibility of each of these ions relative to Ca^{2+} in the calcite structure: Sr^{2+} is not a compatible ion in calcite, as indicated by the fact that end-member $SrCO_3$ forms the aragonite structure, while the other smaller ions form complete solid solutions with calcite and hence are compatible in the calcite structure.

The Lippmann and Roozeboom diagrams in Figure 1 show the case where there is a very large difference in the solubility of the end members and hence a strong tendency to partition the cation of the least soluble end member into the solid phase. For the systems investigated by Lorens (1981), the relative values of the solubility products K_{sp} of the relevant end members are: $CaCO_3$ (calcite) $> CoCO_3 > MnCO_3 > CdCO_3$. The tendency to partition a substituting ion into calcite should therefore follow the sequence Cd > Mn > Co. The relative effect on the distribution coefficient of increasing the growth rate should therefore follow this same sequence. Although Lorens (1981) did not comment on the reason for the different effect of kinetics on each partitioning ion, his data confirm this sequence.

While the general influence of growth rate on the distribution coefficient is well established (Lorens, 1981; Tesoriero & Pankow, 1996; Prieto *et al.*, 1997), the various factors which affect growth rates and the underlying mechanisms of minor and trace element incorporation into a growing crystal are only poorly understood. Part of the problem is that there are many possibly interacting kinetic and thermodynamic factors affecting element partitioning making it very difficult to formulate a general theory. In the rest of this chapter some of these factors will be explored and related to the partitioning phenomena observed in nature and in experimental systems.

3. Supersaturation and nucleation rate

The degree of deviation from equilibrium, expressed as a supersaturation, is the primary driver for nucleation and growth from solution. Initially, we will consider only the supersaturation relative to the pure end-member of a solid solution, as is usually the case in the literature. We will then move on to discuss how the concept of supersaturation applies for solid solution-aqueous solution systems.

To illustrate the principles involved in relating growth rates and mechanisms to supersaturation, we will use the system barite–celestite ($BaSO_4$–$SrSO_4$). The supersaturation of an aqueous solution with respect to the barite end member is defined as

$$\Omega = \frac{\{Ba^{2+}\} \cdot \{SO_4^{2-}\}}{K_{sp}} \tag{2}$$

where $\{Ba^{2+}\}$ and $\{SO_4^{2-}\}$ are the activities of the respective ions in the solution and K_{sp} is the equilibrium solubility product of barite (1.1×10^{-10}, Blount, 1977). A similar expression applies for the supersaturation of a solution with respect to celestite, for which $K_{sp} = 2.4 \times 10^{-7}$ (Reardon & Armstrong, 1987). This large difference in solubilities of the end members results in a Lippmann diagram similar in shape to that in Figure 1.

In classical homogeneous nucleation theory, the formation of a nucleus from a supersaturated solution involves a volume-dependent decrease in free energy due to the precipitation of the stable phase, as well as a surface area-dependent increase in free energy due to the formation of a new surface. The overall free energy change for nucleation can be written:

$$\Delta G_r = \frac{4}{3}\pi r^3 \Delta G_v + 4\pi r^2 \sigma \tag{3}$$

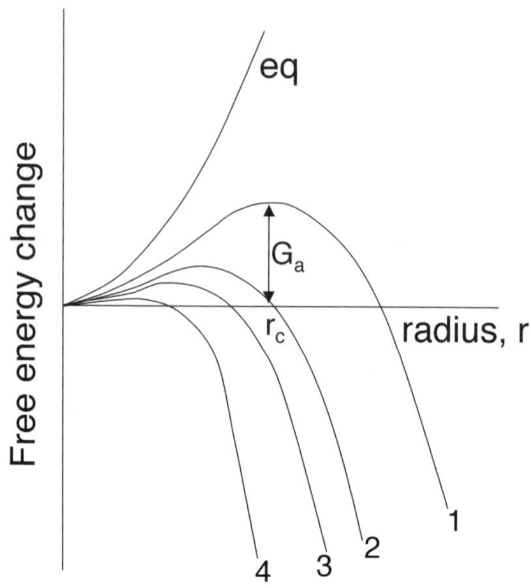

Fig. 2. Schematic diagram showing the change in the free energy for nucleation as a function of the size of the nucleus at increasing departure from equilibrium (supersaturation). At equilibrium (eq) no nucleation can occur. As the supersaturation increases from curves 1–4, the free energy of activation for nucleation, G_a, as well as the critical radius, r_c, decrease.

where ΔG_v is the reduction in free energy per unit volume due to the formation of a spherical nucleus with radius r and σ is the surface energy per unit area. Figure 2 shows the general form of this equation for which there is an activation energy barrier which decreases as the supersaturation increases (curves 1–4). The critical radius of the nucleus is that for which the activation energy barrier is a maximum. As the degree of supersaturation increases the size of the critical radius for nucleation also decreases.

The theoretical nucleation rate, J, *i.e.* the number of nuclei formed per unit volume and time is given by (Nielsen, 1964)

$$J = \Gamma \exp\left(\frac{-\delta\sigma^3\Upsilon^2}{\mathrm{k}^3T^3(\ln S)^2}\right) \tag{4}$$

where δ is a shape related factor, σ is related to the surface energy of the nucleus, Υ is the volume of a growth unit in the nucleus, k is Boltzmann's constant, T is temperature and S is the supersaturation (in this case more simply defined as the ratio of concentration in the solution to the concentration at equilibrium). The pre-exponential factor Γ is related to the rate at which the nucleus can grow to a supercritical size, and hence it involves the diffusion rate of growth units to the surface of the nucleus.

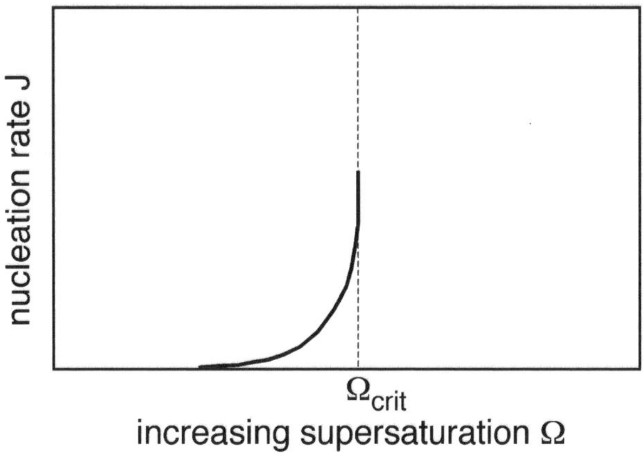

Fig. 3. Schematic representation of equation 4 showing that the nucleation rate is a very sharp function of the supersaturation and at a certain critical value, Ω_{crit}, the nucleation rate becomes very rapid. This corresponds to the situation in Fig. 2 where the activation free energy and critical radius become small.

The main aim of showing equation 4 is to demonstrate that the nucleation rate is a very sharp function of the supersaturation (Fig. 3). Thus there is a value of 'critical supersaturation', Ω_{crit}, at which nucleation takes place at some measurable rate. The value of the critical supersaturation for nucleation is related to the solubility *via* the surface energy term, σ, in equation 3. The greater the solubility, the smaller the surface energy and the lower the critical supersaturation (Mullin, 1993). The pre-exponential term, Γ, in equation 3 is also dependent on the concentration of growth units in the solution at a given supersaturation; in sparingly soluble substances this value is small and the pre-exponential term, and therefore the nucleation rate, are small.

From the above discussion, it follows that the critical supersaturation for nucleation of barite will be much greater than that for celestite. However, measuring the critical supersaturation for nucleation experimentally raises some new questions about these 'classical' concepts of nucleation because the classical theory assumes that supersaturation, and therefore composition, are uniform in both space and time. Gradients in supersaturation can be the result of varying cooling rate, or diffusion-controlled compositional gradients around a growing crystal. In other words, the gradient or rate of change of supersaturation will also determine the actual value of the supersaturation at which nucleation takes place. This supersaturation value for nucleation, termed the 'threshold' supersaturation, Ω_{th}, will increase as the rate of change of supersaturation increases. As it is virtually impossible to avoid supersaturation gradients in real crystal-growth experiments, any experimental determination of supersaturation for nucleation gives the threshold supersaturation. Figure 4 shows the experimentally determined correlation between the threshold supersaturation, Ω_{th}, and the rate of change of supersaturation, R_{Ω}, for a number of simple salts (Prieto *et al.*, 1994; Putnis *et al.*, 1995). The linear ln–ln relationship for each mineral is

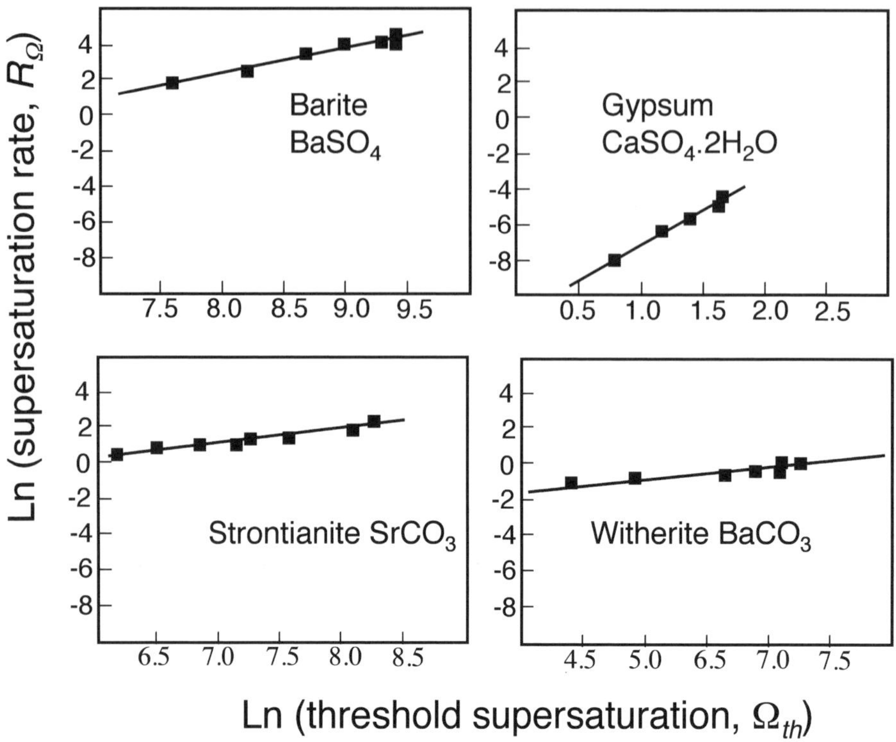

Fig. 4. Experimentally determined relationships between the supersaturation rate and the threshold supersaturation for four different salts (after Prieto *et al.*, 1994).

given by the equation

$$R_\Omega = K(\Omega_{\mathrm{th}})^m \tag{5}$$

where K and m are empirical coefficients. Note the very large difference in the values of the threshold supersaturation for sparingly soluble salts such as barite, compared to gypsum, the most soluble salt in the group. From Figure 4 we could also correctly conclude that witherite ($BaCO_3$) is more soluble than strontianite ($SrCO_3$). This is confirmed by the available data; K_{sp} values for witherite and strontianite are 2.7×10^{-9} (Busenberg & Plummer, 1986a) and 5.4×10^{-10} (Busenberg *et al.*, 1984), respectively.

We began this discussion with the experimental observation that distribution coefficients are a function of the growth rate, and that one of the determining factors in growth rate is the supersaturation. When applied to homogeneous nucleation from aqueous solution we see that the situation is rather more complex than merely relating a growth rate to a particular value of supersaturation, even for pure end-member compounds. The consequences of the differences in threshold supersaturation between end-members of solid solutions has a very significant effect on the growth of solid solutions as we shall see in a later section.

4. Crystal anisotropy and the effect of surface structure

Thus far we have considered the equilibrium distribution coefficient as a single value which may be changed by kinetic effects such as growth rate. However, a crystal growing from an aqueous solution is not generally an isotropic spherical nucleus. Different crystal faces have a different surface structure and therefore the distribution coefficient of any ion may be different on each face (Kinsman & Holland, 1969). The ultimate crystal morphology is also determined by the fact that different crystal faces have different growth rates: those with the slowest growth rates tend to be the dominant faces expressed in the crystal form.

Cross-sections of many minerals show compositional sector zoning (Dowty, 1976; Reeder & Grams, 1987), meaning that there is a differential partitioning of minor elements on different faces of a growing crystal. Reeder & Grams (1987) and Reeder & Paquette (1989) discussed the possibility that this could be due to different growth rates of different faces, but discounted this on two grounds. First, using the data from Lorens (1981) on the dependence of the distribution coefficient of Mn as a function of growth rate to explain the large difference in the Mn/Ca molar ratio in adjacent sectors of natural calcite would require very large differences in the growth rate of adjacent faces, and this is not borne out by the relative thicknesses of the growth zones. Second, an increased growth rate should have the opposite effect on the distribution coefficients for Mn and Sr, decreasing the value for Mn and increasing the value for Sr (see above). Therefore if Mn is lower in one sector than another, the Sr should be higher. The data show this is not the case, and therefore the differential growth rate cannot be the main cause of sector zoning.

The conclusion is that sector zoning is the result of differential adsorption of species into different growth sites for different non-equivalent faces, and that this selectivity must originate from differences in the surface structure (Dowty, 1976; Reeder & Grams, 1987; Reeder & Paquette, 1989). The implications of this statement is that the concept of a unique equilibrium distribution coefficient is not sufficient to explain minor and trace element incorporation into crystals, even when growth rate is not a significant factor. The surface structure can exert significant control on the distribution coefficient.

To account for the observation of sector zoning in crystals even at very slow growth rates, Watson & Liang (1995) presented a model in which the anisotropic surface enrichment is preserved due to low values of diffusivity in the near-surface layers of the growing crystal. The preservation of chemical inhomogeneities thus depends on the competition between the growth rate and lattice diffusion, so that if the diffusivity is sufficiently low, the anisotropic surface enrichment will be preserved and inevitably leads to sector zoning even at small growth rates. For the preservation of sector zoning of a rapidly diffusing species, growth rates would need to be relatively faster.

Watson (1996, 2004) has extended the idea that surface diffusivity plays a major role in preserving the composition of a growing surface to a more general conceptual model for trace-element partitioning. To explain the dependence of partitioning of minor elements on growth rate as observed by Lorens (1981) and Tesoriero & Pankow (1996), Watson (1996, 2004) hypothesized that crystal growth can trap any surface-enriched elements, and that slow ion migration in the near-surface region may not be effective in ridding the structure of these impurities. This 'growth entrapment' model is then applied to any situation where the composition at the mineral surface differs from that in the bulk crystal.

To explain the increase in Sr incorporation into calcite with increasing growth rate, the growth entrapment model requires that the concentration of Sr in the near-surface region of the growing crystal is greater than in the bulk. This was justified by Watson by arguing that Sr is less compatible in the calcite structure and so becomes concentrated in the near-surface region. Slow diffusion of any Sr trapped by the growing crystal ensures preservation of the near-surface chemistry into the crystal structure, and this effect becomes more pronounced with increasing growth rate. For compatible elements such as Cd, Co and Mn an increased growth rate reduces the partitioning into the calcite (*e.g.* Fig. 1b), and the model requires that therefore the near-surface must be depleted in these elements, and that 'growth exclusion' rather than entrapment is called upon.

The Watson model does not explicitly refer to relative solubilities of end-member components or Lippmann diagrams, nor to the observation that the relative effect on the distribution coefficient of increasing the calcite growth rate should therefore follow the sequence Cd > Mn > Co. However, since the relative diffusivities of these elements in the near-surface region are unknown, a robust model combining features of Lippmann's approach, growth rates of different compositions as a function of supersaturation and different surface diffusivities remains a challenge for the future.

Even on a single surface of a mineral various different types of crystallographic sites are exposed which may have a different preference for incorporating ions from solution. The best studied example is the surface structure of monolayer steps on the calcite $\{10\bar{1}4\}$ cleavage plane. Figure 5 shows that the terminations of monolayer growth steps are of two types, referred to as obtuse steps and acute steps. During growth of calcite it has been shown that larger ions such as Sr^{2+} and Ba^{2+} tend to incorporate into the calcite structure at the obtuse steps, whereas ions smaller than Ca^{2+} such as Mg^{2+}, Co^{2+}, Cd^{2+} and Mn^{2+} adsorb preferentially to the more confined acute steps (Paquette & Reeder, 1990, 1995; Reeder, 1996). Paradoxically, Zn^{2+} seems to be an exception, preferring the obtuse steps despite being a smaller ion than Ca^{2+} (Reeder, 1996). The explanation for such preferential partitioning on different sites on the same surface is therefore likely to be more complex than a simple steric model of site confinement. These studies illustrate that a given site in the bulk structure (there is only one type of cation site in the bulk calcite structure) can have multiple surface expressions and that trace-element partitioning is dependent on the detailed surface structure.

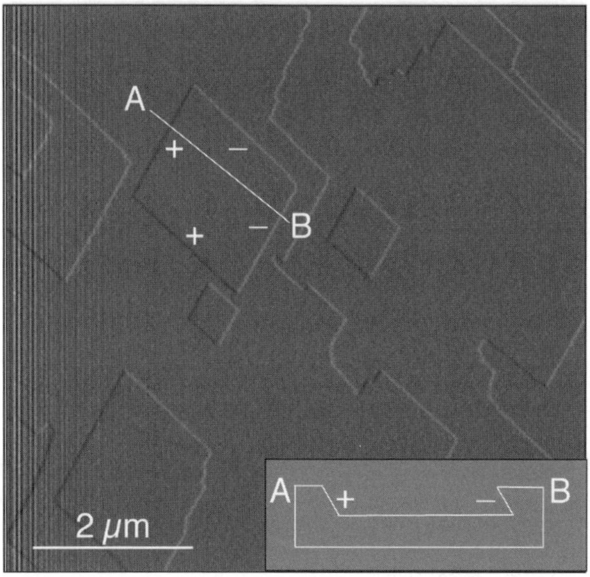

Fig. 5. Atomic force microscope image of a calcite surface after slight dissolution to reveal the characteristic step-edge morphology. The inset shows a cross section between positive or obtuse (+) and negative or acute (−) step edges. (Image: J.M. Astilleros.)

If there can be a variable partition coefficient on the same crystal surface it is likely that trace-element distributions in natural carbonates violate equilibrium partitioning. Furthermore, different calcite growth forms also will have different surface structures and different preferences for trace-element incorporation, so that the details of the calcite habit during growth will also determine the relative trace-element partitioning.

A similar role of surface structure on trace-element partitioning has also been reported in phosphates (Rakovan & Reeder, 1994, 1996), fluorite (Bosze & Rakovan, 2002) and silicates (Northrup & Reeder, 1994). This raises questions about the whole concept of a single equilibrium value of a partition coefficient and suggests that using trace-element compositions as petrogenetic indicators of fluid compositions from which the minerals grew may also be problematic as the trace-element chemistry can vary depending on the growth morphology.

5. Supersaturation and layer growth

With atomic force microscopy (AFM), crystal growth can be continuously imaged *in situ* in a fluid cell, as a function of time and fluid composition. One limitation is that ideally the crystal surface being imaged should be flat, preferably atomically flat, and so most work on crystal growth from solutions has been carried out on phases with perfect cleavage (*e.g.* calcite structures, barite structures, gypsum *etc.*).

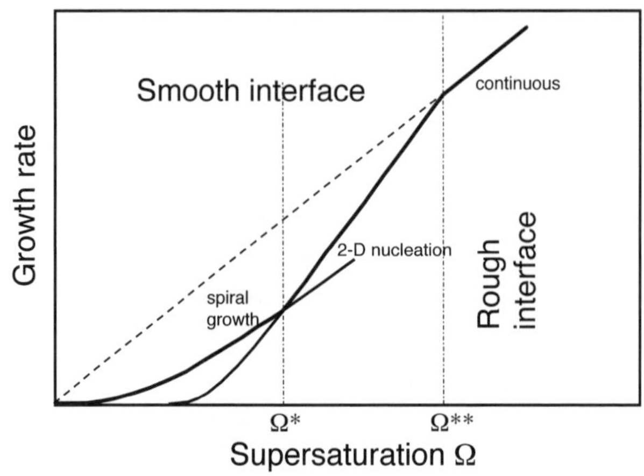

Fig. 6. A schematic Sunagawa diagram showing the relationship between growth rate and supersaturation in three different growth regimes. At small supersaturation, growth is dominated by spiral growth and step advancement. Above a transitional value of supersaturation, Ω^* two-dimensional nucleation takes place and dominates the overall growth rate. Above a further transitional value, Ω^{**}, the interface becomes rough and growth rate is linearly related to supersaturation.

Nevertheless, some important new principles, which could not easily have been obtained from macroscopic growth experiments, have emerged from AFM studies of crystal growth. We will restrict this discussion to those aspects of crystal growth which relate to element distribution coefficients.

In a general way the qualitative relationship between supersaturation and growth mechanism on a flat crystal face has been known for some time from experimental data on macroscopic growth rates (Sunagawa, 1984, 1987, 2005). Figure 6 shows that at small values of supersaturation, the growth rate is dominated by the addition of 'growth units' to existing atomic steps on a crystal surface. Such steps are generated by screw dislocations intersecting the surface and this leads to spiral growth as more material is added to the steps. For spiral growth, the growth rate is a parabolic function of the supersaturation ($\propto \Omega^2$). At a particular transitional value of supersaturation, Ω^*, two-dimensional nuclei form on a flat surface, generating new steps which spread as more growth units are added. This is referred to as the 'birth and spread' model of crystal growth (Ohara & Reid, 1973). The growth rate dominated by two-dimensional nucleation is an exponential function of the supersaturation ($\propto \Omega^n$). At still greater supersaturation, a third growth mechanism becomes dominant, in which the surface becomes rough as growth units are added in direct proportion to the supersaturation ($\propto \Omega$).

From the point of view of element-distribution coefficients during the growth of solid solutions, we consider once again the system barite–celestite ($BaSO_4$–$SrSO_4$) and investigate the value of the transitional supersaturation Ω^* for each end-member. This value can be estimated from AFM observations of the surface nucleation density as the supersaturation is increased *e.g.* Figure 7 which shows the transition from

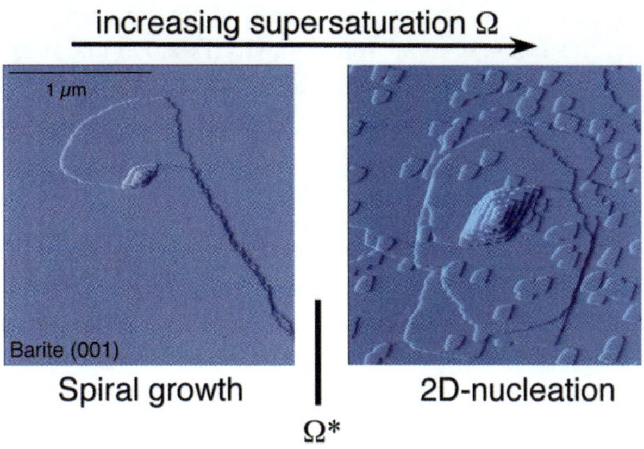

Fig. 7. AFM images showing the transition between spiral growth and two-dimensional nucleation with increasing supersaturation for the case of barite, $BaSO_4$. (Image: C.M. Pina).

spiral growth to two-dimensional nucleation as a function of supersaturation for the case of barite (Pina *et al.*, 1998a,b). For the case of pure barite, Ω^* has been measured at ~7.0, and for celestite, the value is ~2.3 (Pina *et al.*, 2000). This difference is consistent with the fact that barite has a lower solubility than celestite. The value of Ω^* for intermediate compositions of the $BaSO_4$–$SrSO_4$ solid solution will lie somewhere between these two extreme values: if we assume that these depend only on solubility, we can estimate that Ω^* will vary linearly as a function of solid composition (Pina *et al.*, 2000).

6. Supersaturation in binary solid-solution–aqueous-solution systems

Up to this point we have only mentioned the concept of supersaturation when referring to end-member stoichiometric compounds, and so even when an aqueous solution may contain both Ba^{2+} and Sr^{2+} ions, supersaturation, as defined by equation 2, allows us to calculate the value of supersaturation relative to either pure $BaSO_4$ or $SrSO_4$ even when the phase precipitating may be a solid solution. Clearly this is not a satisfactory state of affairs, but the solution to the problem of how to deal with the supersaturation of an aqueous solution relative to any specific solid solution composition is complex and still actively debated (Prieto, 2009).

From the Lippmann diagram, it is clear that any aqueous solution with a value of the total solubility product which plots above the solutus is supersaturated with respect to a whole series of possible compositions of the solid solution, all of which could precipitate. Therefore supersaturation cannot be expressed as a single value, and so we need to determine how the supersaturation of a specific solution varies as a function of the solid-solution composition. Prieto (2009) has discussed in detail the methods proposed for doing this, but for the purposes of this chapter we need only to use the so called "stoichiometric supersaturation" function, Ω_{st} (Prieto *et al.*, 1993). Written for a solid-solution–aqueous-solution such as $BaSO_4$–$SrSO_4$ ($Ba^{2+} = B^{2+}$; $Sr^{2+} = C^{2+}$; $SO_4^{2-} = A^{2-}$).

$$\Omega_{st}(X_{BA}) = \frac{\{B^{2+}\}^x\{C^{2+}\}^{(1-x)}\{A^{2-}\}}{(K_{BA}\gamma_{BA}X_{BA})^x(K_{CA}\gamma_{CA}X_{CA})^{(1-x)}} \quad (6)$$

where $x = X_{BA}$, and $1 - x = X_{CA}$. This equation defines the variation in the supersaturation of an aqueous solution with specific ion activities, with respect to the solid solution composition expressed by X_{BA}. For an ideal solid solution the activity coefficients γ are unity.

To illustrate how equation 6 may apply to a situation where we have a supersaturated solution with very much more Sr^{2+} in solution than Ba^{2+} we calculate the function Ω_{st} for an aqueous solution with $[Ba^{2+}] = 1$ μM and $[Sr^{2+}] = 2000$ μM (Fig. 8) and superimpose this on the plot of the transitional value of supersaturation Ω^* between spiral growth and two-dimensional nucleation (Pina *et al.*, 2000). Figure 8 shows a number of important features (1) the supersaturation function always curves upwards from the end-member values and hence all solutions will be more supersaturated with respect to an intermediate solid solution than to either end-member, (2) even though the solution is very Sr-rich, the maximum supersaturation corresponds to an intermediate composition of the solid solution, (3) although the solution is supersaturated with respect to all solid compositions, some can grow by two-dimensional nucleation and growth, while more Ba-rich compositions could grow by spiral growth or step advancement.

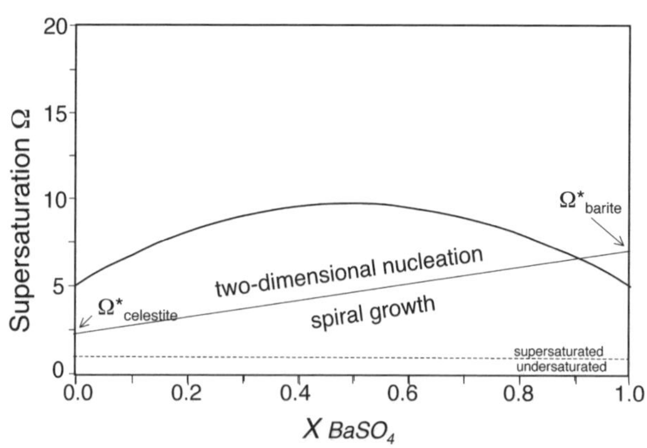

Fig. 8. Supersaturation and growth for the solid-solution system $SrSO_4$–$BaSO_4$. The transitional supersaturation values between spiral growth and two-dimensional nucleation are represented by a straight line between the values determined for the end-members. The bold curve shows the supersaturation of a specific aqueous solution (see text) relative to the complete range of solid-solution compositions (modified from Pina *et al.*, 2000).

In terms of minor-element partitioning, this example has shown that if we have a solution of a salt (in this case $SrSO_4$) with the addition of only 0.05 mole% of Ba then we could in theory grow any composition of the solid, and not even necessarily the composition for which the supersaturation is maximum.

Pina & Putnis (2002) attempted to apply this same principle to three-dimensional nucleation using equation 3 above by estimating values for the relevant constants and then calculating nucleation rates for different solid-solution compositions. When nucleation rates were compared with the supersaturation functions for various solution compositions, they concluded that the maximum supersaturation did not coincide with the maximum nucleation rate, *i.e.* the more soluble solid composition may be kinetically favoured for nucleation than less soluble compositions for which the fluid is more supersaturated. This conclusion is in accord with the discussion above that the more soluble compositions of a solid solution have a lower threshold value of supersaturation.

Pina & Putnis (2002) also calculated Roozeboom diagrams for increasing departures from equilibrium and these were consistent with the experimental evidence that the greater the supersaturation, the greater the deviation from equilibrium partitioning. The solid phase becomes less enriched in the less soluble phase and hence more enriched in the more soluble phase. This is the same as the observation made by Lorens at the beginning of this chapter, *i.e.* that increasing the growth rate of calcite decreased the partitioning of Cd, Mn and Co into the solid, but increased the partitioning of Sr.

7. Oscillatory zoning in SS–AS systems

The conclusion above, that the nucleation and growth of the more soluble compositions of a solid solution may be kinetically preferred, even when the precipitation of less soluble compositions is thermodynamically predicted, provides one key to understanding the common phenomenon of oscillatory compositional zoning formed during growth in minerals (Shore and Fowler, 1996). For binary solid solutions it has been found experimentally that oscillatory zoning occurs when the solubility products of the two end-members vary widely (Prieto *et al.*, 1997) *e.g.* in cases such as $(Ba,Sr)SO_4$ and $(Ca,Cd)CO_3$ (K_{sp} $BaSO_4 = 1.1 \times 10^{-10}$, Blount, 1977; K_{sp} $SrSO_4 = 2.4 \times 10^{-7}$, Reardon & Armstrong, 1987; K_{sp} $CaCO_3 = 3.3 \times 10^{-9}$, Busenberg & Plummer, 1986b; K_{sp} $CdCO_3 = 7.9 \times 10^{-13}$, Stipp *et al.*, 1993). Figure 9 shows an image of oscillatory zoning in a crystal of $(Ba,Sr)SO_4$ (Putnis *et al.*, 1992).

The Lippmann diagrams in such SS–AS systems (such as Fig. 1) show that there is a very strong preferential partitioning of the less soluble component into the solid, and only a very narrow range of aqueous solutions from which an intermediate-composition solid solution can form. At a certain threshold value of the supersaturation, there is a kinetic tendency to nucleate the more soluble component, and its growth will rapidly concentrate the mineral-fluid interface in the other component which will then precipitate. In turn, this depletes the interface in this composition resulting in the precipitation of the more soluble component. The Roozeboom diagram of such systems (*e.g.* Fig. 1b)

Fig. 9. Electron micrograph of a cross section through an experimentally grown crystal of composition $(Ba,Sr)SO_4$. The dark regions are Sr-rich while the paler regions are Ba-rich. The crystal shows both oscillatory and sector zoning of Ba and Sr. The length of the crystal is ~150 μm (from Putnis *et al.*, 1992).

shows that a very small variation in the fluid composition at the interface results in a large swing in the composition of the solid solution with which it is in equilibrium. Thus the system can fluctuate between two widely different compositions and the partitioning of each component into the solid obeys no simple rules.

Even this level of complexity in the description of the mechanism ignores many other potentially important factors such as the strain involved in precipitating one composition onto the surface of another, and how this may influence the composition of the solid solution precipitating. This is discussed in more detail below.

The feedbacks between the precipitation of a solid rich in one component, the resultant diffusion gradients at the solid-fluid interface, and the precipitation of the other component have been explored theoretically for the case of oscillatory zoning in $(Ba,Sr)SO_4$ by Katsev & L'Heureux (2002), L'Heureux & Katsev (2006), Kalischewksi *et al.* (2007) and Lubashevsky *et al.* (2008).

When the end-member compositions of a binary solid solution have similar solubilities, as in the case of $(Ba,Sr)CO_3$ there is a much smaller partition coefficient and a wide range of fluid compositions can exist in equilibrium with intermediate solid compositions. At increased supersaturation the distribution coefficient is further lowered such that the composition of the solid solution and aqueous solution from which it precipitates are almost the same. In such a case, as the fluid composition at the interface varies during crystal growth, the composition of the solid solution precipitating can evolve gradually and a virtually compositionally homogeneous crystal can form.

8. The role of surface chemistry and interfacial strain in the composition of a growing layer

Studies by AFM of the rate of monolayer growth on calcite $\{10\bar{1}4\}$ cleavage steps generally relate changes in growth rate to the adsorption of whichever trace and minor elements have been added to the growth solution. As the minor element composition of calcite has been used as a proxy for various physicochemical factors, such as water temperature, chemistry, growth rate, presence of organics *etc.*, it is of considerable interest to understand the mechanisms of such incorporation. The Mg^{2+}/Ca^{2+} ratio in carbonates is a commonly used proxy for palaeotemperature and so the partitioning of Mg into calcite has been the subject of many studies (Broecker *et al.*, 1979; Mucci & Morse, 1983; Mucci, 1987; Burton & Walter, 1987; Hartley & Mucci, 1996; Davis *et al.*, 2000; Wasylenki *et al.*, 2005; Stephenson *et al.*, 2008).

The partitioning of Mg into calcite surfaces and the resultant decrease in growth rate has usually been explained by considering two 'impurity models' (de Yoreo & Velikov, 2003, and references therein).

(1) The step-pinning model assumes that Mg^{2+} is adsorbed at surface sites and then causes a pinning of the growth monolayers as they spread over the surface, hindering step advancement. At high Mg^{2+} concentration in solution, growth may be completely stopped but can be resumed by increasing the supersaturation, as the amount of impurities required to stop growth needs to be greater at greater supersaturation.

(2) The impurity incorporation model assumes that Mg^{2+} forms a solid solution with the growing calcite layer, and that as this phase is less stable than calcite it will be more soluble. This increased solubility is then equated with a reduced supersaturation of the solution relative to this phase, and hence a reduced growth rate.

In both models there is a direct relationship between the concentration of Mg^{2+} in the aqueous solution and the amount of Mg^{2+} which eventually incorporates into the calcite lattice. However, detailed AFM studies of step-advancement rates of calcite as a function of Mg concentration in the fluid (Astilleros *et al.*, 2010) shows that the growth rate of each monolayer step is dependent on the nature of the substrate on which it grows, so that if an Mg-bearing layer grows over pure calcite, it will have a different growth rate than a subsequent layer growing over this Mg-bearing layer. The main point of the Astilleros *et al.* (2010) paper is to demonstrate that when monolayers grow from a multicomponent solution, with a composition that may differ from that of the substrate, there is no unique value for the rate of step advancement even for specific supersaturation and impurity ion concentration values.

As we have seen from Figure 8, in SS–AS systems there will be a range of compositions of a solid solution for which a specific solution is supersaturated, and the composition which actually grows may not be that for which the supersaturation is greatest. The successive growth of monolayers is known to depend on the degree of misfit between one layer and the next, due to the strain resulting from different ionic sizes and hence bond lengths for the substituted ions on the surface. It is likely that such compositional features of successive growth layers will play a role in determining the actual composition of successive layers. This is especially relevant for the case of oscillatory zoning, also seen in calcites grown in Mg-rich solution (Fernández-Díaz *et al.*, 1996).

The relationship between the solubility of a solid solution, supersaturation functions and growth rates is often confused in the literature. For example, in the impurity incorporation model above, the argument used to explain the decreased growth rate of calcite when Mg^{2+} is added to a solution containing Ca^{2+} and CO_3^{2-} ions goes thus: (1) with Mg^{2+} in solution, the growing calcite becomes richer in Mg; (2) Mg-calcite is less stable and hence more soluble than pure calcite; therefore (3) the solution is less supersaturated with respect to the Mg-calcite, and so the growth rate of Mg-calcite from this solution is smaller. However, as pointed out in this chapter and recently by Astilleros *et al.* (2010) it does not follow that a solution is less supersaturated with respect to the more soluble composition of a solid solution. It all depends on the fluid composition.

9. The role of fluid chemistry in minor-element incorporation during crystal growth

Most of the literature on the effect of 'foreign ions' or organic molecules on crystal growth and minor-element incorporation is focused on adsorption of these ions on the growing crystal surface. However, it is also well known that the incorporation of an ion from solution into the crystal structure involves desolvation, *i.e.* removal of the water of hydration around the ion. In the case of $BaSO_4$ growth, the desolvation of the cation has been shown to be the rate-determining step (Piana *et al.*, 2006), and it therefore follows that additives in the solution which may affect the strength of the cationic hydration will change the activation energy for desolvation. This was the explanation for the increased nucleation rate of barite when a small amount of methanol was added to a supersaturated barium sulphate solution (Piana *et al.*, 2006; Kowacz *et al.*, 2007).

The concept that the chemistry and hence the structure of the aqueous solution affects ion hydration led Kowacz & Putnis (2008) and Kowacz *et al.* (2010) to examine the effect of the addition of a systematic series of simple ionic salts (KCl, NaCl, LiCl, CsCl, NaF, $NaNO_3$) on the nucleation and growth rate of barite at constant supersaturation and constant ionic strength. The results showed that these 'background electrolytes' affected the nucleation and growth rates in a systematic way, depending on how the specific salts modified the activation energy barrier of expelling water molecules from the hydration shells of Ba^{2+} ions in the solution.

The conclusion that different ionic salts affect the structure of an aqueous solution in different ways has important potential implications to element partitioning. Kowacz & Putnis (2008) found that Na^+ could be incorporated into the barite structure in the presence of SO_4^{2-} or F^- ions in solution but not when the anion was Cl^-. The explanation was that SO_4^{2-} and F^- ions strengthen the H-bonded structure of water and so promote 'exclusion' of weakly hydrated cations such as Na^+ from the fluid phase into the solid.

Mg^{2+} is known to be a strongly hydrated ion in solution, and therefore its incorporation as a minor element in calcite during growth will also be affected by the presence of any additives to the solution which change the dynamics of water exchange around the Mg^{2+} ion in solution. Although the effect of solution chemistry on Mg incorporation has not yet been explored in detail, Stephenson *et al.* (2008) found experimentally that the addition of peptides (short chain polymers linking amino acid groups) enhanced the incorporation of Mg into calcite by up to 3 mole%. This suggests that the peptide reduces the activation energy for desolvation of Mg^{2+} more than for Ca^{2+}. This could be due to the fact that Mg^{2+} is more strongly hydrated in solution than Ca^{2+} and therefore a change in the water dynamics associated with the presence of the peptide might have a greater relative effect on Mg. Whatever the ultimate explanation, it is clear that the presence of organic molecules can have a significant effect on the incorporation of 'foreign ions', and that proxies based on such incorporation will need to take this into account.

10. Conclusions

In this chapter we have reviewed some of the ways in which the partitioning of a minor or trace element between an aqueous solution and a growing crystal can be modified from that which might be expected at equilibrium. The growth rate of the crystal can have a very significant effect, increasing the distribution coefficient of some elements while reducing it for others, depending on the compatibility of the element in the solid. However, many factors control the growth rate and it is not clear which of these has a direct effect on the mechanism of partitioning. The supersaturation is a primary driver for nucleation and growth, but as we have seen in this chapter, even this apparently simple concept has complications when applied to solid-solution–aqueous-solution systems. The degree of supersaturation controls the growth mechanism and hence the surface structure, which in turn controls the mechanism of adsorption and incorporation of minor and trace elements into the growing crystal. For a solid solution, the transitional supersaturation for different growth mechanisms can be very significantly different for the two end members. When combined with the fact that a given aqueous solution composition has a different supersaturation for each composition of a solid solution, the crystal growth and partitioning of elements between solution and solid can become a very complex phenomenon, leading to local inhomogeneities in composition, as well as sector and oscillatory zoning. Finally, background electrolytes and organic molecules which do not appear to have any direct chemical relationship to the phase precipitating can also have a significant effect on growth rates and element partitioning by changing the structure of the aqueous solution. This modifies the activation energy for desolvation of the partitioning ions and hence the ease with which they can be incorporated into the solid. We are still some way from a comprehensive understanding of most of these issues, but the papers referred to in this chapter have at least outlined some directions for future research.

Acknowledgements

This paper has greatly benefited from many years of collaborations with José Manuel Astilleros, Lurdes Fernández-Díaz, Carlos Pina and Manuel Prieto. Financial support for these collaborations has been through various EU Marie Curie Actions in Framework Programmes FP5, FP6 and FP7.

References

Astilleros, J.M., Fernández-Díaz, L. & Putnis, A. (2010) The role of magnesium in the growth of calcite: an AFM study. *Chemical Geology*, **271**, 52 – 58.

Blount, C.W. (1977) Barite solubilities and thermodynamic quantities up to 300°C and 1400 bars. *American Mineralogist*, **62**, 942–957.

Bosze, S. & Rakovan, J. (2002) Surface-structure-controlled sectorial zoning of the rare earth elements in fluorite from Long Lake, New York and Bingham, New Mexico, USA. *Geochimica et Cosmochimica Acta*, **66**, 997–1009.

Broecker, W.S., Takahashi, T., Simpson, H.J. & Peng T.H. (1979) Fate of fossil fuel carbon dioxide and the global carbon cycle. *Science*, **206**, 409–418.

Burton, E.A. & Walter, L.M. (1987) Relative precipitation rates of aragonite and Mg calcite from seawater: Temperature or carbonate ion control? *Geology*, **15**, 111–114.

Busenberg, E. & Plummer, L.N. (1986a) The solubility $BaCO_3$(cr)(witherite) in CO_2–H_2O solutions between 0 and 90°C, evaluation of the association constants of $BaHCO_3(aq)^+$ and $BaCO_3(aq)^o$ between 5 and 80°C, and a preliminary evaluation of the thermodynamic properties of $Ba(aq)^{2+}$. *Geochimica et Cosmochimica Acta*, **50**, 2225–2233.

Busenberg, E. & Plummer, L.N. (1986b) A comparative study of the dissolution and crystal growth kinetics of calcite and aragonite. In: *Studies in Diagenesis* (F.A. Mumpton, editor). U.S. Geological Survey Bulletin, **1578**, 139–168. U.S. Geological Survey, Washington, D.C.

Busenberg, E., Plummer, L.N. & Parker, V.B. (1984) The solubility of strontianite ($SrCO_3$) in CO_2-H_2O solutions between 2 and 91°C, the association constants of $SrHCO_3(aq)^+$ and $SrCO_3(aq)^0$ between 5 and 80°C, and an evaluation of the thermodynamic properties of $Sr(aq)^{2+}$ and $SrCO_3$(cr) at 25°C and 1 atm total pressure. *Geochimica et Cosmochimica Acta*, **48**, 2021–2035.

Davis, K.J., Dove, P.M. & De Yoreo, J.J. (2000) The role of Mg^{2+} as an impurity in calcite growth. *Science*, **290**, 1134–1137.

De Yoreo, J.J. & Vekilov, P.G. (2003) Principles of crystal nucleation and growth. In: *Biomineralization* (P.M. Dove, J.J. De Yoreo & S. Weiner, editors). Reviews in Mineralogy and Geochemistry, **54**, 57–93. Mineralogical Society of America, Washington, D.C.

Dowty, E. (1976) Crystal structure and crystal growth II: Sector zoning in minerals. *American Mineralogist*, **61**, 460–469.

Fernández-Díaz, L., Putnis, A., Prieto, M. & Putnis, C.V. (1996) The role of magnesium in the crystallization of calcite and aragonite in a porous medium. *Journal of Sedimentary Research*, **66**, 482–491.

Hartley, G. & Mucci, A. (1996) The influence of p(CO_2) on the partitioning of magnesium in calcite overgrowths precipitated from artificial seawater at 25°C and 1 atm total pressure. *Geochimica et Cosmochimica Acta*, **60**, 315–324.

Kalischewksi, F., Lubashevsky, I. & Heuer, A. (2007) Boundary-reaction-diffusion model for oscillatory zoning in binary crystals grown from solution. *Physical Review E*, **75**, 021601, 1–12.

Katsev, S. & L'Heureux, I. (2002) Autocatalytic model of oscillatory zoning in experimentally grown (Ba,Sr)SO_4 solid solution. *Physical Review E*, **66**, 066206, 1–12.

Katsikopoulos, D., Fernández-González, A. & Prieto, M. (2008) Crystallization of the (Cd,Ca)CO_3 solid solution in double diffusion systems: The partitioning behaviour of Cd^{2+} in calcite at different supersaturation rates. *Mineralogical Magazine*, **72**, 433–436.

Kinsman, D.J.J. & Holland, H.D. (1969) The co-precipitation of cations with $CaCO_3$ IV – The coprecipitation of Sr^{2+} with aragonite between 16°C and 96°C. *Geochimica et Cosmochimica Acta*, **33**, 1–17.

Kowacz, M. & Putnis, A. (2008) The effect of specific background electrolytes on water structure and solute hydration: Consequences for crystal dissolution and growth. *Geochimica et Cosmochimica Acta*, **72**, 4476–4487.

Kowacz, M., Putnis, C.V. & Putnis, A. (2007) The effect of cation/anion ratio in solution on the mechanism of barite growth at constant supersaturation: role of the desolvation process on the growth kinetics. *Geochimica et Cosmochimica Acta*, **71**, 5168–5179.

Kowacz, M., Prieto, M. & Putnis, A. (2010) Kinetics of crystal nucleation in ionic solutions: electrostatics and hydration forces. *Geochimica et Cosmochimica Acta*, **74**, 469–481.

L'Heureux, I. & Katsev, S. (2006) Oscillatory zoning in a (Ba,Sr)SO_4 solid solution: Macroscopic and cellular automata models. *Chemical Geology*, **225**, 230–243.

Lorens, R.B. (1981) Sr, Mn and Co distribution coefficients in calcite as a function of the precipitation rate. *Geochimica et Cosmochimica Acta*, **45**, 553–561.

Lubashevsky, I., Mues, T. & Heuer, A. (2008) Different routes towards oscillatory zoning in the growth of solid solutions. *Physical Review E*, **78**, 041606, 1–16.

Mucci, A. (1987) Influence of temperature on the composition of magnesian calcite overgrowths precipitated from seawater. *Geochimica et Cosmochimica Acta*, **51**, 1977–1984.

Mucci, A. & Morse, J.W. (1983) The incorporation of Mg^{2+} and Sr^{2+} into calcite overgrowths. *Geochimica et Cosmochimica Acta*, **47**, 217–233.

Mullin, J.W. (1993) *Crystallization*. 3rd edition. Butterworth, Oxford, 594 pp.

Nielsen, A.E. (1964) *Kinetics of Precipitation*. Pergamon, Oxford, 151 pp.

Northrup, P.A. & Reeder R.J. (1994) Evidence for the importance of growth surface structure to trace element incorporation in topaz. *American Mineralogist*, **79,** 1167–1175.

Ohara, M. & Reid, P.C. (1973) *Modelling Crystal Growth Rates from Solution.* Prentice Hall, Englewood Cliffs, New Jersey, USA, 272 pp.

Paquette, J. & Reeder, R. (1990) New type of compositional zoning in calcite: insights into crystal growth mechanisms. *Geology*, **18**, 1244–1247.

Paquette, J. & Reeder, R. (1995) Relationship between surface structure, growth mechanism, and trace incorporation in calcite. *Geochimica et Cosmochimica Acta*, **59**, 735–749.

Piana, S., Jones, F. & Gale, J.D. (2006) Assisted desolvation as a key kinetic step for crystal growth. *Journal of the American Chemical Society*, **128**, 13568–13574.

Pina, C.M. & Putnis A. (2002) The kinetics of nucleation of solid solutions from aqueous solutions: A new model for calculating non-equilibrium distribution coefficients. *Geochimica et Cosmochimica Acta*, **66**, 185–192.

Pina, C.M., Becker, U., Risthaus, P., Bosbach, D. & Putnis A. (1998a) Molecular-scale mechanisms of crystal growth in barite. *Nature*, **395**, 483–486.

Pina, C.M., Bosbach, D., Prieto, M. & Putnis A. (1998b) Microtopography of the barite (001) face during growth: AFM observations and PBC theory. *Journal of Crystal Growth*, **187**, 119–125.

Pina, C.M., Enders, M. & Putnis A. (2000) The composition of solid solutions crystallising from aqueous solutions: the influence of supersaturation and growth mechanisms. *Chemical Geology*, **168**, 195–210.

Prieto, M. (2009) Thermodynamics of solid solution-aqueous solution systems. In: *Thermodynamics and Kinetics of Water-rock Interaction* (E.H. Oelkers & J. Schott, editors). Reviews in Mineralogy and Geochemistry, **70**, 47–85. Mineralogical Society of America, Washington, D.C.

Prieto, M., Putnis, A. & Fernández-Díaz, L. (1993) Crystallization of solid solutions from aqueous solutions in a porous medium; zoning in (Ba,Sr)SO_4. *Geoogical Magazine*, **130**, 289–299.

Prieto, M., Putnis, A., Fernández-Díaz, L. & López-Andrés, S. (1994) Metastability in diffusing-reacting systems. *Journal of Crystal Growth*, **142**, 225–235.

Prieto M., Fernández-Gonzalez, A., Putnis, A. & Fernández-Díaz, L. (1997) Nucleation, growth and zoning phenomena in crystallizing (Ba,Sr)CO_3, Ba(SO_4,CrO_4), (Ba,Sr)SO_4, and (Cd,Ca)CO_3 solid solutions from aqueous solutions. *Geochimica et Cosmochimica Acta*, **61**, 3383–3397.

Putnis, A., Fernández-Díaz, L. & Prieto, M. (1992) Experimentally produced oscillatory zoning in the (Ba,Sr)SO_4 solid solution. *Nature*, **358**, 743–745.

Putnis, A., Prieto, M. & Fernández-Díaz, L. (1995) Supersaturation and crystallisation in porous media. *Geological Magazine*, **132**, 1–13.

Rakovan, J. & Reeder, R.J. (1994) Differential incorporation of trace elements and dissymmetrization in apatite: The role of surface structure during growth. *American Mineralogist*, **79,** 892–903.

Rakovan, J. & Reeder, R.J. (1996) Intracrystalline rare earth element distributions in apatite: surface structural influences on zoning during growth. *Geochimica et Cosmochimica Acta*, **60**, 4435–4445.

Reardon, E.J. & Armstrong, D.K. (1987) Celestite ($SrSO_4(s)$) solubility in water, sea water and NaCl solution. *Geochimica et Cosmochimica Acta*, **51**, 63–72.

Reeder, R.J. (1996) Interaction of divalent Co, Zn, Cd, and Ba with the calcite surface during layer growth. *Geochimica et Cosmochimica Acta* **60**, 1543–1552.

Reeder, R.J. & Grams, J.C. (1987) Sector zoning in calcite cement crystals: Implications for trace element distributions in carbonates. *Geochimica et Cosmochimica Acta*, **51,** 187–194.

Reeder, R.J. & Paquette, J. (1989) Sector zoning in natural and synthetic calcites. *Sedimentary Geology*, **65**, 239–247.

Shore, M. & Fowler, A.D. (1996) Oscillatory zoning in minerals: a common phenomenon. *The Canadian Mineralogist*, **34**, 1111–1126.

Stephenson, A.E., DeYoreo, J.J., Wu, L., Wu, K.J., Hoyer, J. & Dove, P.M. (2008) Peptides enhance magnesium signature in calcite: Insights into origins of vital effects. *Science*, **322**, 724–727.

Stipp, S.L.S., Parks, G.A., Nordstrom, D.K. & Leckie, J.O. (1993) Solubility-product constant and thermodynamic properties for synthetic otavite, cadmium carbonate and aqueous association constants for the cadmium (II)–carbon dioxide–water system. *Geochimica et Cosmochimica Acta*, **57**, 2699–2713.

Sunagawa, I. (1984) Growth of crystals in nature. In: *Materials Science of the Earth's Interior* (I. Sunagawa, editor), 63–105. Terra Scientific Publishing, Tokyo.

Sunagawa, I. (1987) Morphology of minerals. In: *Morphology of Crystals Vol. 2* (I. Sunagawa, editor), 511–587. Terra Scientific Publishing, Tokyo.

Sunagawa, I. (2005) *Crystals: Growth, Morphology and Perfection.* Cambridge University Press, Cambridge, UK, 295 pp.

Tesoriero, A.J. & Pankow, J.F. (1996) Solid solution partitioning of Sr^{2+}, Ba^{2+}, and Cd^{2+} to calcite. *Geochimica et Cosmochimica Acta*, **60**, 1053–1063.

Wasylenki, L.E., Dove, P.D. & De Yoreo, J.J. (2005) Effects of temperature and transport conditions on calcite growth in the presence of Mg^{2+}: Implications for paleothermometry. *Geochimica et Cosmochimica Acta*, **69**, 4227–7236.

Watson, E.B. (1996) Surface enrichment and trace-element uptake during crystal growth. *Geochimica et Cosmochimica Acta*, **60**, 5013–5020.

Watson, E.B. (2004) A conceptual model for near-surface kinetic controls on the trace-element and stable isotope composition of abiogenic calcite crystals. *Geochimica et Cosmochimica Acta*, **68**, 1473–1488.

Watson, E.B. & Liang, Y. (1995) A simple model for sector zoning in slowly grown crystals: Implications for growth rate and lattice diffusion, with emphasis on accessory minerals in crustal rocks. *American Mineralogist*, **80,** 1179–1187.

EMU Notes in Mineralogy, Vol. 10 (2010), Chapter 3, 65–138

Geochemical thermodynamic modelling of ion partitioning

DMITRII A. KULIK

Laboratory for Waste Management, Paul Scherrer Institute, CH-5232 Villigen PSI, Switzerland
e-mail: dmitrii.kulik@psi.ch

The partitioning of an ionic species between two or more co-existing phases is closely related to chemical speciation, solubility, sorption and solid-solution formation in the geochemical system. The aim of chemical thermodynamic modelling is to find the stable or metastable chemical speciation in such a system; from that, any possible ion partition or distribution coefficient can be retrieved. This chapter aims to highlight some typical features of modelling the equilibrium partitioning of inorganic ionic species between aqueous electrolyte and solid phases, including adsorption and ion exchange on their surfaces. Some guidance is also given on methods and computer codes that can be used in helping us understand and predict ion partitioning in complex scenarios, where several competing solids are involved in a minor element uptake by different mechanisms.

1. Introduction

Partitioning of a chemical element between coexisting phases is a phenomenon of major interest, studied in many branches of geochemistry, mineralogy and petrology. Some related fundamental issues include the understanding of distribution of elements in coexisting rock-forming minerals, melts and fluids at hydrothermal or high *P-T* conditions (Albarède, 1995; Spera *et al.*, 2007, and references therein); the uptake of trace elements by coprecipitation with carbonates in (sea)water (*e.g.* Kulik *et al.*, 2000; Gaetani & Cohen, 2006; Watson, 2004, and references therein); the partitioning of radionuclides between aqueous solution, clays, carbonates, sulphates by various mechanisms such as adsorption, ion exchange, and solid-solution formation (*e.g.* Kersten & Boettcher, 1997; Curti, 1999; Turner *et al.*, 2006; Payne *et al.*, 2006; Bruno *et al.*, 2007; Bradbury & Baeyens, 2005, 2009).

This contribution is limited by scope to low-temperature aquatic systems; by relevance mainly to applications in marine and waste geochemistry; and by objective to highlight some typical features of models of the equilibrium partitioning of inorganic ionic species between aqueous electrolyte and solid (solution) phases, adsorption on their surfaces, and ion exchange on permanent charge surfaces.

The emphasis will be not so much on definitions and equations, but rather on providing some guidance about the methods and codes for geochemical thermodynamic

DOI: 10.1180/EMU-notes.10.3

modelling, and on showing how to use them in order to help understand and predict ion partitioning in complex scenarios, where several competing solids are involved in (minor) element uptake by different mechanisms. It is assumed that a sufficient process understanding is available, *i.e.* the justification from the atomistic, structural, spectroscopic side, *etc.* is provided elsewhere.

In the context of geochemical equilibria, partitioning is closely related to such phenomena as chemical speciation, solubility, sorption, and formation of variable-composition phases (mixtures or solutions). Quantitative representation of these phenomena in stable or metastable states is the aim of thermodynamic modelling, usually done with help of computer codes. From the given bulk composition of the system and the thermodynamic database, such codes can compute the (meta)stable chemical speciation in all coexisting phases at temperature and pressure of interest. Because at least some phases are non-ideal solutions, and the geochemical system may contain hundreds of chemical species arranged in scores of potentially-present phases, this cannot be done by hand or using a simple spreadsheet. Conversely, there is no simple way of predicting the partition or distribution coefficients without computing the full chemical speciation in all phases.

In the present context, some relevant applications of geochemical modelling include:

(1) trace-metal partitioning in aqueous–solid solution carbonate systems (*e.g.* Sr, Cd, Mn, Zn,... in calcite; Sr, Ba in aragonite), important for development of palaeoclimate proxies or for understanding the environmental fate of heavy metals;

(2) CO_3^{2-} and HCO_3^- ion partitioning between aqueous, gaseous, and mineral phases;

(3) trace ion partitioning between (sea)water and particulate Fe and Mn hydroxides (mainly adsorption or surface precipitation as in the Fe-Mn crusts and nodules); and

(4) uptake of radionuclides (Cs, Sr, Ra, An^{3+}, An^{6+}) into repository host minerals (carbonates, sulphates, clays, cement phases).

Such processes occurring at or near ambient temperature are to some extent irreversible or metastable. Hence, the account for metastability and simulation of irreversible mass transfer (mineral dissolution/precipitation, water-rock interaction, weathering *etc.*) comprises an important component of geochemical modelling.

2. Calculation of equilibrium ion speciation

Partitioning of an ion (element) *M* involves at least two chemical components (species) of it, present in at least two different phases, at least one of which is a solution. The typical case is when one or more aqueous species (*M* ion and its complexes) compete for *M* with its end member in a solid-solution phase and/or with the *M* surface species on mineral particles. Thus, an essential step in modelling the equilibrium partitioning of *M* consists of calculation of the chemical speciation (*i.e.* amounts of all components) in all relevant phases in the chemical system. The speciation is usually augmented by non-ideal effects that must be considered in each phase-solution in the form of activity coefficients of its components. The 'forward thermodynamic modelling problem' consists of finding mole amounts, concentrations, and activities of chemical species in aqueous and other phases present in a complex chemical system in a complete equilibrium defined by temperature, *T*, pressure, *P*, and the system bulk composition *b*. From these results, any possible partition and distribution coefficients can be obtained. Conversely, if some of these coefficients are known from the experiment or from geochemical observations, then some input thermodynamic constants or interaction parameters of the model for non-ideal mixing can be retrieved by solving the 'inverse thermodynamic modelling problem'. Speciation calculations are needed in both cases.

2.1. Methods

2.1.1. Law of Mass Action (LMA)

Problems involving aqueous speciation are most frequently solved using a computer program based on the reaction-law of mass action (LMA) formalism and the Newton-Raphson numerical method. A good introduction into modern LMA methods can be found in textbooks (Van Zeggren & Storey, 1970; Bethke, 1996) and in the PHREEQC manual (Parkhurst & Appelo, 1999). Below, a brief overview of the setup of LMA speciation calculations is provided, and its shortcomings in solving aqueous–solid solution (Aq-SS) equilibria are discussed. Actually, there are two LMA setups: one used in the GWB (*Geochemists Workbench*) commercial code (Bethke 1996), and another used in PHREEQC, CHESS and similar codes. As discussed in detail by Bruno *et al.* (2007), the GWB setup is not capable of calculating solution phases other than aqueous electrolyte and therefore is of very limited applicability to model the ion partitioning. So, our discussion will be limited to the second setup, following Reed (1982). In this setup, minerals, gases, and surface complexes are treated as 'product species' in addition to 'master species' also forming the stoichiometry basis for setting up the bulk composition of the chemical system.

The set of equations, to be solved numerically for the speciation with the Newton-Raphson technique, includes $n(N)$ mole balance equations for 'master' species belonging to the set N (usually aqueous ions, water and SiO_2), and $n(\Phi)$ mass action equations for 'phases', *i.e.* saturated minerals and gases of the set Φ. The bulk composition of the system is set in terms of total amounts $n_i^{(T)}$ of master components only, subject to $n(N)$ balance equations:

$$n_i^{(T)} = m_w\left(m_i + \sum_{j\in Q} \nu_{ij} m_j\right) + \sum_{k\in\Phi} \nu_{ik} n_k, \quad i \in N, \tag{1}$$

where Q is now a set of the product aqueous species with the water mass m_w(in kg) and molalities m_j substituted by the LMA equations written in terms of master species only:

$$m_j = \frac{1}{K_j\gamma_j}\cdot\prod_{i\in N}(\gamma_i m_i)^{\nu_{ij}}, \quad j \in Q \tag{2}$$

where K_j is the equilibrium constant of the jth product species formation reaction from master species, and γ stands for activity coefficients; n_k is the amount of kth pure mineral or gas, the stability of which is described by a LMA equation:

$$K_k = \frac{1}{a_k} \cdot \prod_{i \in N} (\gamma_i m_i)^{\nu_{ik}}, \quad k \in \Phi, \tag{3}$$

where K_k is the equilibrium constant (*e.g.* solubility product), and a_k is the activity of kth mineral or the fugacity of a kth gas. For a pure mineral or a gas at standard pressure (1 bar), $a_k = 1$. For a kth solid solution end member, equation 3 takes the form:

$$K_k = \frac{1}{f_k x_k} \cdot \prod_{i \in N} (\gamma_i m_i)^{\nu_{ik}}, \quad k \in \Phi_{ss},\ \Phi_{ss} \in \Phi,\ x_k = n_k \Big/ \sum_{k \in \Phi_{ss}} n_k \tag{4}$$

where Φ_{ss} is a subset of solid-solution end members, x_k is the mole fraction, and f_k is the activity coefficient of the kth end member. In the case of simple ideal mixing, $f_k = 1$; otherwise, f_k must be computed using an appropriate mixing model as a function of end-member mole fractions and interaction parameters (see below). This is usually done between the iterations of Newton-Raphson algorithm along with the calculation of ionic strength and aqueous activity coefficients. The analytical derivatives of all equations such as 1 to 4, as well as those for activity coefficients, are needed (Parkhurst and Appelo, 1999).

In principle, the LMA method described can handle multiple solid solutions and a gas mixture in equilibrium with the aqueous phase (and pure substance phases), subject to limitations of the Gibbs phase rule at fixed T and P:

$$n(\Phi) = n(N) - n(F) \tag{5}$$

where $n(F)$ is the number of degrees of freedom, $n(N)$ is the number of independent components (in the present case, the number of mass balance equations), and $n(\Phi)$ is the number of phases. In the PHREEQC code (Parkhurst & Appelo, 1999), an ideal solid solution may contain two or more end members, but a non-ideal solid solution is limited to two end members. In general, the strong non-ideality introduced by the aqueous activity coefficients in concentrated solutions, as well as by the solid activity coefficients, may cause bad convergence or failure in the LMA speciation algorithms. In recent codes, this problem has been avoided to some extent by inventing smart numerical procedures for picking up the initial guesses of n_k and m_i values.

As stated by Reed (1982), the real challenge in LMA lies in the selection of mineral phases, especially solid solutions, to be included in the mass balance through equation 1 with boundary conditions (equation 4) such that the Gibbs phase rule (equation 5) is satisfied. Selection of pure phases is normally done upon checking the saturation indexes, *SI*, over the whole list of minerals in the process of solving the LMA speciation task many times after inserting the most supersaturated phase in a semi-empirical 'swap' procedure. Selection of a solid solution in this way is considerably complicated by the fact that pure end members may be undersaturated while their solid solution is supersaturated. There is no easy way to check this because to do it, the end-member mole fractions x_k (and end-member activity coefficients, f_k, if the mixture is non-ideal) must be known. However, the amount and composition of the solid solution depends on that of the coexisting aqueous electrolyte, and both amounts are initially unknown.

To work around this limitation of the LMA method, some more or less successful semi-empirical procedures have been suggested (Reed, 1982). The problematic point in such a procedure is where the estimated mole fractions $x_k^{(\varphi)}$ or activities $a_k^{(\varphi)}$ of the potentially present solid-solution end members are chosen, based upon the approximated molalities, m_i, or activities, a_i, of aqueous ions. A simple way would be to guess mole fractions $x_k^{(\varphi)}$ in the solid as being equal to ionic molality fractions $x_{k,aq}^{(\varphi)} = m_{i,k}^{(\varphi)} / \sum m_{i,k}^{(\varphi)}$ or to ionic activity fractions. However, if the solubility products of end members are different, equilibrium mole fractions in solid solution can be very different from ionic activity fractions in the aqueous phase due to partitioning effects. Therefore, another possibility is to guess the activity $a_k^{(\varphi)}$ of each end member from the current saturation index *SI* of this end member taken as pure substance. For a multi-component ideal solid solution, this yields directly the guess mole fraction values $x_k^{(\varphi)} = a_k^{(\varphi)}$, but there is no easy way to get $x_k^{(\varphi)}$ values from activities in the case of non-Raoultian or non-ideal mixing. For the above reasons, use of Lippmann functions (see chapter 1 this volume, Prieto, 2010) appears to be the only simple and theoretically sound method for estimating activities or mole fractions of solid-solution end members in equilibrium with the aqueous solution in the context of LMA approach. Solving the Aq-SS equilibria with LMA codes is, in general, an inefficient and awkward process because a single aqueous speciation calculation turns into a series of many runs controlled by non-rigorous test-and-swap procedures. The only case which is easy to solve is the stoichiometric saturation state (Glynn & Reardon, 1990), when the solid mixture is effectively 'frozen' and behaves as a pure solid of known stoichiometry. This solid can be included directly in the LMA setup as a 'pure phase' with the mixture solubility product $K^{(x)}$, if the end member solubility products and non-ideal interaction parameters are known.

2.1.2. Gibbs Energy Minimization (GEM)

In Gibbs Energy Minimization setup, the total mole amounts of chemical elements, n_i, and (zero) charge n_Z comprise the input bulk composition, *b*, of the chemical system expressed using $n(N)$ independent components (IC). All $n(L)$ stoichiometrically feasible chemical species or dependent components (belonging to a set *L*) are taken into the mass balance through their chemical formulae and mole amounts $n_j, j \in L$. A dependent component belongs to a single- or multicomponent aqueous, gaseous, fluid, liquid, solid or sorption phase. The stability of the *j*th chemical species at *T*, *P* of interest is defined by its input standard molar Gibbs energy function g_j^o. Activities and concentrations of dependent components are treated separately in each phase (using subsets of indexes $L_\alpha, \alpha \in \Phi$), taking into account the appropriate standard and reference states, concentration scales, and mixing models.

The GEM algorithms can solve complex Aq-SS equilibria in one GEM run, in a straightforward way, without Lippmann functions and semi-empirical iterative procedures required in LMA Aq-SS speciation models. The applicability of GEM is limited only by the knowledge of standard-state molar properties of end-members and the interaction parameters for phases. In the following description, the variant of GEM known as 'convex programming approach' (Karpov *et al.*, 1997; Karpov *et al.*, 2001) will be mainly considered as implemented in the GEM-Selektor code package (Kulik *et al.*, 2004), http://gems.web.psi.ch. More details and comparison with LMA were published in work by Bruno *et al.* (2007).

The GEM algorithm computes simultaneously the 'primal' $\hat{n}^{(x)}$ (speciation – amounts of dependent components) and the 'dual' $\hat{u}$ (chemical potentials of independent components) results, from which other thermodynamic quantities and criteria of equilibrium can be obtained. Solving the GEM forward problem is equivalent to finding a vector of mole amounts of dependent components $n^{(x)} = \{n_j^{(x)}, j \in L\}$ (vector x in Karpov *et al.*, 1997) such that

$$G(n^{(x)}) \Longrightarrow \min \qquad \text{subject to } \mathrm{A}n^{(x)} = n^{(b)} \tag{6}$$

where $n^{(b)} = \{n_i^{(b)}, i \in N\}$ is the input vector of mole amounts of independent components belonging to the set N; $\mathrm{A} = \{\mathrm{a}_{ij}, i \in N, j \in L\}$ is a matrix made of the formula stoichiometry coefficients of dependent components; and $G(n^{(x)})$ is a total Gibbs energy function of the system:

$$G(n^{(x)}) = \sum_j n_j^{(x)} \upsilon_j, \quad j \in L. \tag{7}$$

In equation 7, υ_j is the normalized chemical potential of jth dependent component, which can be written in a simplified dimensionless form as

$$\upsilon_j = g_j^o / RT + \ln C_j + \ln \gamma_j + \Xi, \quad j \in L \tag{8}$$

where g_j^o is the standard molar Gibbs energy function at temperature T (K) of interest (pressure dependence omitted for simplicity); $\mathrm{R} = 8.3145\ \mathrm{J\,K^{-1}\,mol^{-1}}$ is the universal gas constant; $C_j = f(n_j^{(x)})$ is the concentration that depends on the choice of standard concentration scale. C_j equals the mole fraction

$$x_j = n_j^{(x)} / \sum_j n_j^{(x)}, \quad j \in L_\alpha \tag{9}$$

for components of αth gas or condensed mixture phases, and for the water-solvent; C_j equals the molality m_j for aqueous electrolyte species, and unity for the single component of a pure phase. The activity coefficient γ_j of jth dependent component in its respective phase is taken in the chosen standard concentration scale and reference state (unity in a simple ideal mixture or in the single-component pure phase). The term Ξ depends on the chosen standard state: $\Xi = \ln P$ for gases; $\Xi = 1 - x_w$ for aqueous species (x_w is the mole fraction of water in aqueous phase); $\Xi = 2 - x_w - 1/x_w$ for the water-solvent; $\Xi = 0$ for (solid) mixture end-members and pure-substance phases.

Activity coefficients γ_j, as functions of the respective phase composition, are calculated between GEM iterations according to the model of non-ideal mixing within each phase.

The non-linear GEM algorithm finds simultaneously two vectors, the 'primal' $\hat{n}^{(x)}$ and the 'dual' u solutions of the problem (equation 6), using the 'Karpov-Kuhn-Tucker' necessary and sufficient conditions (Karpov *et al.*, 1997):

$$\begin{aligned} &\upsilon - \mathrm{A}^T u \geq 0; \\ &\mathrm{A}\hat{n}^{(x)} = n^{(b)}; \quad \hat{n}^{(x)} \geq 0; \\ &\hat{n}^{(x)}(\upsilon - \mathrm{A}^T u) = 0 \end{aligned} \tag{10}$$

where ^{T} is the transpose operator. The first condition in equations 10, rewritten with indices,

$$g_j^o/\mathrm{R}T + \ln C_j + \ln \gamma_j + \Xi - \sum_i \mathrm{a}_{ij} u_i \geq 0, \quad j \in L, \;\; i \in N \tag{11}$$

implies that for any *j*th species present at some equilibrium concentration C_j in its phase, the primal chemical potential, υ_j, numerically equals the dual chemical potential

$$\eta_j = \sum_i \mathrm{a}_{ij} u_i, \quad j \in L, \; i \in N \tag{12}$$

where u_i has the same value in all coexisting phases. The last condition of orthogonality in equations 10 zeroes off the molar amounts of unstable species and phases. Conditions in equations 10 have also been extended for the case when the molar quantities sought, $\hat{n}_j^{(x)}$, of some metastable species are constrained from below and/or above to model 'partial equilibrium' states and kinetically controlled processes (Karpov *et al.*, 2001).

The phase-stability criteria (based on equations 10 and 11) are used for checking whether a phase should be present or absent from the equilibrium assemblage. If an internally consistent set of thermodynamic data is used, the GEM algorithm will always find a complete (or partial) equilibrium phase assemblage and speciation, regardless of how many ideal or non-ideal solution phases were included in the initial system definition.

From equation 11, the activities of gaseous, aqueous, solid-solution and sorption species, the activity functions such as *pH, pe, Eh*, and the saturation indices of single- and multi- component condensed phases can be computed (Karpov *et al.*, 2001; Kulik, 2006b). In particular, the activity of a *j*th component in its phase in equilibrium is found as

$$\ln a_j = \sum_i \mathrm{a}_{ij} u_i - g_j^o/\mathrm{R}T, \quad j \in L, \; i \in N. \tag{13}$$

For comparison, the traditional calculation of activities is:

$$\ln a_j = \ln C_j + \Xi + \ln \gamma_j. \tag{14}$$

As both C_j and γ_j are functions of the primal solution (speciation) $\hat{n}^{(x)}$ vector, computed 'primal' activities are subject to some numerical limitations. For instance, GEM codes usually cannot handle mole amounts below 10^{-25}; smaller values are not physically realistic and are automatically zeroed off for numerical convergence reasons. For 'zeroed-off' species, equation13 is the only way to compute activities, also applicable to stoichiometrically feasible minor species not even included in the mass balance. A good case is the 'aqueous electron' e_{aq}, a hypothetical species used in the definition of pe, a measure of redox potential. For the e_{aq} species (charge -1 and $g^o = 0$ by convention), equation 13 takes the form $\ln a_e = -1 \cdot u_{\text{Charge}}$. By defining $pe = -\log_{10} a_e$,

$$pe = -\frac{1}{\ln 10}(-u_{\text{Charge}}) \tag{15}$$

Using the fundamental relation $pe \cdot \text{R}T \cdot \ln 10 = \text{F} \cdot Eh$ (F $= 96485$ C mol^{-1} is the Faraday's constant), pe can be further converted into Eh (in Volts):

$$Eh = \frac{\text{R}T}{\text{F}} \cdot T \cdot u_{\text{Charge}}. \tag{16}$$

In a similar way, pH can be calculated for any aquatic system, even if the H^+ species is not explicitly included or was eliminated from the mass balance:

$$pH = -\frac{1}{\ln 10}(u_{\text{H}} + u_{\text{Charge}}). \tag{17}$$

In this equation, the activity coefficient of H^+ ion is not required. Another useful case of equation 13 is the calculation of fugacity f_j of a gaseous component, which also avoids knowledge of a fugacity coefficient.

If a solid-solution phase is present in equilibrium in a positive amount then equation 13 directly yields the activity of its any end member. For a single-component (pure substance) phase, equation 13 yields instead the numerical value of the saturation index, SI, in $\log_{10}$ (see proof in Bruno *et al.* (2007), pp. 102–103):

$$\Omega_{S,j} = \frac{1}{\ln 10}\left(\sum_i \text{a}_{ij} u_i - g_j^o/\text{R}T\right), \quad j \in L, \ i \in N \tag{18}$$

For example, $\Omega_{\text{S,Barite}} = 1/\ln 10[u_{\text{Ba}} + u_{\text{S}} + 4u_{\text{O}} - (-549.48)]$ (at 25°C, $g^o/(\text{R}T) = -549.48$ taken from the GEMS version of the Nagra-PSI database; Hummel *et al.*, 2002a). For any pure phase in equilibrium, SI is numerically zero. In the GEM IPM algorithm, negative values of SI are always computed for unstable solids eliminated from the mass balance; positive values (supersaturation) can be obtained only for compounds with the amount $n_j^{(x)}$ set by a metastability constraint from above. Equation 18 even applies to stoichiometrically feasible single-component solids not included in the mass balance. Note that equation 18 involves neither ion activity nor solubility product, as would be required if the classic definition of *SI* had been used.

As discussed below, the *SI* is a crucial value in many kinetic equations for mineral dissolution or precipitation. Hence, having *SI* also for solid solutions would be desirable. Using equations 13 or 18, the saturation index of the αth condensed solid or liquid non-electrolyte solution phase (relative to the rest of the chemical system) can be defined as

$$\Omega_{SS,\alpha} = \log \sum_{j \in l_\alpha} x_j^* = \log_{10} \sum_{j \in l_\alpha} (10^{\Omega_{s,j}} / f_j^*)$$

where x_j^* is the 'expected' mole fraction, and f_j^* is the 'expected' activity coefficient of *j*th end member. The word 'expected' applies here in the cases where the solution phase is (a) not stable or (b) is metastable, *i.e.* amounts of its end members have metastability constraints. In case a, the phase is not present in the mass balance, so its composition is unknown, and 'true' activity coefficients cannot be calculated. In case b, the phase is present in the mass balance, the metastable mole fractions, x_j, and the respective activity coefficients f_j^* can be calculated, but they may differ from x_j^* and f_j^* values expected from equation 19. In this case, the sum of 'real' mole fractions will be equal to 1.0, but the sum of 'expected' mole fractions computed using equation 19 with activity coefficients calculated from 'real' mole fractions will be >1.0 (supersaturation) or <1.0 (undersaturation). Note that in GEM calculations, the saturation indexes, *SI*, are the only output data because they are not used in determining which phases should be present in the mass balance, as done in LMA computations.

One consequence of equations 11 and 12 is that the equilibrium chemical potential of any element or compound partitioned between different phases is the same in all phases. Because of this, for exchange reactions involved in ion partitioning, the chemical potentials of elements and charge largely cancel out. Consider a distribution coefficient of components *B* and *C* between the aqueous electrolyte and the simple ionic *(B,C)L* solid solution:

$$D = \frac{x_{BL}}{\{B^z\}} \bigg/ \frac{x_{CL}}{\{C^z\}} = \frac{K_{CL} f_{CL}}{K_{BL} f_{BL}} \tag{20}$$

where $\{\}$ stands for the aqueous ion activity, and K_{CL}, K_{BL} are the end-member solubility products (*e.g.* $K_{BL} = \{B^z\}\{L^{-z}\}/a_{BL}$). From equations 20 and 13, we obtain, after rearrangement:

$$D = \exp\left(-g_{C^z}^{(n)} - g_{BL}^{(n)} + g_{B^z}^{(n)} + g_{CL}^{(n)}\right) \frac{f_{CL}}{f_{BL}} = K_{B,C}^{ex} \frac{f_{CL}}{f_{BL}} \tag{21}$$

where $g_j^{(n)} = g_j^o/(RT)$, and $K_{B,C}^{ex}$ is the equilibrium constant of the exchange reaction $CL + B^z(aq) \Leftrightarrow BL + C^z(aq)$. If the solid solution is ideal then $D = K_{B,C}^{ex}$ is independent of the composition. If not, the speciation must be known to estimate activity coefficients of the end members. The middle fraction term in equation 20 can be obtained directly from the (computed) chemical speciation. A special case occurs when B is a trace component ($x_{BL} \ll 1$ and $x_{CL} \approx 1$). Then $f_{CL} \approx 1$ and $f_{BL} \approx f_{BL}^{\infty} = const$, hence $D \approx D^{\infty} = K_{B,C}^{ex}/f_{BL}^{\infty}$ is (nearly) constant and corresponds to the Nernst distribution law, which also holds if both BL and CL are trace components in the ternary solid solution with a major end member AL.

2.2. Thermodynamic databases

The modelling code (LMA or GEM) alone makes little sense without an internally consistent thermodynamic database covering a range of applications in a certain T, P and composition region. The compilation of such a database usually costs many person-years of highly qualified work. At present, web-based resources become state-of-the-art; they can rapidly supply thermodynamic data that can be used in various modelling codes. These web resources reuse the data from earlier-published NIST references (http://webbook.nist.gov/chemistry), the NEA TDB project (http://www.nea.fr/dbtdb/), and many other literature sources.

It is not surprising that the geochemical codes available are distributed with one or several thermodynamic data sets. In particular:

The CHESS LMA code (http://chess.ensmp.fr) is related to the CTDP database project (http://www.ctdp.org) containing data suitable for both LMA and GEM calculations at ambient T, P for ~1500 aqueous, 180 gaseous, 1470 solid and 1750 organic species.

A new web project, THEREDA (http://www.thereda.de) has been recently started in Germany, aimed at providing a database for the geochemical modelling of the near-field and far-field processes occurring in repositories for radioactive wastes in different rock formations. This resource is expected to export data in PHREEQC and other code formats.

The THERMODDEM database (http://thermoddem.brgm.fr) is a fast-growing web-based resource having an emphasis on applications in modelling the alteration of mineral waste and its environmental impact. This database, suitable for both LMA and GEM calculations, contains (imported) a large proportion of the SUPCRT data set developed over 25 years by H.C. Helgeson and co-workers (see Shock *et al.*, 1997; http://www.predcent.org).

The SUPCRT (OBIGT) database (http://www.predcent.org) became so widespread and popular among geochemists that it is by now a 'de-facto standard' in modelling related to hydrothermal geochemistry and water-rock interaction. The declared P, T range of applicability of SUPCRT is from ambient to 1000°C and up to 5 kbar (see Oelkers *et al.*, 2009 for details). The imported SUPCRT database, more or less complete, comes together with some GEM codes – GEM-Selektor (http://gems.web.psi.ch), HCh (http://www.ga.gov.au/minerals/research/methodology/geofluids/HCh.jsp), ChemApp (http://www.gtt-technologies.de). An adapted and extended variant of SUPCRT (restricted to saturated H_2O vapour pressures) known as llnl.dat is available for PHREEQC (wwwbrr.cr.usgs.gov/projects/GWC_coupled/phreeqc). Some data derived from SUPCRT are also available for geochemical modelling at temperatures below the freezing point of water using the FREZCHEM code (Marion *et al.*, 2010).

The LMA geochemical code PHREEQC (perhaps, the most popular and widespread of all) comes with four databases (wateq.dat, phreeqc.dat, minteq.dat, and llnl.dat) that can be used alternatively, but should not be mixed together. The ChemApp GEM modelling code has recently been merged into an integrated thermochemical databank system F*A*C*T; the resulting commercial product is now called *FactSage* (http://www.crct.polymtl.ca/FACT/). The Nagra-PSI 01/01 thermodynamic database for nuclear waste applications (http://les.web.psi.ch/TDBbook/) is distributed with the GEM-Selektor code package and is also available in PHREEQC format.

In the context of application, it often turns out that no thermodynamic database is complete, and some extensions should be made for missing species and phases, or for a different T, P range. This must be done in such a way that the internal consistency of the data is preserved; by no means should the data from different data sets be mixed without checking for consistency and adjusting when necessary. This methodology is complex and requires experience and diligence; an overview is available *e.g.* in Nordstrom & Munoz (1994), and many examples can be found in the NEA TDB volumes (*e.g.*Guillaumont *et al.*, 2003).

In general, LMA codes require fewer thermodynamic data (log$K$ for the product species only) than GEM codes (G^{o} values for all species in all phases). When the modelling has to be performed at non-ambient conditions, these data must be corrected from reference T_r, P_r to the T, P of interest. For condensed substances, this correction requires the standard molar properties (Gibbs energy G^{o}, enthalpy H^{o}, entropy S^{o}, volume V^{o}) at T_r, P_r, as well as the polynomial coefficients of the heat capacity function $C_P = f(T)$. At moderate pressures, the molar volume of a condensed substance can be assumed constant (the same as at T_r, P_r). The required calculations are usually built into the geochemical modelling code package. The T, P-corrected log$K$ values for product species used in LMA codes are given as polynomial coefficients of $\log K = f(T)$ function, or simply as look-up arrays for data interpolation.

The T, P corrections for partial molal thermodynamic properties of aqueous ions, complexes and neutral species from the SUPCRT database are performed using the so-called HKF coefficients (see details and references in Anderson, 2005) and the SUPCRT92 program (Johnson *et al.*, 1992). The source code of SUPCRT92 has been freely available, which led to its incorporation in several codes (also after translation from fortran to C and other programming languages). The extension of SUPCRT data set with new aqueous complexes can be done using the PRONSPREP97 algorithm (Sverjensky *et al.*, 1997; Shock *et al.*, 1997), or by constructing simple temperature extrapolations of equilibrium constants of certain isoelectric/isocoulombic reactions.

In the GEMS code package, such consistent extensions are facilitated by maintaining data for chemical species in two complementary formats: DComp (thermochemical) and ReacDC (reaction-defined). In the latter, the properties of a 'new' chemical species are defined via the effects ($\Delta_r G^{o}$ or $\log K$, $\Delta_r H^{o}$ or $\Delta_r S^{o}$, $\Delta_r Cp^{o}$ and $\Delta_r V^{o}$) of its reaction with the 'old' species already represented in the database records, as well as *via* properties of those 'old' species given either in DComp or in ReacDC format.

This possibility of database extension is especially useful for aqueous, surface complexes and some minerals of trace elements with only their formation equilibrium constants known at ambient *T*, *P* (*cf.* Smith & Martell, 1989; Hummel *et al.*, 2002b; NEA TDB books). However, the reactions depicting the direct complex formation ($M^{z+} + L^{y-} \Leftrightarrow MZ^{z-y}$) or mineral dissolution ($ML \Leftrightarrow M^{z+} + L^{z-}$) are expected to have (very) large entropy, heat capacity, and volume effects due to ion solvation and other phenomena. However, for a common-cation or common-ligand 'isoelectric' reaction (*e.g.* $M1L + M2^{z+} \Leftrightarrow M2L + M1^{z+}$), the entropy and heat capacity effects become small and may be ignored, which provides quite a satisfactory temperature extrapolation to 100–150°C, sometimes even greater (Gu *et al.*, 1994; Kulik, 2002a), for thermodynamic constants actually known only at T_r, P_r. This is an issue for several actinides for which no HKF parameters are yet available.

A distinct separation between the relatively accurate core database and the uncertain extensions provided by the user is crucial for assessing the reliability of modelling results. All published internally consistent databases can be subdivided into subsets of data of different uncertainty and reliability. The 'kernel' data of highest reliability do not cover even the relevant inorganic speciation in many environmental geochemical applications. In such cases, additional, less certain thermodynamic data must be used. Such values (mostly $\log K$ of formation reactions) are derived from the room-temperature experimental data, from geochemical observations, or estimated from linear free energy relationships (LFER).

For minerals, the known low-temperature solubility products often contradict with independently measured calorimetric data. Many weak inorganic aqueous complexes can be considered as unreliable because they are used in 'ion association' aqueous models, but ignored in 'specific interaction' models. Hence, careful thinking about uncertainties in the input thermodynamic data, as well as about the potential effects of the uncertainty propagation to the results of geochemical modelling, is mandatory.

3. Aqueous activity models

This contribution is mainly focused on ion partitioning in aquatic systems. Therefore, the ion speciation in the aqueous electrolyte is of major importance. Thermodynamic calculation of the aqueous speciation must realistically reproduce (1) the effect of ion-ligand complexation and (2) the effect of ionic strength on the ion activity-concentration relationships.

3.1. The effect of complexation

This effect is most directly reproduced in the so-called 'ion association' model of aqueous electrolyte (*e.g.* implicit in the SUPCRT data set and in the Nagra-PSI database). In this model, the specific interactions between cations *M* and major inorganic ligands *L* (OH^-, CO_3^{2-}, HCO_3^-, Cl^-, SO_4^{2-}, HS^- *etc.*) are mainly represented by assigning the stoichiometry and stability to 'strong' or 'weak' *ML* complexes; the residual non-specific interactions are described via the activity coefficients calculated using various forms of the Debye-Hückel equations as functions of the ionic strength (see below). Actually, the discrimination of 'strong' and 'weak' aqueous complexes is to some extent arbitrary; complexes such as $NaCl^0$, very weak (often neglected) at ambient temperature, become very significant at hydrothermal conditions. A question may be asked: 'at what temperature does such a "weak" complex becomes "strong" and can no longer be ignored?' On the other hand, many polyvalent ions practically exist as complexes and not free ions in a wide range of (even ambient) conditions. For example, at near-neutral *pH*, many trivalent cations occur as hydroxocomplexes and not as free ions, which are stable only at relatively low *pH* (Fig. 1).

Aqueous electrolyte models of another kind ('specific ion interaction' models) ignore the existence of 'weak' complexes, replacing them with the activity coefficients involving a 'specific ion interaction' term in addition to the long-range electrostatic Debye-Hückel term. Such models are capable of accurate prediction of salt-concentration effects up to high ionic strength (3–10 M and more) at ambient conditions. However, they become ambiguous at elevated temperature where the interactions between strong acid anions and strong base cations become much stronger, which leads to dramatic activity coefficients with significant (empirical) dependence on composition and temperature. Whatever the aqueous model, it should correctly reproduce the ratio of free *M* ion activity to total *M* concentration in the whole range of relevant conditions.

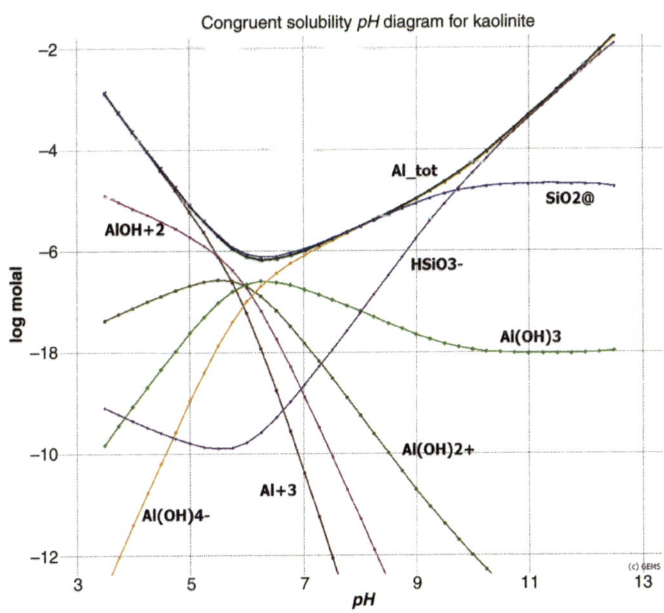

Fig. 1. Aqueous speciation of Al and Si at 1 bar and 25°C as function of *pH* in the system composed of 1000 g H_2O, 30 g $Al_2Si_2O_5(OH)_4$, 1 mol NaCl and titrated with HCl or NaOH. GEM-Selektor screenshot, Nagra-PSI database complexation constants, Davies equation used. Solid kaolinite is present throughout; precipitation of any other Si- and Al-containing solids is suppressed. Free Al^{+3} ion is dominant only at $pH < 6$.

3.2. The effect of ionic strength

This effect on the activity of aqueous species is described with activity coefficients whose numerical values depend on the choice of standard and reference states. The standard state of dissolved species is normally taken as 1 m ($m^{\mathrm{o}} = 1$ mol per 1 kg of H_2O solvent) concentration referenced to the hypothetical ideal state of infinite dilution. Another standard state (pure substance with mole fraction $x^{\mathrm{o}} = 1$) and reference state (activity equals mole fraction, $a_{\mathrm{w}} = x_{\mathrm{w}}$) are used for the water-solvent. Note that the 1 M standard state is used for aqueous species too, but the drawback of molarity M is that it depends on the composition, T and P of aqueous solution. At reference state, the activity coefficient of any chemical species in its phase is unity. The activity of jth aqueous species, in general, is represented as

$$a_j = m_j \gamma_j \tag{22}$$

where γ_j is the molal activity coefficient. In ion-association aqueous speciation models, the activity coefficient is represented as a function of the effective ionic strength:

$$I = \frac{1}{2} \sum_j m_j z_j^2, \tag{23}$$

where z_j is the formula charge and m_j is the molality of j-th aqueous species. At very low $I < 1 \times 10^{-3} m$, activity coefficients of ions are well described by the Debye-Hückel limiting law:

$$\log_{10} \gamma_j = -\frac{1}{2} z_j^2 \sqrt{I}. \tag{24}$$

At moderate ionic strength ($0.1 < I < 0.4$ m), the semi-empirical Davies equation

$$\log_{10} \gamma_j = -A_\gamma z_j^2 \left(\frac{\sqrt{I}}{1 + \sqrt{I}} + 0.3I \right) \tag{25}$$

is used, especially at near-ambient conditions. At the ionic strength below 1.0 m, the extended Debye-Hückel equation with a common linear term $b_\gamma I$ (where b_γ is a parameter specific to the major electrolyte, *e.g.* $b_\gamma =$ 0.064 for NaCl at 25°C) is the most popular one:

$$\log \gamma_j = \frac{-A_\gamma z_j^2 \sqrt{I}}{1 + B_\gamma \dot{a}_j \sqrt{I}} + b_\gamma I \tag{26}$$

Here, $\dot{a}_j$ is the 'ion-size' parameter (in 10^{-8} cm), either ion-specific (Kielland's), or common to all aqueous species (as in HKF EoS); A_γ and B_γ are T, P dependent parameters (at 25°C, $A_\gamma \approx 0.5$ and $B_\gamma \approx 0.3$). According to (Kielland, 1937), the ion size parameter $\dot{a}_j$ is thought to be a function of ion hydration, roughly proportional to the ratio of ionic radius to charge; the values of $\dot{a}_i$ for inorganic ions range from 2.5 to 4.5 Å for monovalent (except Li^+ and H^+), 4 to 8 Å for divalent, 4 to 9 for trivalent, and 5 to 11 for tetravalent ions; many tabulated values are available (Kielland, 1937; Langmuir, 1997). More about the Debye-Hückel equation and calculation of activity coefficients can be found in textbooks (Anderson, 2005; Grenthe *et al.*, 1997; Bruno *et al.*, 2007). For neutral aqueous species and dissolved gases, equation 26 reduces to the 'salting-out' equation,

$$\log \gamma_j = b_\gamma I \tag{27}$$

Individual values of $b_{\gamma j}$ are available for some dissolved gases (Millero & Schreiber, 1982).

At high ionic strength $I > 1\ m$, the aqueous electrolyte models based on the Debye-Hückel equation and ion association become very inaccurate because the short-range interactions begin to dominate the unspecific long-range electrostatic effects encompassed by equation 26. For concentrated aqueous electrolytes, the specific ion interaction models use a Debye-Hückel term describing long-range electrostatic forces, and specific interaction terms that account for short-range molecular interactions as function of the electrolyte composition, temperature and pressure. These models can calculate single-ion activity coefficients and allow the equations describing the activity coefficients in simple electrolytes be extended to multicomponent solution systems. Two models have gained a particular popularity: Brønsted-Guggenheim-Scatchard (SIT) and Pitzer (see Grenthe *et al.*, 1997 and references therein). It is worth mentioning also the relatively new Extended UNIQUAC model (Thomsen, 1997; Thomsen, 2005). Below, a very brief summary of those models is given.

The SIT model is recommended for ionic strength corrections throughout the NEA TDB project (Grenthe & Puigdomenech, 1997; Grenthe *et al.*, 1997), and it is also recommended by IUPAC (Pettit *et al.*, 2003). The SIT model is based on two assumptions: (1) The molal activity coefficient of an ion of charge z_i in solution is given by

$$\log_{10}\gamma_i = -\frac{z_i^2 A_\gamma \sqrt{I}}{1 + 1.5\sqrt{I}} + \sum_k \varepsilon(i, k)\, m_k, \tag{28}$$

where A_γ is the limiting Debye-Hückel slope coefficient, and $\varepsilon(i, k)$ is a binary interaction parameter which describes the specific short-range interaction between the ionic species i and k. The summation extends over all species k with molality m_k. (2) The interaction parameters $\varepsilon(i, k)$ for ions of the same sign are set to zero. The interaction coefficient for uncharged species ($z_i = 0$) is usually assumed to be zero; however, this is not necessarily feasible.

Shortcomings of the SIT model, as discussed by Grenthe *et al.* (1997) and by Spahiu (2002), consist mainly of its very limited temperature and pressure applicability range. Although SIT is widely used in the correction of complexation constants to zero ionic strength, its usage in forward modelling of aqueous equilibria is still very limited. The SIT model is relatively simple, it has only binary interaction parameters, but it seems not to be valid for high ionic strength (above 2–3 *m*), and it needs to be improved at lower ionic strength for highly charged ions.

The Pitzer model was formulated by Pitzer and co-workers in 1970s and has been discussed in many publications (Grenthe *et al.*, 1997). In the Pitzer approach, the long-range interactions are accounted for by a Debye-Hückel term, and the short-range interactions by a virial expansion. A general form of the Pitzer equation is (Anderson, 2005):

$$\frac{G^{EX}}{kg_W RT} = f(I) + \sum_i^{L_i} \sum_j^{L_j} \lambda_{ij} m_i m_j + \sum_i^{L_i} \sum_j^{L_j} \sum_k^{L_k} \mu_{ijk} m_i m_j m_k \qquad (29)$$

where G^{EX} is the excess Gibbs energy of aqueous electrolyte; kg_W is the number of kilograms of water; *m* indicates molalities, and indexes *i*, *j*, *k* represent aqueous species of any kind. The first virial term is a Debye-Hückel term $f(I)$. The next term accounts for the binary interactions, and the last term for ternary interactions. The binary interaction coefficient λ_{ij} is a function of ionic strength *I*; the ternary interaction coefficient μ_{ijk} is taken independent of *I*, and set to zero if indexes *i*, *j*, *k* belong to ions with charge of the same sign.

Harvie *et al.* (1984) showed that the Pitzer approach can be extended to construct accurate models of solubility in complex brines, successfully applied to natural fluids at ambient temperatures using a specially written GEM code (Harvie *et al.*, 1987). This model has also been extended to a broad temperature range (0–250°C) (Greenberg & Møller, 1989; Christov & Møller, 2004). The Pitzer model has been used in marine chemistry (Millero & Pierrot, 1998) and in research related to nuclear-waste disposal in salt rock (Fanghänel, 2002). The Pitzer model is able to describe data for high ionic-strength systems (>1 *m*) accurately, also in mixed electrolytes. The main shortcoming of the Pitzer approach is that a relatively large amount of accurate experimental data is needed for parameterization; there are correlations among the interaction parameters; the latter are difficult in physical interpretation and cannot be extrapolated beyond the experimental range. The Pitzer model parameters are virial coefficients that do not seem to have a fundamental temperature or pressure dependence (Königsberger, 2002).

The Extended UNIQUAC model (Thomsen, 2005 and references therein) combines a Debye-Hückel term for long-range electrostatic interactions with the UNIQUAC local composition model (Abrams & Prausnitz, 1975) for short-range interactions. Water is considered the only solvent; ions, non-electrolytes including alcohols, and dissolved gases are considered solutes. This rather new model has been successfully applied to predict solid–liquid equilibria in several systems involving heavy metal cations (Mn^{2+}, Fe^{2+}, Co^{2+}, Ni^{2+}, Cu^{2+} or Zn^{2+}) and a number of anions (Iliuta *et al.*, 2002). The model parameters were found to be generally valid in the temperature range from the cryohydratic point to the boiling point of the respective solutions. The pressure dependencies were also added to the EUNIQUAC model (Villafáfila García *et al.*, 2006), which was then used for correlation and prediction of solid–liquid equilibria involving scaling minerals (Ca, Ba, Sr sulphates and carbonates) at geothermal conditions. This model is able to represent binary, ternary and quaternary solubility data within the experimental accuracy in the temperature range from −20 to 300°C, at pressures up to 1000 bar. The geochemical applications of EUNIQUAC are still in their infancy, but this model appears to be promising because it has only binary interaction parameters (*i.e.* close to SIT model in complexity), but has a predictive power close to that of the Pitzer model. If properly re-parameterized and combined with a GEM solver (as already implemented in GEM-Selektor v.3 code) and the appropriate solid solution models, the EUNIQUAC model could become serious competition for both the Pitzer and SIT approaches in the future development of thermodynamic models of ion partitioning, including those related to nuclear-waste geochemistry.

With these approaches to ionic strength corrections, the inorganic speciation of the majority of elements can be modelled in seawater and in fresh surface- and groundwaters in a wide range of salinities and temperatures (Millero, 1996; Turner & Whitfield, 1987; Byrne, 2002). In turn, this is a prerequisite for correct calculation of equilibrium partitioning of elements between two and more (non-ideal) solution or sorption phases.

There is a potential ambiguity in values of distribution coefficients computed for the same solid solution or sorption phase using different models of aqueous electrolyte in otherwise the same system (especially at medium to high ionic strength). In a typical distribution coefficient such as for cations A and B between the solid solution $(A, B)L$ and aqueous phase

$$D_{BA} = \frac{x_{BL}\, m_{A^{z+}}}{x_{AL} m_{B^{z+}}} = \frac{K_{AL} f_{AL} \gamma_{B^{z+}}}{K_{BL} f_{BL} \gamma_{A^{z+}}} \quad (30)$$

the free-ion molalities and activity coefficients are used (here K_{AL}, K_{BL} are solubility products of solid solution end members, and f_{Al}, f_{BL} are their activity coefficients). In ion-association aqueous models, the free-ion molality m_A is a difference between the total molality $m_{A,\mathrm{tot}}$ of A and the sum of molalities of all its complexes. The respective free ion molality fraction $m_A/m_{A,\mathrm{tot}}$ may be different for ions of the same charge. But in simple aqueous models based on the Debye-Hückel equation, the ratio of their activity coefficients will be close to unity because they depend mainly on charge and ionic strength. However, in the SIT or Pitzer aqueous models, the aqueous complexes are largely ignored, and in many cases $m_A/m_{A,\mathrm{tot}}$ and $m_B/m_{B,\mathrm{tot}}$ will both be close to unity, while the difference in aqueous behaviour of A and B will be largely accounted for in their free-ion activity coefficients, so the numerical value of D_{BA} is likely not to be the same as that in the case when the aqueous ion-association model was employed. This consideration shows that a direct calculation of full chemical speciation in all phases of the system may be more reliable and may lead to a better process understanding than the reasoning based on simplified standalone estimates of distribution coefficients.

4. Ion partitioning into solids and solid solutions

The simplest trivial case of M ion partitioning is the equilibrium between a multi-component solution phase (*e.g.* aqueous electrolyte, gas mixture) and a pure M-containing solid phase. This is usually called 'solubility' and described with reactions like $ML \Leftrightarrow M^{z+} + L^{z-}$ having an equilibrium constant $K_{ML} = \{M^{z+}\}\{L^{z-}\}/a_{ML} \approx \{M^{z+}\}\{L^{z-}\}$ referred to as a 'solubility product' because the activity of a pure solid present in equilibrium $a_{ML} = 1$. Formally, one can write a 'partition coefficient' using the free ion molality also in this case:

$$k_d(M) = \frac{x_{ML}}{m_{M^{z+}}} = \frac{1 \cdot \gamma_{M^{z+}}}{\{M^{z+}\}} = \frac{m_{L^{z-}}\gamma_{L^{z-}}\gamma_{M^{z+}}}{K_{ML}} \tag{31}$$

In fact, this $k_d(M)$ is mainly dependent on the molality of the ligand L, hence equation 31 is an expression of the so-called 'ligand concentration effect'. In hydrolysis solubility reactions, the main ligand is OH^-, and this explains in part the strong pH dependence of dissolved M concentration in the presence of the solid phase of M. For very simple reactions $M(s) \Leftrightarrow M(aq) \pm H_2O$, such as SiO_2 solubility at non-alkaline pH, the k_d value will be (almost) constant, close to the equilibrium constant of the reversed reaction $M(aq) \Leftrightarrow$ M(s), because no ligand is involved.

Sulphates of divalent cations present a good example of the ligand concentration effect and variations of k_d according to equation 31. In Table 1, the solubility products and g^o_{298} values of some sulphate minerals are collected; they differ by more than eight orders of magnitude. If a system is composed of an excess amount of a $M1SO_4$ mineral composition and water, the calculated equilibrium speciation will correspond to a case of congruent solubility of the sparingly soluble $M1SO_4$ solid, and the dissolved molalities of $M1$ and S will be about equal. However, if a more soluble $M2SO_4$ phase stoichiometry is added in an amount in excess of its solubility then the $M2SO_4$ solid will define the total molality of dissolved sulphate.

According to equation 31, this should dramatically decrease the molality of $M1$ and increase the k_d of $M1$. Table 2 demonstrates that this happens indeed: the greater dissolved sulphate concentrations maintained by the presence of a more soluble phase

Table 1. Thermodynamic data for some crystalline sulphates[1] (at 1 bar 298.15 K).

Mineral	log K_{SP}	g^o_{298} kJ mol^{-1}	Ref.
$RaSO_4$	−10.26	−1364.52	[1]
$BaSO_4$ (barite)	−9.97	−1362.15	[1]
$SrSO_4$ (celestite)	−6.63	−1346.15	[1]
$PbSO_4$ (anglesite)	−7.85	−813.17	[2]
$CaSO_4$ (anhydrite)	−4.36	−1322.12	[1]
$CaSO_4 \cdot 2H_2O$ (gypsum)	−4.58	−1797.76	[1]
$MgSO_4 \cdot 7H_2O$ (epsomite)	−2.15	−2871.00	[3]

[1] SO_4^{2-} ion: $g^o_{298} = -744.459$ kJ mol^{-1}

[1] GEMS version of Nagra-PSI chemical thermodynamic database (Hummel *et al.*, 2002a)

[2] SUPCRT92 database, Slop98.dat file (Johnson *et al.*, 1992; Shock *et al.*, 1997)

[3] Grevel & Maizlan (2009).

Table 2. Calculated[1] total dissolved molalities of metals and sulphur in presence of various combinations of stable crystalline sulphate phases.

Phases	**Log_{10} k_d(Ba) Equation 31**	**log_{10} m (S)**	**log_{10} m (Ba)**	**log_{10} m (Sr)**	**log_{10} m (Ca)**	**log_{10} m (Mg)**	***I, m***
Aq + barite	4.972	−4.970	−4.970				4.e−5
Aq + barite + celestite	6.560	−3.190	−6.483	−3.190			0.002
Aq + barite + celestite + gypsum	7.339	−1.781	−7.004	−3.839	−1.785		0.045
Aq + barite + gypsum	7.338	−1.784	−7.0035	−10.00	−1.784		0.045
Aq + barite + epsomite[(2)]	8.051	0.3825	−7.388			0.3825	3.633

[1] Calculated using GEM-Selektor with built-in Nagra-PSI database, data from Table 1, and using the Debye-Hückel equation 26 with common $\dot{a} = 3.72$ and $b_\gamma = 0.064$.

[2] This calculation is only illustrative because of incompatibility of equation 26 with high ionic strength; the Pitzer model is needed to obtain quantitative results.

effectively 'repartition' Ba^{2+} cations from the aqueous into the barite phase. This effect is more pronounced at low ionic strength, where the non-ideality and aqueous complexation effects are less significant. The presence of 'intermediately soluble' solid such as celestite with barite and gypsum has little impact on the partitioning of barium. A less-than-expected ligand concentration effect due to the presence of highly soluble epsomite phase can be explained by strong ionic strength and aqueous complexation effects on the cation aqueous speciation.

Another system with pure barium (barite) and lead (anglesite) sulphate phases is a good illustration of the 'ionic strength effect' and the 'cation speciation effect' on ion partitioning, in enhancement of the 'common ligand effect' shown in Table 2. In order to follow the cation speciation effect, let us introduce a total element molality-based partition coefficient

$$k_d^*(M) = \frac{x_{ML}}{m_{M,\text{tot}}}. \tag{32}$$

Comparison with equation 31 shows the difference: equation 32 is an inverse of total molality of all aqueous species of M, whereas equation 31 involves the molality of free ion only. If the latter is predominant at all conditions, the two partition coefficients will be nearly equal. However, if a large part of M is bound into aqueous complexes,

$$k_d(M^{z+}) > k_d^*(M,\text{tot})$$

In a system composed of 1000 g of H_2O, 0.1 mol $BaSO_4$, 0.001 or 0.1 mol NaCl, trace air, and trace $PbSO_4$, the barite is expected to be stable in a wide range of pH (adjusted by adding NaOH or HCl). A model titration of that system to calculate equilibria with pH between 3.0 and 10.0 with step 0.2 (performed using GEM-Selektor) results in a diagram presenting both k_d (equations 31 and 32) as a function of pH.

As seen in Figure 2, the partition coefficients are almost independent of pH and they are not subject to a significant complexation effect. The only difference is that the k_d at low ionic strength is ~0.4 $\log_{10}$ units (2.5 times) greater than at moderate ionic strength.

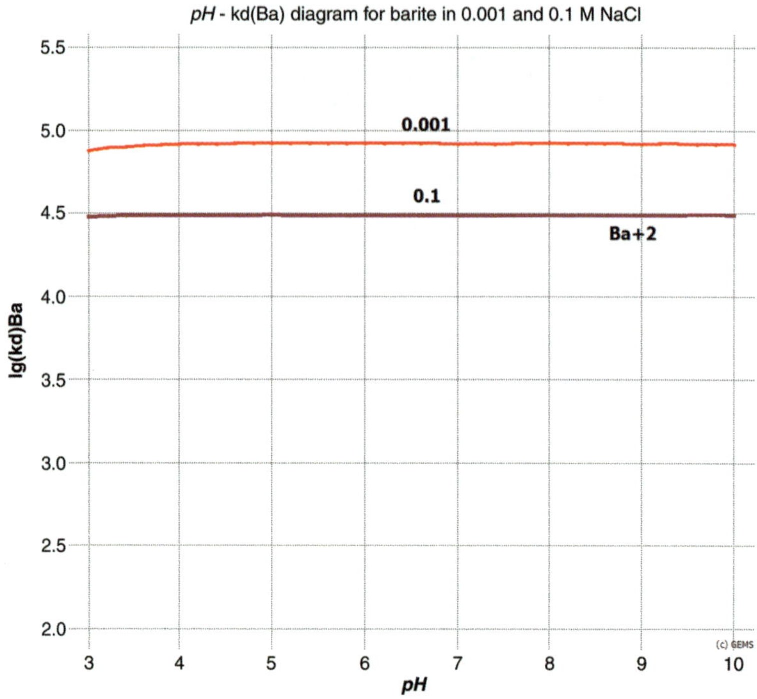

Fig. 2. Partition coefficients of Ba in barite in 0.001 M and 0.1 M NaCl as function of *pH*. Curves for k_d (equation 31) and k_d^* (equation 32) cannot be discriminated by eye.

Indeed, the ratio of calculated activity coefficients of Ba^{2+} ion at 0.001 m and 0.1 m ionic strength is $0.8636/0.3453 = 2.501$. So, Figure 2 shows the ionic strength effect only: the bulk partitioning of Ba^{2+} into barite is weaker at greater ionic strength.

Similar calculations for the system $PbSO_4$-NaCl-H_2O (with trace Ba) show a much less trivial behaviour of partition coefficients (Fig. 3a,b). At low ionic strength and $pH \leq 5$, both k_d and k_d^* values are nearly the same and independent of *pH*. However, at greater *pH*, $k_d(Pb^{2+})$ dramatically increases with *pH*, whereas the k_d^*(Pb) decreases, indicating much less efficient partitioning of Pb into anglesite at elevated *pH* due to a strong aqueous complexation effect. At higher ionic strength (0.1 *m* NaCl), the complexation effect is also present at $pH < 5$, as seen in a separation of curves for the two partition coefficients. Details about the Pb aqueous speciation in the same model titration runs are shown in Figure 3c,d. There are two speciation patterns of lead: hydroxocomplexes, predominant at $pH > 6$, and chloride complexes prevailing at I(NaCl) > 0.01 m and $pH > 6$.

As seen in Figure 3, there are strong aqueous complexation effects affecting the partitioning of some transitional cations (also trivalent and tetravalent cations) which are easily modelled using the appropriate thermodynamic database. In such cases, the

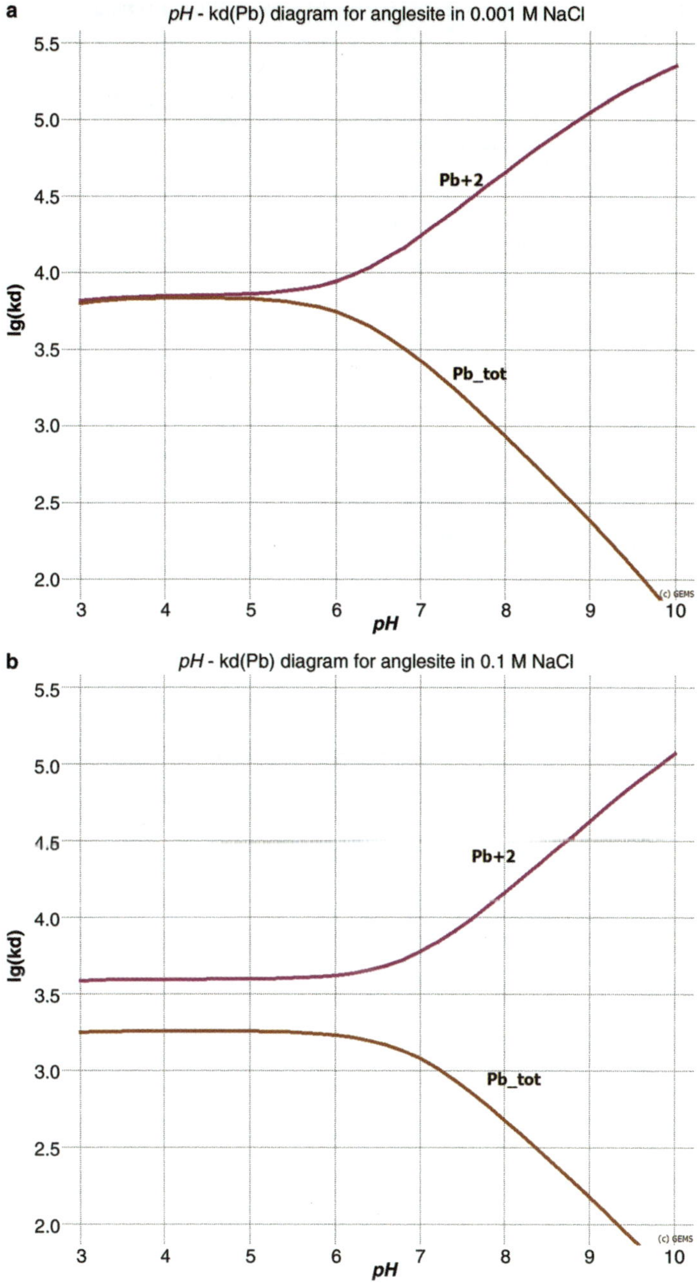

Fig. 3. Equilibrium partitioning coefficients of Pb into anglesite in 0.001 M NaCl (***a***) and 0.1 M NaCl (***b***) as function of *pH*, with the respective aqueous speciation diagrams (***c***) and (***d***). Pb+2 stands for k_d (equation 31), Pb_tot – for k_d^* (equation 32). Pb_tot curves are in the $\log_{10}(m_{\mathrm{tot}})$ scale; other curves are in the $\log_{10}$(activity) scale.

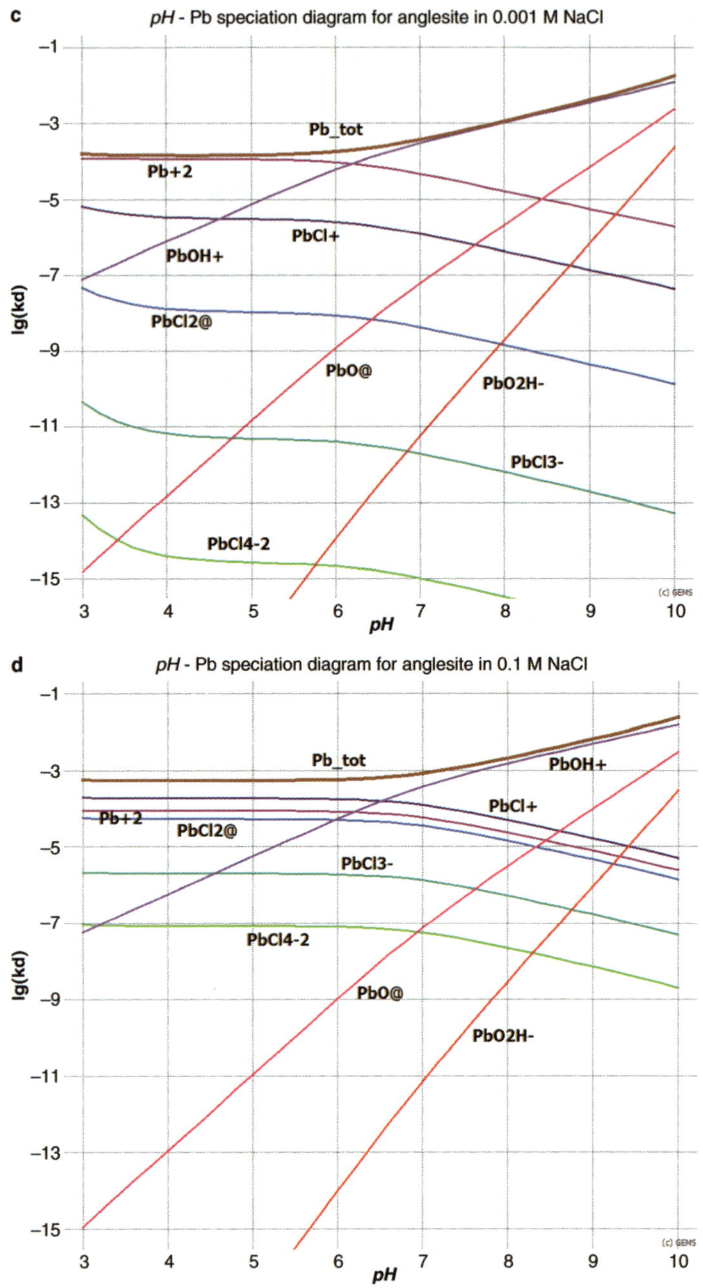

Fig. 3. (*Continued*).

partition coefficients like equation 31 are of little or no value because the free ion aqueous concentration may be very different from the total dissolved element concentration. Equation 32, which is a theoretical analogue of the experimental distribution ratio *Rd*, is then more appropriate, but hardly predictable if the aqueous speciation of the cation is not known in detail and a full chemical speciation model is not available. Otherwise, the relevant effects can be predicted, and all possible partition and distribution coefficients can be obtained from the computed speciation and aqueous activity coefficients.

The speciation effects on ion partitioning can be very complex, and different cations may influence each other through the common ligand concentration effect. To illustrate this, let us add 0.1 mol $BaSO_4$ to the system considered in generating Figure 3. Both lead and barium sulphate solids will be stable at the same time. As seen in Figure 4, the presence of barite does not have much influence on the partitioning of Pb into anglesite. However, the partitioning of Ba into barite is strongly affected by the Pb speciation at $pH > 6$. The reason is the enhanced solubility of anglesite due to the hydroxo and chloride complexation with the associated aqueous sulphate ligand concentration effect on k_d(Ba) in less soluble barite.

In a closed system, the temperature increase alone may cause significant speciation effects in ion partitioning, as shown in Figure 5a. This diagram is plotted for the same system as in Figure 4b with a composition yielding $pH \approx 7$ at 25°C. Variation of temperature from 0 to 350°C at pressures of saturated H_2O vapour results in a variation of *pH* within one unit, as well as in steadily increasing aqueous complexation effects at elevated temperatures.

Very strong complexation effects can be triggered by even a small addition of the oxidant or reductant to a redox-unbuffered bulk system composition. In Figure 5b, such a reduction effect is modelled at fixed $P = 1$ bar, $T = 25$°C by adding milligrams of H_2 to the initial system recipe made of 1000 g H_2O, 0.001 mol $BaSO_4$, 0.001 mol $PbSO_4$, 0.1 mol NaCl, 0.00047 mol NaOH, 1 mg of air, and 0.009 g O_2. In this initially oxic system, $pH = 7.00$, $Eh = 0.804$, and both barite and anglesite are present. Upon addition of small amounts of H_2, complex redox speciation effects on partitioning of Ba and Pb occur; they are manifested in a dramatic increase in *pH*, decrease in *Eh*, and a complete switch in the stable Pb mineral from anglesite $PbSO_4$ to galena PbS, with a jump in $\log_{10} k_d^*$(Pb_tot) of >10 orders of magnitude. The reduction of this system results in a conversion of sulphate into sulphide aqueous species. When sulphate becomes unstable, the barite phase disappears, and all the barium dissolves in the aqueous phase while raising *pH*. This illustrative system shows that very dramatic redox effects on ion speciation and partitioning are readily possible. In reality, such redox transformations may be smeared by the kinetics of redox- and solid surface reactions, which can make the experimental validation of modelled redox profiles a very difficult task.

4.1. Solid solutions

The aqueous equilibria involving only pure crystalline mineral phases are often oversimplified. In geochemistry and mineralogy, it is well known that no pure minerals exist; at least to a small extent, most minerals can be regarded as solid solutions. In

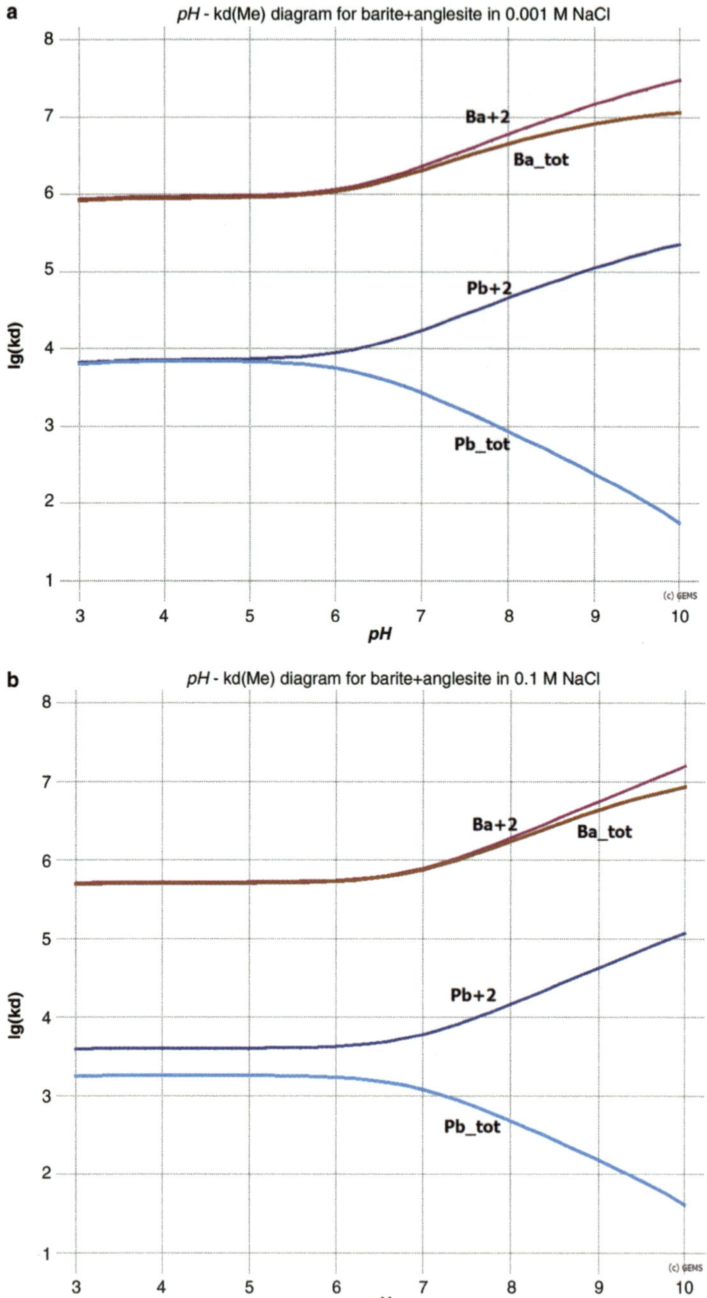

Fig. 4. Partition coefficients of Ba in barite and Pb in anglesite in 0.001 *m* NaCl (***a***) and 0.1 *m* NaCl (***b***) as function of *pH*. *M*+2 stands for k_d (equation 31), *M*_tot stands for k_d^* (equation 32).

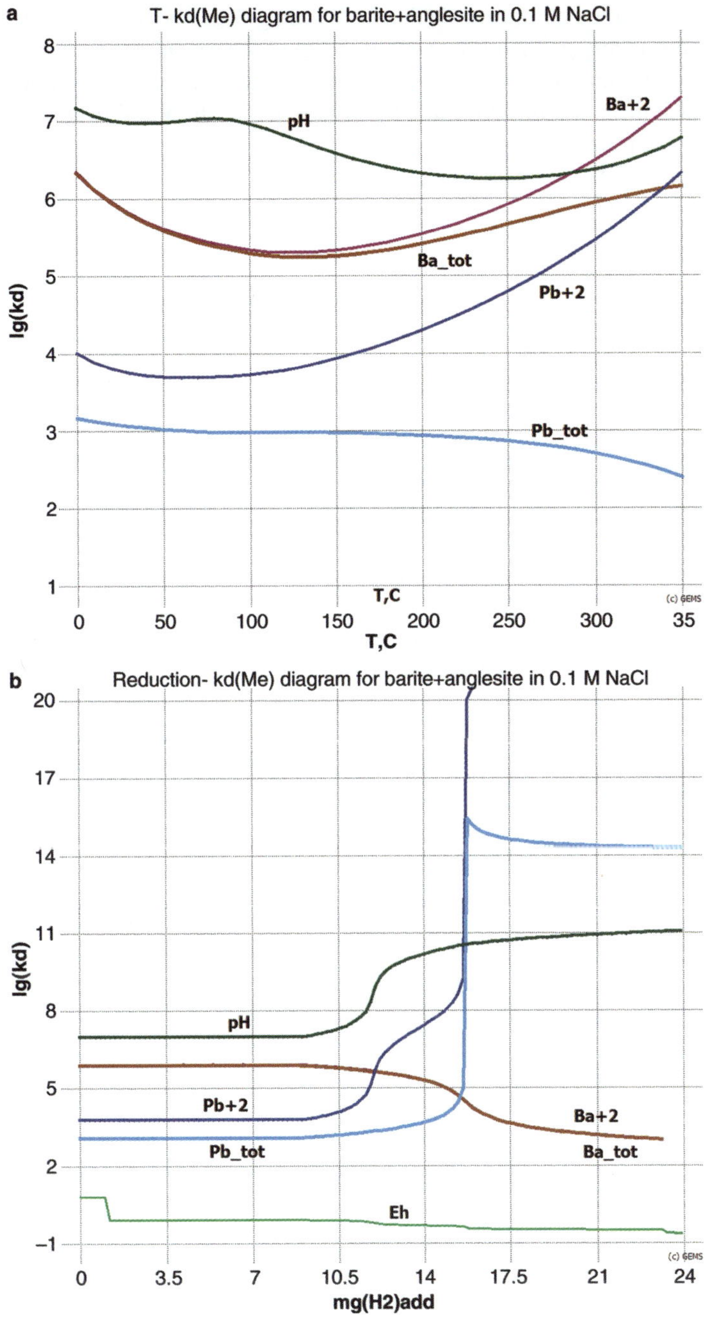

Fig. 5. Effects of temperature (***a***) and redox (***b***) on speciation and partitioning of Ba and Pb into their pure solid phases. See text for explanations.

aqueous-solid solution equilibria, the ion partitioning depends on the end-member solubility products, common ligand concentration, aqueous speciation (temperature) effects, plus also on the properties of mixing between the end members. The simplest case is a $(B,C)L$ ideal solid solution of BL and CL end members with solubility constants K_{BL} and K_{CL}. Activity of each end member in this case equals its mole fraction (x_{BL} or x_{CL}), and the end member activity coefficients are $f_{BL} = f_{CL} = 1$. Two partition coefficients can now be written:

$$k_{d,B} = \frac{x_{BL}}{m_{B^{z+}}} = \frac{m_{L^{z-}}\gamma_{L^{z-}}\gamma_{B^{z+}}}{K_{BL}f_{BL}} \tag{33}$$

$$k_{d,C} = \frac{x_{CL}}{m_{C^{z+}}} = \frac{m_{L^{z-}}\gamma_{L^{z-}}\gamma_{C^{z+}}}{K_{CL}f_{CL}} \tag{34}$$

and combined into a single 'equilibrium distribution coefficient'

$$D_{BC} = \frac{k_{d,B}}{k_{d,C}} = \frac{x_{BL}\, m_{C^{z+}}}{x_{CL} m_{B^{z+}}} = \frac{K_{CL}f_{CL}\gamma_{B^{z+}}}{K_{BL}f_{BL}\gamma_{C^{z+}}} \tag{35}$$

We see that the partitioning of either B or C between the aqueous and the solid solution is subject to the speciation and ligand concentration effects, but it is independent of the solid composition, and the distribution coefficient is (nearly) constant. However, the distribution of B and C between the two phases is very strongly affected by the value of D_{BC}, *i.e.* mainly by the effect of ratio of the end member solubility products because the ratio of aqueous activity coefficients for ions of the same charge is usually close to unity. Let us rewrite equation 35 after defining $x_{C,aq} = \dfrac{m_{C^{z+}}}{m_{B^{z+}} + m_{C^{z+}}}$ leading to $x_{BL}x_{C,aq} = D_{BC}x_{CL}x_{B,aq}$:

$$x_{CL} = x_{C,aq}\frac{1}{D_{BC} - D_{BC}x_{C,aq} + x_{C,aq}} \tag{36}$$

When $D_{BC} = 1$, *i.e.* the solubility products are about the same, the denominator is almost unity, so the ionic fraction of C in the solid phase will be the same as in the aqueous phase; *i.e.* there is no preference for C at any $x_{C,aq}$. If, for example, $D_{BC} = 1000$ (*i.e.* K_{CL} is much greater than K_{BL}) and $x_{C,aq} = 0.9$, then according to equation 36, $x_{CL} = 9.91 \times 10^{-3} x_{C,aq}$, *i.e.* the ion C will occur mainly in the aqueous solution, while the ion B will be strongly concentrated in the SS phase ($x_{B,aq} = 0.1$, $x_{BL} = 0.9901$). Of course, the opposite will happen if $D_{BC} \ll 1$ ($K_{BL} \gg K_{CL}$): now the C ion will be strongly concentrated in the solid solution. At very different end-member solubility products, this 'K_{SP} ratio effect' will override other effects due to solid solution and aqueous electrolyte non-ideality.

4.2. Solid solution non-ideality effects

How significant are the solid solution non-ideality effects? The non-ideality in the $(B,C)L$ solid solution means that the activity of end member is a product such as

$a_{BL} = x_{BL} f_{BL}$ where f_{BL} is the activity coefficient, a function of composition with a limiting behaviour $f_{BL} \to 1$ at $x_{BL} \to 1$ (pure end member reference state) and $f_{BL} \to const \to f_{BL}^{\infty}$ at $x_{BL} \to 0$ (dilute state). In equations 33–35, it can be seen that the impact of non-ideality on the ion partitioning is accounted for by the end-member activity coefficients (together with the aqueous ion activity coefficients). If they all are computed within a chemical speciation model then all possible partition and distribution ratios can be obtained directly from the molalities and mole fractions in the equilibrium state, without explicit consideration of solubility products.

For a simple (solid) solution, the molar Gibbs energy of mixing can be represented as

$$G_m = G_m^{id} + G_m^{ex} \tag{37}$$

where

$$G_m^{id} = RT \sum_j x_j \ln x_j \tag{38}$$

is a contribution from the simple ideal mixing, and G_m^{ex} is the molar excess Gibbs energy.

Description of thermodynamic properties of a non-ideal solid solution, and hence of the solid-phase activity coefficient *vs.* composition relations, requires a model for calculating G_m^{ex} and other properties of non-ideal mixing. Many such models, sometimes quite complex, have been used in petrology for the rock-forming minerals (Ganguly, 2001; Holland & Powell, 1998; Holland & Powell, 2003; Ghiorso & Sack, 1995).

In low-temperature geochemistry, most popular models for simple binary solid solutions are based on the Margules expression (*e.g.*Anderson, 2005; Bruno *et al.*, 2007):

$$G_m^{ex} = \mathrm{R}T(x_1 \ln f_1 + x_2 \ln f_2) = x_1 x_2 W_{12} x_2 + x_1 x_2 W_{21} x_1 \tag{39}$$

which becomes symmetric (regular) if the interaction parameters are equal ($W_{12} = W_{21} = W$). A more general approach is based on the Guggenheim (Redlich-Kister) polynomial

$$G_m^{ex} = \mathrm{R}T(x_1 \ln f_1 + x_2 \ln f_2) = x_1 x_2 \sum_{r=0}^{r} G_r (x_1 - x_2)^r \tag{40}$$

where G_r are interaction parameters (J mol^{-1}) which can be functions of temperature and pressure, but not of composition. In practice, such models of the order $r > 3$ are seldom used. At $r = 3$, equation 40 takes the form

$$G_m^{ex} = x_1 x_2 [G_0 + G_1 (x_1 - x_2) + G_2 (x_1 - x_2)^2] \tag{41}$$

Equations for the end-member activity coefficients

$$\ln f_1 = \frac{x_2^2}{RT}[G_0 + G_1(3x_1 - x_2) + G_2(x_1 - x_2)(5x_1 - x_2)] \quad (42)$$

$$\ln f_2 = \frac{x_1^2}{RT}[G_0 - G_1(3x_2 - x_1) + G_2(x_2 - x_1)(5x_2 - x_1)] \quad (43)$$

can be obtained by differentiating equation 41. Equations for the 'sub-regular' (asymmetric) mixing model can easily be obtained by setting $G_2 = 0$, and those for the 'regular' (symmetric) model – by setting both $G_1 = 0$ and $G_2 = 0$. It can be shown that for the sub-regular case, the interaction parameters in the Margules and Guggenheim models are inter-related:

$$W_{12} = G_0 - G_1; \quad W_{21} = G_0 + G_1 \quad (44)$$

Both models can also be generalized to multi-component solutions (*e.g.* Helffrich & Wood, 1989; Hillert, 1998), which is beyond the scope of this chapter. The usual assumptions in applying mixing models to solid solutions are that the same crystallographic structure exists through the whole compositional range, and that the enthalpy of mixing is equal to the excess Gibbs energy ($G_m^{ex} = H_m$).

In carbonate or sulphate solid solutions, the excess Gibbs energies are positive, which may result in a miscibility gap. In the regular mixing model ($G_m^{ex} = x_1x_2G_0$), the binodal gap is predicted to occur when $G_0/(RT) > 2.0$ (Glynn, 2000; Glynn, 1991). If the bulk composition of the system falls within a binodal miscibility gap then (in equilibrium) two solid-solution phases of different composition will coexist with a certain composition of the aqueous solution. The width of miscibility gap usually decreases with increasing temperature, until the gap disappears at a 'critical' temperature T_{cr}. Below T_{cr}, two solid solution phases instead of one must be included in the system definition for the GEM calculation of correct stable equilibrium speciation (more in Bruno *et al.*, 2007).

If pure end members have different structures, *e.g.* *Cal* $CaCO_3$ (calcite, $R\bar{3}c$) and *Str* $SrCO_3$ (strontianite *Pmcn*), their mixing behaviour can be described using the DQF model (Darken's quadratic formalism). The DQF is an extension of the regular mixing model applied to two structural/compositional regions (Powell, 1987). In the *Cal-Str* case (Kulik *et al.*, 2010), in the Ca-rich region 1 (calcite structure), the activity coefficients will be given as

$$RT \ln f_{Cal} = W_{CalStr} x_{Str}^2 \quad (45a)$$

$$RT \ln f_{Str} = \Delta g_{SrCO_3}^{DQF} + W_{CalStr} x_{Cal}^2 \quad (45b)$$

In the Sr-rich region 2 (aragonite structure), the analogous equations will apply:

$$RT \ln f_{Cal} = \Delta g_{CaCO_3}^{DQF} + W_{StrCal} x_{Str}^2 \quad (46a)$$

$$RT \ln f_{Str} = W_{StrCal} x_{Cal}^2 \quad (46b)$$

The Darken's constants appearing in equations 45b and 46a, respectively, can be defined as

$$\Delta g^{DQF}_{CaCO_3} = G^o_{Arg} - G^o_{Cal} \tag{47a}$$

$$\Delta g^{DQF}_{SrCO_3} = G^o_{SrCO_3R3c} - G^o_{Str} \tag{47b}$$

where G^o_j stands for the standard molar Gibbs energy of jth end member at the temperature of interest. Two Darken's constants, as well as two interaction parameters W_{CalStr} and W_{StrCal}, must not necessary have the same value. A cornerstone assumption made in the DQF model is that between the terminal compositional regions 1 and 2, a transitional region 3 exists where one structure changes to another. The position and the width of this region are somewhat arbitrary; however, the width should be smaller than that of the miscibility gap. If an incomplete miscibility is expected, two SS phases with the same end members and the same DQF mixing model are necessary for calculation of the complete equilibrium state.

Values of interaction parameters in Margules or Guggenheim equations can be predicted for simple ionic crystal structures (*e.g.* carbonates, sulphates) using a Madelung-Vegard approach in calculating electrostatic excess energies of the mixed crystals (Lippmann, 1980; Urusov, 1977, 2001), estimated semi-empirically from the experimental or geochemical data (Glynn, 2000; Kulik *et al.*, 2000), or simulated using the cluster expansion atomistic methods (Vinograd & Winkler, 2010; Vinograd *et al.*, 2009b).

4.3. Lippmann functions

A convenient way to visualize Aq-SS equilibria, saturation states, and ion partitioning for a binary ionic solid solution with a common ligand, such as $(B,C)L$ with end members BL and CL, is based on plotting the Lippmann total solubility product (Lippmann, 1980; Glynn, 2000; Glynn & Reardon, 1990; Prieto, 2009) defined as

$$\Sigma\Pi = \{L^-\}(\{B^+\} + \{C^+\}) \tag{48}$$

where braces denote ion activities in the aqueous solution. Note that at the end-member compositions, equation 48 reduces to a usual solubility product K_{BL} (reaction $BL = B^{z+} + L^{z-}$) or K_{CL} (reaction $CL = C^{z+} + L^{z-}$). The Lippmann diagram usually consists of two plots with a common ordinate $\log_{10}\Sigma\Pi$ and two superimposed abscissas x_{CL} and $x_{C,aq}$, where the aqueous ion activity fraction is defined as

$$x_{C,aq} = \frac{\{C^+\}}{\{B^+\} + \{C^+\}}; \quad x_{B,aq} = 1 - x_{C,aq} \tag{49}$$

The plot of $\log_{10}\Sigma\Pi_{eq}$ (equation 48) against the mole fraction in the solid x_{CL} is called the 'solidus', and that against the aqueous ion activity fraction (equation 49) is called the 'solutus'. In the case of a miscibility gap, segments of solidus and solutus curves are metastable. The boundary points between stable and metastable segments show the compositions of two SS phases in equilibrium with the 'eutectic' composition of the aqueous phase. Lippmann diagrams can be plotted easily using the MBSSAS code (Glynn, 1991) or directly from results of GEM speciation calculations at varying composition of the system. More about the Lippmann and other diagrams for Aq-SS systems can be found in chapter 1 of this volume (Prieto, 2010). Example Lippmann diagrams for the sulphate aqueous–solid solution systems are provided in Figures 6 and 7.

In the $(Ba,Ra)SO_4$ system (Fig. 6), the end-member solubility products are similar and the mixing is probably not far from ideal. Figure 6a shows only a little partitioning of Ra^{2+} into the solid solution, which is somewhat enhanced by a moderate non-ideality of mixing (the non-ideal case is shown in dashed curves) at $x_{Ra} > 0.4$, with no partitioning around $x_{Ra} = 0.3$ (alyotropic point). As seen in Figure 6b, in the ideal mixing case, D_{RaBa}, is constant, and k_d(Ra), k_d(Ba) are nearly constant throughout the whole composition range. Introduction of the non-ideality of mixing makes D_{RaBa} a function of composition, k_d(Ra) decreasing in the Ba-rich region, and k_d(Ba) decreasing in the Ra-rich region. One can conclude that positive non-ideality impedes the ion partitioning into solid solution especially at trace concentrations.

The celestite-barite system (Fig. 7a,b) shows a strong partitioning of Ba into the solid and of Sr into aqueous solution due to very different end-member solubility products (see equations 35 and 36) resulting in large values of k_d(Ba) and D_{BaSr}. The non-ideal mixing is accounted for using equations 41, 42 and 43 with interaction parameters $G_0 = 7.5804$; $G_1 = -0.1066$; $G_2 = 1.0538$ (in kJ mol^{-1}) predicted using the double-defect method of atomistic modelling (Vinograd *et al.*, 2009a). This non-ideality leads to a miscibility gap between $x(BaSO_4) = 0.05$ and 0.94 in equilibrium where two phases of (different) fixed composition coexist with a certain constant composition of the aqueous solution. However, at ambient temperatures and relatively short reaction times, a single metastable 'disordered' solid solution phase may form throughout the whole composition range even though the enthalpy of mixing is great enough for a miscibility gap (*e.g.*Katsikopoulos *et al.*, 2009). For this reason, both stable and metastable diagrams are plotted in Figure 7a,b. Thick lines in Figure 7a represent stable parts of the (very asymmetric) Lippmann solutus and solidus curves with a miscibility gap between where the equilibrium aqueous composition is set by a peritectic point (PP) (see inset) at $x(Ba,aq) = 0.00046$ and $\log_{10} \Sigma\Pi = -6.647$. In Figure 7b, the horizontal dash-dotted lines show partition and distribution coefficients in two phases coexisting within the miscibility gap. Thin solid lines in Figure 7a,b represent a metastable case with a single solid-solution phase forming.

Examples provided in Figures 6 and 7 give some idea about the magnitudes of the effects of mixing on the cation partitioning between aqueous and solid solutions. The major effect is due to the ratio of end-member solubility products (involved in the distribution coefficient D, equation 35). If this ratio is close to unity (see Fig. 6), the partitioning is rather small at any composition, and can be significantly changed or reversed by even a moderate non-ideality. The impact of non-ideality is more pronounced at minor-to-trace concentrations (Fig. 6b); the partition and distribution coefficients can be considered constant only when the mixing is simple and ideal. If the K_{SP} ratio is large (Fig. 7) then it becomes a major factor in ion partitioning, whereby the ion,

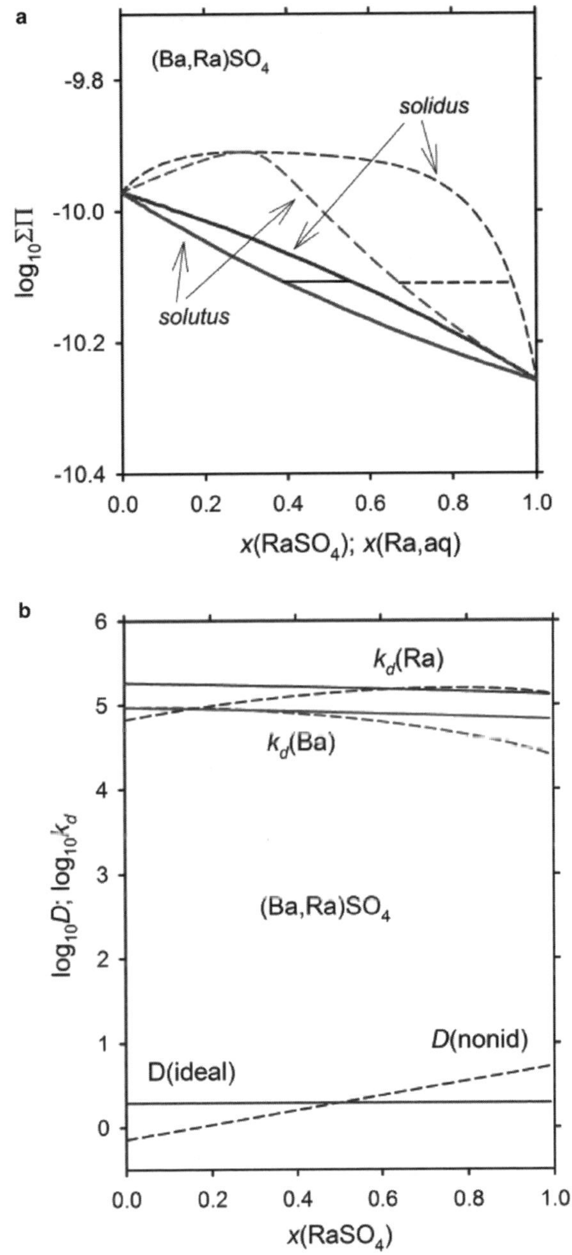

Fig. 6. Diagrams for the barite-$RaSO_4$ aqueous–solid solution system: (***a***) Lippmann diagrams for the ideal mixing (solid lines) and for a regular model with a guessed interaction parameter $G_0 = 4.0$ kJ mol^{-1} (dashed curves). (***b***) Variations of partition and distribution coefficients for the cases of ideal and non-ideal mixing. Calculated in GEM-Selektor using the built-in Nagra-PSI database.

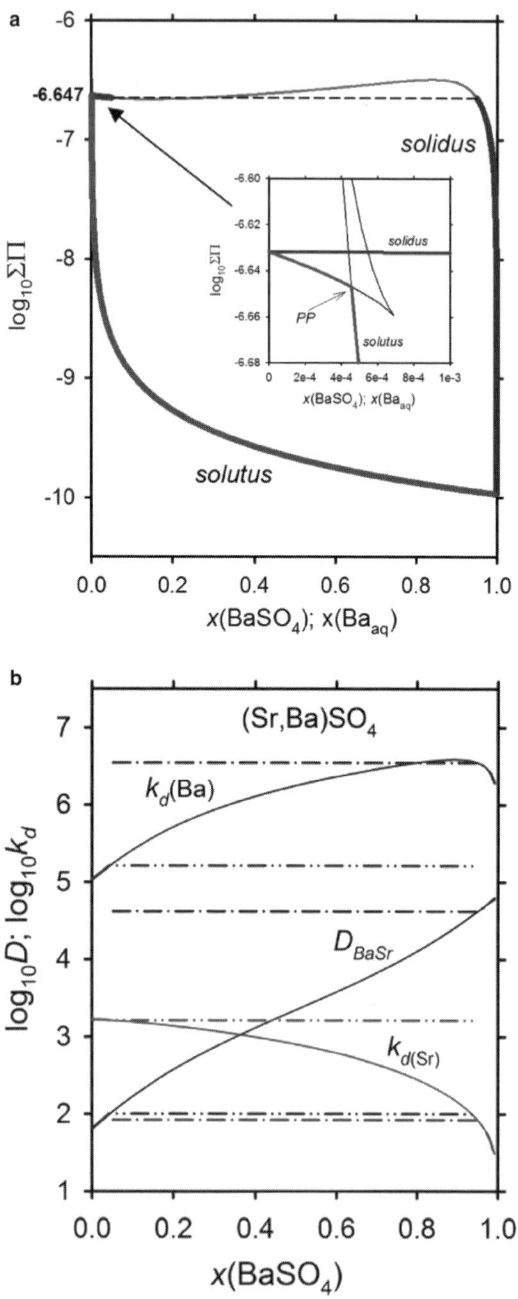

Fig. 7. Diagrams for the celestite-barite aq-SS system: Lippmann functions (***a***) and partition/distribution coefficients (***b***) at 1 bar and 25°C. Computed in GEM-Selektor; thermodynamic data were taken from the built-in Nagra-PSI database (see also Table 1). See text for explanations.

having a smaller solubility product (Ba^{2+}) is strongly concentrated in the solid phase, and the non-ideality of mixing (even a miscibility gap) is a second-order effect on that. In such systems, the partition and distribution coefficients may be highly variable with composition and, thus, difficult to predict without a complete thermodynamic speciation model.

4.4. Trace-ion partitioning into solid solutions

Trace-ion partitioning in solid solutions is important in geochemistry, *e.g.* of marine carbonates or of (radio) toxic waste. Lippmann (Figs 6 and 7) or Roozeboom diagrams (see chapter 1, this volume, Prieto, 2010) are not very helpful because of the linear scale in phase composition. The experimental data on trace-element *M* uptake in host solids are often represented as the equilibrium 'uptake isotherms', where the concentration of *M* (*e.g.* mole fraction) in the solid is plotted against the concentration of *M* in the aqueous phase, both usually in $\log_{10}$ scale. Figure 8 shows the Ba isotherm (a) and the partition and distribution coefficients (b) for the same aqueous-solid solution system as in Figure 7, but plotted in logarithmic scale from the Sr-rich end.

Figure 8 shows that the non-ideal isotherms are linear, and the k_d and D are constant at $x(BaSO_4) < 0.001$ (m(Ba,tot) $< 10^{-8}$); the non-ideality affects curve shapes only at greater Ba concentrations. In the trace region, the only impact of non-ideal mixing is manifested in a vertical shift of the isotherm, k_d and D plots down by ~1.5 orders of magnitude relative to the ideal case. This Henry-law-type behaviour can be understood by looking at equations 42 and 43 or any similar ones for the end-member activity coefficients. Suppose that end-member 2 is in trace concentration ($x_2 < 0.001$), then $x_1 > 0.999$. Substitution of $x_2 = 0.001$ and values of interaction parameters given above into equations 42 and 43 (at $T = 298.15$ K) yields $\ln f_1 \approx 5 \times 10^{-6}$ ($f_1 \approx 1.0001$) and $\ln f_2 \approx 3.43$ ($f_1 \approx 30.88$). Obviously, in the dilution limit $x_2 \to 0$, $\ln f_1 \to 0$, and $\ln f_2 \to (G_0 + G_1 + G_2)/(RT) \to 3.44$. We see that below the mole fraction 0.001 of the trace end member, both activity coefficients become practically constant. The activity of the major solid-solution end-member is $a_1 \approx x_1$ in both ideal and non-ideal cases, whereas the activity of trace end member is $a_2 = x_2$ in the ideal and $a_2 \approx 31.187x_2$ in the non-ideal case. Because the activity of the $BaSO_4$ end member is proportional to the aqueous activity (molality) of the dominant Ba^{2+} ion, the limiting value of trace end-member activity coefficient $f_2^{\infty} = 31.187$ is exactly the shift on Figure 8 of non-ideal from ideal curves for the Ba isotherm, k_d and D values.

This kind of reasoning can be applied to any aqueous–solid solution system. If solubility products of major and trace end members are known, and measured trace ion uptake isotherms, k_d or D values are available, they can be used in retrieval of the activity coefficient of trace end member f_2^{∞}, which then yields immediately at least the regular interaction parameter value G_0. In some systems where trace ions are incorporated into the host mineral such as calcite, some trace-element end-members are not known to exist as pure phases, therefore only the product $K_2 f_2^{\infty}$ can be retrieved from the uptake isotherm data. Europium in calcite is a good example (Curti *et al.*, 2005).

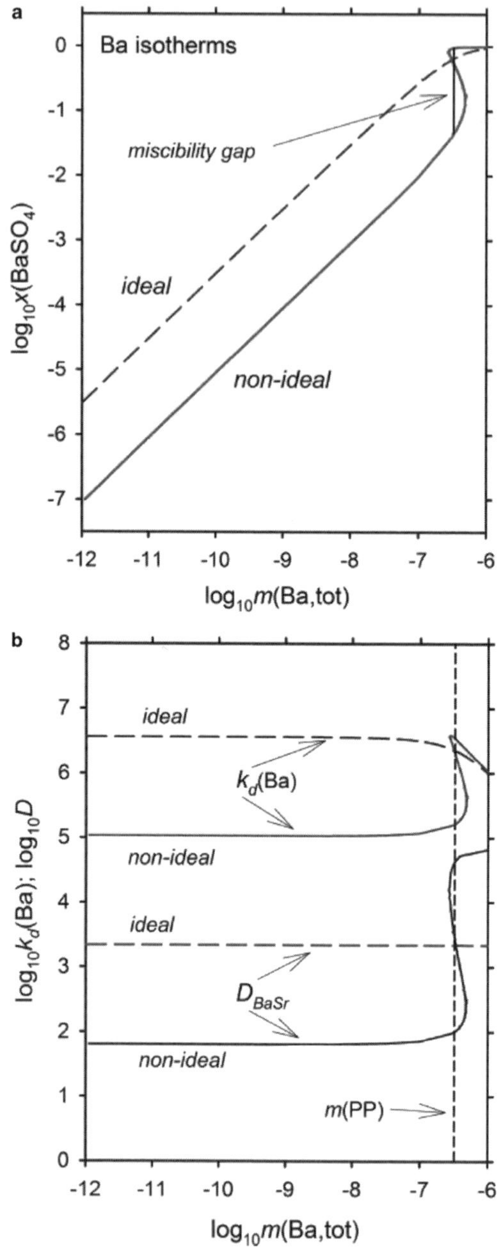

Fig. 8. Ba isotherm (***a***) and Ba partition and distribution coefficients (***b***) for the aqueous–solid solution system $(Sr,Ba)SO_4$ at 1 bar and 25°C. Computed with GEM-Selektor using the same data as in Figure 7. Solid curves represent the metastable non-ideal solid-solution model (PP is the aqueous composition at the peritectic point, defining the position of miscibility gap). Dashed curves, for comparison, show the same runs for the ideal solid-solution model.

4.5. Sublattice solutions

In many minerals, ionic substitutions may occur simultaneously in two or more different structural sites. If so then simple solution models cannot be used even in the case of ideal mixing on each site independent of the other site. Treatment of such 'multi-site' solutions requires the identification of 'sublattices', *i.e.* sets of equivalently coordinated sites (and charges) in the crystal structure (Hillert, 1998). The next step is to consider which species (atoms, ions or vacancies) can be substituted in a given sublattice site. A single species in each sublattice is a trivial case of pure ionic solid such as calcite $[Ca^{2+}]_1{:}[CO_3^{2-}]_1$.

Random mixing in a given sublattice (site) is described by the configurational entropy

$$S_{\mathrm{conf}} = -R\sum_{i\in M} y_i \ln y_i \tag{50}$$

where y_i is the site fraction of the ith substituting species (one of $n(M)$ possible species). If there is only one sublattice with two or more $n(L)$ substituting species then the S_{conf} reduces to

$$S_{\mathrm{simple}} = -R\sum_{j\in L} x_j \ln x_j \tag{51}$$

where x_j is the jth end-member mole fraction, $x_j = y_i, j = i$. The molar Gibbs energy of mixing in any (solid) solution is

$$G_m = H_m - TS_m \tag{52}$$

In the case of simple ideal solid solution, $H_m = 0$ and $S_m = S_{\mathrm{simple}}$. On the other hand, G_m can be expressed in chemical potentials of end members:

$$G_m = \sum_{j\in L} x_j(\mu_j - \mu_j^o) = \mathrm{R}T\sum_{j\in L} x_j \ln a_j \tag{53}$$

where μ_j is the chemical potential of jth end member, μ_j^o is its standard chemical potential (standard molar Gibbs energy), and a_j is its activity. Because in the simple ideal solid solution, $G_m = -TS_{\mathrm{simple}}$, from equation 53 it follows that

$$a_j = x_j \tag{54}$$

which leads to equation 38. The chemical potential of the jth end member is then

$$\mu_j = \mu_j^o + RT\ln(x_j \cdot 1) \tag{55}$$

where 1 stays in the place of the activity coefficient f_j. These relations hold for any number of end members, *i.e.* for any number of substituting species in a single sublattice.

When the energetic non-ideality is absent ($H_m = 0$), but different substitutions may occur simultaneously in several sub-lattices, it can be assumed (as a first approximation) that the mixing in one sublattice is independent of that in another; the total entropy S_{conf} is therefore the sum of S_{conf} for each sublattice. For instance, consider an $[A,B]_1{:}[C,D]_1$ solid solution (Na,K)(Br,Cl). This system can be described with four end members: NaBr, KBr, NaCl and KCl. According to equation 50 applied to two sublattices, (c) for cations and (a) for anions,

$$S_{\text{conf}} = -\mathrm{R}(y_{Na}^{(c)} \ln y_{Na}^{(c)} + y_{K}^{(c)} \ln y_{K}^{(c)} + y_{Br}^{(a)} \ln y_{Br}^{(a)} + y_{Cl}^{(a)} \ln y_{Cl}^{(a)}) \tag{56}$$

By applying $S_m = S_{\text{conf}}$ in equations 52 and 53, one obtains

$$a_{NaBr} = y_{Na}^{(c)} y_{Br}^{(a)};\; a_{KBr} = y_{k}^{(c)} y_{Br}^{(a)};\; a_{NaCl} = y_{Na}^{(c)} y_{Cl}^{(a)};\; a_{KCl} = y_{K}^{(c)} y_{Cl}^{(a)} \tag{57}$$

Ionic fractions on cationic (c) and anionic (a) sites can now be found from end-member mole fractions:

$$y_{Na}^{(c)} = x_{NaBr} + x_{NaCl};\; y_{k}^{(c)} = x_{KBr} + x_{KCl}; \tag{58a}$$
$$y_{Br}^{(a)} = x_{NaBr} + x_{KBr};\; y_{Cl}^{(a)} = x_{NaCl} + x_{KCl} \tag{58b}$$

In equation 57, for instance, $a_{NaBr} = y_{Na}^{(c)} y_{Br}^{(a)} = (x_{NaBr} + x_{NaCl})(x_{NaBr} + x_{KBr})$, which is a quadratic function of x, obviously not equal to just x_{NaBr} in the simple ideal model (equation 54). The Gibbs energy of this (still energetically ideal) mixing equals to $G_m^{id} = -TS_{\text{conf}}$, where S_{conf} is given by equations 56–58.

The system such as (Na,K)(Br,Cl) has a 'reciprocity' – a property manifested by the possibility of writing a 'reciprocal reaction' between end-members, KBr + NaCl ⇔ NaBr + KCl, with the Gibbs energy effect:

$$\Delta G_{rcp} = G_{NaBr}^{o} + G_{KCl}^{o} - G_{KBr}^{o} - G_{NaCl}^{o} \tag{59}$$

This reaction shows that only three of four end members are compositionally independent. The chemical potential of an ideal end member can be expressed as (*e.g.* for KBr)

$$\mu_{KBr} = \mu_{KBr}^{o} + (1 - y_{K}^{(c)})(1 - y_{Br}^{(a)})\Delta G_{rcp} + RT \ln (y_{K}^{(c)} y_{Br}^{(a)}) \tag{60}$$

A consequence of reciprocity is that the bulk composition of the solid solution phase cannot be split to a unique combination of mole amounts of the end members without minimizing the total Gibbs energy of the system. More about thermodynamics of reciprocal solid solutions can be found in Hillert (1998) and Wood & Nicholls (1978). Such reciprocal solid solutions can, of course, have energetic non-ideality in addition to complex configurational entropy, especially if the short- or long-range ordering is present.

Yet another complication is due to the sublattice site stoichiometry ratios (Bruno *et al.*, 2007). For example, in the garnet structure $A_3M_2T_3O_{12}$, independent ion substitutions are possible on three different sublattices: the large 8-coordinated A sites for divalent cations, the octahedral M sites hosting trivalent cations, and the tetrahedral T sites occupied with highly charged cations such as Si^{4+}. In a binary SS $Mg_3Al_2Si_3O_{12}$–$Fe_3Al_2Si_3O_{12}$, the random mixing of Mg^{2+} and Fe^{2+} ions in A sublattice sites results in the ideal configurational entropy of mixing $S_m = -3\mathrm{R}\ (x_1 \ln x_1 + x_2 \ln x_2)$, leading to end-member activities in solid solution $a_1 = x_1^3$; $a_2 = x_2^3$ which again are different from that in equations 54 and 55 implicitly used for simple ideal SS end members in most LMA and GEM equilibrium speciation codes. There are two ways to model this in a chemical speciation code with the implicit simple ideal mixing model. One is to use a fictive activity coefficient, λ_j,

$$\mu_j = \mu_j^o + \mathrm{R}T(\ln x_j + \ln \lambda_j) \quad \text{where } \lambda_j = \frac{a_j}{x_j} \tag{61}$$

For the above SS system, $\lambda_j = x_j^2$. This method also leaves the definitions of partition and distribution coefficients unchanged (when 'energetic' activity coefficients f_j are replaced with λ_j, see equations 33 to 35). Another way is to re-scale the end-member formulae to $Mg_1Al_{2/3}Si_1O_4$ and $Fe_1Al_{2/3}Si_1O_4$ (also dividing the standard chemical potentials of end members μ^o by 3). In this case, the configurational entropy of mixing formally becomes the same as in the simple ideal mixing model: $S_\mathrm{m} = -\mathrm{R}(x_1 \ln x_1 + x_2 \ln x_2)$ and $a_1 = x_1$; $a_2 = x_2$.

However, in natural garnets $(Mg,Fe,Ca,Mn)_3(Al,Cr,Fe)_2Si_3O_{12}$, cation substitutions occur in A and M sublattices at the same time, so the activity of one end member must be expressed as (*e.g.* for the $Mg_3Al_2Si_3O_{12}$ end member) $a(Mg_3Al_2Si_3O_{12}) = (y_{Mg}^{(A)})^3 (y_{Al}^{(M)})^2$. These site fractions can be calculated as products of sums of mole fractions of Mg- and Al-containing end members (similarly to equations 56–58); hence the first method (equation 61) can still be rigorously applied. However, the second method (*i.e.* downscaling of end-member stoichiometries) is no more feasible.

Given the sublattice formula such as $[A,B]_x{:}[C,D]_y$ a list of solid-solution end members can be created by combining one species per sublattice site (*e.g.* A_xC_y, B_xC_y, A_xD_y and B_xD_y). Since not all such end members may exist in a pure state, there will be a problem of estimating or fitting their thermodynamic properties. The site fraction for ith species will be a sum of mole fractions of end members that contain this ionic species. In the absence of non-ideal (excess) terms, the activity of the end member will be equal to a product of site fractions of species that it contains (*e.g.* for A_xC_y, $a_{AxCy} = y_A^x \cdot y_C^y$).

Specific complications are expected for the ordering effects, as well as for the heterovalent substitutions, *e.g.* in $(Na,Eu)_2(CO_3)_3$. The latter involve vacancies and are interdependent in two sublattices because of charge balance; a careful bookkeeping of charges when converting end member mole fractions into ionic (equivalent) fractions will be necessary. Currently, this seems to be a main challenge in further applications of solid solutions to model ion partitioning in geochemical systems. In the context of waste geochemistry and trace radionuclide incorporation, some relevant solids that need more careful consideration in terms of sublattice models include clay minerals (montmorillonite, illite), LDH phases (hydrotalcites, Ca-sulphoaluminates), and calcium silicate hydrates (C-S-H).

5. Sorption: surface complexation and ion exchange

Sorption on mineral-water interfaces (MWI) controls the partitioning and retention of many trace elements and organic substances of environmental concern, as well as the kinetics of mineral dissolution/precipitation (Sherman, 2009; Kulik, 2009; Schott *et al.*, 2009 and references therein). A minor element (M) can be retained on MWI by a variety of sorption mechanisms.

At short reaction times, the aqueous complexation competes for M with 'ion exchange' on permanently-charged surfaces (such as basal planes of clay minerals), as well as with the 'outer-sphere surface complexation' on (hydr)oxides. At longer reaction times, the 'specific (multi-site) adsorption' on (hydr)oxides or clay particle edges takes place; this kind of 'inner-sphere' binding starts already at low total M concentrations in the system and becomes progressively irreversible with time. At even longer reaction times, the incorporation of (ad)sorbed species due to recrystallization of the host mineral may lead to formation of a 'surface solution' layer. At a rather large dissolved $[M_{aq}]$, various 'surface (co)precipitation' phenomena may also be observed. In all these cases of sorption, the experimental ion-partitioning data can be represented as partition ratios Rd:

$$Rd_M = \frac{\text{sorbed concentration of } M \text{ in moles per unit mass of sorbent}}{\text{dissolved concentration of } M \text{ (molar or molal)}} \tag{62}$$

The aim of thermodynamic sorption models (TSM) is to predict Rd_M values, *i.e.* partitioning of M between dissolved aqueous and (ad)sorbed states, in the wide range of composition, temperature and pressure of interest. In particular, such models are important in assessing the retention and mobility of M in the environment or in (radio)toxic waste repositories because at trace M concentrations, sorption is often more efficient than other retention mechanisms such as solid solution formation or pure solid precipitation. In principle, ion partition by sorption is subject to ligand concentration, aqueous speciation (including acid-base and redox changes), ionic strength, and temperature effects, similar to those discussed above for the ion partitioning into solids and solid solutions. However, there are effects specific and relevant for sorption: the 'electrical interface layer' (EIL) effects; the 'surface site saturation and competition' effects, and the 'sorption site heterogeneity' effects. The size of the present contribution precludes a detailed account on TSMs on which a vast literature exists in interface chemistry and geochemistry. Instead, some parallels will be drawn between the ion partitioning by sorption and that by other (already considered) mechanisms.

Thermodynamic sorption models are inherently more complex than the solid solubility or aqueous-solid solution models because the specific surface and site heterogeneity of the sorbent must be considered together with the aqueous complexation and surface binding subject to the electrostatic phenomena on variable-charge amphoteric surface. There are three kinds of thermodynamic sorption models: (1) based on site (or charge) mole balances; (2) based on the surface-density scaling without site balances; and (3) representing sorption by ion exchange as a (solid) solution. They, of course, can be combined with each other and with other solution models (gases, solid solutions) and pure solids in the same chemical thermodynamic system in order to model all possible competition and partitioning effects on the *M* speciation in various phases.

5.1. Site-balance-based TSMs

Site-balance-based TSMs use a concept of site ligand, denoted '$>S$OH', '≡OH' or simply '≡'. The surface binding of *M* is represented as a complexation reaction with the site ligand, analogous to the aqueous complexation: $M_{aq} + >S\text{OH} = >SOM + \text{H}^+$ or $M_{aq} + \equiv \Leftrightarrow \equiv M$. Such reactions are referred to as 'monodentate' in contrast to 'δ-dentate' ones where one *M* species binds to δ surface ligands: $M_{aq} + \delta\equiv \Leftrightarrow \equiv_{\delta}M$. Charged surface species may be subject to the electrostatic (Coulombic) activity coefficient terms taken from the EIL model (Lützenkirchen, 2002a; Kulik, 2009). To run a TSM, a total mole amount or molarity of each site ligand $[>S\text{OH}]_{\text{Tot}}$ or $[\equiv]_{\text{Tot}}$ must be provided for setting up an additional balance constraint for all species containing the site ligand in their formula. The surface complexes are treated similarly to aqueous complexes, but with EIL Coulombic terms applied instead of Debye-Hückel activity coefficients. In other words, a species like $>SOM^+$ is involved in three balances: that for M^{2+}, for H^+, and for the $>S$OH group. The stability of $>SOM^+$ is defined by an 'intrinsic' equilibrium constant of its formation reaction, referenced to 1 M standard concentration, infinite dilution, and zero surface potential. This typical TSM setup is implemented in many LMA speciation codes (MINEQL/FITEQL, MINTEQA2, PHREEQC) and has been described extensively in textbooks (Stumm, 1992; Stumm & Morgan, 1996; Morel & Hering, 1993; Benjamin, 2002). Recently, such TSMs have been implemented in GEM codes HCh and GEM-Selektor (Gaskova & Bukaty, 2008; Kulik *et al.*, 2010, in prep.).

An important feature of the site-balance approach is that no sorbent as such is included in the chemical system definition; only sorption sites are represented in site ligand balances. The total molarity of a site ligand is estimated as a product $[\equiv]_{\text{Tot}} = c_S A_S \Gamma_C$ where c_S is the given sorbent concentration (in kg dm^{-3}), A_S is the known specific surface area (in m^2 kg^{-1}) (usually measured by BET), and Γ_C is the monolayer surface site density parameter (in mol m^{-2}), fixed to some conventional value or taken from the crystallography data. Often, only the sorbent concentration, c_S, is known, but the specific surface area is not. In this case, the product $A_S\Gamma_C$ is replaced by a sorption capacity q_C (mol kg^{-1}) parameter of the solid. This is justified because in many systems with internal porosity, such as soils, clays, and cements, it is difficult to measure reactive specific surfaces separately in pores, interlayers, and on particle surfaces.

Further, site-balance based TSMs depend on what is known about A_S. If A_S is given, then adsorption models can be developed using the surface density concentration scale.

If only the sorption capacity q_C is known, then sorption or ion-exchange models with site ligands can be constructed, as done *e.g.* for clay minerals. The experimental data expressed as a partition ratio *Rd* (equation 62) are normalized by the sorbent mass (or volume) defined at standard conditions (*e.g.* a homoionic Na form of clay at certain relative humidity).

Conversely, only the mass of the sorbent $M = c_S \cdot V_{aq}$ is known in the system; the sorbed *M* concentration q_M (in mol kg^{-1}) can be defined as $q_M = n_{M,sorb}/M = [\equiv M]/c_S$ and then compared with the measured distribution ratio *Rd* value via the modelled partition coefficient (often called distribution coefficient) *Kd* (in dm^3 kg^{-1}):

$$Kd_M = \frac{q_M}{[M_{aq}]} \tag{63}$$

The sorption cannot exceed the sorption capacity q_C as expressed by a site molar balance

$$[\equiv M] + [\equiv] = [\equiv]_{\text{Tot}} \tag{64}$$

where $[\equiv]_{\text{Tot}} = c_S q_C$. Each surface complex of *M* is represented as a product of reaction such as $M_{aq} + \equiv\text{OH} \Leftrightarrow \equiv\text{O}M + \text{H}^+$ with the intrinsic equilibrium constant

$$K_M^{\text{int}} = \frac{[\equiv \text{O}M]\{\text{H}_{aq}^+\}}{\{M_{aq}\}[\equiv \text{OH}]} \tag{65}$$

where braces denote the aqueous species activity (Stumm, 1992; Benjamin, 2002). It is often assumed that the ratio of activity coefficients of surface species in equation 65 is unity, and the only effective correction is due to the (optional) Coulombic term obtained from the chosen EIL model which determines the relative electrical potential acting on a surface species as function of the surface-charge density. The latter, in turn, is mainly created by the (de)protonation of surface $\equiv$OH groups according to reaction such as $\equiv\text{OH}^{z-} + \text{H}^+ \Leftrightarrow \equiv\text{OH}_2^{1-z}$ with the LMA expression

$$K_H^{\text{int}} = \frac{[\equiv \text{OH}_2^{1-z}]}{\{\text{H}_{\text{aq}}^+\}[\equiv \text{OH}^{z-}]};\ \log_{10} K_H = pH_{\text{PZC}} \tag{66}$$

Equations 65 and 66 show why surface speciation is strongly *pH*-dependent, and why the total surface charge density can be positive, negative, or zero. Each solid surface has its characteristic *pH* point of zero charge (PZC) which is known experimentally for many solids (Kosmulski, 2009) and can also be predicted reasonably (Sverjensky & Sahai, 1996). Thermodynamic insights into site-balance-based TSMs were summarized by Kulik (2006a, 2009). In particular, each balance (equation 64) for monodentate surface species simulates a (competitive) Langmuir isotherm. When several surface site types are involved in a multisite model, the adsorption of *M* is the sum of amounts bound to each site, subject to the site balance and to the overall mass balance for *M*. Typically, the 'strong' sites (present in the smallest amount) get occupied first, then weaker sites are populated upon increasing total *M* concentration, leading to complex sorption isotherms with the slope less than one at large M loadings. In the case of δ-dentate adsorption, the Langmuir isotherm is no more appropriate, and the values of intrinsic adsorption constants dependent on the site density parameter, solid concentration, and other factors. Application of EIL Coulombic terms makes the TSM obeying a form of the (competitive) Frumkin isotherm (Stumm, 1992).

5.2. TSMs without site balances

The TSMs without site balances were implemented in the GEM framework (Kulik, 2002b, 2006c) using a concept of the 'sorption phase' where the mineral sorbent (pure or solid solution) with known specific surface area A_S keeps the surface species located on different site types on different surface patches (faces) of sorbent particles. In fact, such a sorption phase is a combination of two or more 'solutions': one for the sorbent, and one or more for surface species on each site type on each surface patch. Surface species are treated in the surface-density concentration Γ_M (mol m^{-2}) scaled to the standard density $\Gamma_o = 2 \times 10^{-5}$ mol m^{-2}, the same for all surfaces on all sorbents. This approach provides the automatic scaling of amounts of surface species to the actual surface area (and amount) of the sorbent. No sites are directly involved in adsorption reactions such as $M_{aq} \Leftrightarrow M_{ads}$ or $M_{aq} + \mathrm{O}_{0.5}\mathrm{H}_{ads} = \mathrm{O}_{0.5}M_{ads} + \mathrm{H}^+_{aq}$ used in defining the surface species. Elemental stoichiometry of surface hydroxyl groups and surface complexes can be determined according to certain rules based on the Pauling bond-valence principle; stability of surface species having such stoichiometry is then given by its standard partial molal Gibbs energy at temperature of interest (Kulik, 2000, 2006c).

The activity of an M surface species involves its relative concentration Γ_M/Γ_o and two activity coefficient terms: the optional EIL Coulombic term (the same as in the LMA TSMs), and the surface activity coefficient term (SACT) derived from the appropriate adsorption isotherm equation (such as Langmuir) to reproduce site saturation and competition effects. The site density parameter, Γ_C, and denticity, δ, are involved only in the SACT, but not in the thermodynamic properties of the adsorbed species referenced to infinite dilution, standard surface density, and zero surface potential. Conversely, linear isotherms are reproduced on any site if M concentration is low enough; the non-linearity due to site saturation and denticity effects becomes significant only at greater concentrations. The advantage of this approach is that it can model not only competitive Langmuir or Frumkin isotherms (as in site-balance based TSMs), but also isotherms for δ-dentate binding or up to surface condensation/precipitation. A shortcoming is that the SACT functions sometimes behave stiffly at dense site coverage and may impede the convergence of the GEM algorithm. Another advantage is that the sorbent is represented with its real chemical stoichiometry and stability (even as a solid solution) and with known specific surface, not only as a collection of abstract site balances. This makes the whole TSM more chemically plausible.

5.3. Choice of TSM and EIL model

This choice is mainly dictated by research objectives. If the emphasis is to reproduce the potentiometric titration and zeta-potential data measured in dilute suspensions of well pre-conditioned (hydr)oxide with known specific surface area at various background electrolyte concentrations, then any non-electrostatic (NE), double layer (DL), or constant capacitance (CC) EIL model must be discarded in favour of the basic Stern (BS), triple layer (TL) or charge-distribution three-plane (CD) model. The final quality of results will depend very much on the accuracy of experimental procedures and methods of reduction of the raw titration data (Lützenkirchen, 2002b; Lefevre *et al.*, 2006).

If the goal is to model *pH* adsorption edges or isotherms of strongly adsorbing trace ions measured in dilute suspensions of hydroxide such as goethite or gibbsite at relatively high background electrolyte concentration, then a CC or DL model may be sufficient and an NE model may be satisfactory, especially for model adsorption isotherms in *pH*-buffered systems. A multi-site TSM may be more adequate to account for the site heterogeneity over a wide range of *M* total concentrations (Dzombak & Morel, 1990; Benjamin & Leckie, 1981); at high sorbate loadings, an extension of the model to surface precipitation may be needed (Zhu, 2002). In multi-site sorption models, the total Kd_M value is actually split into 'partial' site-specific *Kd*'s, each with its own dependence on *pH*, solid concentration, and dissolved *M* and other ions concentration. When a full chemical speciation model (including TSMs) is available for the system, any such partial *Kd* can be retrieved from the speciation calculation results and plotted together with the overall Kd_M *e.g.* to be compared with experimental Rd_M. Representing adsorption data (and modelling results) as $\log_{10} Rd_M$ is of advantage especially when the adsorbed fraction of *M* is >95%; the latter quantity is then subject to a large uncertainty and cannot be used *e.g.* in fitting the adsorption constants.

In the geochemistry of natural waters, the emphasis may be put on the sorption of trace elements and radionuclides on mineral particles present in river, lake, and sea water (Honeyman & Santschi, 1992; Brown & Parks, 2001) as the main form of ion partitioning. The *pH* of natural waters is relatively limited because of buffering mineral associations or the atmosphere; the suspended particulate matter is very dilute and heterogeneous, containing detrital and secondary mineral particles, organic colloids, and bacteria, hence the large difficulties in separating the particles and characterizing their surface areas and composition. These and other factors lead users of direct applications of electrostatic TSMs to favour simple multi-site non-electrostatic sorption models based on empirical binding constants or Rd_M values instead. The affinity of natural particle surfaces to transitional elements appears to be not less than that of synthetic goethite, HFO or gibbsite particles. Hence, the generally accepted idea is that the sorption on natural particulate matter is probably responsible for the present (very low) trace-element concentrations in seawater (Krauskopf, 1956; Brown & Parks, 2001).

In the geochemistry of waste disposal, the interest is shifted from dilute laboratory systems to compacted porous mineral aggregates such as soils, clay rocks, bentonites, cements, where pores are filled with aqueous electrolyte. In such systems, the *pH* of pore solutions is usually buffered by reactive minerals such as carbonates, sulphates, and clays present together in a high solid-to-liquid ratio; it is difficult to characterize reactive specific surfaces in such compact mineral aggregates. In modelling such complex systems, the electrostatic TSMs in their present form seem to have little chance for success, and new TSM approaches need to be developed. There are two paradigms in modelling sorption on heterogeneous natural mineral mixtures, known as the 'Component Additivity' or CA ('bottom-up') and the 'Generalized Composite' GC ('top-down') approach (Payne *et al.*, 2006; Turner *et al.*, 2006). The CA approach represents the mixture by proportions of several reactive mineral phases, each with individual surface properties and appropriate (electrostatic or not) TSM parameters obtained separately for each phase; the TSMs are then subjected to the chemical mass transfer in the whole system. In contrast, the GC approach assumes that the assemblage of mineral surfaces in the aggregate is too complex to be described by adding contributions from individual mineral components. Instead, one or more generic site types (with site capacities) are introduced, and sorption affinities to those sites are represented *via* surface binding reactions with empirical binding constants, commonly developed in the non-electrostatic framework because of uncertainty in defining the appropriate charge-potential relationships. While extrapolation beyond experimental ranges used in fitting the model parameters is questionable, the GC approach may provide better fits with fewer adjustable parameters compared to the CA approach for the same system (Payne *et al.*, 2006).

5.4. Ion exchange

Ion exchange is considered in geochemistry as an important mechanism of partitioning and retention of aqueous ions onto clays and zeolites, *i.e.* minerals that possess permanent structural charge. Clay minerals (montmorillonite, illite, kaolinite) have received a lot of recent attention because of the prospective use of clay rocks and bentonites in nuclear-waste disposal (Bradbury & Baeyens, 2006, 2009).

Clays possess strong retention properties for (radio)toxic cations, and are ubiquitous materials in repository systems. As in other cases of ion partitioning, the uptake of ions on clay materials is expressed with Rd_M (equation 62) values or $Kd_{M,tot}$ as their theoretical counterparts compiled into 'sorption databases' applicable at certain conditions. The total Kd_M value comprises several partial Kd values related to different sorption mechanisms on clay particle edges (multi-site adsorption on amphoteric $>S$OH groups) and permanent-charge basal planes (ion exchange): $Kd_{M,tot} = Kd_{M,edge} + Kd_{M,iex}$. This complexity is due to the sheet-silicate structure and 'platy' morphology of clay minerals (Fig. 9).

The main difference between them is the structural origin of permanent charge. Kaolinite has almost no substitutions and little or no permanent charge. In montmorillonite and other smectites, the negative permanent charge is mainly due to the cation substitutions (Mg^{2+} for Al^{3+}, *etc.*) in the octahedral "O" layer, and in illite and other mica-like clays due to the substitutions (Al^{3+} for Si^{4+}, *etc.*) mainly in tetrahedral 'T' layers. Clay minerals in clay rocks may have significant substitutions in both T and O layers, or be a 'mixed-layer' intercalation of smectite- and illite-like domains. The negative permanent charge is compensated in the interlayer and outer basal (001) planes by various cations (Na^+, K^+, NH_4^+, Ca^{2+}, Mg^{2+}, *etc.*); it can be measured as CEC (cation

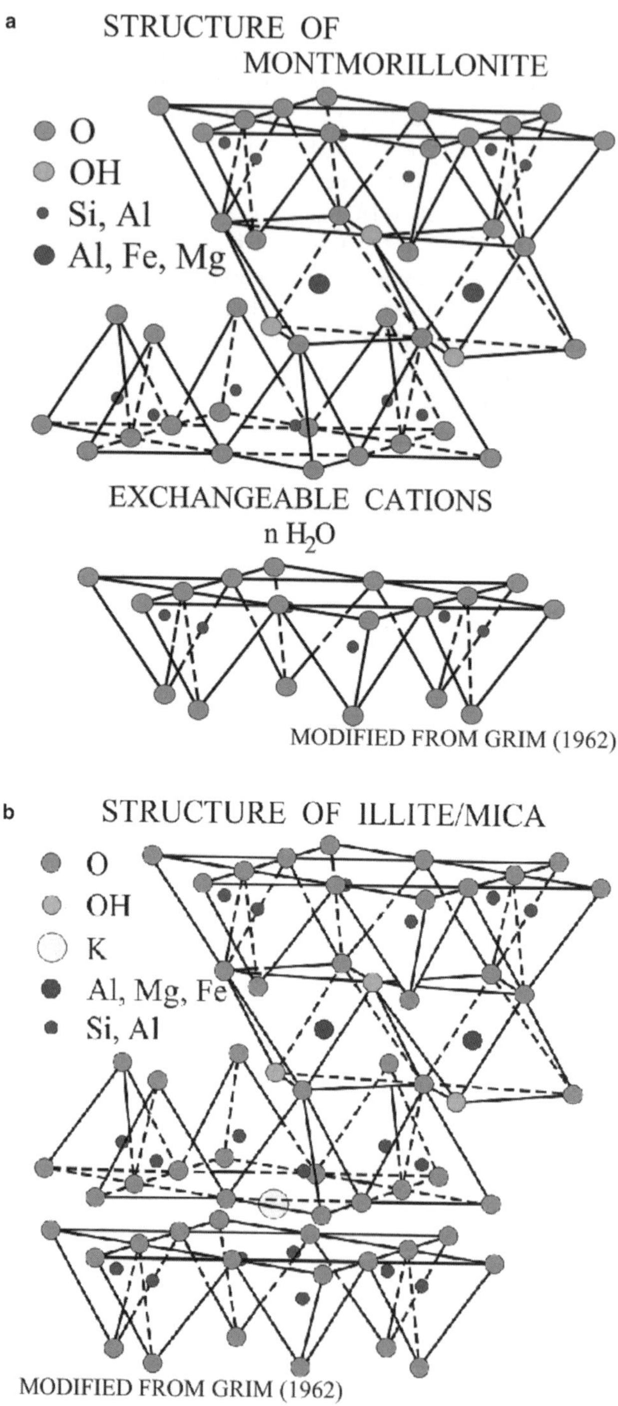

Fig. 9. Structures of montmorillonite and illite (from http://pubs.usgs.gov/of/2001/of01-041/htmldocs/clay.htm, part *a* can be found under 'Smectite Group' and part *b* under 'Illite').

exchange capacity per unit mass of clay) or estimated from the chemical composition formula.

The thickness of the smectite platelets may correspond to a single T-O-T layer where the T layer is made up of SiO_4 tetrahedra, and the O layer of Al^{3+} ions in octahedral coordination. The CEC of montmorillonite is $\sim q_z = 0.9$ eq kg^{-1} (corresponding to charge density on siloxane planes $\Gamma_{ch} \approx 1 \times 10^{-6}$ mol m^{-2}). In illite, the permanent charge originating from T layers is mainly compensated with K^+ ions. Because of high stacking in particles (10–50 T-O-T packages), only some part of the interlayer surface in illite is accessible to ion exchange, hence smaller measured CEC values of ~0.2 eq kg^{-1} (Grim, 1968; Meunier, 2005).

Due to their structure, clay platelets have sorption properties of amphoteric (hydr)oxides such as gibbsite and silica on the edges, and properties of ion exchangers on the basal planes at the same time. This results in a strong dependence of M^{2+}or M^{3+} cation sorption (*e.g.* Zn^{2+}, Eu^{3+}) on electrolyte concentration in the acidic *pH* region (due to the competition with major cations for the ion exchange sites), and in a weak dependence in the alkaline *pH* region (where the amphoteric >*S*OH edge sites play a major role).

There is a large uncertainty in characterizing the surface area of clay samples (typically 20–90 m^2 g^{-1}) and in discriminating between the edge and the basal surfaces, even though this directly influences the sorption modelling results. While the edge surface may be predominant for kaolinite particles, it is certainly minor for smectite platelets whose total reactive area reaches 800 m^2 g^{-1}. However, a recent work (Tournassat *et al.*, 2003) has led to a consistent value of $A_{ES} \approx 8$ to 9 m^2 g^{-1} for the specific area of montmorillonite particle edges; the known discrepancies are largely due to the stacking of platelets at dry conditions (*ca.* 20 T-O-T layers) compared to their aqueous suspension (1 or 2 layers). These findings are important because they provide a consistent link between the bulk-sorption capacity on edges q_{EC} and the geometric site densities per unit area which can be found from the edge-surface structure data.

5.5. The B&B sorption model for clays

The most comprehensive experimental data set for sorption of di- and trivalent cations on smectite, illite and bentonite, with the most extensive database of sorption constants is due to Bradbury & Baeyens (2006, 2009 and references therein). Their (B&B) non-electrostatic multi-site sorption model combines several site balances for amphoteric groups on the clay particle edge with one balance for the basal-surface permanent charge sites with the cation exchange.

Surface complexation reactions on edge sites are represented in the B&B model as:

$$\equiv OH^0 \iff \equiv O^- + H^+ \qquad K_{A2} \qquad (67a)$$

$$\equiv OH^0 + H^+ \iff \equiv OH_2^+ \qquad K_{A1} \qquad (67b)$$

$$\equiv OH^0 + M^{z+} + yH_2O \iff \equiv OM(OH)_y^{z-y-1} + H^+ \qquad K_{y,M} \qquad (67c)$$

Table 3. Parameters fixed in multisite B&B sorption model for Na-montmorillonite (Bradbury & Baeyens, 2006).

Site type	Site capacity mol kg^{-1} (eq kg^{-1})	$\log_{10} K_{A2}$	$\log_{10} K_{A1}$
$\equiv^{W1}OH$	$4.0 \cdot 10^{-2}$	4.5	7.9
$\equiv^{W2}OH$	$4.0 \cdot 10^{-2}$	6.0	10.5
$\equiv^{S}OH$	$2.0 \cdot 10^{-3}$	4.5	7.9
-clay	0.87		

Cation-exchange reactions on basal permanent charge sites are given as

$$z_B A^{zA}\text{-clay} + z_A B^{zB} \Longleftrightarrow z_A B^{zB}\text{-clay} + z_B A^{zA} \qquad {}^{B}_{A}K_c \tag{68}$$

with the selectivity coefficients ${}^{B}_{A}K_c$ defined in the Gaines-Thomas convention, *i.e.* on an equivalent (charge) fraction basis for sorbed and in activities for aqueous ions. The protolysis reaction constants were adjusted originally together with the sorption capacity parameters for two 'weak' edge site types, W1 and W2, separately against the potentiometric data for pre-conditioned Na-montmorillonite (Table 3) and Na-illite, and then kept constant in fitting all the metal sorption data (*e.g.* for Eu(III), Table 4). The B&B model was recently extended to account for the influence of carbonate on radionuclide sorption (Marques Fernandes *et al.*, 2008). The power of the non-electrostatic B&B sorption model consists largely of its ability to describe the *pH* edges and sorption isotherms of a number of strongly sorbing polyvalent cations (see an example for Eu(III) in Fig. 10). Some mono- and divalent cations such as Cs^+, Ca^{2+}, Sr^{2+}, Ba^{2+}, Ra^{2+} are considered only on the cation-exchange sites.

Many cations sorbing on clay edges show stronger sorption at trace concentrations, which is accounted for by using a small capacity of 'strong' sites $\equiv^{S}OH$ with protolysis constants taken the same as for one of the 'weak' sites ($\equiv^{W1}OH$) (see examples in Fig. 10c,d). The fitted intrinsic constants for metal complexes on $\equiv^{S}OH$ sites on both montmorillonite and illite show good correlations with the respective aqueous hydrolysis constants, leading to important LFERs (Bradbury & Baeyens, 2006, 2009) useful in predicting cation sorption constants in the absence of experimental data. These LFERs show that metal cations sorb as 'inner-sphere' complexes on clay edges in a very similar

Table 4. B&B model parameters for the sorption of Eu(III) on Na-montmorillonite.

Surface complex	Reaction	$\log_{10} K^{int}$ at 25°C	Analogous aqueous hydroxocomplex	$\log_{10} \beta_y$[1] (I = 0, 25°C)
$\equiv^{S}OEu^{+2}$	(67)	+1.6	$Eu(OH)^{+}$[2]	−7.2
$\equiv^{S}OEuOH^{+}$	(67)	−5.9	$Eu(OH)_2^{+}$	−15.1
$\equiv^{S}OEu(OH)_2^0$	(67)	−14.2	$Eu(OH)_3^0$	−26.2
$\equiv^{W1}OEuOH^{+}$	(67)	−0.5	$Eu(OH)_4^-$	−36.2[2]
$Clay_3Eu^0$	(68)	$\log K_{ex,Na} = 7.1$[3]		

[1] For the hydrolysis reaction: $M^{z+} + yH_2O \Leftrightarrow M(OH)_y^{z-y} + yH^+$.
[2] Value from Nagra-PSI data base, used in GEM-Selektor calculations only.
[3] From ${}^{B}_{A}K_c = 29$. Equilibrium constants including ${}^{B}_{A}K_c$ are from (Marques Fernandes *et al.*, 2008).

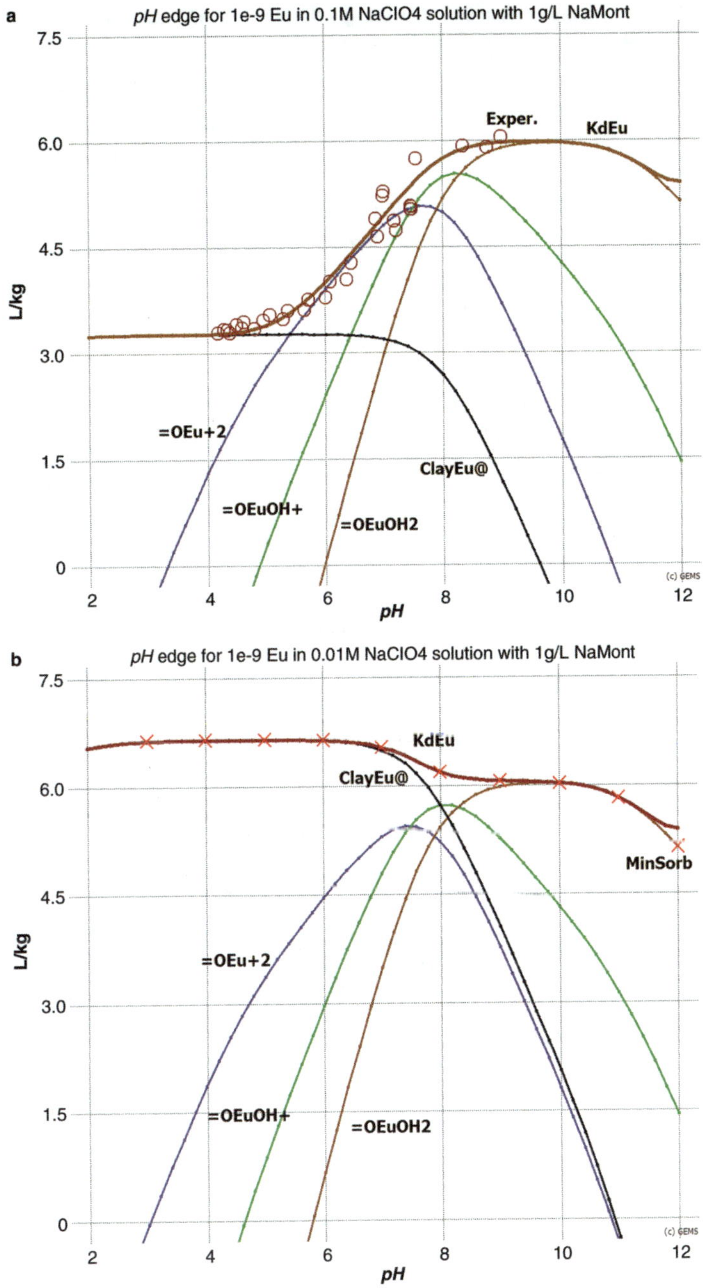

Fig. 10. GEM-Selektor screenshots for modelled sorption *pH* edges of trace Eu(III) (1×10^{-9} M) on 1 g/L Na-montmorillonite in 0.1 M $NaClO_4$ electrolyte (***a***) and 0.01 M $NaClO_4$ (***b***), compared with the experimental data (Bradbury & Baeyens, 2002) (*a*), and the same TSM results computed with the MINSORB code (crosses) (*b*).

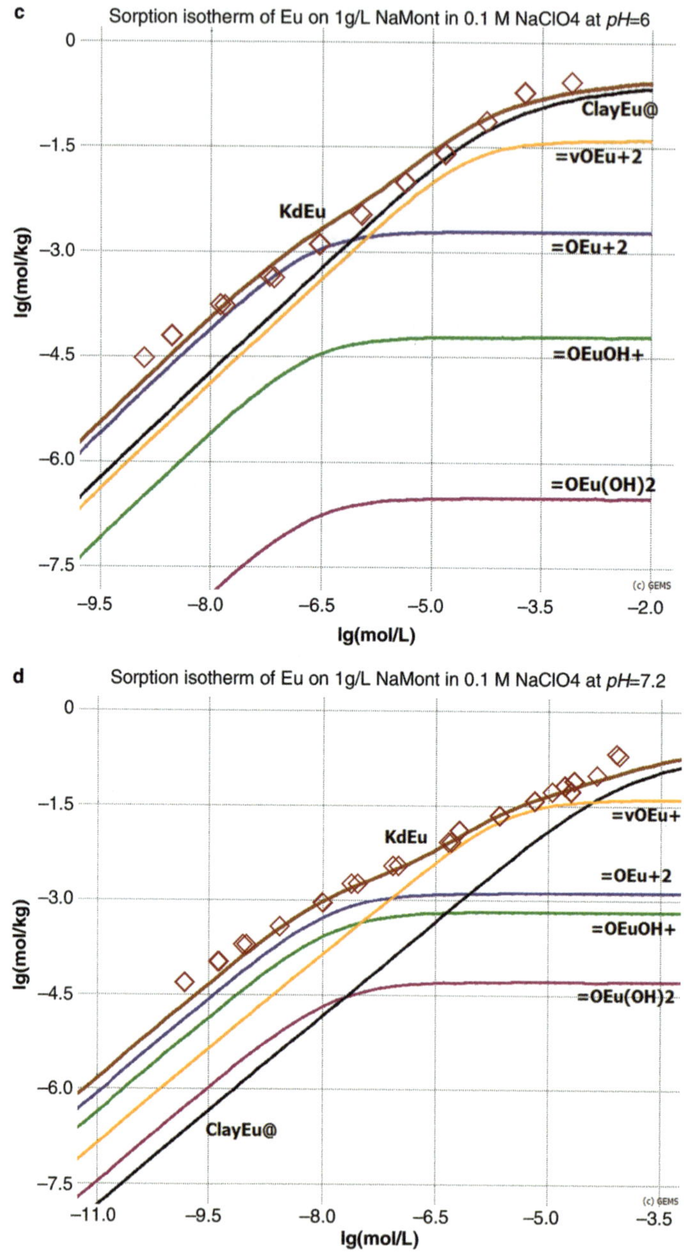

Fig. 10. (*Continued*) Sorption isotherms of Eu(III) on 1 g/L Na-montmorillonite in 0.1 M $NaClO_4$ at $pH = 6.0$ (***c***) and $pH = 7.2$ (***d***), compared with the experimental data (Bradbury & Baeyens, 2002). Input data from Tables 3, 4 and built-in Nagra-PSI thermodynamic data base. The =vOEu+2 surface species belongs to the $\equiv^{W1}OH$ edge site, other surface species of europium to $\equiv^{S}OH$ sites.

way to adsorption on (hydr)oxide surfaces, where such correlations were established previously (Schindler & Stumm, 1987; Dzombak & Morel, 1990). The sorption of transitional cations on clay edges is strong and does not seem to be much affected by the variable edge surface charge.

The TSM example in Figure 10 shows some ion-partitioning effects typical of multi-site sorption models. At relatively high electrolyte concentration, there is a modest contribution from the ion exchange, only visible at $pH < 6$ (Fig. 10a). However, at low electrolyte concentration (Fig. 10b), there is much weaker competition from Na^+ on ion exchange, and the contribution of $Clay_3Eu^0$ surface complex into Kd_{Eu} becomes dominant at $pH < 8$. Contributions of 'inner-sphere' Eu surface complexes on edge $\equiv^S$ sites are not affected by the bulk electrolyte concentration; they appear in the same order as the analogous aqueous hydroxocomplexes. The same pH sorption edges are modelled both in GEM and LMA (MINSORB code) implementations of the B&B TSM (Fig. 10b).

The relative impact of various sites and surface complexes into the sorption isotherms of Eu(III) on montmorillonite is seen in Figure 10c,d. At greater pH, the 'trace concentration' part of the isotherm is shifted to lower dissolved Eu molarities because the surface hydroxo-species become more abundant. However, due to the limited availability of strong edge $\equiv^S$ sites, the upper part of the isotherm has a slope of <1 and is dominated by the $\equiv^{W1}OEuOH^+$ on more abundant weak edge sites (not visible in pH edges at trace Eu concentrations), and, finally, by the $Clay_3Eu^0$ ion exchange species. Overall, the B&B sorption model is capable of reasonably reproducing the effects of speciation, ionic strength, surface-site competition, and site heterogeneity onto the partitioning of Eu(III) between aqueous solution and montmorillonite particles. Competition from the side of major cations is the most pronounced partitioning effect in ion exchange because the selectivity constants are rather small, but the number of exchange sites is limited.

There are two interesting thermodynamic aspects in modelling the aqueous-clay ion partitioning due to cation exchange on permanent-charge sites: (1) representation of heterovalent ion exchange in site-balance sorption models; and (2) representation of clay ion exchangers as multi-component solid solutions. These issues are discussed briefly below.

5.6. Representation of heterovalent ion exchange

Representation of heterovalent ion exchange used in the B&B model (equation 68) is one of several possible; the related Gaines–Thomas selectivity coefficient ${}^B_AK_c$ actually depends on composition and is not a true equilibrium constant. It may not be trivial to compare values of selectivity coefficients for the heterovalent ion exchange data without knowing exactly the conventions and concentration scales behind them. This issue has been discussed in soil chemistry (Sposito & Mattigod, 1979; Fletcher & Sposito, 1989), in chemical engineering (Grant & Fletcher, 1993; Ioannidis *et al.*, 2000), and in studies of ion exchange in clays (Townsend, 1984) and zeolites (Townsend & Coker, 2001; Fletcher *et al.*, 1984; Pabalan & Bertetti, 1994). Without going deep into theory, let us compare some basic definitions. The generic ion exchange reaction is

$$\nu_j M_j^{z_j}(\text{sorbed}) + \nu_i M_i^{z_i}(\text{aq}) \Longleftrightarrow \nu_j M_j^{z_j}(\text{aq}) + \nu_i M_i^{z_i}(\text{sorbed}) \qquad (69)$$

where j and i are indexes of ions, and ν are the appropriate reaction stoichiometry coefficients. The permanent-charge solid–water interface is traditionally represented as a hypothetical exchanger ligand X^- (for anion exchange X^+). The sorbed species (end members) are constructed in two alternative ways. A species of Vanselow-type (V) has composition $M_i\mathrm{X}_{z_i}$, *i.e.* one mole of ion i combined with part of the sorbent carrying z_i moles of charge. A species of Gapon-type (G) with the composition $(M_i)_{z_i^{-1}}\mathrm{X}$ represents one mole of charge of ith ion combined with the part of the sorbent carrying one mole of the opposite charge. Reactions of these components with ions in aqueous solution, drawn from equation 69, have different stoichiometry coefficients, thus leading to different exchange equilibrium constants:

$$ {}^{i}_{j}K_{\mathrm{V}} = \frac{a^{z_i}_{j(\mathrm{aq})}\, a^{z_j}_{i(\mathrm{V})}}{a^{z_j}_{i(\mathrm{aq})}\, a^{z_i}_{j(\mathrm{V})}} \tag{70} $$

$$ {}^{i}_{j}K_{\mathrm{G}} = \frac{a^{1/z_j}_{j(\mathrm{aq})}\, a_{i(\mathrm{G})}}{a^{1/z_i}_{i(\mathrm{aq})}\, a_{j(\mathrm{G})}} \tag{71} $$

related as ${}^{i}_{j}K_{\mathrm{V}} = {}^{i}_{j}K_{\mathrm{G}}^{z_i z_j}$. In turn, these equilibrium constants are connected to measurable selectivity coefficients through the ratios of activity coefficients of V-type and G-type end members, respectively:

$$ {}^{i}_{j}K^{a}_{\mathrm{V}} = \frac{a^{z_i}_{j(\mathrm{aq})}\, x^{z_j}_{i(\mathrm{V})}}{a^{z_j}_{i(\mathrm{aq})}\, x^{z_j}_{j(\mathrm{V})}} \quad \text{and} \quad {}^{i}_{j}K_{\mathrm{V}} = {}^{i}_{j}K^{a}_{\mathrm{V}} \frac{f^{z_i}_{i(\mathrm{V})}}{f^{z_i}_{j(\mathrm{V})}} \tag{72} $$

(Vanselow selectivity coefficient)

$$ {}^{i}_{j}K^{a}_{\mathrm{G}} = \frac{a^{1/z_j}_{j(\mathrm{aq})}\, x_{i(\mathrm{G})}}{a^{1/z_i}_{i(\mathrm{aq})}\, x_{j(\mathrm{G})}} \quad \text{and} \quad {}^{i}_{j}K_{\mathrm{G}} = {}^{i}_{j}K^{a}_{\mathrm{G}} \frac{\gamma_{i(\mathrm{G})}}{\gamma_{j(\mathrm{G})}} \tag{73} $$

(Gapon selectivity coefficient)

Here, x stands for the end-member mole fraction; hence both selectivity coefficients (and the respective activity coefficients) refer to the mole fraction concentration scale. For an ideal ion exchanger, all activity coefficients $f_{i(\mathrm{V})}$ or $\gamma_{i(\mathrm{G})}$ remain unity over the whole range of the ion exchange isotherm. Alternatively, the V-type end members can be used to define a selectivity coefficient $\left({}^{B}_{A}K_c \text{ in B\&B model}\right)$ in an ionic charge equivalent fraction scale (Gaines & Thomas, 1953):

$$ {}^{i}_{j}K^{a}_{\mathrm{GT}} = \frac{a^{z_i}_{j(\mathrm{aq})}\, E^{z_j}_{i}}{a^{z_j}_{i(\mathrm{aq})}\, E^{z_i}_{j}} \quad \text{where } E_i = \frac{z_i x_{i(\mathrm{V})}}{z_i x_{i(\mathrm{V})} + z_j x_{j(\mathrm{V})}} \tag{74} $$

leading to

$$ {}^{i}_{j}K_{\mathrm{V}} = {}^{i}_{j}K^{a}_{\mathrm{Gt}} \frac{y^{z_j}_{i(\mathrm{V})}}{y^{z_i}_{j(\mathrm{V})}} \tag{75} $$

where the activity coefficients y are also taken in the equivalent fraction scale.

Equations 72 to 75 show that for the same set of heterovalent ion exchange experimental data, the values of selectivity- and activity coefficients will be different depending on the underlying conventions for ion-exchange components and reactions, and on the applied concentration scales. Hence, one needs to be careful in compiling selectivity coefficients from different sources into a data base for geochemical speciation codes where they should correspond to the same kind of exchange equilibrium constant.

5.7. Representation of ion exchange as solid solution

Representation of ion exchange as solid solution becomes desirable, *e.g.* in modelling the retention of radionuclides upon clay-cement interaction in the engineering barrier in waste repository. Upon interaction with highly alkaline cement pore water, clay minerals may become unstable and partially dissolve, desorbing major cations and radionuclides and providing matter for the formation of other phases such as kaolinite, zeolites, or C-S-H (*e.g.* Gaucher & Blanc, 2006). These complex processes are expected to occur at very high solid/water ratios, which makes the classic sorption models with site balances (like the B&B model) not very helpful in modelling the overall reactive mass transport.

Let us return to the equilibrium distribution coefficient (equation 35) for ions B and C between the aqueous and the solid solution. Replacing the molalities of species by their activities yields an equilibrium constant

$$K_{D,BC} = \frac{a_{BL} a_{C^{z+}}}{a_{CL} a_{B^{z+}}} = \frac{x_{BL} f_{BL} m_{C^{z+}} \gamma_{C^{z+}}}{x_{CL} f_{CL} m_{B^{z+}} \gamma_{B^{z+}}} \tag{76}$$

This equilibrium distribution constant is clearly a particular case of either Vanselow (equation 70) or Gapon (equation 71) constant for the homovalent ion exchange ($z_i = z_j = 1$). However, for carbonate or sulphate solid solution, the 'exchanger' is represented by the divalent ligand L^{2-} and equation 76 takes the form

$$K_{D,BC} = \frac{a_{BL} a_{C^{z+}} a_{L^{z-}}}{a_{CL} a_{B^{z+}} a_{L^{z-}}} = \frac{K_{CL}}{K_{BL}} \tag{77}$$

which for $|z| > 1$ is not a G-type exchange constant valid for a single monovalent site ligand. In the case of homovalent exchange (of 2, 3, ... valent cations), the activities of dissolved aqueous ligand in equation 77 will cancel out regardless of the ligand charge and stoichiometry coefficient in the end member formula (compare CaF_2 and $CaSO_4$).

Therefore, the numerical values of $K_{D,BC}$ will be equal to that of the Vanselow exchange constant (equation 70) and the mixing behaviour (*i.e.* activity coefficients) will be the same as in the Vanselow model. Conversely, ignoring the kinetics, the aqueous–solid solution systems involving carbonates or sulphates of divalent cations can be modelled alternatively using the classic solid-solution models (see Figs 6–8) or simulated using the Vanselow ion-exchange constants and the same mixing models (by setting the ligand amount in the solid as balance constraint). Accordingly, the homovalent ion exchange equilibria such as those on clays can be modelled in a straightforward way using solid solutions like (*B*,*C*)X where X is the 'anionic part' of the mineral, for instance, $[MgAl_5(Si_4O_{10})_3(OH)_6]_2^{2-}$ for theoretical stoichiometry of montmorillonite, and *B* and *C* are divalent cations.

Replacement of the balance-constraint based on CEC by the 'real' clay mineral stoichiometry will require, however, the determination of stability (G^o value or solubility product) of at least one end member (*e.g.* $Ca[MgAl_5(Si_4O_{10})_3(OH)_6]_2$). The G^o values of any similar end members can then be obtained from cation exchange reactions and their Vanselow equilibrium constants.

The definition (equation 77), however, does not allow us to consider the heterovalent exchange substitutions in a so simple way because of an ambiguity in defining the ligand charge and the cation stoichiometry. For instance, in the Vanselow convention, the exchange between mono-, di- and trivalent cations bound to a monovalent ligand will be written in terms of end members such as $M1^{1+}L^-$; $M2^{2+}L_2^-$; and $M3^{3+}L_3^-$. The respective K_D expressions will be

$$K_{D,12} = \frac{a_{M1L} a_{M2^{2+}} a_{L^-}^2}{a_{M2L_2} a_{M1^+} a_{L^-}} = \frac{K_{M2L_2}}{K_{M1L}} \tag{78a}$$

or

$$K_{D,13} = \frac{a_{M1L} a_{M3^{3+}} a_{L^-}^3}{a_{M3L_3} a_{M1^+} a_{L^-}} = \frac{K_{M3L_3}}{K_{M1L}} \tag{78b}$$

The system can also be represented in the Gapon convention, using the end members $M1L$; $M2_{0.5}L$; and $M3_{0.3333}L$ and the equilibrium exchange constants

$$K_{G,12} = \frac{a_{M1L} a_{M2^{2+}}^{0.5} a_{L^-}}{a_{M2_{0.5}L} a_{M1^+} a_{L^-}} = \frac{K_{M2L_2}^{0.5}}{K_{M1L}} \tag{79a}$$

or

$$K_{G,13} = \frac{a_{M1L} a_{M3^{3+}}^{0.3333} a_{L^-}}{a_{M3_{0.3333}L} a_{M1^+} a_{L^-}} = \frac{K_{M3L_3}^{0.3333}}{K_{M1L}} \tag{79b}$$

In this case, the ideal mixing behaviour of cations in the solid phase will be different from the previous (Vanselow) case where the end member activity would equal its mole fraction. To reproduce the Vanselow-type mixing using the Gapon-type end members, their activities must be represented as (see reasoning at equation 61):

$$a_{M1L} = x_{M2_{0.5}L}^1; \quad a_{M2_{0.5}L} = x_{M2_{0.5}L}^{0.5}; \; a_{M3_{0.3333}L} = x_{M3_{0.3333}L}^{0.3333} \tag{80}$$

Coming back to the solid-solution representation of ion exchange in clays or zeolites, we must admit that even if the energetic non-ideality of mixing between exchange cations on montmorillonite is insignificant (*cf.* Fletcher & Sposito, 1989), the ideal mixing will be different in models based on Vanselow and Gapon conventions. It seems still to be an open question as to which of the two conventions would lead to a more chemically plausible thermodynamic model of ion partitioning. Shao *et al.* (2009b) implemented a solid solution model of cation exchange on montmorillonite (Table 5) based on the Gapon convention (G-type end members) and assuming the simple ideal mixing. The Gapon exchange constants were obtained from Gaines-Thomas selectivity coefficients using

$$K_{G,ex} = K_{V,ex}^{0.5} = \left({}^{M}_{Na}K^{c}_{GT} \cdot \frac{\Sigma n}{2\Sigma zn} \right)^{0.5} \tag{81}$$

where $\sum n$ is the sum of mole amounts, and $\sum zn$ is the sum of charge equivalents of ions present on the exchanger at the composition used in determining the value of ${}^{M}_{Na}K^{c}_{GT}$.

The initial guesses of end-member G^o values were adjusted in GEM-Selektor calculations simultaneously such that minor amounts of kaolinite appeared together with montmorillonite and quartz, while gibbsite remained undersaturated. In addition, a sensitivity analysis was performed by varying the end member G^o values simultaneously by -30 kJ mol^{-1}. Only concentrations of Al aqueous species were affected while kaolinite disappeared and gibbsite remained undersaturated. Hence, the ion distribution between ion exchanger (clay) and aqueous solution is largely unaffected by the solubility of the montmorillonite phase as the whole, as long as it remains stable relative to gibbsite and kaolinite (Shao *et al.*, 2009b).

As discussed above, an alternative representation of this solid solution model of ion exchange is possible in the Vanselov convention, and it is expected to result in a different ideal mixing behaviour between mono- and divalent cations compared with the original G-type model. The V-type end member stoichiometries of monovalent cation end members remain the same as the G-type ones; the divalent cation end members can be obtained just by doubling the stoichiometries and the G^o_{298} values, as done in Table 5. In order to see how different the two model variants are, the equilibria should be modelled along a systematic change in the clay composition (*e.g.* from almost pure Na*Mont* until almost pure $Ca_{0.5}$*Mont* or Ca$Mont_2$), either using the G or

Table 5. Solid-solution models of montmorillonite phase (after Shao *et al.*, 2009b).

G-type end members	$K_{G,ex}$	G^o_{298} (kJ · mol^{-1})	V-type end members[2]	$K_{V,ex}$	G^o_{298} (kJ · mol^{-1})
Na*Mont*[1]		−15908.27[3]	Na*Mont*		−15908.27
K*Mont*	10.6	−15934.70	K*Mont*	10.6	−15934.70
$Ca_{0.5}$*Mont*	2.2[4]	−15924.74	Ca$Mont_2$	4.84[2]	−31849.47
$Mg_{0.5}$*Mont*	2.0[4]	−15875.10	Mg$Mont_2$	4.00[2]	−31750.19
$Sr_{0.5}$*Mont*	2.5[4]	−15930.58	Sr$Mont_2$	6.25[2]	−31861.15
$Ba_{0.5}$*Mont*	2.5[4]	−15929.05	Ba$Mont_2$	6.25[2]	−31858.10
$Ra_{0.5}$*Mont*	2.5[4]	−15929.40	Ra$Mont_2$	6.25[2]	−31858.81

[1] *Mont* stands for $[MgAl_5(Si_4O_{10})_3(OH)_6]$.
[2] Reaction $M^{2+} + 2\text{Na}Mont \Leftrightarrow 2\text{Na}^+ + MMont_2$.
[3] Adjusted in trial GEM runs.
[4] Reaction $\frac{1}{2}M^{2+} + \text{Na}Mont \Leftrightarrow \text{Na}^+ + M_{0.5}Mont$

the V variant of the montmorillonite solid solution in otherwise the same chemical system. The results of this exercise are shown in Figure 11a.

The system recipe was composed of 1 kg H_2O, 10 kg 'atmospheric air', 0.001 mol NaCl, 6 mol SiO_2, and the initial montmorillonite composition 0.5 mol Na*Mont*, 0.5 mol $Ca_{0.5}$*Mont*, 0.001 mol K*Mont* and 0.0001 mol $Mg0_{.5}$*Mont*. 0.07 mol of HCl was added to obtain an equilibrium $pH = 8.55$ when the V-type solid solution model was used ($pH = 8.77$ when the G-type solid solution model was used with the same bulk composition of the system). The ionic strength was ~ 0.075 *m*. The bulk composition of the system was then varied systematically using the process extent variable, η, changed from 0.001 to 1.001 in 0.01 steps. The addition of $Ca_{0.5}$*Mont* was equal to η and that of Na*Mont*, to $1.001 - \eta$ at each step, then the equilibrium states were computed and the results plotted as shown in Figure 11.

The only difference between the solid and dashed curves in Figure 11a is that the V-type model was replaced by the G-type model of the montmorillonite solid solution (Table 5). Conversely, only the mixing behaviour was different, while solubilities of pure end members remained the same. It is seen that the change from V to G ion exchange model has very little effect on aqueous molalities of monovalent cations Na^+ (major) and K^+ (minor) at $E(\text{Ca}) < 0.7$; this effect becomes visible (0.3 $\log_{10}$ units) at $E(\text{Ca}) = 0.95$.

However, for divalent cations, the differences between G- and V-type ion exchange models are significant in the whole composition range, especially for minor cations (0.7 to 1 $\log_{10}$ units for Mg^{2+} or for Ca^{2+} at $E(\text{Ca}) < 0.1$). This shows that the choice of convention for ion exchange end members is important in describing the retention of trace di- and trivalent cations on ion exchange as affected by the competition with major cations. The Vanselow convention leads to more conservative predictions (*i.e.* less retention of the minor cation under more competition from the side of major cations) in comparison with the Gapon convention. This kind of discrepancy may probably be resolved by introducing a non-ideality of mixing, *i.e.* activity coefficients of the end members, and fitting the interaction parameters against the appropriate set of experimental data, making the TSM far more complicated. In any case, the above modelling exercise calls for careful investigations.

Figure 11b shows that this issue applies to implementations of ion-exchange models either based on the site balance or not. The B&B site-balance ion-exchange model (see equation 68) is compared here with the V-type ion-exchange solid-solution model. Although the reasons for small deviations (<0.2 $\log_{10}$ units) between curves for Ca and Mg are not quite clear, it is evident that both V-type SS and B&B models show a very similar mixing behaviour of cations on montmorillonite ion exchange. Hence, LMA site-balance TSMs for ion exchange can be reproduced by rather simple solid-solution models such as those given in Table 5.

5.8. Solid/water ratio effects on ion partitioning between several phases

Partitioning of some radionuclide cations at repository conditions may occur between aqueous solution and several solid solution, adsorption and ion-exchange phases. In this case for the 'Component Additivity' approach, all phases relevant for radionuclide

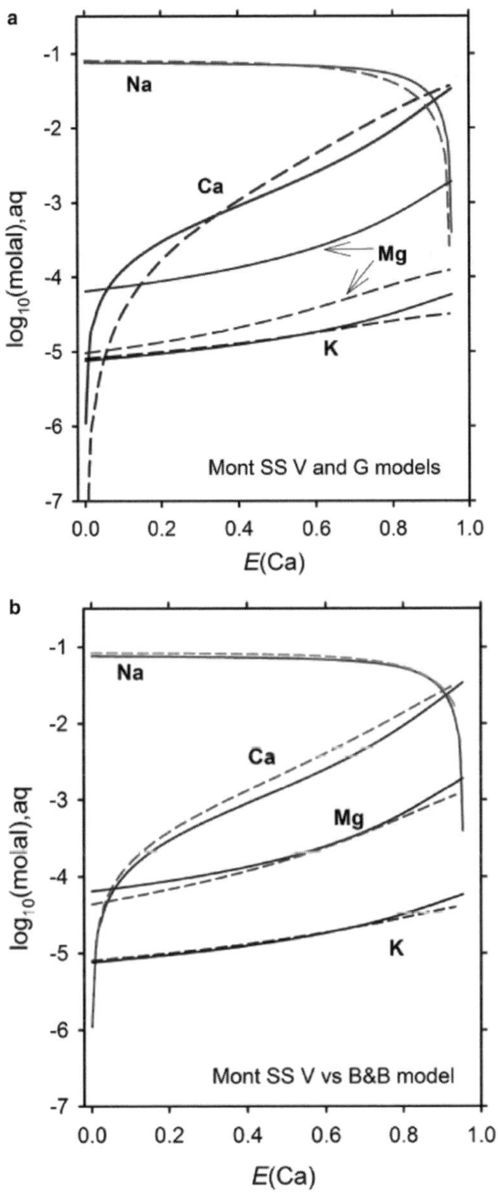

Fig. 11. Total dissolved molalities of some mono- and divalent cations in equilibrium with the amount of montmorillonite having ~1 mol CEC in 1 kg H_2O and excess air at 1 bar and 25°C as a function of the equivalent fraction of Ca^{2+} on ion exchange *E*(Ca). (***a***) Comparison of V-type (solid curves) and G-type (medium-dashed curves) ideal solid-solution models from Table 5. (***b***) Comparison of the GEM V-type solid-solution model (solid curves) with the site-balance B&B sorption model (cation exchange only according to reaction in equation 68, with exchange constants $K_{V,ex}$ from Table 5) for approximately the same conditions. All models were calculated using the GEM-Selektor package with built-in Nagra-PSI database.

Table 6. Regular interaction parameters[1] $G_0/(RT)$ (ternary interactions set to 0).

End-member binary	Carbonate SS	Sulphate SS
Ba-Sr	5.3	2.34
Ba-Ra	0	0
Sr-Ra	5.3	2.34

[1] From Table A4 in Shao *et al.* (2009a)

retention must be included in a thermodynamic model in order to predict the dissolved aqueous-ion concentrations which, at the end, determine its migration. In this respect, radium retention in a clay-rich barrier system is a particularly good example.

Shao *et al.* (2009a) considered the reactive transport of Ra involving aqueous-solid solution chemical systems with both non-ideal Ba-Sr-Ra carbonate and Ba-Sr-Ra sulphate phases. They concluded that the re-partitioning of radium from carbonate to sulphate may cause a strong retardation effect on Ra. A shortcoming of that study was that the cation exchange in clay was not considered in their reactive transport model. In further work (Shao *et al.*, 2009b), those authors extended their model to investigate how the reactive transport of radium may be influenced by the competition between radium-containing solid solutions and the cation exchange in bentonite clay minerals.

Two simple reaction-path scenarios can illustrate complex effects on partitioning of Ra, Ba and Sr between aqueous and solid phases that may occur in this system (results of the full reactive mass transport modelling with the OpenGeosys-GEM coupled code can be found in two papers cited above). Scenario 1 shows how the ion partitioning is influenced by the changing clay/water ratio at constant loadings of cations and sulphate in the whole system. Scenario 2, at the constant clay/water ratio, simulates the effect of removal of sulphate on the redistribution of radium, strontium and barium between the sulphate and carbonate solid solution phases in equilibrium with aqueous electrolyte. Both simulations can be run easily using the GEM-Selektor code and database (using Tables 5 and 6).

5.9. Impact of increasing montmorillonite/water (m/w) ratio (scenario 1)

Similar to the model system in (Shao *et al.*, 2009b), the initial recipe included 1 kg of H_2O, 500 g of air, 0.0163 mol of $BaSO_4$, 1.14 mol of $CaCO_3$, 0.155 mol of $CaSO_4 \cdot 2H_2O$, 0.42 mol of NaCl, 6.0 mol of SiO_2, 0.0206 mol of $SrSO_4$, 2×10^{-6} mol $RaSO_4$, and 1 kg of *MontBent* stoichiometry composed of 0.330905 mol of $Ca_{0.5}$*Mont*, 0.331905 mol of $Mg_{0.5}$*Mont*, 0.023 mol of K*Mont* and 0.2695 mol of Na*Mont* (see Table 5). In the process simulator (reaction-path model), addition of *MontBent* to the bulk-system composition was changed stepwise as $10^{p\xi}$ mol, where the process extent variable $p\xi$ ran from -2.7 to 1.3 in steps of 0.1. The maximum value of $p\xi$ corresponds to highly compacted bentonite. Carbonate solid solutions were represented by non-ideal $(Ca,Ba,Sr)CO_3$ (mainly calcite) and $(Ra,Ba,Sr)CO_3$ (mainly witherite) phases; sulphate solid solutions were represented as $(Ra,Ba,Sr)SO_4$ (mainly barite) and $(Ra,Sr,Ba)SO_4$ (mainly celestite) phases, with solubility products from the Nagra-PSI database and binary regular interaction parameters given in Table 6. The results are shown in Figure 12.

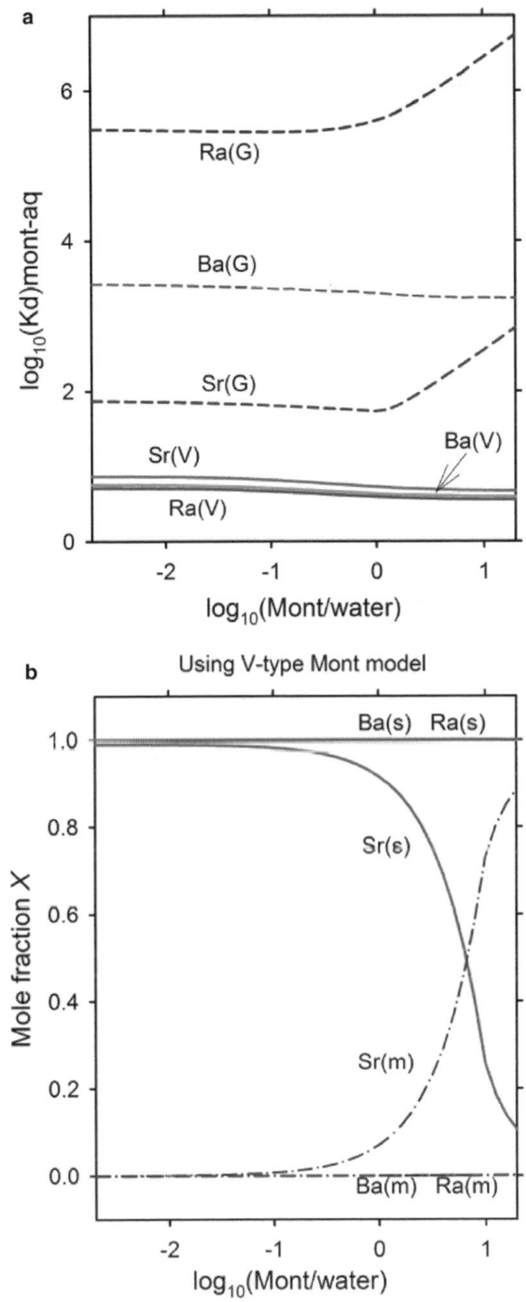

Fig. 12. The impact of m/w (*Mont*/water) mass ratio (in kg clay per 1 kg H_2O) on the partitioning of Ba, Sr and Ra between aqueous, clay and sulphate + carbonate solution phases. (***a***) Profiles of *Kd* (in L kg^{-1}) for cations in montmorillonite solid solution of V-type (V) and G-type (G). (***b***) Fraction of total moles of cations in the system bound in sulphates + carbonates (s) and clay (m) solid solution phases, modelled with the V-type Mont phase.

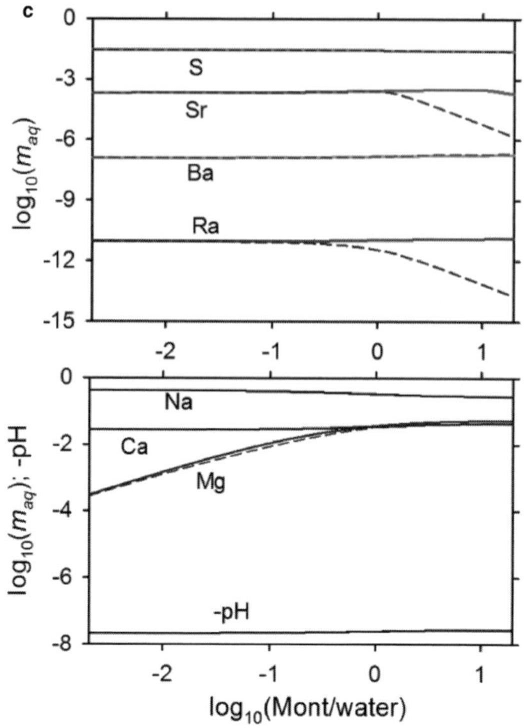

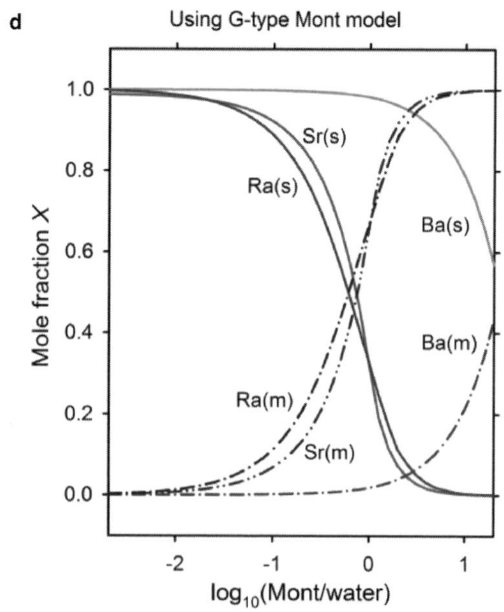

Fig. 12. (*Continued*) (***c***) Profiles of total dissolved cation concentrations and *pH*: solid curves using the V-type, and dashed curves using the G-type Mont model. (***d***) The same as (*b*), but using the G-type Mont solid solution-model instead.

The calculations were performed at 1 bar and 25°C using either the V- or the G-type *Mont* solid solution models from Table 5. Kaolinite and quartz were present in equilibrium throughout, together with the *Mont*, two sulphate, one carbonate (mainly calcite) solid solution phases, gas phase, and aqueous electrolyte. The carbonate solid solution with mainly Ba and Sr was not stable. As seen in Figure12c, the *pH* and total dissolved molalities of Na, Ca, Mg, Ba and S calculated with both V-type and G-type *Mont* models are very similar. However, the G-type variant results in significantly smaller Sr and Ra concentrations at $m/w > 1.0$, and in very much greater *Kd* values in clay for Ra, Ba and Sr compared with the V-type variant (Fig. 12a).

Moreover, the two *Mont* phase variants predict very different patterns for the equilibrium distribution of Ba, Ra and (to lesser extent) Sr between *Mont* and sulphate solid solutions. For the V-type model (Fig. 12c), Ba and Ra remain throughout in barite, whereas Sr is found first in a celestite phase and then, at $m/w > 1.0$, is gradually repartitioned into the *Mont* phase (up to 80%). In G-type model runs (Fig. 12d), both Ra and Sr begin to re-partition from sulphates into *Mont* phase at $m/w = 0.01$ until completion at $m/w > 8.0$. Barium remains in the barite phase until $m/w = 1.0$, and then up to 40% of it goes into the clay ion exchange.

This modelling exercise shows how strong the host solid solution amount effect can be on the trace ion partitioning between different phases. The increasing availability of ion-exchange sites on montmorillonite may lead to the almost complete removal of a cation (*e.g.* Sr^{2+}) from sulphate and its accumulation in the clay phase, if its sulphate phase is not stable enough.

Comparison of two ideal *Mont* model variants with V- and G-type end members shows another effect of mixing behaviour in the predominant solution phase, which is potentially important for minor/trace polyvalent cations. It seems that the V-type ion-exchange model tends to underestimate, and the G-type model to overestimate the uptake of minor divalent cations into clay in this system. As seen in Figure 12a, the V-type model predicts rather small and similar *Kd* values of Sr, Ba, and Ra (<7 L kg^{-1}) in the whole range of m/w ratios. Compared to the V-type model, the G-type model predicts seven times greater *Kd* for Sr, 500 times greater *Kd* for Ba, and 70,800 times greater *Kd* for Ra, in the inverse order to the aqueous dissolved cation concentrations shown in Figure 12c. Repartitioning of Ra and Sr from sulphates into clay leads to further increase in *Kd* (about one order of magnitude), associated with the decrease in total dissolved Ra and Sr (more than two orders). A question arises here: which of the two hypothetical model profiles is more compatible with the chemical reality? In other words, which of clay ion exchange models (the G- or the V-type) is more chemically plausible? Do these model profiles correspond to the experimental reality, keeping in mind that the kinetics of dissolution/precipitation of sulphates and carbonates is completely ignored?

5.10. Impact of decreasing total amount of sulphate (scenario 2)

In this modelling exercise, the same setup of chemical system is taken, but with a fixed 20 g addition of the *MontBent* stoichiometry (initial $m/w = 0.02$). This m/w ratio was selected because in both V-type and G-type variants, minor cations (Ba, Ra, Sr) are still mainly present in sulphate phases. The process extent variable, η, was defined as an addition of $CaSO_4 \cdot 2H_2O$ to the system bulk composition, decreased in steps of -0.0004 mol from the initial 0.155 mol down to -0.036901 mol, which results in 1.0×10^{-6} mol total S remaining in the system when no more sulphate phase is stable. Actually, at $0.155 > \eta > 0.063$ mol nothing interesting happened, so the model runs were limited to $0.063099 > \eta > -0.036901$ mol. The results are presented in Figure 13 as functions of the total amount of sulphur S_{tot} in the bulk chemical composition of the system.

Again, the choice of the *Mont* model convention (V *vs.* G) has almost no influence on the profiles of total dissolved elements and *pH*, except Ra at $S_{tot} < 0.015$ mol (Fig. 13c). However, below this point, the *Kd*(mont) profiles of Ba and Ra decrease sharply, and that of Sr increase and level off in G-variant, whereas all three profiles increase slightly and the V-variant becomes constant (Fig. 13a). Again, the amount of Sr, Ba and Ra in the clay is negligible in the V-type and minor in the G-type model at $S_{tot} > 0.015$ mol (Fig. 13b,d). In both cases, Sr is initially distributed between the sulphate (celestite) and some carbonate (mainly calcite with some Sr) solid solutions. However, at $S_{tot} < 0.063$, gypsum disappears, and another carbonate solid solution (mainly strontianite) begins to form, causing Sr to re-partition from celestite into a growing amount of this new phase. In the G-variant, a minor amount of Sr goes, at the same time, into *Mont* (Fig. 13d), whereas in the V-model it does not.

Dramatic changes in the cation distribution begin from $S_{tot} < 0.015$ when the celestite phase disappears completely. In G-type model runs, the fraction of Ba in the barite phase decreases linearly with S_{tot}, the fraction of Ra decreases non-linearly, and both cations go into a new (mainly witherite) carbonate solid solution which co-exists with mainly calcite and mainly strontianite. Only minor amounts of Ba and Ra go into the *Mont* ion exchange (Fig. 13b). A different pattern develops for Ba and Ra in G-type model profiles (Fig. 13d). Barium still goes into the witherite (carbonate) solid solution, with only a minor amount of it sorbed into the *Mont* phase, whereas radium is strongly re-partitioned into the *Mont* phase, up to completion when the barite phase is gone.

The profiles of total dissolved concentrations (Fig. 13c) are all (except Ra at $S_{tot} < 0.015$ mol) controlled by the solubility of sulphate and carbonate solid solutions. The Sr profile is almost flat because of a decreasing sulphate concentration effect on the solubility of celestite, caused by the removal of $CaSO_4$ from the bulk composition. The Ba profile increases suddenly because of that effect and also due to much greater solubility of witherite than that of barite. The increase in dissolved Ra profile is due to its incorporation in both barium sulphate and carbonate phases (remember that the ideal mixing between Ra and Ba was assumed), but in the G-type model, the increase is less because the larger part of Ra is sorbed by the clay.

These striking differences in the cation partitioning between sulphate and carbonate solid solutions and the Mont solid solution set up in Vanselow or Gapon end members, as modelled in three exercises above, of course, do not prove that either of these two *Mont* models is correct or incorrect. Rather, these model profiles call for more attention to be paid to analysis of thermodynamic model setup in relation to our 'mechanistic

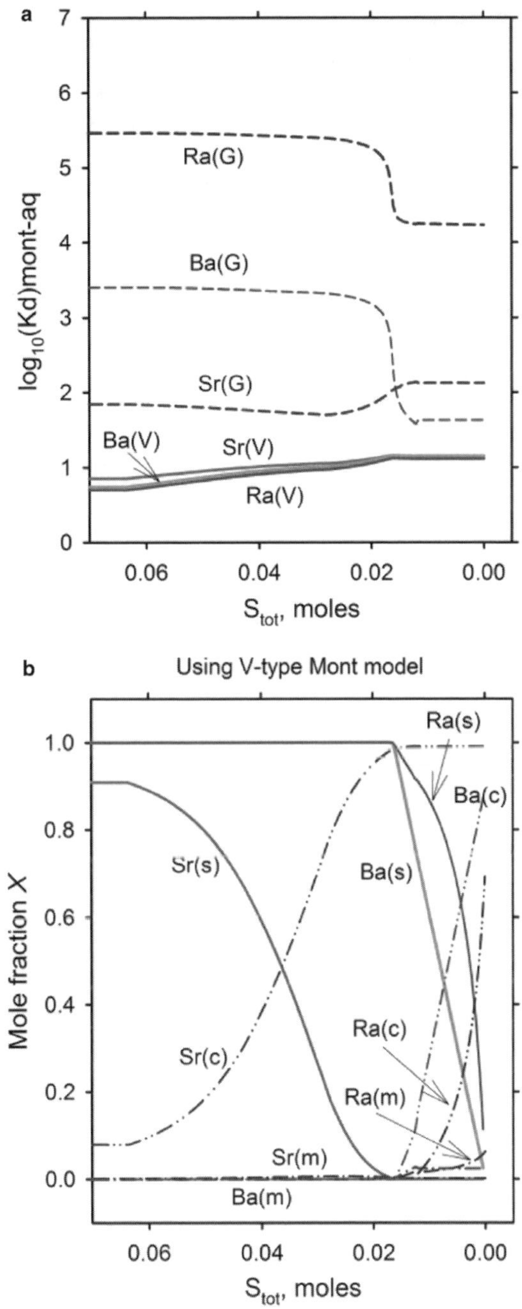

Fig. 13. Impact of the total amount of sulphur S_{tot} on the cation patitioning between sulphate, carbonate, and *Mont* solid solutions at constant *Mont*/water ratio 0.02. (***a***) Profiles of *Kd* (in L kg^{-1}) for cations in montmorillonite solid solution of V-type (V) and G-type (G). (***b***) Fraction of total moles of cations in the system bound in sulphates (s) and clay (m) solid solution phases, modelled with the V-type *Mont* phase.

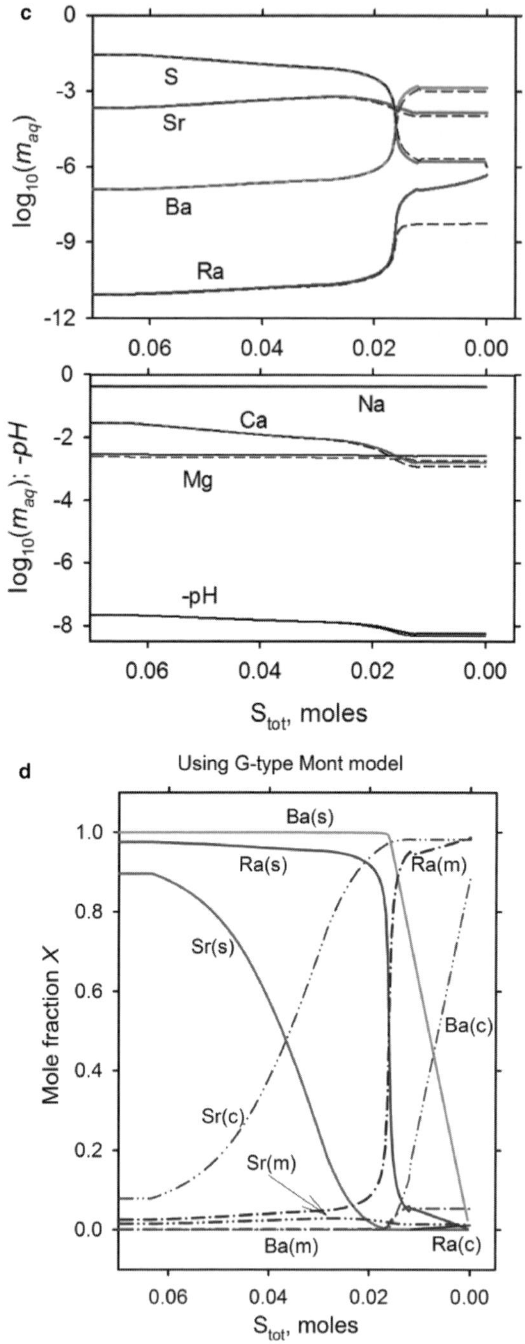

Fig. 13. (*Continued*) (***c***) Profiles of total dissolved cation concentrations and *pH*: solid curves using the V-type, and dashed curves using the G-type *Mont* model. (***d***) The same as (b), but using the G-type *Mont* solid-solution model instead.

understanding' of the underlying process, in this case, heterovalent ion exchange on permanent charge surfaces of montmorillonite. In the B&B model, it is assumed that the ion exchange follows the Vanselow convention (Bradbury & Baeyens, 1995; see also equation 68; Fig. 11b). In the mechanistic sense, this is similar to the bidentate binding to surface sites for a divalent cation, tridentate binding for a trivalent cation, and so on. The mechanistic interpretation of the Gapon convention is not so straightforward, although this setup seems to be more natural from the clay mineralogy viewpoint. Assuming that the ideal mixing in Mont solid solution should follow the Vanselow configurational entropy, the activity of a z_i-charged cation G-type end member should be set as $a_i = x_i^{1/z_i}$ (see equation 80), and the solid-solution model should be calculated using fictive activity coefficients analogous to equation 61:

$$\lambda_i = \frac{a_i}{x_i} = x_i^{1/z_i - 1} \tag{82}$$

In real systems, the energetic non-ideality may also need to be considered in addition. Of course the ultimate answer can only be found on the basis of careful modelling of experimental data. There are different opinions in the literature concerning the issue of Vanselow *vs.* Gapon conventions in modelling ion exchange on clays (Evangelou & Phillips, 1988; Shainberg *et al.*, 1987; Sposito, 1989; McBride, 1994).

6. Concluding remarks

In modelling geochemical systems, ion partitioning is closely related to chemical speciation, solubility, sorption, and formation of variable-composition phases (mixtures or solutions). Quantitative representation of stable or metastable states in such systems is the goal of thermodynamic modelling, performed with help of computer codes that compute the (meta)stable chemical speciation in all coexisting phases at temperature and pressure of interest. The required input data include the 'recipe' (*i.e.* bulk composition of the system), the list of phases and chemical species, and the internally consistent thermodynamic data base covering relevant *T*, *P* and composition regions. Description of equilibrium ion-partitioning phenomena in geochemical thermodynamic models is closely related to the concept of aqueous solubility. Therefore, the ion activity/speciation models for the aqueous electrolyte are of major importance.

The simplest case of *M* ion partitioning is between a multi-component aqueous solution phase and a pure *M*-containing solid phase. This kind of aqueous solubility is defined by the thermodynamic equilibrium constant K_{SP}; the related *M* partition coefficient k_d values are subject to ligand concentration, ion aqueous speciation, and ionic strength effects.

When solid solutions involving *M* are present in the system, in addition to abovementioned effects, the major impact on *M* partitioning between all *M*-containing phases is expressed by the ratio of end member solubility products (K_{SP} ratio involved in the distribution coefficient *D*, equation 35). If this ratio is close to unity, the partitioning will be rather small, and it may be significantly changed or reversed by even a moderate non-ideality of mixing, especially at minor-to-trace M concentrations. The partition and distribution coefficients will be constant only when the mixing is simple and ideal. If the K_{SP} ratio is large then the ion whose end member has the lower K_{SP} will tend to repartition into the solid solution phase; the non-ideality of mixing effect (even causing a miscibility gap) is then only subordinate. In such non-ideal cases, the k_d and D coefficients will vary with the system composition and are difficult to predict without a thermodynamic model of speciation in all phases.

The trace ion partitioning into solid solutions is important in geochemistry of *e.g.* marine carbonates or of (radio)toxic waste. At trace M concentrations, the M isotherms are linear, and k_d and D coefficients are constant; the non-ideality affects the isotherm, k_d and D curve shapes only at high M concentrations. In the trace region, the impact of non-ideal mixing in solid phase is manifested by a vertical shift of the isotherm, k_d and D plots down relative to the ideal mixing case (Henry-law-type behaviour). All these effects can be modelled using modern GEM codes and thermodynamic databases.

So far, mainly simple solid solutions (carbonates, sulphates) of divalent cations were considered in marine- and waste-geochemistry applications (Bruno *et al.*, 2007). However, the uptake of radionuclides into host phases such as calcite, LDH, C-S-H, or clay minerals, often involves heterovalent substitutions (*e.g.* An^{3+} in calcite) or complex multisite substitutions (*e.g.* U(VI) in C-S-H). In addition to experimental and spectroscopic studies, such systems call for development of more sophisticated models of sublattice solid solutions.

Sorption on mineral-water interfaces (MWI) controls the retention of many trace elements of environmental concern, as well as the kinetics of mineral dissolution/precipitation. Various sorption mechanisms – ion exchange on permanent-charge sites, multi-site outer- or inner-sphere surface complexation on amphoteric OH groups, and surface (co)precipitation – all cause specific effects on ion partitioning with aqueous solutions, including electrostatic effects on variable-charge amphoteric surfaces as a function of pH and temperature. Advanced multi-site sorption models incorporated in modern chemical speciation codes can reasonably reproduce the effects of speciation, ionic strength, site competition and saturation, and site heterogeneity into the ion partitioning aqueous solution and mineral sorbents. Competition from the side of major electrolyte ions is the most pronounced factor of ion partitioning by the ion exchange of minor/trace ions, reproduced differently in Vanselow or Gapon representations.

Homovalent clay ion exchange equilibria can be modelled in a straightforward way using solid solutions such as $(B,C)X$ where X is the 'anionic part', *e.g.* $[MgAl_5(Si_4O_{10})_3(OH)_6]^-$ of montmorillonite, and B and C are divalent cations. Extension to heterovalent exchange is ambiguous. In the Vanselow convention, the end members will be written as $M1^{1+}L^-$; $M2^{2+}L_2^-$; and $M3^{3+}L_3^-$; in the Gapon convention, as $M1L$; $M2_{0.5}L$; and $M3_{0.3333}L$. In these two cases, the ideal mixing between mono- and divalent cations will be different, resulting in different predictions especially for the retention of minor/trace (radiotoxic) di- and trivalent cations. The Vanselow convention leads to more conservative predictions (*i.e.* less retention of the cation) in comparison with the Gapon convention. This kind of discrepancy may be resolved in dedicated studies involving experimental sorption data.

Regardless of the uptake mechanism and the kind of the solid phase, the partitioning of M between the aqueous and the solid parts can be represented with the empirical ratio

$$Rd_M = \frac{M \text{ concentration in solid}}{M \text{ aqueous concentration}}.$$

Most 'retention on something' experimental data are represented as Rd values. The theoretical counterpart of Rd is the partition coefficient Kd or k_d:

$$Kd_M^{(w-s)} = \frac{C_M^{(s)}}{C_M^{(w)}} \text{ in} \left(\frac{mol/kg}{\text{mol}/m^3} \right)$$

or

$$k_{d_M}^{(w-s)} = \frac{x_M^{(s)}}{m_M} \text{ in} \left(\frac{mol/mol}{mol/kg\text{H}_2\text{O}} \right)$$

where $C_M^{(s)}$ is the concentration of M in the solid; $C_M^{(w)}$ is the concentration of M in the aqueous phase; $x_M^{(s)}$ is the mole fraction of M end member in the solid (solution), and m_M is the aqueous molality of M. Thus, a partition coefficient measures a relative solubility of M in two coexisting phases in the equilibrium state. The former (mass-based) coefficient is used for clays and other substrates, whereas the latter (k_d) is applied to solid solutions. The ratio of two partition coefficients for different ions in the same solid phase is termed the distribution coefficient (D). All possible such coefficients can be extracted from the equilibrium speciation computed in all phases in the system, and then compared with experimental data.

In complex systems such as soil, sediment, or repository backfill, the ion M can be distributed between aqueous and several solid phases involving different incorporation mechanisms (ion exchange, sorption, solid solution, precipitation). To predict the overall effect on aqueous concentration of M and its migration, two approaches are in use (Payne *et al.*, 2006): the Generalized Composite GC ('top-down'), and Component Additivity CA ('bottom-up').

In the GC approach, the retention of M on the solid phase aggregate is described using one or more generic site types (with site capacities), with sorption affinities represented via non-electrostatic surface binding reactions with empirical equilibrium constants. In contrast, the CA approach represents the aggregate by proportions of several reactive mineral phases, each with individual thermodynamic model parameter set obtained separately for each phase, subject to the chemical mass transfer in the whole system. Conversely, the overall retention of M (the bulk Kd) can be represented as the weighted sum of model *Kd*s for each solid phase, in turn, obtained from the computed speciation:

$$Kd_{\Sigma_M}^{(w-s)} = w_{s1} Kd_M^{(w-s1)} + w_{s2} Kd_M^{(w-s2)} + \cdots$$
$$= \frac{w_{s1} C_M^{(s1)} + w_{s2} C_M^{(s2)} + \cdots}{C_M^w} = \frac{C_M^{(bulk)}}{C_M^{(w)}}$$

where w_{s1}, w_{s2}, ... are the mass fractions of solid phases $s1$, $s2$, ... in the aggregate.

The use of equilibrium thermodynamics in estimating individual *Kd* for phases is often justified by long reaction times (100 years or more) and slow transport processes. The *Kd* values can be extrapolated along composition and temperature, if allowed by the thermodynamic model. Some complex geochemical systems seem to approach equilibrium closely enough to verify this thesis. Co-precipitated rhodochrosite $(Mn,Ca,Mg,Sr,Fe,Ba)CO_3$ solid solution phases in anoxic Baltic Sea sediments provide a good example (Kulik *et al.*, 2000; Neumann *et al.*, 2002; Kersten & Kulik, 2005), also in the context of thermodynamic modelling of elemental speciation and ion partitioning.

However, in the majority of cases, the chemical evolution of subsurface systems is irreversible; some phases never reach equilibrium and remain metastable throughout. Therefore, the account for metastability, kinetic controls, and irreversible mass transfer (mineral dissolution/precipitation, water-rock interaction, weathering, cement hydration, *etc.*) comprises an important component of geochemical modelling. In most codes, this kind of simulations is based on the assumptions of partial and local equilibrium states, and implemented as calculation of many constrained equilibria in a sequence driven by changes in the system bulk composition, temperature, pressure, time, or any combination of them.

Local equilibrium means that an inhomogeneous 'megasystem' with fluxes of matter or heat, existing as the whole far from equilibrium, can be divided into 'nodes' or domains small enough such that a single node can be considered uniformly mixed within a time step, thus reaching a partial equilibrium state. This concept underpins the 'operator-splitting' approach to reactive mass transport simulation in coupled codes (van der Lee & De Windt, 2001).

Partial equilibrium means that at least one phase or component in the system is metastable, *i.e.* not in equilibrium with the (equilibrated) rest of the system. This allows simulation of kinetically dependent chemical evolution of the system if the amount(s) of metastable component(s) can be defined as a function of time, kinetic rate, specific surface, saturation index, and other parameters (some computed from partial equilibrium speciation at previous step). This vast topic makes a bridge between mineral dissolution/precipitation, sorption/desorption kinetics on one side, and chemical speciation in phases of the system on another side; unfortunately, it goes far beyond the scope and size of this chapter. Some comments on it can be found in Kersten & Kulik (2005), and more information on kinetics in Schott *et al.* (2009) or in companion chapters in this volume. Just a few general remarks on the relation between metastability and ion partitioning are given below.

In general, the partitioning of M will be different from the complete equilibrium case if at least one M-containing component (of a solid phase) is metastable, *i.e.* present in mass balance in the amount different from that prescribed by complete equilibrium. The simplest way of introducing metastability is to exclude some (relevant) components from the mass balance; the profound potential impact of this 'brutal force' operation, which can make the whole modelling effort misleading, is usually underestimated. However, some geochemical modelling codes allow the user to impose metastability constraints (from below, from above, or from two sides) on the amounts or concentrations sought of some components in some phases (*e.g.* Karpov *et al.*, 2001). Thus, induced metastability is measured quantitatively relative to the complete equilibrium state by calculating *the saturation index* (see equations 18 and 19) which can be used in many kinetic equations together with other parameters (rate constant, specific surface area, *pH*, *etc.*) to reproduce the irreversible change as function of time. Two cases are mineral dissolution (for mixed solids, usually stroichiometric) and precipitation.

To simulate the dissolution of a (undersaturated) solid phase, amounts of all its end members must be constrained from below, and changed with time (step) proportionally to the initial mole fractions. Other phases containing M may or may not appear as stable in the mass balance at some step; accordingly, the dissolution process will be either incongruent or congruent. Precipitation of the (oversaturated) solid phase can be modelled by imposing constraints from above on its end members and letting them evolve with time step according to known kinetic equation. However, this simulation is much less certain than that for dissolution because it is usually unclear how the mole fractions of end members should change with time, and what supersaturation is needed to reach the nucleation and to control the growth of particles of the newly-formed phase. In closed systems, an additional complication is that upon growth, some ions will be preferentially incorporated into the solid (*e.g.* when $D \gg 1$), and some will be accumulated in aqueous solution, leading to change in the (precipitated) solid composition with time, to the formation of crystal zoning, and so on. Good speciation codes to model such chemical evolution await development.

Trace ion uptake into a ubiquitous host mineral such as calcite seems to be a particularly complex and intriguing interplay between thermodynamics of solid solution and adsorption and transport phenomena on a local scale. The uptake of trace cations depends on the growth rate of the host mineral, resulting in experimental k_d or D values deviating from those expected in equilibrium aqueous - solid solution ion partitioning (Rimstidt *et al.*, 1998; Curti, 1999; Gaetani & Cohen, 2006; Böttcher, 1998). It seems that at very large growth rates, the distribution coefficients, D, of various trace ions co-precipitated in calcite tend to unity, whereas at very small growth rates, they tend to their equilibrium values.

This bias can be modelled semi-quantitatively in the surface entrapment model (Watson & Liang, 1995; Watson, 2004) based on a postulate that a growing crystal assumes the composition of its surface, unless diffusion in the near-surface region is effective during growth. The surface enrichment or depletion results from (metastable) adsorption equilibrium between the MWI and the aqueous solution. Upon crystal growth, the surface layer is 'buried' and trapped in the newly-formed mineral volume, resulting in ion concentrations different from those required by the aqueous-solid solution equilibrium. The efficiency of this process depends on the interplay between the growth rate (*i.e.* burial rate of the entrapped M) and the diffusivity of M ions in the near-surface layer (determining how efficiently M can escape to the fluid). The growth entrapment of trace cations becomes highly efficient at high ratios of bulk growth rate to diffusivity, when the newly-grown crystal mainly preserves the surface composition within the bulk lattice. Several recent studies (Watson, 2004; Gaetani & Cohen, 2006; Tang *et al.*, 2008) corroborated the validity of this model (implemented as computer code) combining diffusive transport with thermodynamics of sorption- and aqueous-solid solution ion partitioning, thus providing an amazing connection between different ion-uptake mechanisms.

A future research challenge might lie in the areas of introducing a realistic TSM without site balances in order to provide an enhanced multi-species sorption thermodynamic model for the surface layer. Next, a simple theoretical approach needs to be formulated which would link different 'layer solutions' in the surface layer and in the underlying 'entrapment layer' with the inner particle core. In case of success, this would eventually lead to a simpler but sound understanding of trace elements partitioning within the 'sorption continuum' of ion retention in space and time.

Acknowledgments

The author is grateful to B. Baeyens, U. Berner, D. Bosbach, M. Bradbury, L. Charlet, E. Curti, O. Gaskova, W. Hummel, N. Kallay, M. Kersten, L. Lakshtanov, J. Lützenkirchen, E. Oelkers, T. Payne, M. Prieto, V. Sinitsyn, G. Sposito, J. Tits, V. Vinograd, T. Wagner, E. Wieland, and many others for stimulating discussions. Partial financial support by Nagra, Wettingen, is much appreciated. Special thanks go to the organizers of the EMU School in Oviedo, 2010, for getting the author involved.

References

Abrams, D.S. & Prausnitz, J.M. (1975) Statistical thermodynamics of liquid mixtures: A new expression for the excess Gibbs energy of partly or completely miscible systems. *American Institute of Chemical Engineers Journal*, **21**, 116–128.

Albarède, F. (1995) *Introduction to Geochemical Modelling*. Cambridge University Press, Cambridge, UK, 543 pp.

Anderson, G.M. (2005) *Thermodynamics of Natural Systems*. Cambridge University Press, Cambridge, UK, 648 pp.

Benjamin, M.M. (2002) *Water Chemistry*. McGraw Hill, New York, 668 pp.

Benjamin, M.M. & Leckie, J.O. (1981) Multiple-site adsorption of Cd, Cu, Zn, and Pb on amorphous iron oxyhydroxide. *Journal of Colloid and Interface Science*, **79**, 209–221.

Bethke, C.M. (1996) *Geochemical Reaction Modeling: Concepts and Applications*. Oxford University Press, New York, 414 pp.

Böttcher, M.E. (1998) Manganese(II) partitioning during experimental precipitation of rhodochrosite-calcite solid solutions from aqueous solutions. *Marine Chemistry*, **62**, 287–297.

Bradbury, M.H. & Baeyens, B. (1995) Sorption by cation exchange. Incorporation of a cation exchange model info geochemical computer codes. *Paul Scherrer Institut Bericht*, **94-07**, 1–32.

Bradbury, M.H. & Baeyens, B. (2002) Sorption of Eu on Na- and Ca-montmorillonites: Experimental investigations and modelling with cation exchange and surface complexation. *Geochimica et Cosmochimica Acta*, **66**, 2325–2334.

Bradbury, M.H. & Baeyens, B. (2005) Modelling the sorption of Mn(II), Co(II), Ni(II), Zn(II), Cd(II), Eu(III), Am(III), Sn(IV), Th(IV), Np(V) and U(VI) on montmorillonite: Linear free energy relationships and estimates of surface binding constants for some selected heavy metals and actinides. *Geochimica et Cosmochimica Acta*, **69**, 875–892.

Bradbury, M.H. & Baeyens, B. (2006) A quasi-mechanistic non-electrostatic modelling approach to metal sorption on clay minerals. In: *Surface Complexation Modelling* (J. Luetzenkirchen, editor). Elsevier, Amsterdam, pp. 518–538.

Bradbury, M.H. & Baeyens, B. (2009) Sorption modelling on illite. Part II: Actinide sorption and linear free energy relationships. *Geochimica et Cosmochimica Acta*, **73**, 1004–1013.

Brown, G.E.J. & Parks, G.A. (2001) Sorption of trace elements on mineral surfaces: Modern perspectives from spectroscopic studies, and comments on sorption in the marine environment. *International Geology Review*, **43**, 963–1073.

Bruno, J., Bosbach, D., Kulik, D. & Navrotsky, A. (2007) *Chemical thermodynamics of solid solutions of interest in radioactive waste management. A state-of-the-art report.* OECD NEA, Paris, 288 pp.

Byrne, R.H. (2002) Inorganic speciation of dissolved elements in seawater: the influence of pH on concentration ratios. *Geochemical Transactions*, **3**, 11–16.

Christov, C. & Møller, N. (2004) Chemical equilibrium model of solution behaviour and solubility in the H-Na-K-OH-Cl-HSO_4-SO_4-H_2O system to high concentration and temperature. *Geochimica et Cosmochimica Acta*, **68**, 1309–1331.

Curti, E. (1999) Coprecipitation of radionuclides with calcite: estimation of partition coefficients based on a review of laboratory investigations and geochemical data. *Applied Geochemistry*, **14**, 433–445.

Curti, E., Kulik, D. & Tits, J. (2005) Solid solutions of trace Eu(III) in calcite: thermodynamic evaluation of experimental data over a wide range of pH and pCO_2. *Geochimica et Cosmochimica Acta*, **69**, 1721–1737.

Dzombak, D.A. & Morel, F.M.M. (1990) *Surface Complexation Modeling: Hydrous Ferric Oxide.* Wiley Interscience, New York, 393 pp.

Evangelou, V.P. & Phillips, R.E. (1988) Comparison between the Gapon and Vanselow exchange selectivity coefficients. *Soil Science Society of America Journal*, **52**, 379–382.

Fanghänel, T. (2002) Application of Pitzer's ion-interaction approach for the calculation of excess properties of trace radionuclides in electrolyte solutions. In: *The Use of Thermodynamic Databases in Performance Assessment* (NEA TDB International Workshop, Barcelona, Spain, 29–30 May 2001). OECD, Paris, pp. 45–60.

Fletcher, P. & Sposito, G. (1989) The chemical modelling of clay/electrolyte interactions for montmorillonite. *Clay Minerals*, **24**, 375–391.

Fletcher, P., Franklin, K.R. & Townsend, R.P. (1984) Thermodynamics of binary and ternary ion exchange in zeolites: The exchange of sodium, ammonium, and potassium ions in mordenite. *Philosophical Transactions of the Royal Society of London A*, **312**, 141–178.

Gaetani, G.A. & Cohen, A.L. (2006) Element partitioning during precipitation of aragonite from seawater: A framework for understanding paleoproxies. *Geochimica et Cosmochica Acta*, **70**, 4617–4634.

Gaines, G.L. & Thomas, H.C. (1953) Adsorption Studies on Clay Minerals. II. A Formulation of the Thermodynamics of Exchange Adsorption. *Journal of Chemical Physics*, **21**, 714–718.

Ganguly, J. (2001) Thermodynamic modelling of solid solutions. In: *Solid Solutions in Silicate and Oxide Systems of Geological Importance* (C.A. Geiger, editor). EMU Notes in Mineralogy, vol. **3**, 37–69. Eötvös University Press, Budapest.

Gaskova, O.L. & Bukaty, M.B. (2008) Sorption of different cations onto clay minerals: Modelling approach with ion exchange and surface complexation. *Physics and Chemistry of the Earth*, **33**, 1050–1055.

Gaucher, E.C. & Blanc, P. (2006) Cement/clay interactions – A review: Experiments, natural analogues, and modeling. *Waste Management*, **26**, 776–788.

Ghiorso, M.S. & Sack, R.O. (1995) Chemical mass transfer in magmatic processes IV. A revised and internally consistent thermodynamic model for the interpolation and extrapolation of liquid-solid equilibria in magmatic systems at elevated temperatures and pressures. *Contributions to Mineralogy and Petrology*, **119**, 197–212.

Glynn, P.D. (1991) MBSSAS: A code for the computation of Margules parameters and equilibrium relations in binary solid-solution aqueous-solution systems. *Computer and Geosciences*, **17**, 907–966.

Glynn, P.D. (2000) Solid-solution solubilities and thermodynamics: Sulfates, carbonates and halides. In: *Sulfate Minerals: Crystallography, Geochemistry and Environmental significance* (C.N. Alpers, J.L. Jambor & D.K. Nordstrom, editors) Reviews in Mineralogy and Geochemistry, **40**, 481–511. Mineralogical Society of America, Washington, D.C.

Glynn, P.D. & Reardon, E.J. (1990) Solid-solution aqueous solution equilibria: thermodynamic theory and representation. *American Journal of Science*, **290**, 164–201.

Grant, S.A. & Fletcher, P. (1993) Chemical thermodynamics of cation exchange reactions: theoretical and practical considerations. *Ion Exchange and Solvent Extraction*, **11**, 1–108.

Greenberg, J.P. & Møller, N. (1989) The prediction of mineral solubilities in natural waters: A chemical equilibrium model for the Na-K-Ca-Cl-SO_4-H_2O system to high concentration from 0 to 250°C. *Geochimica et Cosmochimca Acta*, **53**, 2503–2518.

Grenthe, I. & Puigdomenech, I. (1997) Symbols, standards, and conventions. In: *Modelling in Aquatic Chemistry* (I. Grenthe & I. Puigdomenech, editors). NEA OECD, Paris, pp. 35–67.

Grenthe, I., Plyasunov, A.V. & Spahiu, K. (1997) Estimation of medium effects on thermodynamic data. In: *Modelling in Aquatic Chemistry* (I. Grenthe & I. Puigdomenech, editors). NEA OECD, Paris, pp. 325–426.

Grim, R.E. (1968) *Clay Mineralogy*. McGraw Hill, New York, 596 pp.

Grevel, K.-D. & Majzlan, J. (2009) Internally consistent thermodynamic data for magnesium sulfate hydrates. *Geochimica Cosmochimica Acta*, **73**, 6805–6815.

Gu, Y., Gammons, C.H. & Bloom, M.S. (1994) A one-term extrapolation method for estimating equilibrium constants of aqueous reactions at elevated temperatures. *Geochimica et Cosmochimca Acta*, **58**, 3545–3560.

Guillaumont, R., Fanghänel, T., Fuger, J., Grenthe, I., Neck, V., Palmer, D.A. & Rand, M.H. (2003) *Update on the Chemical Thermodynamics of Uranium, Neptunium, Plutonium, Americium and Technetium*. Elsevier, Amsterdam, 970 pp.

Harvie, C.E., Møller, N. & Weare, J.H. (1984) The prediction of mineral solubilities in natural waters: The Na-K-Mg-Ca-H-Cl-SO_4-OH-HCO_3-CO_3-CO_2-H_2O system to high ionic strengths at 25°C. *Geochimica et Cosmochimca Acta*, **48**, 723–751.

Harvie, C.E., Greenberg, J.P. & Weare, J.H. (1987) A chemical equilibrium algorithm for highly non-ideal multiphase systems: Free energy minimization. *Geochimica et Cosmochimca Acta*, **51**, 1045–1057.

Helffrich, G. & Wood, B. (1989) Subregular model for multicomponent solutions. *American Mineralogist*, **74**, 1016–1022.

Hillert, M. (1998) *Phase Equilibria, Phase Diagrams and Phase Transformations: Their Thermodynamic Basis*. Cambridge University Press, Cambridge, UK, 510 pp.

Holland, T.J.B. & Powell, R. (1998) An internally consistent thermodynamic data set for phases of petrological interest. *Journal of Metamorphic Geology*, **16**, 309–343.

Holland, T. & Powell, R. (2003) Activity–composition relations for phases in petrological calculations: an asymmetric multicomponent formulation. *Contributions to Mineralogy and Petrology*, **145**, 492–501.

Honeyman, B.D. & Santschi, P.H. (1992) The role of particles and colloids in the transport of radionuclides and trace metals in the oceans. In: *Environmental Particles* (J. Buffle & H. van Leeuwen, editors). Lewis Publishers, London, pp. 379–423.

Hummel, W., Berner, U.R., Curti, E., Pearson, F.J. & Thoenen, T. (2002a) *Nagra-PSI chemical thermodynamic data base, version 01/01*. Universal Publishers, New York, 589 pp.

Hummel, W., Berner, U.R., Curti, E., Pearson, F.J. & Thoenen, T (2002b) Nagra/PSI Chemical Thermodynamic Data Base 01/01. *Radiochimica Acta*, **90**, 805–813.

Iliuta, M.C., Thomsen, K. & Rasmussen, P. (2002) Modeling of heavy metal salt solubility using the Extended UNIQUAC model. *American Institute of Chemical Engineers Journal*, **48**, 2664–2689.

Ioannidis, S., Anderko, A. & Sanders, S.J. (2000) Internally consistent representation of binary ion exchange equilibria. *Chemical Engineering Science*, **55**, 2687–2698.

Johnson, J.W., Oelkers, E.H. & Helgeson, H.C. (1992) SUPCRT92: A software package for calculating the standard molal thermodynamic properties of minerals, gases, aqueous species, and reactions from 1 to 5000 bar and 0 to 1000°C. *Computer and Geosciences*, **18**, 899–947.

Karpov, I.K., Chudnenko, K.V. & Kulik, D.A. (1997) Modeling chemical mass-transfer in geochemical processes: Thermodynamic relations, conditions of equilibria and numerical algorithms. *American Journal of Science*, **297**, 767–806.

Karpov, I.K., Chudnenko, K.V., Kulik, D.A., Avchenko, O.V. & Bychinskii, V.A. (2001) Minimization of Gibbs free energy in geochemical systems by convex programming. *Geochemistry International*, **39**, 1108–1119.

Katsikopoulos, D., Fernandez-Gonzalez, A., & Prieto, M. (2009) Precipitation and mixing properties of the "disordered" (Mn,Ca)CO_3 solid solution. *Geochimica et Cosmochimica Acta*, **73**, 6147–6161.

Kersten, M. & Böttcher, M. (1997) Geochemische Modellierung von Partikel/Wasser-Wechselwirkungen. *Geowissenschaften*, **15**, 34–39.

Kersten, M. & Kulik, D.A. (2005) Thermodynamic modeling of trace element partitioning in the environment: New concepts and outlook. In: *Handbook of Elemental Speciation II: Species in the Environment, Food, Medicine and Occupational Health* (R. Cornelis, J. Caruso, H. Crews & K. Heumann, editors). John Wiley & Sons, London, pp. 651–689.

Kielland, J. (1937) Individual activity coefficients of ions in aqueous solutions at 25 °C and 1 atmosphere. *Journal of the American Chemical Society*, **59**, 1675–1678.

Königsberger, E. (2002) Solubility equilibria. From data optimization to process simulation. *Pure and Applied Chemistry*, **74**, 1831–1841.

Kosmulski, M. (2009) *Surface Charging and Points of Zero Charge*. CRC Press, Boca Raton, Florida, USA, 1064 pp.

Krauskopf, K.B. (1956) Factors controlling the concentrations of thirteen rare metals in sea-water. *Geochimica et Cosmochimica Acta*, **10**, 1–26.

Kulik, D.A. (2000) Thermodynamic properties of surface species at the mineral-water interface under hydrothermal conditions: A Gibbs energy minimization triple layer model of rutile in NaCl electrolyte to 250°C. *Geochimica et Cosmochimica Acta*, **64**, 3161–3179.

Kulik, D.A. (2002a) Minimizing uncertainty induced by temperature extrapolations of thermodynamic data: A pragmatic view on integration of thermodynamic databases into geochemical computer codes. In: *The Use of Thermodynamic Databases in Performance Assessment*. NEA OECD, Paris, pp. 125–137.

Kulik, D.A. (2002b) Sorption modelling by Gibbs energy minimisation: Towards a uniform thermodynamic database for surface complexes of radionuclides. *Radiochimica Acta*, **90**, 815–832.

Kulik, D.A. (2006a) Classic adsorption isotherms incorporated in modern surface complexation models: Implications for sorption of actinides. *Radiochimica Acta*, **94**, 765–778.

Kulik, D.A. (2006b) Dual-thermodynamic estimation of stoichiometry and stability of solid solution end members in aqueous–solid solution systems. *Chemical Geology*, **225**, 189–212.

Kulik, D.A. (2006c) Standard molar Gibbs energies and activity coefficients of surface complexes (thermodynamic insights). In: *Surface Complexation Modelling* (J. Lützenkirchen, editor). Elsevier, Amsterdam, pp. 171–250.

Kulik, D.A. (2009) Thermodynamic concepts in modeling sorption at the mineral-water interface. In: *Thermodynamics and Kinetics of Water-rock Interaction* (E.H. Oelkers & J. Schott, editors). Reviews in Mineralogy and Geochemistry, **70**, pp. 125–180. Mineralogical Society of America, Washington, D.C. and the Geochemical Society, St. Louis, Missouri, USA.

Kulik, D.A., Kersten, M., Heiser, U. & Neumann, T. (2000) Application of Gibbs energy minimization to model early-diagenetic solid-solution aqueous-solution equilibria involving authigenic rhodochrosites in anoxic Baltic Sea sediments. *Aquatic Geochemistry*, **6**, 147–199.

Kulik, D., Berner, U. & Curti, E. (2004) Modelling chemical equilibrium partitioning with the GEMS-PSI code. In: *PSI Scientific Report 2003 / Volume IV, Nuclear Energy and Safety* (B. Smith & B. Gschwend, editors). Paul Scherrer Institut, Villigen, Switzerland, pp. 109–122.

Kulik, D.A., Vinograd, V.L., Paulsen, N. & Winkler, B. (2010) (Ca,Sr)CO_3 aqueous- solid solution systems: From atomistic simulations to thermodynamic modeling. *Physics and Chemistry of the Earth* (in press). (doi: 10.1016/j.pce.2010.04.011).

Langmuir, D. (1997) *Aqueous Environmental Geochemistry*. Prentice Hall, New Jersey, 600 pp.

Lefevre, G., Duc, M. & Fedoroff, M. (2006) Accuracy in the determination of acid-base properties of metal oxides surfaces. In: *Surface Complexation Modelling* (J. Luetzenkirchen, editor). Elsevier, Amsterdam, pp. 35–66.

Lippmann, F. (1980) Phase diagrams depicting aqueous solubility of binary mineral systems. *Neues Jahrbuch für Mineralogie Abhandlungen*, **139**, 1–25.

Lützenkirchen, J. (2002a) Surface complexation models of adsorption. In: *Encyclopedia of Surface and Colloid Science* (A. Hubbard, editor). Marcel Dekker, New York, pp. 5028–5046.

Lützenkirchen, J. (2002b) Surface complexation models of adsorption: A critical survey in the context of experimental data. In: *Adsorption: Theory, modeling, and analysis* (J. Tòth, editor). Marcel Dekker, New York, pp. 631–710.

Marion, G.M., Mironenko, M.V. & Roberts, M.W. (2010) FREZCHEM: A geochemical model for cold aqueous solutions. *Computer and Geosciences*, **36**, 10–15.

Marques Fernandes, M., Baeyens, B. & Bradbury, M.H. (2008) The influence of carbonate complexation on lanthanide/actinide sorption on montmorillonite. *Radiochimica Acta*, **96**, 691–697.

McBride, M.B. (1994) *Environmental Chemistry of Soils*. Oxford University Press, Oxford, UK, 416 pp.

Meunier, A. (2005) *Clays*. Springer, Berlin, 472 pp.

Millero, F.J. (1996) *Chemical Oceanography*. CRC Press, Boca Raton, Florida, USA. 469 pp.

Millero, F.J. & Pierrot, D. (1998) A chemical equilibrium model for natural waters. *Aquatic Geochemistry*, **4**, 153–199.

Millero F.J. & Schreiber, D.R. (1982) Use of the ion pairing model to estimate the activity coefficients of the ionic components of natural waters. *American Journal of Science*, **282**, 1508–1540.

Morel, F.M.M. & Hering, J.G. (1993) *Principles and Applications of Aquatic Chemistry*. Wiley Interscience, New York, 588 pp.

Neumann, T., Heiser, U., Leosson, M.A. & Kersten, M. (2002) Early diagenetic processes during Mn-carbonate formation: Evidence from the isotopic composition of authigenic Ca-rhodochrosite of the Baltic Sea. *Geochimica et Cosmochimica Acta*, **66**, 876–879.

Nordstrom, D.K. & Munoz, J.L. (1994) *Geochemical Thermodynamics*. Blackwell Scientific, Boston, Massachusetts, USA, 493 pp.

Oelkers, E.H., Bénézeth, P. & Pokrovski, G.S. (2009) Thermodynamic databases for water-rock interaction. In: *Thermodynamics and Kinetics of Water-rock Interaction* (E.H. Oelkers & J. Schott, editors). Reviews in Mineralogy and Geochemistry, **70**, 1–46. Mineralogical Society of America, Washington, D.C. and the Geochemical Society, St. Louis, Missouri, USA.

Pabalan, R.T. & Bertetti, F.P. (1994) Thermodynamics of ion-exchange between Na^+/Sr^{2+} solutions and the zeolite mineral clinoptilolite. *Material Research Society Symposium Proceedings*, **333**, 731–738.

Parkhurst, D.L. & Appelo, C.A.J. (1999) *User's guide to PHREEQC (Version 2) – A computer program for speciation, batch-reaction, one-dimensional transport and inverse geochemical calculations*. U.S. Geological Survey Water Resources Investigations Report 99–4259, U.S. Geological Survey, Denver, 312 pp.

Payne, T.E., Davis, J.A., Ochs, M., Olin, M., Tweed, C.J., Altmann, S. & Askarieh, M.M. (2006) Comparative evaluation of surface complexation models for radionuclide sorption by diverse geologic materials. In: *Surface Complexation Modelling* (J. Luetzenkirchen, editor). Elsevier, Amsterdam, 605–633.

Pettit, L.D., Puigdomenech, I. & Wanner, H. (2003) *Ionic Strength Corrections using Specific Interaction Theory*. IUPAC.

Powell, R. (1987) Darken's quadratic formalism and the thermodynamics of minerals. *American Mineralogist*, **72**, 1–11.

Prieto, M. (2009) Thermodynamics of solid solution-aqueous solution systems. In: *Thermodynamics and Kinetics of Water-rock Interaction* (E.H. Oelkers & J. Schott, editors). Reviews in Mineralogy and Geochemistry, **70**, 47–85. Mineralogical Society of America, Washington, D.C. and the Geochemical Society, St. Louis, Missouri, USA.

Reed, M.H. (1982) Calculation of multicomponent chemical equilibria and reaction processes in systems involving minerals, gases and aqueous phase. *Geochimica et Cosmochimica Acta*, **46**, 513–528.

Rimstidt, J.D., Balog, A. & Webb, J. (1998) Distribution of trace elements between carbonate minerals and aqueous solutions. *Geochimica et Cosmochimica Acta*, **62**, 1851–1863.

Schindler, P.W. & Stumm, W. (1987) The surface chemistry of oxides, hydroxides, and oxide minerals. In: *Aquatic Surface Chemistry* (W. Stumm, editor). Wiley-Interscience, New York, pp. 83–110.

Schott, J., Pokrovsky, O.S. & Oelkers, E.H. (2009) The link between mineral dissolution-precipitation kinetics and solution chemistry. In: *Thermodynamics and Kinetics of Water-rock Interaction* (E.H. Oelkers & J. Schott, editors). Reviews in Mineralogy and Geochemistry, **70**, pp. 207–258. Mineralogical Society of America, Washington, D.C. and the Geochemical Society, St. Louis, Missouri, USA.

Shainberg, I., Alperovitch, N.I. & Keren, R. (1987) Charge density and Na-K-Ca exchange on smectites. *Clays and Clay Minerals*, **35**, 68–73.

Shao, H., Dmytrieva, S.V., Kolditz, O., Kulik, D.A., Pfingsten, W. & Kosakowski, G. (2009a) Modeling reactive transport in non-ideal aqueous–solid solution system. *Applied Geochemistry*, **24**, 1287–1300.

Shao, H., Kulik, D.A., Berner, U., Kosakowski, G. & Kolditz, O. (2009b) Modelling the competition between solid solution formation and cation exchange on the retardation of aqueous radium in an idealized bentonite column. *Geochemical Journal*, **43**, e37–e42.

Sherman, D.M. (2009) Surface complexation modeling: Mineral fluid equilibra at the molecular scale. In: *Thermodynamics and Kinetics of Water-rock Interaction* (E.H. Oelkers & J. Schott, editors). Reviews in Mineralogy and Geochemistry, **70**, 181–205. Mineralogical Society of America, Washington, D.C. and the Geochemical Society, St. Louis, Missouri, USA.

Shock, E.L., Sassani, D.C., Willis, M. & Sverjensky, D.A. (1997) Inorganic species in geologic fluids: Correlations among standard molal thermodynamic properties of aqueous ions and hydroxide complexes. *Geochimica et Cosmochimica Acta*, **61**, 907–950.

Smith, R.M. & Martell, A.E. (1989) *Critical Stability Constants: Second Supplement*. Plenum Press, New York.

Spahiu, K. (2002) The use of SIT in the NEA TDB project – advantages, drawbacks and comparison with the approaches used in some geochemical codes. In: *The Use of Thermodynamic Databases in Performance Assessment* (NEA TDB International Workshop, Barcelona, Spain, 29–30 May 2001). OECD, Paris, pp. 61–68.

Spera, F.J., Bohrson, W.A., Till, C.B. & Ghiorso, M.S (2007) Partitioning of trace elements among coexisting crystals, melt, and supercritical fluid during isobaric crystallization and melting. *American Mineralogist*, **92**, 1881–1898.

Sposito, G. (1989) *The Chemistry of Soils*. Oxford University Press, Oxford, UK, 277 pp.

Sposito, G. & Mattigod, S.V. (1979) Ideal behavior in Na^+-trace metal cation exchange on Camp Berteau montmorillonite. *Clays and Clay Minerals*, **27**, 125–128.

Stumm, W. (1992) *Chemistry of the Solid-water Interface*. Wiley-Interscience, New York, 428 pp.

Stumm, W. & Morgan, J.J. (1996) *Aquatic Chemistry*. Wiley Interscience, New York, 1022 pp.

Sverjensky, D.A. & Sahai, N. (1996) Theoretical prediction of single-site surface-protonation equilibrium constants for oxides and silicates in water. *Geochimica et Cosmochimica Acta*, **60**, 3773–3797.

Sverjensky, D.A., Shock, E.L. & Helgeson, H.C. (1997) Prediction of the thermodynamic properties of aqueous metal complexes to 1000 C and 5 kb. *Geochimica et Cosmochimica Acta*, **61**, 1359–1412.

Tang, J., Köhler, S.J. & Dietzel, M. (2008) Sr^{2+}/Ca^{2+} and $^{44}Ca/^{40}Ca$ fractionation during inorganic calcite formation: I. Sr incorporation. *Geochimica et Cosmochimica Acta*, **72**, 3718–3732.

Thomsen, K. (1997) *Aqueous electrolytes: model parameters and process simulation*. PhD thesis, Technical University of Denmark, Denmark, 130 pp.

Thomsen, K. (2005) Modeling electrolyte solutions with the Extended UNIQUAC model. *Pure and Applied Chemistry*, **77**, 531–542.

Tournassat, C., Neaman, A., Villieras, F., Bosbach, D. & Charlet, L. (2003) Nanomorphology of montmorillonite particles: Estimation of the clay edge sorption site density by low-pressure gas adsorption and AFM observations. *American Mineralogist*, **88**, 1989–1995.

Townsend, R.P. (1984) Thermodynamics of ion exchange in clays. *Philosophical Transactions of the Royal Society of London A*, **311**, 301–314.

Townsend, R.P. & Coker, E.N. (2001) Ion exchange in zeolites. *Studies in Surface Science and Catalysis*, **137**, 467–524.

Turner, D.R., Bertetti, F.P. & Pabalan, R.T. (2006) Applying surface complexation modeling to radionuclide sorption. In: *Surface Complexation Modelling* (J. Luetzenkirchen, editor). Elsevier, Amsterdam.

Turner, D.R. & Whitfield, M. (1987) An equilibrium speciation model for copper in sea and estuarine waters at 25°C including complexation with glycine, EDTA and NTA. *Geochimica et Cosmochimica Acta*, **51**, 3231–3239.

Urusov, V.S. (1977) *Theory of Isomorphous Miscibility*. Moscow State University, Moscow (in Russian).

Urusov, V.S. (2001) The phenomenological theory of solid solutions. In: *Solid Solutions in Silicate and Oxide Systems of Geological Importance* (C.A. Geiger, editor). EMU Notes in Mineralogy, vol. **3**, 121–153. Eötvös University Press, Budapest.

van der Lee, J. & De Windt, L. (2001) Present state and future directions of modeling of geochemistry in hydrogeological systems. *Journal of Contaminant Hydrology*, **47**, 265–282.

Van Zeggren, F. & Storey, S.H. (1970) *The Computation of Chemical Equilibria*. Cambridge University Press, Cambridge, UK, 176 pp.

Villafáfila García, A., Thomsen, K. & Stenby, E.H. (2006) Prediction of mineral scale formation in geothermal and oilfield operations using the Extended UNIQUAC model. Part II. Carbonate-scaling minerals. *Geothermics*, **35**, 239–284.

Vinograd, V.L. & Winkler, B. (2010) An efficient cluster expansion method for binary solid solutions: application to the halite-silvite, NaCl-KCl, system. In: *Theoretical and Computational Methods in Mineral Physics: Applications to Geophysics* (R. Wentzovitch & L. Stixrude, editors). Mineralogical Society of America and Geochemical Society, Washington, D.C. and the Geochemical Society, St Louis, Missouri, USA, pp. 413–436.

Vinograd, V.L., Kulik, D.A., Pina, C.M., Fernandez Gonzalez, A., Prieto, M. & Winkler, B. (2009a) $(Ba,Sr)SO_4$ solid solution–aqueous solution system: Atomistic calculations and experimental constraints. In: *Migration 2009, 12th International Conference on the Chemistry and Migration Behavior of Actinides and Fission Products in the Geosphere*. Kennewick, Washington, USA, pp. 155.

Vinograd, V.L., Sluiter, M.H.F. & Winkler, B. (2009b) Subsolidus phase relations in the $CaCO_3$–$MgCO_3$ system predicted from the excess enthalpies of supercell structures with single and double defects. *Physical Review B*, **79**, 10421–10429.

Watson, E.B. (2004) A conceptual model for near-surface controls on the trace-element and stable isotope composition of abiogenic calcite crystals. *Geochimica et Cosmochimica Acta*, **68**, 1473–1488.

Watson, E.B. & Liang, Y. (1995) A simple model for sector zoning in slowly grown crystals: Implications for growth rate and lattice diffusion, with emphasis on accessory minerals in crustal rocks. *American Mineralogist*, **80**, 1179–1187.

Wood, B.J. & Nicholls, J. (1978) The thermodynamic properties of reciprocal solid solutions. *Contributions to Mineralogy and Petrology*, **66**, 389–400.

Zhu, C. (2002) Estimation of surface precipitation constants for sorption of divalent metals onto hydrous ferric oxide and calcite. *Chemical Geology*, **188**, 23–32.

EMU Notes in Mineralogy, Vol. 10 (2010), Chapter 4, 139–187

Metal-ion partitioning during low-temperature precipitation and dissolution of anhydrous carbonates and sulphates

MICHAEL E. BÖTTCHER[1,*] and MARTIN DIETZEL[2,*]

[1]*Geochemistry and Isotope Geochemistry Unit, Marine Geology Section, Leibniz Institute for Baltic Research, Seestrasse 15, D-18119 Warnemünde, Germany, e-mail: michael.boettcher@io-warnemuende.de*
[2]*Institute of Applied Geosciences, Graz University of Technology, Rechbauerstrasse 12, A-8010 Graz, Austria, e-mail: martin.dietzel@tugraz.at*
**Corresponding authors*

Anhydrous carbonate and sulphate-bearing solids are formed on Earth under very different environmental conditions and are the most abundant constituents in so-called chemical sediments. At low temperatures, both biogenic and abiotic formation mechanisms are observed and their occurrence ranges from freshwater, *via* open marine to hypersaline aquatic environments. To understand individual mechanisms of mineral formation and the impact of boundary environmental conditions such as reaction temperature, composition of the aqueous mother fluid and precipitation rate, geochemical tracers such as trace metals or stable isotopes are needed. The thermodynamics of element partitioning in solid solutions–aqueous solutions (SS–AS) systems provides a frame for the evaluation of *e.g.* the impact of reaction kinetics, vital activity, or ionic strength. The presence of metastable phases may influence further reactions by providing the necessary reactants and surfaces for dissolution, adsorption, and re-precipitation reactions, thereby changing reaction kinetics in comparison to reactions in homogeneous solutions. The evaluation of trace-element substitution requires the combination of modelling approaches, in particular with experimental calibrations, but also empirical relationships. In this communication, the authors focus on selected binary carbonate and sulphate SS–AS systems at low temperatures. Systems involving the partitioning of Mg^{2+}, Mn^{2+} and Sr^{2+} and others are discussed in more detail. They give a presentation of relevant current topics related to low-temperature metal-ion partitioning during complex precipitation and dissolution behaviour. The aim of this contribution is to stimulate future research in the highly relevant, low-temperature carbonate and sulphate SS–AS systems.

1. Introduction

Carbonate- and sulphate-bearing solids can be formed under very different environmental conditions and are the most abundant constituents in the so-called 'chemical sediments' on Earth. At low temperatures, their occurrence ranges from freshwater to hypersaline aquatic environments. Formation sites range from terrestrial speleothems in direct contact with the atmosphere (Fairchild & Treble, 2009), *via* the water

DOI: 10.1180/EMU-notes.10.4

column or the seafloor of the ocean (Averyt & Paytan, 2003), down to the realm of the deep biosphere several hundred metres below the seafloor (Arndt *et al.*, 2006). Besides carbonates, calcium sulphate minerals are formed under near-surface conditions during the evaporation of seawater in arid marginal seas. Whereas abiotic formation mechanisms were dominant on the early Earth, most modern sedimentary carbonates are directly or indirectly influenced by biological activity (Lippmann, 1973; Morse & Mackenzie, 1990; Morse *et al.*, 2007). Microbial activity has been found to be involved in the formation and transformation of carbonate and sulphate minerals in marine and freshwater environments (*e.g.* Aller & Rude, 1988; Gonzales-Munoz *et al.*, 2003; Mitchell & Ferris, 2005; Bisset *et al.*, 2008).

To understand individual mechanisms of mineral formation and the impact of boundary environmental conditions such as reaction temperature, the composition of the aqueous mother fluid and precipitation rate, geochemical tracers are needed. Luckily, carbonate and sulphate minerals are generally not pure substances in natural 'dirty' systems. Impurities can be caused by isomorphic substitution of a host ion by foreign ions in a given crystal structure (McIntire, 1963). Substitution of metal ions in a crystal may result in solid-solution formation such as, for example, the incorporation of Sr into barite or aragonite, and Mg into calcite (Church, 1979; Speer, 1983; Veizer, 1983; Goldsmith, 1983; Hanor, 2000; Dietzel, 2010). However, isomorphic substitution of divalent cations in carbonate and sulphate minerals depends on crystal structure, the cation charge and radius, which further accounts for the solubility and reactivity with aqueous solutions of the respective minerals. For instance, the large Sr and Ba ions are easily incorporated into aragonite, where the divalent cations are coordinated to nine oxygen ions (Speer, 1983). In contrast, less substitution of Ca by the large metal ions occurs in calcite, where cations such as Sr are coordinated to six oxygen ions (Pingitore *et al.*, 1992). Throughout this chapter, we will ignore charge signs when dealing with aqueous or incorporated divalent ions. Equilibrium element partitioning in SS–AS systems can be calculated from thermodynamic data (Glynn and Reardon, 1990; Gamsjäger *et al.*, 2000; Prieto, 2009) which provide the frame for further evaluation of, *e.g.* the impact on kinetics or ionic strength influences. The low-temperature formation of metastable phases is of particular importance as they are the first limits for the accumulation of aqueous species (Steefel and van Cappellen, 1990). Such solids may form the crystalline seeds that can further react with interstitial fluids upon sedimentary diagenesis towards more stable phases (*e.g.* Lippmann, 1973; Morse & Mackenzie, 1990). Prominent examples are the biotic and abiotic formation of aragonite and high-magnesian calcites (HMC) that can further react to low-magnesian calcites (LMC; Mackenzie *et al.*, 1983; Morse *et al.*, 2007). Another prominent example is the complete calcium-manganese(II) carbonate solid-solution series that forms at low temperatures instead of the potentially thermodynamic stable Mn-bearing calcite, Ca-bearing rhodochrosite with an intermediate ordered, dolomite-type $CaMn(CO_3)_2$ phase (Jakobsen & Postma, 1989; Mucci, 2004; Böttcher, 1998, 2010; Katsikopoulos *et al.*, 2009). The presence of metastable phases influences further reactions by providing the necessary reactants and surfaces for dissolution, adsorption, and re-precipitation reactions, thereby changing reaction kinetics in comparison to reactions in homogeneous solutions. In natural, low-temperature systems, the complex mechanisms of metal ion substitution of metal ions into carbonate and sulphate minerals can further be complicated by a number of factors (*e.g.* Morse & Bender, 1990). Accordingly, trace-element substitution estimated from modelling approaches combined with experimental calibrations and empirical relationships, may be used as an archive for environmental conditions during primary mineral formation or to decipher transformation conditions during alteration and diagenesis. For instance, Mg/Ca and Sr/Ca ratios in calcium carbonate minerals of foraminifera, bivalves or in speleothems are used to estimate paleotemperature and environmental records as well as diagenetic alterations (*e.g.* Purton *et al.*, 1999; Huang & Fairchild, 2001; Fairchild & Treble, 2009). On the other hand, substitution of metal ions not only affects the composition of the precipitates, but can also regulate crystallization mechanisms, precipitation rates, and individual shapes of the crystals as has been observed for Ca replacement in calcite by Mg (Devery & Ehlmann, 1981; Davis *et al.*, 2004), Mn, Sr and Ba (Franke & Hofer, 1986; Paquette & Reeder, 1995; Astilleros *et al.*, 2002), and Co (Freij *et al.*, 2004).

In the present chapter we focus on selected binary carbonate and sulphate SS–AS systems at low temperatures. We aim to present helpful, relevant, current topics related to low-temperature metal-ion partitioning and to stimulate future research in the highly relevant carbonate and sulphate SS–AS systems.

2. General background

Upon crystallization from or re-crystallization in aqueous metal ion-bearing solutions, carbonate and sulphate minerals may incorporate foreign ions depending on a number of different factors such as the solution composition, temperature, precipitation rate, the presence of crystal seeds, inorganic and organic reaction inhibitors and promoters, solid–liquid ratios, *etc.* Metal-ion content is a very important tool in deducing mineral-formation and transformation conditions (Curtis *et al.*, 1986; Pingitore & Eastman, 1985), and water-rock ratios (Banner & Hanson, 1990). Therefore, metal-ion partitioning has been a focus of geochemical research for >50 y.

In the case of calcium carbonate, a foreign element is incorporated into the crystal structure replacing Ca and building a solid solution. This can be expressed by the overall equation

$$x\text{Ca}^{2+} + (1-x)Me^{2+} + \text{CO}_3{}^{2-} \Longleftrightarrow \text{Ca}_x Me_{(1-x)}\text{CO}_3 \qquad (1)$$

Under conditions of thermodynamic equilibrium and considering the respective ratios of the activity coefficients of both solid and aqueous components to be of the same value, a thermodynamic distribution constant K_D can be calculated by the application of the Berthelot-Nernst distribution law (Usdowski, 1975; Morse & Bender, 1990). Under non-ideal conditions, the apparent non-thermodynamic partition coefficient defines the isomorphic substitution of an element in a solid during its precipitation as a function of the composition of the aqueous solution. Divalent metal-ion incorporation into, as an example, calcite can be followed by the distribution law according to the expression

$$\left(\frac{[Me]}{[\mathrm{Ca}]}\right)_{\mathrm{calcite}} = D_{Me} \cdot \left(\frac{[Me]}{[\mathrm{Ca}]}\right)_{aq} \tag{2}$$

where, in a first approach, respective concentrations are given in brackets. D_{Me} is the Henderson-Kracek partition coefficient. This is the quantitative information normally measured in natural and synthetic samples. On a rigorous physicochemical base, however, partition coefficients have to be related to respective activities in the solid and in the aqueous solution (*e.g.* Lippmann, 1980; Glynn, 2000; Glynn & Reardon, 1990) and to the dependence from the degree of disequilibrium conditions.

In general, the complete achievement of exchange equilibrium conditions at low temperatures is limited to SS–AS systems including very soluble salts (Glynn & Reardon, 1990; Glynn, 2000). In the case of carbonates and sulphates, however, a reservoir ('Rayleigh') effect may occur, when the solution composition and/or the reaction boundary conditions (*e.g.* reaction rates) change upon reaction progress. At closed-system conditions with limited amounts of aqueous *Me* and Ca, therefore, heterogeneous distribution of *Me* may occur in the crystals during the precipitation. In this latter case, the overall evolution of the *Me* distribution with respect to the bulk calcite can be followed by the expression

$$\left(\frac{[Me]}{[\mathrm{Ca}]}\right)_{\mathrm{bulk\,calcite}} = \left(\frac{[Me]}{[\mathrm{Ca}]}\right)_{aq,\mathrm{o}} \cdot \frac{1 - \left(\frac{[\mathrm{Ca}]}{[\mathrm{Ca}]_{\mathrm{o}}}\right)_{aq}^{D_{Me}}}{1 - \left(\frac{[\mathrm{Ca}]}{[\mathrm{Ca}]_{\mathrm{o}}}\right)_{aq}} \tag{3}$$

(see Usdowski, 1975; Morse & Bender, 1990). $([Me]/[\mathrm{Ca}])_{\mathrm{bulk\ calcite}}$ is the molar Me/Ca ratio of the bulk calcite, $([Me]/[\mathrm{Ca}])_{aq,\mathrm{o}}$ is the initial molar Me/Ca ratio in the aqueous solution, and $([\mathrm{Ca}]/[\mathrm{Ca}]_{\mathrm{o}})_{aq}$ denotes progress of calcite precipitation in terms of instantaneous aqueous Ca concentration to the initial Ca concentration. In a number of experimental studies it has been found that the magnitude of D_{Me} depends on precipitation rate, R', of the solid solution (*e.g.* McIntire, 1963), and can be expressed by

$$\log D_{Me} = A + B \log R' \tag{4}$$

at constant temperature (A and B are regression coefficients; *e.g.* Dromgoole & Walter, 1990a).

2.1. Modelling approaches

Metal-ion partitioning for thermodynamic equilibrium SS–AS systems has been treated theoretically in a number of studies (*e.g.* Lippmann, 1980; Sverjensky, 1984; Glynn & Reardon, 1990; Rimstidt *et al.*, 1998; Prieto *et al.*, 2000; Wang & Xu, 2001). The concept is based on the stabilities of the end-member solids and aqueous species. For equilibrium conditions, the ideal exchange equilibrium constants can be calculated according to (written for divalent trace metal ion incorporation into calcite)

$$K_{eq} = \exp\left(\frac{\Delta G_{f^0,Me^{2+}} - \Delta G_{f^0,MeCO_3} - \Delta G_{f^0,Ca^{2+}} + \Delta G_{f^0,CaCO_3}}{RT}\right)$$
$$= \frac{K_{sp,CaCO_3}}{K_{sp,MeCO_3}} \quad (5)$$

ΔG_f^o, the Gibbs free energies of formation of the aqueous species and end-member solids, as well as the solubility product constants, K_{sp}, for minerals can be found in the databases of thermodynamic modelling programs, *e.g.* SUPCRT92 (Johnson *et al.*, 1992).

If the ratio of the activity coefficients of the dissolved free ions equals 1, the relation between the partition coefficient, D_{Me}, and the equilibrium constant under equilibrium conditions is given by

$$D_{Me} = K_{eq}(\gamma_{CaCO_3}/\gamma_{MeCO_3}), \quad (6)$$

where γ values are the activity coefficients of the solid-solution components. If equilibrium is established the activity coefficient for the trace compound can be estimated from (*e.g.* Tesoriero & Pankow, 1996):

$$\gamma_{MeCO_3} = K_{sp,CaCO_3}/(K_{sp,MeCO_3} \cdot D_{Me}) \quad (7)$$

One complication arises from the fact that, for a number of metals, the coordination sphere of the trace-metal cation in a solid solution may be different from that observed for its end member. This is illustrated, for example, by Sr incorporation into the calcite structure; as in the case of Ca, Sr occurs in 6-fold coordination in solid solution (Pingitore *et al.*, 1992), although only orthorhombic $SrCO_3$ is found as a stable, pure solid. This problem has been addressed by (semi-)empirical relations (Sverjensky, 1984; Rimstidt *et al.*, 1998; Zhu, 2004). Sverjensky & Molling (1992) provided estimates of the stabilities of hypothetical solids such as rhombohedral strontium carbonate that was applied to construct Lippmann phase diagrams for equilibrium SS–AS system (*e.g.* Böttcher, 1997c) and predict the equilibrium trace incorporation of a number of trace metal ions into the structure of calcite, aragonite and rhodochrosite (Table 1; Böttcher, 1993; Wang & Xu, 2001). Although the equilibrium constants cannot predict kinetic factors that will influence metal-ion partitioning in real non-equilibrium systems (*e.g.* Fig. 1), they provide a helpful basis for understanding the underlying principle thermodynamics, and can help to design experimental approaches to calibrate new systems, and, therefore, offer the roots for future research directions.

Table 1. Predicted thermodynamic equilibrium constants for trace-element incorporation into calcite, aragonite, siderite and barite at 25°C and 1 atm total pressure, and selected, experimentally derived partition coefficients.

	log K_{eq}									log D_{Me}			
Cation	$CaCO_3$[1] Cc	$CaCO_3$[2] Cc	$CaCO_3$[3] Cc	$MnCO_3$[1]	$MnCO_3$[3]	$FeCO_3$[2]	$CaCO_3$[1] Ar	$CaCO_3$[3] Ar	$BaSO_4$[4]	$CaCO_3$ Cc	$FeCO_3$	$CaCO_3$ Ar	$BaSO_4$
Be	−4.74	−1.33	−6.87	−6.74	−8.11	−3.10	−9.65	−8.09	–	–	–	–	–
Mg	−0.36	−3.06	−1.44	−2.36	−2.40	−4.83	−2.78	−2.06	–	−1.92	−2.60	–	–
Ca	0.00	0.00	0.00	−2.00	−0.67	−1.57	0.00	0.00	−7.02	0.00	−1.08	0.00	−7.72
Mn	2.00	2.15	0.92	0.00	0.00	0.38	1.06	0.90	−11.77	1.71	0.91	−0.07	–
Fe	2.10	2.38	1.20	0.10	0.19	0.00	0.71	1.25	−13.28	0.57	–	–	–
Co	2.01	1.67	0.97	0.01	−0.08	−0.10	0.32	0.98	−14.56	0.92	−1.14	–	–
Ni	1.64	−1.28	0.54	−0.36	−0.55	−3.05	−0.26	0.46	–	–	−0.92	–	–
Cu	2.56	3.18	1.57	0.56	0.48	1.42	1.06	1.74	−14.52	1.40	2.60	–	–
Zn	1.40	3.12	0.59	−0.60	−0.43	0.35	−0.10	0.49	−14.32	1.70	–	–	–
Sr	−0.87	−0.74	−0.36	−2.87	−0.86	–	0.82	0.10	−3.36	−1.68	–	0.08	−4.62
Cd	2.68	3.84	2.52	0.68	1.68	2.07	3.06	3.09	−7.57	3.18	–	–	–
Sn	2.35	–	2.78	0.35	2.08	–	4.33	3.88	–	–	–	–	–
Ba	−3.22	−1.71	−1.90	−5.22	−2.22	–	0.23	−0.86	–	−2.00	–	0.18	0.00
Eu	−0.85	–	−0.30	−2.85	−0.80	–	0.84	0.21	−3.12	–	–	–	–
Hg	3.81	–	3.97	1.81	3.13	–	5.28	5.10	–	–	–	–	–
Pb	1.80	3.42	2.50	−0.20	1.87	–	4.48	3.80	−2.03	–	–	–	−1.59
Ra	−3.71	−1.72	−2.28	−5.71	−2.57	–	0.18	−1.15	0.39	−1.84	–	–	0.26
UO_2	−2.68	–	−3.75	−4.68	−4.52	–	−5.62	−5.00	–	−1.70	–	−0.52	–
PuO_2	–	–	−2.67	–	−3.49	–	–	−3.64	–	–	–	–	–

[1] Böttcher (1993); [2] Rimstidt *et al.*, 1998; [3] Wang & Xu (2001); [4] Zhu (2004). Experimental data were chosen to reflect conditions as close to equilibrium as possible. They were taken from Averyt & Paytan (2003), Dietzel *et al.* (2004), Raiswell & Brimblecombe (1977), and the compilations of Rimstidt *et al.* (1998), and Zhu (2004).

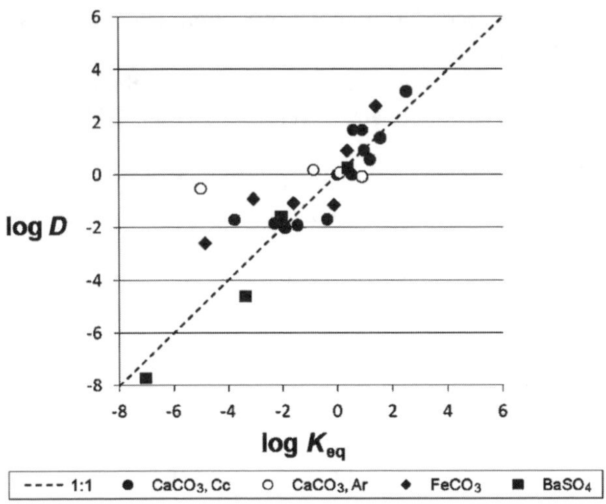

Fig. 1. A comparison of calculated exchange equilibrium constants (25°C) for calcite, aragonite, rhodochrosite, siderite, and barite, with experimentally derived distribution coefficients (for data sources and details see comments in Table 1).

2.2. Empirical relations

The approach closest to the natural environment is the deduction of empirical partition coefficients from field studies. Glynn (2000), for instance, used miscibility-gap data to deduce thermodynamic properties for a number of binary carbonate and (anhydrous and hydrated) sulphate systems. The deduction of partition coefficients from natural environments, however, bears a number of uncertainties since, for instance, different water-rock ratios or dynamic changes in pore-water chemistry are difficult to quantify. In addition, other boundary conditions are not always constant or exactly known (*e.g.* reaction rates, temperature, presence of complexing reactants, co-precipitation of other foreign elements).

Aragonite is the second important modification of calcium carbonate and a common constituent of biogenic and abiogenic carbonates (Lippmann, 1973; Morse & Mackenzie, 1990). Due to its metastability under Earth's surface conditions, aragonite may transform into the more stable rhombohedral calcite by the reaction with aqueous solutions (Pingitore & Eastman, 1985). The redistribution of metal ions during the transformation process has attracted much attention (*e.g.* Katz *et al.*, 1972; Böttcher, 1997a) because of its potential in estimating the reacting solute composition and modelling water-rock interaction processes (*e.g.* Pingitore, 1978; Banner & Hanson, 1990). In the case of natural aragonite-calcite transformations, field applications have been applied successfully (Pingitore & Eastman, 1985; Böttcher & Parafiniuk, 1998). The results also have implications for the use of aragonite to remove metal ions from contaminated aqueous systems (Köhler *et al.*, 2007). In the same way, the stabilization of primary metastable solids *via* dissolution-re-precipitation reactions and the metal ion contents of authigenic minerals have been used to deduce changes in pore-water compositions acting upon secondary mineral formations (Pingitore, 1978; Curtis *et al.*, 1986; Jakobsen & Postma, 1989).

In a prominent sulphate-bearing system, marine barite, the empirical calibration of Sr incorporation into barium sulphate (Averyt & Paytan, 2003) is still a matter of debate (Wang *et al.*, 2007; Averyt & Paytan, 2007). Travertine formation from recent carbonate-precipitating springs or speleothems formation seem to be useful natural laboratories (Bisset *et al.*, 2008; Fairchild & Treble, 2009) to follow metal ion partitioning upon calcium carbonate formation. But this approach is typically limited to redox-insensitve cations. For redox-sensitive metals (*e.g.* Mn and Fe), authigenic mineral formation in anoxic pore water systems, as found in the Baltic Sea surface sediments, seems to be a promising environment (Jakobsen & Postma, 1989; Böttcher, 2010).

2.3. Experimental approaches

The best controlled method for obtaining quantitative information on trace-element incorporation and providing the basis for a mechanistic understanding is an experimental calibration of metal-ion partitioning during the formation and transformation of carbonates. Various experimental approaches have been used to study incorporation of metal cations, ranging from classical free-drift batch-type experiments to the application of better controlled chemo-stat systems. Carbonates, for instance, have been synthesized by: (1) mixing of carbonate and metal bearing solutions (*e.g.* Dromgoole & Walter, 1990a,b; Böttcher, 1998); (2) degassing of CO_2 (*e.g.* Kitano *et al.*, 1971; Oomori *et al.*, 1987; Böttcher, 1998); and (3) decomposition of ammonium carbonate crystals, ion diffusion in gel (*e.g.* Katsikopoulos *et al.*, 2008), replacement of metastable precursors (*e.g.* aragonite; Katz *et al.*, 1972; Böttcher, 1997a), or CO_2 membrane diffusion technique (*e.g.* Dietzel *et al.*, 2004). The distribution coefficients obtained for trace elements in related minerals show largely different behaviour from that proposed by a thermodynamic equilibrium approach (*e.g.* Morse and Bender, 1990; Rimstidt *et al.*, 1998). However, related data sets for trace-element distribution into precipitated solids display systematic relationships mostly referred to saturation states, precipitation rates, and temperatures. If the results of inorganic experiments are related to biogenic-induced mineral formation, individual metabolic effects of organisms may cause additional significant changes of incorporation behaviour (*e.g.* Lea, 2004). A similar diversity is found for experimental dissolution approaches, which covers different types of batch reactors (free-drift batch systems open or closed to a gas phase), flow-through reactors, and hybrid systems (*e.g.* Plummer & Busenberg, 1987, 1989; Böttcher, 1997a,b; Dove & Hochella, 1993). One great advantage of an experimental approach is to follow specific processes with different methods and under different boundary conditions.

3. Metal-ion partitioning during precipitation

3.1. Magnesian calcite formation

The literature dealing with the incorporation of Mg in calcite is voluminous and mostly focused on the mechanisms of formation, transformation, and persistence of low-Mg calcite (LMC), aragonite, high-Mg calcite (HMC), (proto)dolomite and magnesite in seawater and sedimentary environments (*e.g.* Lippmann, 1973; Morse & Mackenzie, 1990; Busenberg & Plummer, 1989; Morse *et al.*, 2007).

Both anhydrous magnesium and calcium carbonate crystallize in the calcite-type structure. Accordingly, precipitation of calcite from Mg-rich solutions causes significant isomorphic substitution of Ca by Mg. Nevertheless, the distribution coefficient of Mg^{2+} between calcite and aqueous solution is less than unity. A limiting value of 4 mol.% $MgCO_3$ is defined to distinguish LMC and HMC (Veizer, 1983). In marine environments, inorganic co-precipitation of Mg generally results in calcite with up to ~10 mol.% $MgCO_3$, whereas in natural skeletal calcite, the Mg content can increase up to ~30 mol.% $MgCO_3$. Uptake of Mg into calcite crystals is stimulated by an increase of the Mg/Ca ratio of the solution. However, the Mg content in inorganically formed calcite is limited due to the formation of aragonite (Mg/Ca ratio > 0.5 to 3, depending on the saturation degree and temperature), as an elevated Mg content inhibits the precipitation of calcite. The Mg content of biogenic-induced calcite is closely related to individual metabolisms, *e.g.* as observed in foraminifera (Elderfield *et al.*, 2006).

The magnitude of Mg substitution in calcite causes changes in physical, chemical, and structural characteristics and properties of the solid. For example: (1) Mg substitution of bulk carbonates can be quantified easily by a shift in X-ray diffraction (XRD) patterns due to a decrease in crystal lattice constants as a function of Mg content. Quantitative Fourier transform infrared (FTIR) spectroscopy was applied successfully to assess the influence of Mg contents on the bonding environment and the vibrational spectra of the carbonate group and a corresponding shift of the IR absorption bands (Böttcher *et al.*, 1997). (2) Mg substitution affects the solubility of Mg calcites, which control stability in contact with aqueous solutions (Fig. 2). A minimum solubility is obtained

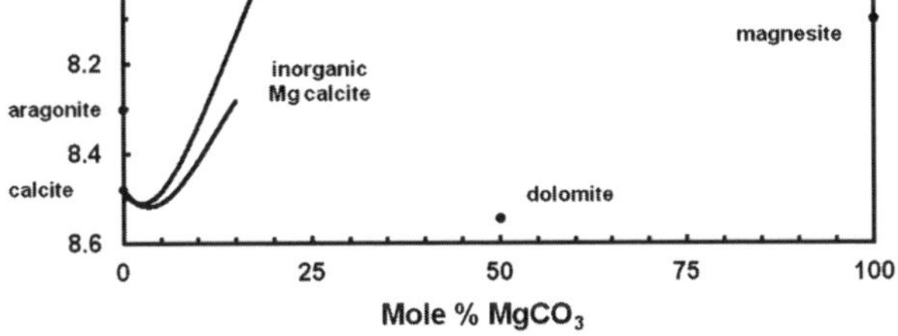

Fig. 2. Stoichiometric solubility constants at 25°C and 1 atm total pressure for calcite, magnesite, abiotic and biogenic Mg calcite solid solutions as a function of Mg content (trends from Morse & Mackenzie, 1990), together with the pK_{ss} values for dolomite and huntite.

at ~3 mol.% $MgCO_3$ and the greatest solubilities are related to HMC (*e.g.* Mackenzie *et al.*, 1983; Morse *et al.*, 2007). The solubilities of biogenic and inorganic Mg calcite differ substantially. Interestingly, the increase in solubility is sufficiently smaller for inorganically precipitated magnesian calcite. Calcites with up to ~15–20 mol.% $MgCO_3$ are all less soluble than aragonite, and calcite containing Mg above this level is less stable than aragonite. It should be noted, however, that the solubility behaviour of Mg calcites is still a matter of debate (Busenberg & Plummer, 1989; Morse *et al.*, 2007). Busenberg & Plummer (1989) attributed the differences in the solubilities of biogenic and low-temperature magnesian calcites (their so-called "group II" carbonates) to the differences in the concentrations of point defects (due to Na and SO_4 incorporation) in the solid solutions rather than to differences in $MgCO_3$ mole fractions. Since multi-element substitution is observed in low-temperature marine HMC, new experimental approaches are required to resolve this question. In this context, Bertram *et al.* (1991) showed experimentally that the solubility of synthetic magnesian calcites decreases with temperature in a manner similar to pure calcite. (3) Mg can directly modify the surface morphology of calcite. Interaction of Mg with crystallographically controlled step directions can explain respective shape modifications of calcite (Devery & Ehlmann, 1981; Davis *et al.*, 2004).

The thermodynamic equilibrium partitioning of divalent metal ions (including Mg) into calcite as defined according to equation 5 are compiled in Table 1. At 25°C it is predicted that Mg is depleted in the solid solution compared to the aqueous solution, in agreement with experimental studies. Although Mg is significantly smaller than Ca, the hydration shell and frequency of the aqueous ion makes it more difficult for it to enter the calcite structure (Lippmann, 1973). It is well known that the Mg content in modern inorganic and biogenic calcite samples is positively correlated with the temperature (*e.g.* Elderfield & Ganssen, 2000). Systematic inorganic co-precipitation experiments indicate a significant increase of the partitioning coefficient of Mg in calcite from ~0.01 to 0.04 as the temperature increases from 5 to 50°C, respectively (Oomori *et al.*, 1987; Mucci, 1987). Consequently, LMC is mainly associated with cool water while HMC is dominantly found in warm-water environments. Note that temperature may be not sufficient to account for the variability of Mg calcite in nature as further parameters may impact the Mg incorporation in calcite. For instance, the Mg incorporation behaviour depends on the Mg/Ca ratio of the solution (Mucci & Morse, 1983) and an increase of D_{Mg} value is obtained with decreasing partial pressure of CO_2 (Burton & Walter, 1991). At a given temperature, the increase in the saturation state enhances the calcite precipitation rate without affecting the Mg content of precipitating calcite (*e.g.* Mucci, 1987; Lopez *et al.*, 2009). Pokrowsky (1998) studied the impact of aqueous Mg/Ca ratio on nucleation and precipitation kinetics of Ca-Mg carbonates from homogeneous solution. Interestingly, in his experiments the main factor for the Mg content in the solid was the saturation index with respect to $MgCO_3$. An early model to account for the kinetically controlled Mg incorporation into calcite was proposed by Lahann & Siebert (1982).

In a recent study, Dissard *et al.* (2010) cultured benthic foraminifera at different salinities to test the reliability of Mg/Ca ratios as a temperature proxy. At constant temperature the amount of Mg incorporation in precipitated calcite increases with increasing salinity. In terms of D_{Mg} values, an increase in salinity of 2 PSU is equivalent to a notable increase of 1°C.

3.2. Sr incorporation during calcite formation

Incorporation of Sr into calcite is involved in a wide range of questions in natural and applied low-temperature environments. The Sr content in calcite, for example, is used in palaeo-oceanographic studies as an indicator of sea-surface temperature (*e.g.* Gagan *et al.*, 1998), for paleo-environmental studies in speleothems or travertine (*e.g.* Huang & Fairchild, 2001), and in the reconstruction of the diagenetic history of fluid composition and authigenic-carbonate formation (*e.g.* Malone & Baker, 1999). Tasks in applied systems are covering, *e.g.* the retention of Sr contaminants in natural ground-water-aquifer systems (*e.g.* Mitchell & Ferris, 2005), and scale formation in drainage systems (e.g. Dietzel *et al.*, 2008).

Various temperature dependencies of Sr incorporation into calcite are reported for different experimental techniques and different biogenic origins. Separate equations for individual organisms are given to estimate palaeo-temperatures from the Sr contents of biogenic calcite (*e.g.* Smith *et al.*, 1979). For biogenic calcite and experimental transformation of aragonite to inorganic calcite a positive or insignificant temperature dependence is observed (Katz *et al.*, 1972; Lea *et al.*, 1999; Malone & Baker, 1999; Stoll *et al.*, 2002a), whereas a negative temperature dependence of Sr incorporation is obtained for inorganically precipitated calcite (*e.g.* Holland *et al.*, 1964; Kinsman & Holland, 1969).

Under equilibrium conditions, a SS–AS system combining calcite and a hypothetical rhombohedral $SrCO_3$ (Sverjensky & Molling, 1992) describes the boundary partitioning of Sr and Ca between solid and solution under standard state conditions (Böttcher, 1997c). These relations may approximate natural systems at very low reaction rates. However, precipitation of a solid solution from aqueous solution requires disequilibrium conditions and, hence, physicochemical supersaturation. Incorporation of Sr into inorganically precipitated calcite is clearly a function of precipitation rate (see Fig. 3), where a constant value of the distribution coefficient is reached at small rates close to thermodynamic equilibrium conditions.

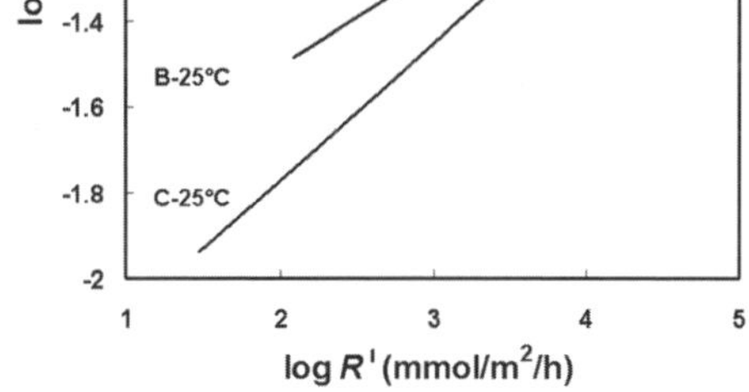

Fig. 3. Partition coefficient for Sr in calcite as a function of precipitation rate obtained from inorganic calcite precipitation experiments. Solid lines: A-25°C for 25°C and *pH* 8.3 (Tang *et al.*, 2008a); B-25°C for 25°C and *pH* 7.5 (Lorens, 1981); C-25°C for 25°C and *pH* 6.1 (Tesoriero & Pankow, 1996). Dashed lines: *A-5°C* and *A-40°C* for 5° and 40°C, respectively (Tang *et al.*, 2008a).

In accordance with the results for inorganically precipitated calcite, Sr incorporation into biogenic and abiotic calcite increases with increasing calcification rate (Lea *et al.*, 1999; Stoll *et al.*, 2002b; Nehrke *et al.*, 2007).

Nevertheless, the results above show that incorporation of Sr into calcite is controlled by both temperature and precipitation rate. A reasonable explanation of the complex behaviour for Sr incorporation during precipitation of inorganic calcite can be given by the 'Surface Entrapment Model' adapted from Watson & Liang (1995). Using this model, Sr is enriched in the surface layer of the growing calcite quantified by the value of the so-called 'Sr entrapment factor'. Incorporation of Sr into the final calcite crystal structure depends on the value of the Sr entrapment factor, crystal-growth kinetics, and ion-diffusion behaviour within the proposed surface layer. Using the smallest available D_{Sr} values from the literature at a given temperature as equilibrium, $D_{Sr,eq}$ values, of 0.01, 0.021, and 0.034 for 5°, 10°, and 25°C, respectively (Stoessell *et al.*, 1987; Tesoriero & Pankow, 1996; Malone & Baker, 1999), the Surface Entrapment Model can simulate Sr incorporation into calcite and calculate D_{Sr} values in the newly formed calcite crystal at specific temperature and the precipitation rate (see Tang *et al.*, 2008a). The Surface Entrapment Model is illustrated in Figure 4 with regard to the effect of precipitation rate on the incorporation of Sr into calcite.

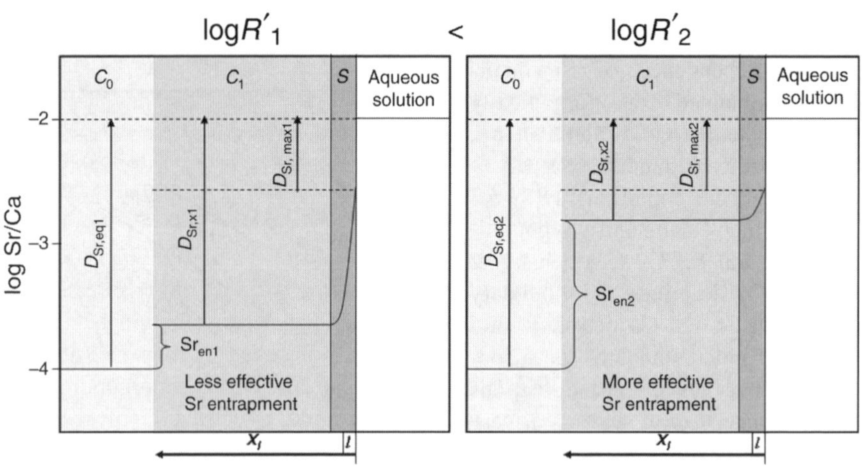

Fig. 4. Illustration of the precipitation rate effect on Sr partitioning during calcite formation at constant temperature ($T = 5$°C, $\log R'_1 = 1.2$ and $\log R'_2 = 2.6$ μmol $m^{-2}\,h^{-1}$) using the surface entrapment approach (Tang *et al.*, 2008a). C_0: crystal at equilibrium; C_1: newly formed crystal structure; *S*: surface layer; *x*: the thickness of the newly formed crystal. $[Sr]_s$ and $[Ca]_s$: Sr and Ca concentrations in regular crystal structure at equilibrium; $(Sr/Ca)_{aq}$: Sr/Ca molar ratio in aqueous solution; F_{Sr}: surface enrichment factor for Sr; $[Sr_{en}]$: entrapment of Sr at newly formed crystal structure; $[Sr]_x$: the enriched Sr content in the newly formed crystal, $[Sr]_x = [Sr]_s + [Sr_{en}]$; $D_{Sr,eq}$: equilibrium distribution coefficient for Sr, $D_{Sr,eq} = ([Sr]_s/[Ca]_s)/(Sr/Ca)_{aq}$; $D_{Sr,x}$: disequilibrium distribution coefficient for Sr, $D_{Sr,x} = ([Sr]_x/[Ca]_s)/(Sr/Ca)_{aq}$. Note that the maximum disequilibrium distribution coefficient is limited by the surface enrichment factor F_{Sr}, *i.e.* $D_{max} = F_{Sr} D_{Sr,eq}$.

An increase in precipitation rate from $10^{1.2}$ to $10^{2.6}$ μmol m^{-2} h^{-1} causes thicker, newly formed crystal structure at a given time. Consequently, more effective entrapment of Sr occurs in the newly formed crystal structure at greater precipitation rates due to a longer distance for Sr to escape by ion diffusion. Thus, Sr is more enriched in the newly formed crystal at elevated precipitation rate. This conceptual model can also rationalize the above-mentioned paradoxical observation of temperature-dependencies of D_{Sr} values for inorganic calcite formation because, at low precipitation rates (transformation reaction for close to thermodynamic equilibrium conditions), a positive temperature-dependence is modelled, whereas at high precipitation rates, a negative temperature dependence is obtained.

Besides temperature and precipitation rate, the apparent D_{Sr} values depend also on the concentrations of additional foreign ions *via* competitive substitution, on chemical composition of the solution (*e.g.* ionic strength, *pH*) and on different partitioning of the Sr on crystallographically non-equivalent faces of calcite (*e.g.* Pingitore, 1986; Paquette & Reeder, 1995). The effect of *pH* can be obtained from Figure 3, where at constant precipitation rate and temperature, D_{Sr} values increase as *pH* increases from 6.1 to 8.3. This relationship can be explained by the Surface Entrapment Model with a change of the entrapment factor of strontium as a function of *pH*, as H^+ activity accounts for surface speciation (*e.g.* Davis & Kent, 1990). In principle, an increase in *pH* results in elevated cation adsorption capacities at the crystal surface, which indicates a higher entrapment factor for Sr. In accordance with Figure 3, elevated D_{Sr} values are expected at greater *pH* values. Different partitioning of Sr on crystallographically non-equivalent faces of calcite was verified by Wasylenki *et al.* (2005a). Sr is highly fractionated between obtuse and acute stepped flanks by a factor of ~2. Note that incorporation of small ions such as Mg shows an inverse behaviour (*e.g.* Davis *et al.*, 2000). Thus, precipitated calcite displays large Sr/Mg ratios at obtuse and low ratios at acute sites, whereas both ions have individual effects on bulk- and site-growing rates. Consequently, the apparent distribution coefficient for the bulk calcite is related to a competition of element fraction at different crystallization sites with individual related distribution coefficients at different crystallization rates. Besides other effects, this might be a reason for a variation in Sr incorporation during calcite formation in the presence of dissolved components (*e.g.* Mg or organic molecules) as may be valid for calcite-forming organisms (vital effect). Moreover, preferential ion incorporation can be completely changed when calcite growth is limited by diffusion of reactants through a boundary layer at the mineral-solution interface instead of surface reaction (*e.g.* McIntire, 1963; for Mg: Wasylenki *et al.*, 2005b).

From the above complex behaviour of Sr incorporation during calcite precipitation, it is concluded that apparent D_{Sr} values in natural and applied systems can be used as an environmental proxy, but only if other relevant environmental parameters are known or reliable assumptions can be made.

In particular, incorporation of Sr into calcite is successfully applied to systems where, besides the chemical composition of calcite, the content of the aqueous solution is known precisely upon calcite formation. A promising case study is the formation of travertine from calcium and DIC-rich spring water. In this case calcite precipitation is induced by CO_2 degassing from the solution into the Earth's atmosphere along the discharge. The corresponding increase in *pH* results in elevated $[CO_3^{2-}]/[DIC]$ ratios and supersaturation with respect to calcite. In Figure 5 the evolution of such solution chemistry is shown for ongoing calcite precipitation from the right to the left. From the slope of the regression line ($D_{Sr} - 1 = -0.83$) a D_{Sr} value of 0.17 is obtained. On the other hand, a mean apparent D_{Sr} value of 0.17 ± 0.02 is estimated from the individual Sr/Ca ratios of the solution and respective exposed precipitated calcite at a given location according to equation 2. As the temperature is $\sim 11 \pm 1$°C and $pH = 8.0 \pm 0.2$, the corresponding precipitation rate with respect to surface area of calcite can be estimated ($R' \approx 10^{3.5}$ μmol m^{-2} h^{-1}; see Fig. 5 and Tang *et al.*, 2008a). Thus, calcite precipitation along a flow path from spring water can be used as an excellent 'in-situ laboratory' for element partition in the terrestrial field.

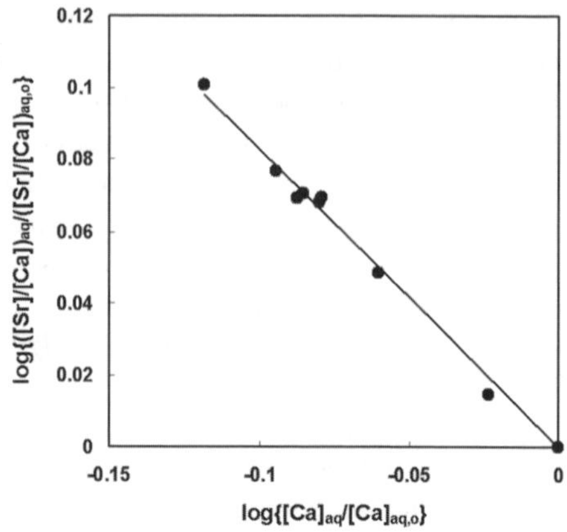

Fig. 5. Evolution of the chemical composition of spring water during ongoing calcite precipitation along a flow path of ~100 m (Carinthia, Austria). Brackets denote concentrations of aqueous Sr and Mg according to an ongoing Rayleigh fractionation process during calcite precipitation from the right to the left. Index o indicates the initial composition of the water at the spring outcrop.

In man-made environments a promising application of Sr incorporation into calcite is related to carbonate scaling. Calcite formation in drainage systems of tunnels, for example, is induced by the dissolution of portlandite ($Ca(OH)_2$) from hydraulic cement used for construction (see Dietzel *et al.*, 2008). In this case, Sr/Ca as well as Mg/Ca ratios in the precipitated calcite can be used to validate sources of dissolved compounds (*e.g.* impact of gypsum dissolution with elevated Sr/Ca ratios), ongoing calcite formation (*e.g.* in analogy to Fig. 5) and *pH* during formation history in carbonate scaling horizons, where low Mg/Ca ratios represent $pH > 10.5$.

3.3. Incorporation of divalent Mn and Fe into calcite

Incorporation of divalent Mn and Fe into calcite is of particular interest for evaluating carbonate diagenesis under low-oxygen conditions, where the redox-sensitive divalent metals are mobile and authigenic carbonates are formed. Formation of Mn(II)-bearing carbonates have been observed in marine and non-marine environments (Jakobsen & Postma, 1989; Friedl *et al.*, 1993), some have been partly influenced by anthropogenic contamination (Lind & Hem, 1993). Most of the authigenic Mn(II) carbonates are not biogenic but have been formed as an indirect consequence of microbial activity from supersaturated pore waters (Jakobsen & Postma, 1989; Friedl *et al.*, 1993). One important general finding is the occurrence of a complete calcite–rhodochrosite solid-solution series in low-temperature environments (Jakobsen & Postma, 1989; Böttcher, 1998), the so-called 'pseudo-kutnahorites' (Mucci, 1988), in spite of the existence of an ordered, dolomite-structured intermediate phase, kutnahorite ($CaMn(CO_3)_2$; Peacor *et al.*, 1987). This phase is stable compared to a solid solution of the same chemical composition (Fig. 6; Mucci, 2004), but has only been described from hydrothermal settings (Peacor *et al.*, 1987). Despite the thermodynamic stability field of kutnahorite, its formation at low temperatures seems to be kinetically inhibited. Due to the solubility products of pure $MnCO_3$ and $FeCO_3$ which are ~2 orders of magnitude smaller than the calcite value (Fig. 6; Dietzel, 2010), Mn/Ca and Fe/Ca tend to be greater in the solid solution compared to the aqueous solution, and authigenic carbonates act as a trap for these redox-sensitive metals. The role of further recrystallization in carbonate composition is still a matter of debate (Jakobsen & Postma, 1989), but its stability in oxygen-depleted interstitial waters makes it a good candidate for preserving, long-term, information on the environmental precipitation conditions (Gingele & Kasten, 1994). The calcite–rhodochrosite SS–AS system has been physico-chemically interpreted from thermodynamic and field relations points of view (Middelburg *et al.*, 1987; Jakobsen & Postma, 1989) but a strict analysis of the SS–AS system requires the consideration of reaction (crystallization and dissolution) kinetics (see below).

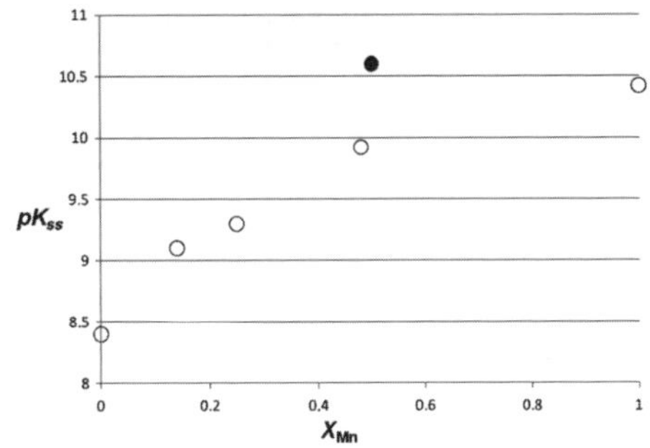

Fig. 6. Stoichiometric solubility constants, K_{ss}, at 20°C and 1 atm total pressure for synthetic calcite-rhodochrosite solid-solutions as a function of Mn content (data from Plummer & Busenberg, 1982; Böttcher, 1993, 1997). Filled symbol: ordered kutnahorite ($Ca_{0.45}Mn_{0.52}Mg_{0.017}Zn_{0.017}Fe_{0.003}CO_3$; Böttcher, 1993; Böttcher *et al., in prep.*) from Sterling Hill mine, USA.

Interactions between calcium carbonate particles and Mn-bearing solutions have been investigated experimentally at low temperatures in a number of studies (*e.g.* Franklin & Morse, 1983; Pingitore *et al.*, 1988; Mucci, 1988; Dromgoole & Walter, 1990a,b; Böttcher, 1993, 1997a,b; Lea *et al.*, 2003; and references therein). It has generally been found, that D_{Mn} shows the preferred incorporation of Mn into the carbonate solid, with a limiting factor of $D_{Mn} = 1$ (Böttcher, 1998). This trend is the opposite to the behaviour of Sr upon calcite formation and mostly due to the differences in metal ionic radii and their hydration behaviour (Burgess, 1978; Nielsen, 1984). No significant influence of Mg in the aqueous solution was found, where a limited co-precipitation of ~5 mol.% into Ca-Mn carbonate solid solutions was observed (Mucci, 1988; Böttcher, 1998). In a detailed study, Dromgoole & Walter (1990a,b) investigated the effects of temperature, PCO_2 and Mn/Ca ratios on the partition coefficients for Mn and Fe. The results are shown in Figures 7 and 8 together with further experimental determinations. At 25°C ($PCO_2 = 0.97$ atm; [Ca] = 10 mM; R' in μmol m^{-2} hr^{-1}) the experimental results follow the relationship

$$\log D_{\mathrm{Mn}} = 1.92 - 0.267 \log R' \tag{8}$$

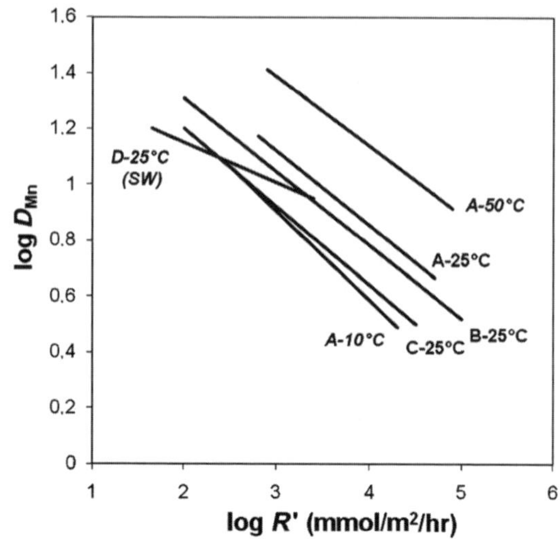

Fig. 7. Partition coefficients of Mn for inorganic calcite precipitation as a function of precipitation rate in seeded chemostat experiments at different temperatures. A: Dromgoole & Walter, 1990; Mn/Ca ratio = 0.01; B: regression line through data of Lorens (1981) after Dromgoole & Walter (1990a); C: Pingitore *et al.* (1988) after Dromgoole & Walter (1990a); D: Mucci (1988) using seawater-like solutions. A slight difference was observed for other CO_2 partial pressures and aqueous Mn/Ca ratios (Dromgoole & Walter, 1990a).

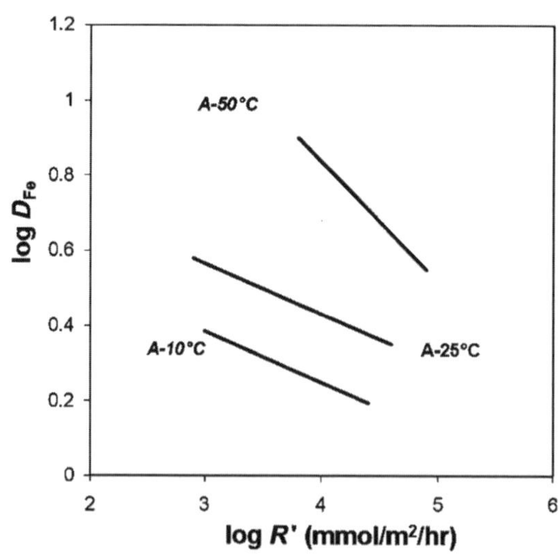

Fig. 8. Partition coefficients of Fe for inorganic-induced calcite precipitation as a function of precipitation rate in seeded chemostat experiments at different temperatures. Fe/Ca ratio = 0.001. Lines according to data in Dromgoole & Walter (1990a). Regression equations presented in the original publication do not always match the experiments and have been refitted from the reported data.

An observed influence of the CO_2 partial pressure may, in analogy to the above discussion on Sr, be explained according to the Surface Entrapment Model by the influence of the solution's *pH* on the surface entrapment factor for divalent Mn. Mn is known to be sorbed effectively to the calcite surface with a dependence on *pH* (*e.g.* Franklin & Morse, 1983). The precipitation of a solid solution from a mixed aqueous solution (written for a Mn-Ca carbonate solid solution)

$$(1-x)\mathrm{Ca}^{2+} + x\mathrm{Mn}^{2+} + 2\mathrm{HCO}_3^- \longrightarrow \mathrm{Ca}_{(1-x)}\mathrm{Mn}_x\mathrm{CO}_3 + \mathrm{H}_2\mathrm{CO}_3^* \quad (9)$$

is, at normal precipitation conditions, controlled by the relative supersaturation that can be related to the total solubility product, $\Sigma\Pi$ (Lippmann, 1980), according to (Kornicker *et al.*, 1991; Böttcher, 1993)

$$R' = \frac{V}{F_t}\frac{dC}{dt} = k'(\Omega_\Sigma^m - 1)^u \quad (10)$$

with

$$\Omega_\Sigma = \{\mathrm{CO}_3^{2-}\}\left(\frac{\{\mathrm{Mn}^{2+}\}}{\gamma_{\mathrm{MnCO}_3}K_{sp,\mathrm{MnCO}_3}} + \frac{\{\mathrm{Ca}^{2+}\}}{\gamma_{\mathrm{CaCO}_3}K_{sp,\mathrm{CaCO}_3}}\right), \quad (11)$$

where C is the concentration, V is the solution volume, k' is the reaction rate constant, m is the number of ionic units of the solid, and u the reaction order (the meaning of F_t is explained below). Based on the experimental data of Dromgoole & Walter (1990a) at 25°C, the dependence of k' (in mol cm^{-2} s^{-1}) from the composition of the aqueous solution (aqueous [Mn]/[Ca] ratios between 0 and 0.05; correlation coefficient, r = 0.99; number of experiments, $n = 5$) follows the relationship (Böttcher, 1993)

$$\log k' = -8.8 \pm 0.2 - \left[17.0 \pm 1.8\left(\sqrt{X_{\mathrm{Mn},aq}}/(X_{\mathrm{Mn},aq}+1)\right)\right] \quad (12)$$

$X_{\mathrm{Mn},aq}$ is the activity fraction in the aqueous solution. In fact, with increasing Mn in the aqueous solution, log k' values decreased from -9.1 to -12.5 mol cm^{-2} s^{-1} and the reaction orders increased from 1.6 to 4.0 (Dromgoole & Walter, 1990a; after Böttcher, 1993). This is consistent with the experimental rate data obtained in seeded experiments for pure calcite (log $k' = -9.1$ mol cm^{-2} s^{-1}; $u = 1.6$; 25°C; Dromgoole & Walter, 1990a; Böttcher, 1993) and pure rhodochrosite (log $k' = -13.5$ mol cm^{-2} s^{-1}; $u = 2.5$; 20°C; Böttcher, 1993). For the individual precipitation rates of Mn and Ca, equation 10 can be separated into (Denis & Michard, 1983; Kornicker *et al.*, 1991; Böttcher, 1993)

$$\frac{V}{F_t}\frac{d[\mathrm{Mn}^{2+}]}{dt} = k_1'(\Omega_\Sigma^m - 1)^u \quad (13)$$

and

$$\frac{V}{F_t}\frac{d[\mathrm{Ca}^{2+}]}{dt} = k_2'(\Omega_\Sigma^m - 1)^u \tag{14}$$

where k_1' and k_2' are the individual rate constants. F_t is the actual reactive surface modified by precipitation considering unchanged particle morphology and Gaussian particle distribution (Barone *et al.*, 1983) according to:

$$F_t = F_0\left(\frac{\mathrm{MG}\cdot V\cdot([C_t]-[C_{t-1}])/2}{\mathrm{M}_0}\right)^{2/3} \tag{15}$$

where F_0 is the initial surface area, M_0 and MG are the mole weight and the initial mass of the solid, respectively, and $[C_t]$ and $[C_t - 1]$ are the dissolved cationic concentrations at times t and $t - 1$ (Böttcher, 1993). Since, $k' = k_1' + k_2'$, combining equations 13 and 14 yields

$$\frac{d[\mathrm{Mn}^{2+}]}{d[\mathrm{Ca}^{2+}]} = \frac{\mathrm{Y}}{1-\mathrm{Y}} = \frac{k_1'}{k_2'} = D_{\mathrm{Mn}}\frac{\{\mathrm{Mn}^{2+}\}}{\{\mathrm{Ca}^{2+}\}} \tag{16}$$

Y and {} are the average Mn mole fractions of the solid and the activities of the dissolved species between the times, t and $t+1$, respectively. Combination of equations 10 and 16 yields

$$R' = \frac{V}{F_t}\frac{dC}{dt} = k_1'(\Omega_\Sigma^m - 1)^u \cdot \frac{1+\{\mathrm{Ca}^{2+}\}}{D_{\mathrm{Mn}}\{\mathrm{Mn}^{2+}\}} \tag{17}$$

At constant precipitation rates, R', the D_{Mn} values increase with temperature (Fig. 9). Results for different experimental studies are comparable for both, diluted and seawater-type solutions, considering the different experimental set-ups used. Therefore, the wide range of experimentally determined D_{Mn} values below 100°C (Fig. 9; Böttcher, 1998) can be explained by the dominant kinetic effects of the precipitation rates. The observed limiting D_{Mn} value of 1 (Pingitore *et al.*, 1988; Böttcher, 1998) corresponds to a very high precipitation rate (Fig. 7) consistent with precipitation from spontaneously mixed, highly supersaturated homogeneous aqueous solutions (Böttcher, 1998). The experimentally observed kinetic partition coefficients are related to fractionation effects at the solid-aqueous solution interface (McIntire, 1963; Lorens, 1981). At very high precipitation rates the overall metal-ion fractionation is probably controlled by ion diffusion from the aqueous solution (McIntire, 1963). Since the diffusion coefficients of dissolved Mn and Ca are very close (Boudreau, 1997), a limiting value of 1 may be approached at very high precipitation rates.

Partition coefficients for Fe values are slightly smaller than the D_{Mn} values at the same precipitation rates with an observed increase with temperature and a decrease with increasing R' (Fig. 8). The incorporation of Fe into calcite, as well as the incorporation of Ca into siderite, seem to be limited at low temperatures (Goldsmith, 1983), but

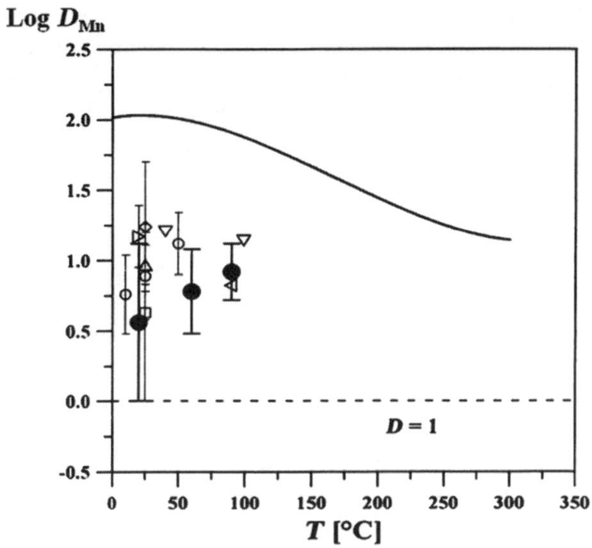

Fig. 9. Range of experimentally determined partition coefficients for Mn(II) incorporation into calcite-rhodochrosite solid solutions as a function of temperature up to 100°C (modified after Böttcher, 1998). Dashed line denotes the limiting partition coefficient of $D_{Mn} = 1$; the solid line is the temperature dependence of the predicted equilibrium constant. Data sources are given in Böttcher (1998).

promoted by the presence of Mg (Romanek *et al.*, 2009). Note that Mn incorporation is easier to interpret than Fe. This is due to the different limiting phases for divalent Fe and Mn fixation under anoxic low-temperature conditions. Whereas Mn can only form carbonates under typical diagenetic conditions (Böttcher, 2010), Fe is also able to react easily with dissolved biogenic sulphide to form iron sulphides. Therefore, the common evaluation of coupled Mn and Fe incorporation is typically limited to suboxic limnic or sulphide-free marine environments.

3.4. Sr and Ba incorporation during aragonite precipitation

Inorganic-induced aragonite precipitation occurs preferentially at elevated temperatures, high aqueous Mg/Ca ratios, and in the presence of certain organic acids (*e.g.* Morse & Mackenzie, 1990; Dove *et al.*, 2003; Morse *et al.*, 2007). At such conditions the formation of calcite, the thermodynamically stable $CaCO_3$ polymorph at low temperature, is inhibited. In the case of organisms, aragonite formation is also referred to individual metabolisms. For both biogenic and inorganic-induced aragonite formation, the respective trace-element content provides an important tool for environmental and palaeoclimate studies.

Strontianite and witherite are isostructural with aragonite and form a complete binary solid-solution series (Speer, 1983). The incorporation of Sr and Ba into aragonite is about two orders of magnitude greater than into calcite (Gaetani & Cohen, 2006;

Dietzel *et al.*, 2004; Lorens, 1981; Tesoriero and Pankow, 1996), which is also predicted by thermodynamic equilibrium calculations (Table 1). The differences between the two elements regarding isomorphic substitution can be explained by the ionic radii of Sr^{2+} and Ba^{2+} and the specific crystal structure of the polymorphs. Sr is the most frequent co-precipitate in aragonite found in natural marine and terrestrial environments, whereas Ba can be highly accumulated in aragonite *vs.* the precipitating solution.

In Figure 10 the partitioning coefficient of Sr in aragonite for inorganic-induced formation is shown as a function of temperature from literature data. A clear negative correlation between the incorporation of Sr in aragonite and temperature is shown (Kinsman & Holland, 1969; Dietzel *et al.*, 2004; Gaetani & Cohen, 2006). In analogy to Sr incorporation during calcite formation (section 3.2), element compatibilities and systematic change in partitioning of Sr and Ba with temperature can be modelled and reasonable explained by the "Surface Entrapment Model" (see Gaetani & Cohen, 2006; Watson and Liang, 1995).

Most of the experimental studies were carried out using seawater-type solutions (*e.g.* Kinsmann & Holland, 1969; Gaetani & Cohen, 2006), whereas Dietzel *et al.* (2004) used Mg and Ca chloride solutions for aragonite precipitation without sodium and sulphate and with a significantly lower ionic strength than normal seawater. As D_{Sr} values from both studies are well correlated to each other, it is concluded that the partitioning coefficient of Sr into aragonite does not depend on sulphate and electrolyte content or salinity, in accord with the experimental results of Zhong & Mucci (1989) and Mucci *et al.* (1989). In several studies it has been proposed that Sr incorporation during aragonite precipitation is rather unaffected by precipitation rate (*e.g.* Zhong & Mucci, 1989). However, comparing data from Zhong & Mucci (1989) and Mucci *et al.* (1989) with those from other studies in Figure 10, the D_{Sr} values for aragonite display a significant precipitation rate effect. The greater the precipitation rate the more Sr is incorporated

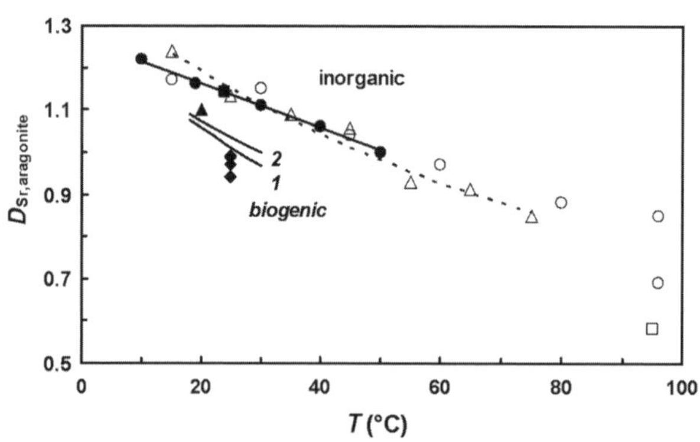

Fig. 10. Partition coefficients for Sr incorporation into inorganic and biogenically induced precipitated aragonite as a function of temperature. Inorganic: (○) Kinsman & Holland (1969); (■) Joschi (1960), (□) Holland *et al.* (1963); (▲) Kitano *et al.* (1971); (◆) Zhong & Mucci (1989) and Mucci *et al.* (1989); solid line, (●) Dietzel *et al.* (2004); dashed line, (Δ) Gaetani & Cohen (2006). Biogenic: solid line (1) De Villiers *et al.* (1995); solid line (2) Beck *et al.* (1992).

during crystal growth. This corresponds to elevated D_{Sr} values for experiments conducted at relatively high growth rates keeping the temperature constant in Figure 10 (*e.g.* data from Dietzel *et al.*, 2004; Gaetani & Cohen, 2006) versus experiments at small growth rates (*e.g.* Zhong & Mucci, 1989). Such growth-rate dependence indicates that Sr partitioning in aragonite is controlled by kinetics other than temperature. The apparent independence of Sr partitioning of aragonite on precipitation rate observed in former studies may be due to rather similar aragonite growth rates in the respective experiments. To evaluate and quantify the proposed impact of kinetics on Sr partitioning in aragonite, further experiments are needed, where incorporation of Sr into aragonite has to be investigated at various but precisely known precipitation rates and, in other respects, at constant physicochemical condition. Such data can be also used to verify geochemical modelling of Sr incorporation into aragonite by the surface entrapment approach for related temperature and precipitation-rate effects.

The trend of D_{Sr} values for biogenic-induced aragonite formation as a function of temperature is similar to that for inorganic precipitation (see solid lines 1 and 2 in Fig. 10). For a given species, incorporation of Sr into aragonite may be used as a proxy to reconstruct temperature or solution chemistry. The respective shift in D_{Sr} values compared to those for inorganic aragonite may be mostly related to the individual metabolisms. Interestingly, De Villiers *et al.* (1994) accounted for variations in extension and/or calcification rate by means of an offset of temperature-related Sr concentration trends in coralline aragonite. This is in accordance with the above conclusions from the inorganic precipitation experiments, where a dependence of D_{Sr} values of aragonite on growth rate is proposed.

Experimental results for inorganic co-precipitation of Ba with aragonite are given in Figure 11. As an analogy to the incorporation of Sr (Fig. 10), D_{Ba} values of aragonite decrease systematically as temperature increases, and Ba incorporation behaviour can also be modelled using the surface entrapment approach. In general, Ba is more compatible in aragonite than Sr. However, at high temperatures both the D_{Sr} and D_{Ba} values for aragonite are close to unity, which represent insignificant discrimination of Sr and Ba during aragonite precipitation. The incorporation of Ba during aragonite formation is significantly more temperature sensitive than that of Sr (Dietzel *et al.*, 2004; Gaetani & Cohen, 2006; Fig. 10). This feature qualifies Ba as a promising proxy for temperature reconstruction if the composition of the precipitating solution is known. Deviation of absolute D_{Ba} values in Figure 11 may be mostly related to individual physicochemical conditions during the precipitation of aragonite. Further experiments

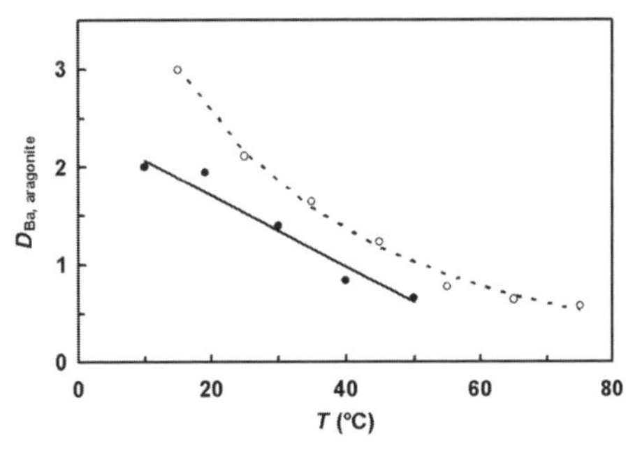

Fig. 11. Partition coefficients for Ba incorporation into aragonite upon inorganic-induced precipitation as a function of temperature. Solid line (●), Dietzel *et al.* (2004); dashed line (Δ), Gaetani & Cohen (2006).

are required to verify if the different behaviour of Ba incorporation into aragonite may be related to *e.g.* ionic strength, sulphate content or/and precipitation rate. It should be noted, that for experimental samples as well as for field samples, precaution should be taken with regard to impurities of witherite ($BaCO_3$) and barite ($BaSO_4$).

3.5. Metal-ion incorporation into carbonates other than calcium carbonate

Much fewer studies have focused on the non-calcium carbonate systems (Glynn, 2000; Morse & Mackenzie, 1990) and experimental studies at low temperatures are scare. However, similar to the calcite and aragonite solid-solution systems, replacement of Ca by metal ions also take place into other rhombohedral and orthorhombic carbonate systems, essentially depending on their cationic radii and hydration behaviour (Rimstidt *et al.*, 1998; Wang & Xu, 2001). Complete solid-solution formation in the carbonate system is, for instance, observed for $BaCO_3$–$SrCO_3$, $ZnCO_3$–$MnCO_3$, $CdCO_3$–$MnCO_3$, $CoCO_3$–$MnCO_3$ and $FeCO_3$–$MnCO_3$ (Goldsmith, 1983; Böttcher, 1993, 1995; Prieto *et al.*, 1997; Glynn, 2000). In natural environments, however, aqueous concentrations must provide very specific conditions to lead to significant solid-solution enrichments of the respective trace metal ions. Glynn (2000) estimated thermodynamic properties from field or experimental miscibility gaps, closing temperatures, and experimental studies. One of the best investigated non-Ca-bearing carbonate systems is the strontianite–witherite SS–AS (Prieto *et al.*, 1997; Noguera *et al.*, 2009; and references therein). Solid-solution formation is limited in a number of binary systems and ordered intermediate phases may be stable as, for instance, observed in the systems $CdCO_3$–$MgCO_3$, $SrCO_3$–$MgCO_3$, $PbCO_3$–$MgCO_3$, $BaCO_3$–$MgCO_3$ (Lippmann, 1973; Goldsmith, 1983), and $BaCO_3$–$MnCO_3$ (Böttcher *et al.*, in prep.). Often the crystal structures of the intermediate phases in these binary systems deviates from the ideal dolomite structure (Lippmann, 1973; Goldsmith, 1983), because of larger differences in the metal ions' radii. Again very little work has been carried out on more complex systems. Therefore, modelling is an important tool in the prediction of trace-metal ion behaviour in non-calcium carbonate systems. Interestingly, the experimental formation of the intermediate phases seems to be less difficult at low temperatures as in rhombohedral systems (*e.g.* dolomite or kutnahorite; Lippmann, 1973; Böttcher, 1993). Equilibrium metal partitioning under standard-state conditions has been predicted for rhodochrosite ($MnCO_3$; Table 1; Böttcher, 1993; Wang & Xu, 2001) and siderite ($FeCO_3$; Table 1; Rimstidt *et al.*, 1998). Trace incorporation into siderite was also investigated experimentally by Thornber & Nickel (1976) and Wersin *et al.* (1989). Theoretical K_{eq} and experimental D_{Me} values are compared in Figure 1. Based on regression results for calcite and siderite, Rimstidt *et al.* (1998) derived a general empirical relation for divalent metal cations that form rhombohedral carbonates (*e.g.* Cd, Zn, Cu, Mn, *etc.*), between the distribution coefficient and the solubility products of the end-member carbonates ($Me\mathrm{CO_3}$ and $Me\mathrm{CO_3^*}$) to follow

$$D_{Me} = 2.25(K_{sp,Me\mathrm{CO}_3^*}/K_{sp,Me\mathrm{CO}_3})^{0.57} \quad (18)$$

For metal-ion incorporation into calcite and siderite, respective pre-factors of 1.6 and 4.1 were reported (Rimstidt *et al.*, 1998). For elements that form stable orthorhombic end-member carbonates (*e.g.* Sr, Ba, Pb, *etc.*), those authors reported a pre-factor of 0.022 for incorporation into calcite. Much more experimental work is needed for these carbonate systems to verify individual incorporation behaviour.

3.6. Sr, Ra and Mg incorporation into calcium sulphates

Although, anhydrite is the anhydrous calcium sulphate mineral in natural environments, the respective substitution behaviour of cations is closely related to the formation of the hydrous form as gypsum is preferentially precipitated at low temperature. Because early precipitated gypsum can be transformed easily into anhydrous anhydrite upon diagenesis, the partitioning of cations into both minerals has to be considered (*e.g.* Kushnir, 1982).

Coprecipitation of Sr with gypsum was studied experimentally as function of temperature, brine concentration (evaporated sea water) and growth rate between 30 and 50°C by Kushnir (1980). The partitioning coefficients increase with precipitation rates from 0.1 to a limit at ~0.3 (30°C). Such a D_{Sr} limit at elevated precipitation rates is in accordance with the "Surface Entrapment Model" (see section 3.2). Maximum entrapment at high growth rates is given by the value of the entrapment factor. Consequently, large discrepancies of the partitioning coefficient of Sr in gypsum from 0.2 to 0.6 at 25°C in former studies (*e.g.* Purkayastha & Chatterjee, 1966; Usdowski, 1973) were mostly attributed to different growth rates. Incorporation of Sr during gypsum formation depends not only on growth rate, but also on aqueous ion content and temperature. The D_{Sr} values tend to increase with brine concentration and decrease with temperature. Thus, Sr incorporation in gypsum can be used to reconstruct *e.g.* evaporation degrees if temperature and growth rate are known. For analysis of natural samples caution has to taken with regard to impurities of celestine in gypsum (*e.g.* Playa & Rosell, 2005).

The Ra incorporation in gypsum is about the same order of magnitude as for Sr at slow precipitation rates and a similar increase in partitioning coefficient at elevated precipitation rate is obtained (Yoshida *et al.*, 2009). As the ionic radii of Sr and Ra are too large for compatibility with the coordination in the monoclinic gypsum structure (celestine and $RaSO_4$ crystallize in the orthorhombic crystal system), the respective partitioning coefficients are <1. In contrast, the smaller Mg ion is well incorporated into gypsum ($D_{Mg} > 1$; Kushnir, 1980).

Co-precipitation of Sr and Mg with anhydrite during the phase transformation of gypsum to anhydrite was studied experimentally in seawater by Kushnir (1982). The partition coefficients of Sr and Mg between anhydrite and the brine are similar to those of gypsum, and similar to the results observed in systems with meteoric solutions. Thus, co-precipitated ions may be not be a promising indicator to distinguish primary from secondary calcium sulphate minerals. However, elemental substitution in gypsum and anhydrite may help to evaluate palaeosalinity and depositional environments. Caution is needed during the dehydration of gypsum rock with respect to celestine formation as the water released contains elevated Sr concentrations (Playa & Rosell, 2005). This phenomenon must also be taken into account as soon as carbonate diagenesis also further involved (Böttcher & Parafiniuk, 1998).

Analyses of Mg and Sr contents in calcium sulphate samples by Lu *et al.* (1997) indicated that fluid and solid inclusions can contribute considerable amounts of elements to bulk analyses. Previous results may need to be revised by a pretreatment of samples to fracture and remove fluid and solid inclusions from calcium sulphate. However, Mg and Sr are less sensitive to the inclusion-stripping compared to Na and the general trends should prevail.

3.7. Sr, Ca and Pb incorporation during barite precipitation

Barite is known to form a continuous solid-solution series with Sr (Church, 1979; Hanor, 2000; Prieto *et al.*, 2000), while substitution by Ca appears to form a discontinuous series (Church, 1979; Hanor, 2000). Although, barite is known to support up to ~8% substitution of Ca for Ba in its crystal structure (Glynn, 2000; Hennessy & Graham, 2002), natural barite samples typically contain Ca only as a trace element (Averyt & Paytan, 2003). However, the amount of Ca incorporation can increase significantly in the presence of Mg ions (Hennessy & Graham, 2002). The formation of low-temperature barite has been observed in the water column and/or at the sediment-water interface (Church, 1979; Paytan *et al.*, 1998; Paytan & Griffith, 2007), which has been introduced as a potential proxy for export productivity (Paytan *et al.*, 2007). Incorporation of both Sr and Ca into marine barite recovered from oxic deep-sea sediments is used to reconstruct Sr/Ca ratios and Sr and Ca concentrations of seawater during geological history. The main focus is on the interpretation of palaeo-oceanographic records and on the understanding the related global geochemical cycles, which were influenced *e.g.* by chemical weathering and hydrothermal activities. Barite from the top of the core was selected from open-ocean sites in order to estimate empirical individual partition coefficients using modern global seawater average Sr and Ca contents and the saturation concentration of Ba for barite at the depth of its formation. The results show that Sr incorporation into barite is greater than of Ca, which is caused by similar ionic radii of Sr and Ba (Averyt & Paytan, 2003). As the natural variability of Sr and Ca contents of the marine barite is significantly smaller than the expected variation in seawater throughout Earth's history, a reliable resolution may be obtained to reconstruct Sr/Ca ratios and Sr and Ca concentrations in the seawater. The advantage of this approach compared to the use of calcium carbonates is that barite is precipitated inorganically at small rates by the release of Ba associated with the decay of organic matter; barite is largely preserved in closed-system conditions, and no diagenetic alteration occurs (*e.g.* Church, 1979; Ganeshram *et al.*, 2003). However, experimental results for precipitation of barite from seawater and modelling of incorporation behaviour indicate that lower temperatures, increasing pressures, and lower precipitation rates decrease respective partition coefficients and thus decrease Sr and Ca incorporation into marine barite (Church, 1979). Although, Averyt & Paytan (2003) concluded that temperature and pressure change during marine barite

Fig. 12. Diagenetic barite band found in sediments recovered during ODP Leg 207 (Site 1258A; photograph courtesy of H.J. Brumsack).

formation are within the variation of apparent empirical partition coefficients, further knowledge of the potential impact of environmental parameters such as growth rate, Sr and Ca content in spatial and temporal marine (micro)environment, where barite formation occurs, are required (see Wang *et al.*, 2007; Averyt & Paytan, 2007). Future studies must consider potential heterogenic trace-element distribution into

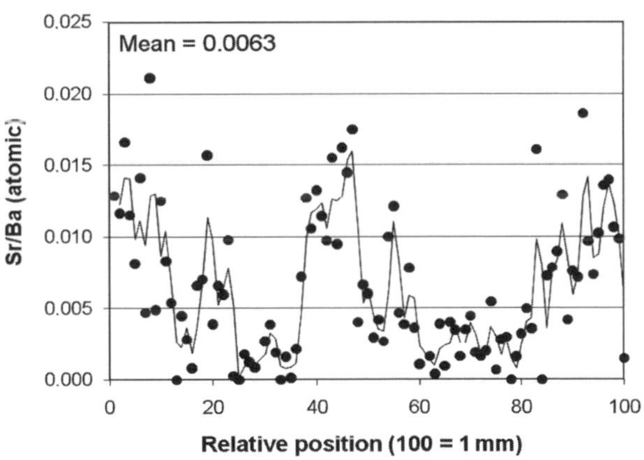

Fig. 13. Variations of Sr/Ba ratios through a diagenetic barite (section 1258A-34R, 103.5–104 cm; ~2 mm thick, clear crystal with idiomorphic faces) recovered from >300 m b.s.f. during ODP Leg 207 (microprobe measurements: M. Prieto) (Böttcher *et al.*, in prep.) demonstrating a sensitive recording of changes in the (local?) pore-water composition. Total bottom scale corresponds to a width of 1 mm. Line indicates running average. A $\delta^{34}S$ value of +53‰ *vs.* V-CDT clearly indicates a diagenetic origin (Böttcher *et al.*, in prep.).

barite crystals and potential sample impurities with respect to fluorite (CaF_2). Figure 12 gives an example for a diaganetic barite recovered from deep marine sediments of ODP leg 207. As an example of trace-element oscillations, Figure 13 shows variable atomic Sr/Ba ratios, from a small section of a diagenetic barite crystal recovered during ODP Leg 207 which formed in deep marine sediments due to diffusion-induced formation of barite fronts (Torres *et al.*, 1996). This process is triggered by black-shale driven anaerobic methane oxidation (Arndt *et al.*, 2006). Different Sr content profiles were found for barites from the same site (Böttcher *et al.*, in prep.) indicating formation from pore-water gradients with different dynamics and/or oscillations upon precipitation as observed in experimental studies (Putnis *et al.*, 1992; Prieto *et al.*, 1997). The latter process may complicate the determination of the compositions of solids with low solubilities that may preserve compositional Ostwald-type zonations (Putnis *et al.*, 1992; Prieto, 2009) and their corresponding mother solutions.

In the ocean-water column, celestine, $SrSO_4$, with several mole percent of barium may form the tests of Acentharia (Church, 1979). Because the world's ocean is undersaturated with respect to pure celestine and only few regions reach saturation with respect to barite (Monnin & Hoareau, 2006, 2010, this volume) the exact (trans)formation conditions are still a matter of debate. Further transformations have been suggested to take place, *e.g.* when Sr-rich barium sulphate recrystallizes to Sr-poor authigenic barite minerals (Church, 1979). Different textures may indicate the place and mechanism of barite formation (Gingele & Dahmke, 1994). Celestine and barite have also been found as the product of replacement reactions where aragonite was transformed to calcite in a sulphate-rich environment, causing the liberation of trace metals such as Sr and Ba for further reaction to form the respective sulphate minerals (*e.g.* Böttcher & Parafiniuk, 1998), thereby linking the low-temperature carbonate and sulphate SS–AS systems.

In addition to Sr, Pb incorporation into barite at low temperatures is also complete (Kornicker *et al.*, 1991). The actual contents, however, depend heavily on the availability in the dissolved phase. Massive celestine precipitation with minor Ba incorporation has been suggested to occur from saline basinal brines containing moderately high levels of dissolved sulphate (Hanor 2000, 2004).

Incorporation of up to 20% Pb has been observed in natural samples (Glynn, 2000), but a complete (metastable?) solid-solution series was prepared at low temperatures (Kornicker *et al.*, 1991). Theoretical trace-element incorporation into barite and celestine has been modelled by Zhu (2004). For the partitioning between barite and aqueous solutions, the sequence of partitioning into barite, from favourable to unfavourable, is Ra^{2+}, Pb^{2+}, Eu^{2+}, Sr^{2+}, Ca^{2+}, Cd^{2+}, Mn^{2+}, Fe^{2+}, Zn^{2+}, Co^{2+}, Cu^{2+}, Mg^{2+}, Ni^{2+} and Be^{2+} (Table 1). The modelled magnitude of trace-element fractionation is in general agreement with the few experimental data available so far (Fig. 1) and allows the prediction of trace-element behaviour as long no experimental calibration is available.

The low-temperature formation of the pure barium and strontium carbonates witherite and strontiante is often described as the replacement of barite and celestine (Hanor,

2000). Incorporation of trace elements will, therefore, depend on the metal-ion contents of the precursor sulphates of the aqueous solutions associated with transformation, and the openness of the system (Hanson & Banner, 1990). Both witherite and strontianite require large concentrations of dissolved Ba or Sr to be stable in the presence of dissolved Ca (Hanor, 2000). The reversed reaction, the incorporation of carbonate-derived trace elements into sulphate (gypsum) minerals was described from groundwater systems by Glynn (1992).

4. Metal-ion partitioning upon congruent and incongruent dissolution

In low-temperature natural or man-made environments the formation of metastable solid solutions provides the potential for further reactions. This may also be induced by changes in the composition of the aqueous solution, temperature, and microbial activity, or the metastability of the early precipitates. The transformation of precursor solid solutions requires a dissolution step before further precipitation of a secondary solid can take place. Further reaction progress or preservation of initially formed solid solutions depends on surface characteristics. Metal-ion partitioning during the dissolution of carbonate and sulphate solid solutions has been investigated in a number of low-temperature experimental studies (Thorstenson & Plummer, 1977; Denis & Michard, 1983; Plummer & Busenberg, 1987; Busenberg & Plummer, 1989; Chou *et al.*, 1989; Galinier *et al.*, 1989; Glynn *et al.*, 1990; Plummer *et al.*, 1992; Bertram *et al.*, 1991; Kornicker *et al.*, 1991; Felmy *et al.*, 1993; Böttcher, 1997b). The reaction paths have been analysed and discussed in detail for a number of carbonate and sulphate SS–AS systems (Glynn *et al.*, 1990; Kornicker *et al.*, 1991; Plummer *et al.*, 1992; Felmy *et al.*, 1993).

4.1. Magnesian calcites

The interest in the dissolution behaviour of metal-ion solid-solution systems started from observations of magnesian calcites and was later extended to other carbonate and sulphate SS–AS systems. In dissolution reactions of metastable or unstable solids the aqueous solution may reach a state of 'stoichiometric saturation' (Thorstenson & Plummer, 1977; Glynn & Reardon, 1990) with respect to the initial solid, if the solids react with fixed composition and the formation of secondary (more stable) solid solutions is kinetically inhibited or the water-solid ratio is sufficiently high (Plummer & Busenberg, 1987, 1989; Glynn *et al.*, 1990; Bertram *et al.*, 1991; Böttcher, 1997b).

Upon interaction of undersaturated aqueous solutions with mono-phase solid solutions, congruent dissolution will take place in the initial phase, as for the example of a general Ca-*Me* carbonate solid solution, according to

$$\mathrm{Ca}_{(1-x)}Me_{x}\mathrm{CO_3} + \mathrm{H_2CO_3^*} \rightarrow (1-x)\mathrm{Ca}^{2+} + xMe^{2+} + 2\mathrm{HCO_3^-} \quad (19)$$

Without further formation of secondary carbonates, the dissolution is controlled until reaching stoichiometric saturation ($\Omega_{ss} = 1$), by the degree of stoichiometric undersaturation (*e.g.* Thorstenson and Plummer, 1977; Kornicker *et al.*, 1991; Glynn *et al.*, 1992), which is calculated according to

$$\Omega_{ss} = IAP_{ss}/K_{ss} \tag{20}$$

IAP_{ss} is the stoichiometric ion activity product. K_{ss}, the stoichiometric solubility constant for the congruent dissolution reaction (19) is given by:

$$K_{ss} = \{Ca^{2+}\}^{(1-x)}\{Me^{2+}\}^{x}\{CO_3^{2-}\} \tag{21}$$

This stoichiometric solubility product differs from the total solubility product as defined for thermodynamic exchange equilibrium by Lippmann (1980). The thermodynamic consistency of both approaches was shown by Glynn & Reardon (1990). The rate of dissolution of a solid solution depends on the composition of the dissolving phase and the saturation state of the aqueous solution, according to (*e.g.* Konicker *et al.*, 1991)

$$R = \frac{V}{F_t}\frac{dC}{dt} = k(1 - \Omega_{ss}^{m})^{n} \tag{22}$$

with k as the reaction rate constant and n as the order of the reaction. The congruent part of the reaction process adds dissolved cations in the proportion controlled by the composition of the initial solid solution, and is independent from the initial composition of the fluid (Plummer *et al.*, 1992). As soon as supersaturation of the aqueous solution with respect to more stable solid solutions occurs, secondary solid solutions may be formed and the reaction becomes incongruent according to

$$\begin{aligned} Ca_{(1-x)}Me_xCO_3 + H_2CO_3^* \longrightarrow\ & zCa_{(1-y)}Me_yCO_3 + (1 - x - z + z \cdot y)Ca^{2+} \\ & + (x - z \cdot y)Me^{2+} + 2(1 - z)HCO_3^- + zH_2CO_3^* \end{aligned} \tag{23}$$

In this case, the secondary Mg calcite is depleted in Mg compared to the initial solid. The congruent part of the reaction has been extrapolated by the use of inverse-time plots (Thorstenson & Plummer, 1977; Bertram *et al.*, 1991; Böttcher, 1997b) to obtain the stoichiometric solubility of the respective solid solution. Using these data, a significant influence of the Mg contents on the solubility behaviour of biogenic and abiotic magnesian calcites was deduced (Fig. 2; Morse & Mackenzie, 1990). Later, however, it was found that the incorporation of Na and SO_4 and associated point defects in the crystal structure may influence the stability of magnesian calcites in aqueous solution, too (Busenberg & Plummer, 1989); contradictory observations that still await a detailed clarification (Morse *et al.*, 2007). In addition, the evaluation of the dissolution kinetics of natural Mg calcites has been found to be complicated by the co-occurrence of different solid solutions or Mg-bearing phases (Mackenzie *et al.*, 1983). Dissolution kinetics of magnesian calcites has been carried out in a number of studies (Morse & Mackenzie, 1990). As recently outlined out by Morse *et al.* (2007), the quantitative evaluation is complicated for natural, especially biogenic, carbonates due to the complex surface microstructures, a multi-element solid-solution composition, and other factors.

Incongruent dissolution behaviour is caused by the attempt of a SS–AS system to reach more stable thermodynamic boundary conditions and is not only the basis for the recrystallization of HMC to LMC but also for the aragonite–calcite transformation (Lippmann, 1973; Morse & Mackenzie, 1990; Morse *et al.*, 2007). In the latter case, the dissolving carbonate has a crystal structure different from the precipitating solid solution. The formation of the secondary solids leads to a slowing of the dissolution process due to the blocking of reactive surfaces and is more effective when isostructural (*e.g.* calcite) substrates react (Böttcher, 1997a; Cubillas *et al.*, 2005; Köhler *et al.*, 2007). With respect to the partition coefficients, the slow transformation of aragonite to calcite-type solid solutions, which is essentially controlled by the dissolution kinetics of the precursor phase aragonite, leads to D_{Me} values that are consistent with results from seeded precipitation experiments at low precipitation rates (Katz *et al.*, 1972; Böttcher, 1997a).

4.2. Manganese(II)-calcium carbonate dissolution

Compared to the complex natural magnesian calcites, the Ca-Mn carbonate SS–AS system is less complicated since the solid solutions found in nature seem not to be of biogenic origin and they are easy to prepare in the laboratory (*e.g.* Böttcher, 1998; Mucci, 2004; Katsikopoulos *et al.*, 2009). Regarding the inability to synthesize the intermediate ordered double carbonate kutnahorite, however, the system seems to be similar to or even more complicated than the dolomite system. The dissolution behaviour of synthetic Ca-Mn carbonate solid solutions was investigated at between 20° and 25°C by Böttcher (1993, 1997b) and Mucci (2004). For example, it was found at 20°C that carbonates with X_{Mn} values of 0.14, 0.25 and 0.48 initially dissolved congruently (analogue to equation 19; Fig. 14). The results were extrapolated using inverse-time plots to obtain the stoichiometric solubility constants compiled in Figure 6. The replacement of Ca by Mn in the calcite structure leads to an enhanced stability of the solid solutions compared to pure calcium carbonate. The congruent dissolution rate of Mn calcites #1 and #4 (Fig. 14) was evaluated using equation 22 and yields reaction rate constants, k, between the data found for pure calcite and rhodochrosite (Fig. 15), and reaction orders of 2.7 ± 0.2 (#1) and 3.5 ± 0.7 (#4) (Böttcher, 1993).

After the initial phase, incongruent dissolution behaviour was observed for solids #1 and #2 (Figs 14 and 16). The overall changes in the concentrations of the individual dissolved metal ions upon the reaction

$$Ca_{(1-x)}Mn_xCO_3 + H_2CO_3^* \longrightarrow zCa_{(1-y)}Mn_yCO_3 + (1-x-z+z\cdot y)Ca^{2+} + (x-z\cdot y)Mn^{2+} + 2(1-z)HCO_3^- + zH_2CO_3^* \quad (24)$$

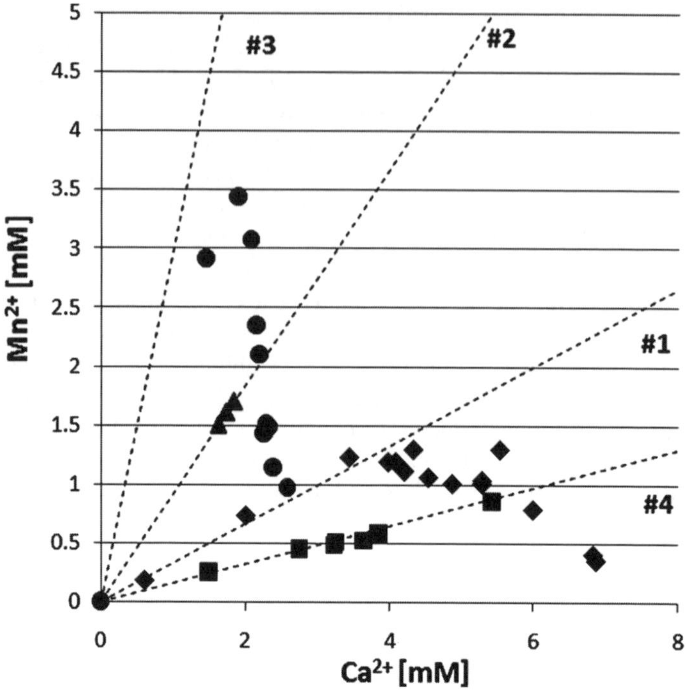

Fig. 14. Illustration of congruent and incongruent dissolution behaviour of Ca-Mn carbonate solid solutions as a function of reaction progress. Experimental data points are for synthetic solid solutions with X_{Mn} fractions of 0.25 (#1; ◆), 0.48 (#2; ▲), and 0.75 (#3; ●), *cf.* Fig. 16. Solid #4 (■) is a synthetic hydrothermal manganoan calcite ($X_{Mn} = 0.14$). Results are given for batch experiments open with respect to CO_2 gas (20°C, $PCO_2 = 0.98$ atm; Böttcher, 1993, 1997b), lasting up to 370 h. Only the experiment with the most concentrated solution carried out with #4 is from a batch experiment closed with respect to CO_2. Degassing of the final solution of the #1 experiment for 25 h with N_2 gas led to the precipitation of a secondary solid solution with a mean X_{Mn} of 0.47.

can be described by (Böttcher, 1993, modified after Kornicker *et al.*, 1991)

$$d[\mathrm{Mn}^{2+}] = \frac{\delta[\mathrm{Mn}^{2+}]}{\delta\mathrm{Ca}_{(1-x)}\mathrm{Mn}_x\mathrm{CO}_3} d\mathrm{Ca}_{(1-x)}\mathrm{Mn}_x\mathrm{CO}_3 + \frac{\delta[\mathrm{Mn}^{2+}]}{\delta\mathrm{Ca}_{(1-y)}\mathrm{Mn}_y\mathrm{CO}_3} d\mathrm{Ca}_{(1-y)}\mathrm{Mn}_y\mathrm{CO}_3 \tag{25}$$

and

$$d[\mathrm{Ca}^{2+}] = \frac{\delta[\mathrm{Ca}^{2+}]}{\delta\mathrm{Ca}_{(1-x)}\mathrm{Mn}_x\mathrm{CO}_3} d\mathrm{Ca}_{(1-x)}\mathrm{Mn}_x\mathrm{CO}_3 + \frac{\delta[\mathrm{Ca}^{2+}]}{\delta\mathrm{Ca}_{(1-y)}\mathrm{Mn}_y\mathrm{CO}_3} d\mathrm{Ca}_{(1-y)}\mathrm{Mn}_y\mathrm{CO}_3 \tag{26}$$

The changes as a function of time are quantitatively described by

$$\frac{d[\mathrm{Mn}^{2+}]}{dt} = \frac{kF_t X_{\mathrm{Mn}}}{V}(1 - \Omega_{ss}^m)^n - \frac{k_2' F_t}{V}(\Omega_\Sigma^m - 1)^u \frac{D_{\mathrm{Mn}}\{\mathrm{Mn}^{2+}\}}{\{\mathrm{Ca}^{2+}\}} \tag{27}$$

and

$$\frac{d[\mathrm{Ca}^{2+}]}{dt} = \frac{kF_t(1 - X_{Mn})}{V}(1 - \Omega_{ss}^m)^n - \frac{k_1' F_t}{V}(\Omega_\Sigma^m - 1)^u \frac{\{\mathrm{Ca}^{2+}\}}{D_{\mathrm{Mn}}\{\mathrm{Mn}^{2+}\}} \tag{28}$$

The composition of the aqueous solutions followed the trend predicted by the stoichiometric saturation curves towards the solutus curve, indicating the precipitation of a Mn-rich carbonate phase (Figs 14 and 16). The evaluation of the incongruent part of the dissolution of Mn calcite #1 gave a precipitation rate constant of $\log k' = -11.5$ mol cm^{-2} s^{-1} ($n = 2.7$; Böttcher, 1993), between the results for the pure end members and in general agreement with the influence of Mn on the precipitation kinetics in the SS–AS system (Dromgoole & Walter, 1990a,b).

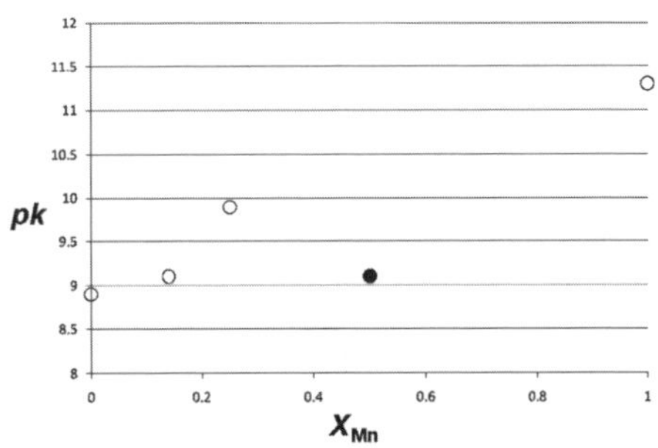

Fig. 15. Dependence of the dissolution constant *k* for congruent dissolution as a function of the solid-solution (open symbols) composition in the system calcite – kutnahorite (solid symbol) – rhodochrosite (data from Böttcher, 1993).

At low temperatures, complex dissolution-reprecipitation behaviour of ordered kutnahorite in pure CO_2-saturated water was suggested previously (Mucci, 2004). To look in more detail into this process, the dissolution behaviour of ordered natural kutnahorite ($Ca_{0.45}Mn_{0.52}Mg_{0.017}Zn_{0.017}Fe_{0.003}CO_3$; Sterling Hill mine, USA) was investigated with a focus on the behaviour of trace and minor metal ions in free-drift batch-type experiments, open and closed with respect to gaseous carbon dioxide, and flow-through reactors. The low-temperature dissolution in aqueous solutions was found to be generally congruent with respect to major, minor and trace elements (Fig. 17). Extrapolation of experimental results gives a solubility product at 20°C (pK_{ss}) of 10.6 ± 0.1 (Fig. 6), close to results reported by Mucci (2004) for ordered kutnahorite. Upon stoichiometric dissolution, none of the possible stoichiometric saturation curves nor the solutus curve for the solid solutions or pure end members (Fig. 16) was reached or passed. This suggests that kutnahorite should be stable compared to a solid solution of the same chemical composition (Fig. 6). The experimental data upon congruent dissolution followed the rate law depending on the relative undersaturation according to equation 20. Values of −8.6 and −9.1 were found for log *k* in experiments open and closed (20°C; initial $PCO_2 = 0.98$ atm), with respective empirical reaction orders, *n*,

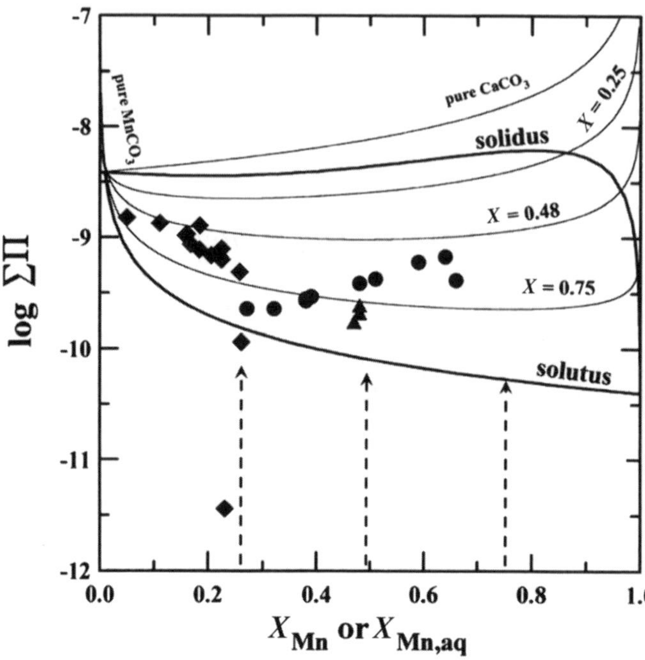

Fig. 16. Lippmann phase diagram for the binary $(Ca,Mn)CO_3$ system at 25°C and 1 atm total pressure (modified from Böttcher, 1997b). X_{Mn} and $X_{Mn,aq}$ are the manganese mole fraction and activity fraction of the solid solution and aqueous solution, respectively. The stability of ordered kutnahorite is not considered here. Experimental data points are calculated for the dissolution of synthetic solid solutions precipitated at room temperature with X_{Mn} fractions of 0.25 (#1; ◆), 0.48 (#2; ▲) and 0.75 (#3; ●) (see stoichiometries of the process in Fig. 15). Dashed arrows indicate theoretical initial congruent dissolution paths. Stoichiometric saturation curves are calculated for solid solutions and end members. $\Sigma\Pi$ is the 'total solubility product' according to $\Sigma\Pi = (\{Ca^{2+}\} + \{Mn^{2+}\})\ \{CO_3^{2-}\}$. The equations to calculate solidus, solutus, and stoichiometric saturation curves can be found in Glynn & Reardon (1990). The non-dimensional Guggenheim parameter considering a regular solid-solution system ($a_0 = 3.23$) was taken from Lippmann (1980). It should be noted that a sub-regular model probably has to be considered to describe the binary system (Böttcher, 1997b; Katsikopoulos *et al.*, 2009).

of 11 and 7. The dissolution rate of kutnahorite decreased far from equilibrium, as previously shown in experimental studies of ordered dolomite dissolution (Busenberg & Plummer, 1982). The rate constants for kutnahorite are in between those rates found for pure calcite and rhodochrosite, or selected calcite-rhodochrosite solid solutions under the same experimental conditions, but the reaction rates are much higher. Under the same conditions, pure rhodochrosite had a dissolution rate constant of $pk = 11.3 \pm 0.3$ with a reaction order of 2 ± 0.5 (Fig. 13; Böttcher, 1993).

4.3. Dissolution of barium sulphates

The dissolution behavior of Sr- and Pb-bearing barite solid solutions was investigated in a number of experimental studies (Denis & Michard, 1983; Galinier *et al.*, 1989;

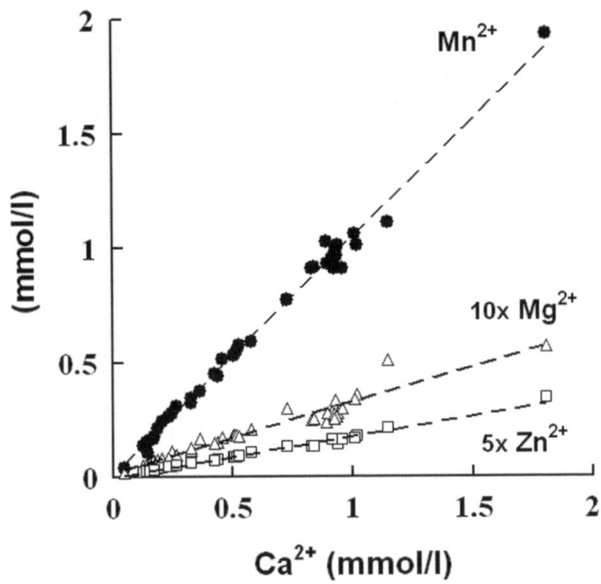

Fig. 17. Major, minor and trace-element behaviour upon congruent dissolution of ordered natural kutnahorite between 20 and 30°C (Böttcher, 1993). Congruency of the dissolution reaction was also reflected by Ba and Sr liberation that followed the bulk kutnahorite stoichiometry (Böttcher *et al.*, in prep.).

Kornicker *et al.*, 1991; Felmy *et al.*, 1993). The barite SS–AS systems arc of particular importance due to their unresolved occurrence in seawater (Monnin, 2006) but their occurrence as scaling minerals and their ability to trap relevant anthropogenic trace-metal ions such as Pb and Ra are also of interest (Zielinski & Budahn, 2007).

In experiments at 25°C, Kornicker *et al.* (1991) observed initially congruent dissolution behaviour of Ba-Pb sulphate solid solutions. This turned incongruent after ~30 min. The authors obtained the stoichiometric solubility constants of the solid solutions and applied a modified rate law (Denis & Michard, 1983) to evaluate the dissolution kinetics. Probably because of modifications of the reacting solid surfaces, the model based on a similar set of equations to equations 24–28 was not able to fit the concentration evolutions of both dissolved metal ions. At 60°C, dissolution by incongruent reaction had already started within 3 min. As a function of time the aqueous solution moved towards the two-component thermodynamic equilibrium as defined by the solutes curve (Lippmann, 1980) while maintaining undersaturation with respect to stoichiometric saturation (Kornicker *et al.*, 1991). This indicates that the precipitation rate of the secondary sulphate was greater than the dissolution rate of the initial solid, probably due to covering of the precursor by the newly formed solid solution. Interestingly, the aqueous solutions maintained states closer to the stoichiometric solubility curves at 60°C. Note however, that the quantitative position of the stoichiometric saturation curve may depend on the thermodynamic model used for the respective SS–AS system (Glynn & Reardon, 1990). Denis & Michard (1983), Galinier *et al.* (1989) and Felmy *et al.*

(1993) carried out dissolution experiments in the low-temperature barite–celestine SS–AS system. The reaction paths of the experimental results of the first study were re-evaluated by Glynn *et al.* (1990). Together with a discussion of dissolution experiments with strontianite–aragonite solid solutions, the authors demonstrated that stoichiometric solubility controlled the congruent part and at least the initial section of the incongruent part of the dissolution reaction. Similar to the Ba–Pb sulphate system the formation of secondary Ba-rich sulphates was observed leading to shifts in the reaction paths of the aqueous solutions towards the theoretical solutus curve. Thermodynamic equilibrium of the complete SS–AS system was not reached within 285 days. Felmy *et al.* (1993) observed reversible equilibrations (steady state) after 315 days, but complex behaviour with respect to the solutus and stoichiometric saturation curves was found. They observed the formation of a set of solid solutions that influenced the aqueous composition and argued for different solid phases to control the activities of dissolved species in experiments lasting up to 3 y. Clearly much longer times are required to reach a complete equilibration in a given SS–AS system *via* continuous dissolution recrystallization steps. The reaction times will depend on solid-to-liquid ratios, the particle sizes, the differences in the solubility products and other factors. Time is an advantage of nature and the experimental results mentioned argue for an inclusion of well chosen natural systems to study the equilibration pathways in low-temperature SS–AS systems.

The few examples clearly show the enormous complexity in the kinetic behaviour of carbonate and sulphate SS–AS systems at low temperatures. Further studies are required to resolve the complex behaviour upon congruent and incongruent dissolution of binary and multi-component solid solutions at low temperatures to understand the pathways and rates of natural processes in SS–AS systems.

4.4. Dissolution of carbonates by phosphoric acid

We also want to address the influence of metal-ion partitioning on a well studied congruent dissolution process of metal carbonates. Research related to the stoichiometric dissolution of carbonates by phosphoric acid has already been carried out since the 1950s as it is the standard analytical procedure for CO_2 production from metal carbonates for further precise stable carbon and oxygen isotope analysis *via* mass spectrometry (*e.g.* McCrea, 1950; Guo *et al.*, 2009). The unidirectional dissolution reaction proceeds with concentrated (nominal 105%) H_3PO_4 according to the overall equation

$$\begin{aligned} Ca_xMe_{(1-x)}CO_3 + H_3PO_4 \longrightarrow x\mathrm{Ca}^{2+} + (1-x)Me^{2+} + \mathrm{HPO}_4^- \\ + H_2O + CO_2\uparrow \end{aligned} \tag{29}$$

Due to the small pH value, the reaction proceeds generally congruently with respect to metal cations. With respect to oxygen isotopes, however, a fractionation is observed. The magnitude depends on temperature as well as on the structure and composition of the carbonate in question (Sharma & Sharma, 1969; Böttcher, 1996; Gilg *et al.*, 2003). It has recently been speculated that carbonic acid formation as an intermediate

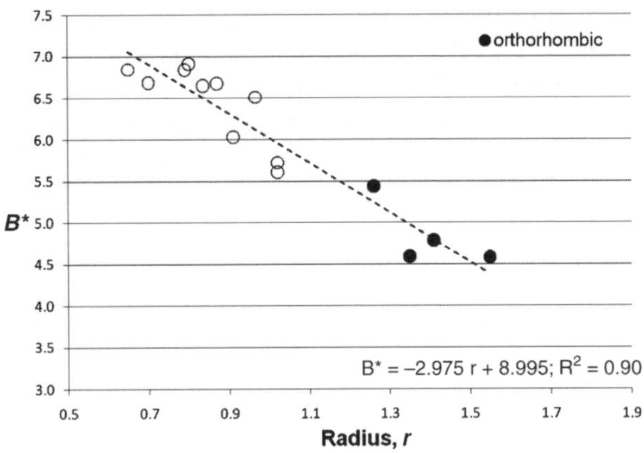

Fig. 18. Variation of the temperature-dependence coefficient B^* (equation 30) with the mean cation radii of pure and mixed carbonates. Radii are taken from Zhu (2004).

species will have significant influence (Guo *et al.*, 2009), but theoretical considerations have so far been unable to explain the experimental results (Guo *et al.*, 2009). Metal-ion partitioning has been shown to influence this unidirectional reaction significantly (Böttcher, 1996). In a number of studies it was found that the temperature dependence of reaction in equation 29 is described by

$$1000 \ln \alpha = A^* + B^* 10^5/T^2 \qquad (30)$$

where T is the temperature in Kelvin, and A^* and B^* are empirical regression coefficients relating the fractionation factor, α, between the solid mixed carbonate and gaseous CO_2 to temperature. Figure 18 shows, for a set of published experimental B^* values for rhom bohedral and orthorhombic carbonates, that there is a linear relationship to the radii of the metal ions. Figure 18 includes not only pure end-member solids, but also dolomite-structured double carbonates and a Mn-calcite solid solution (Böttcher, 1996). The same relation was observed for the A^* values for rhombohedral carbonates, but a reverse trend was observed for aragonite-structured solids (data not shown). These results clearly indicate that (partial) substitution by metal ions must be taken into account upon the evaluation of the kinetic oxygen isotope fractionation according to equation 28. This is of particular importance during the evaluation of the effects of metal ions on oxygen isotope fractionation between carbonate solid solution and water, as, *e.g.* in the case of Mg calcites (*e.g.* Tarutani *et al.*, 1969).

5. Stable metal-isotope partitioning

5.1. Carbonates

Stable isotope distribution of cations in carbonates and sulphates is a promising tool for deciphering kinetics and mechanisms of individual exchange reactions and the source of components for ion partitioning (*e.g.* Johnson *et al.*, 2004). Stable isotopes can be considered as a kind of trace or foreign component in the solid. For instance, Ca, Sr and Mg isotope distribution in related carbonate minerals are used to reconstruct the variability of sea water and terrestrial solution content during the geological history. From knowledge of temporal and spatial variations in global geochemical cycles, changes in weathering conditions and environmental conditions are anticipated.

Several studies exist for Ca isotopes fractionations during carbonate precipitation, wherein Ca isotope fractionation was explained by a kinetic diffusion model (Gussone *et al.*, 2003), a rate-controlled model (Lemarchand *et al.*, 2004), an equilibrium fractionation model (Marriott *et al.*, 2004), an adsorption-controlled steady-state calcite-growth model (Fantle & DePaolo, 2007) and a diffusion-controlled surface-entrapment model (Tang *et al.*, 2008b). Experimental results show that in analogy to trace-element partitioning, the apparent Ca isotope fractionation factor between calcite and aqueous calcium ions according to equation

$$\alpha_{\text{calcite-}aq} = (^{44}\text{Ca}/^{40}\text{Ca})_{\text{calcite}}/(^{44}\text{Ca}/^{40}\text{Ca})_{aq} \tag{31}$$

depends on both temperature and precipitation rate (*e.g.* Lemarchand, *et al.*, 2004; Tang *et al.*, 2008b). Different fractionation behaviour is obtained for individual experimental conditions, which may be caused by respective ion diffusion condition in the solution and in the solid surface layer, ionic strength and chemical content of the solution. However, Fantle & DePaolo (2007) found that calcite equilibrated with marine pore fluids for time periods of millions of years indicates an isotope equilibrium $\alpha_{\text{calcite-}aq}$ value of 1. Interestingly, a good correlation is obtained for the fractionation of Ca-isotopes and Sr incorporation during calcite precipitation, which does not depend on precipitation rate and temperature (Tang *et al.*, 2008b). Accordingly, both signatures may be used in a multi-proxy approach to evaluate Sr/Ca ratios or calcium isotopic composition of the primary precipitating solution.

The distribution of stable Mg isotopes in calcium carbonates shows that ^{26}Mg is discriminated *vs.* ^{24}Mg during carbonate formation (Young & Galy, 2004; Buhl *et al.*, 2007). In this aspect Mg isotopes behave similarly to Ca isotopes. In contrast to Ca isotopes, Mg isotope fractionation is assumed to be insensitive to low temperatures, which is concluded from the small amount of Mg-calcite formation of speleothems (Galy *et al.*, 2002). A significant variability of Mg isotope fractionation is attributed to differences in mineralogy concerning marine skeletal carbonates (Hippler *et al.*, 2009). However, further systematic studies are required to evaluate and decipher the parameters and mechanisms which control stable Mg isotope incorporation during carbonate formation. For coral aragonite the ^{88}Sr/^{86}Sr ratio may serve as a promising novel palaeo-temperature proxy as Rüggeberg *et al.* (2008) showed that ^{88}Sr/^{86}Sr ratios of cold-water corals display ambient seawater temperature.

In the case of stable Fe isotope fractionation between aqueous Fe(II) and siderite, Schauble *et al.* (2001) predicted a large equilibrium fractionation from spectroscopic data. However, a significant smaller isotopic fractionation is typically obtained from experiments, where siderite is precipitated from Fe(II) solutions (*e.g.* Wiesli *et al.*,

2004). For the complex behaviour of iron isotopes, $^{56}Fe/^{54}Fe$ fractionation between aqueous ferrous and ferric species and Fe isotope fractionation upon sorption of Fe(II) onto mineral surfaces also have to be considered (Welch *et al.*, 2003; Mikutta *et al.*, 2009).

5.2. Sulphates

For sulphate minerals, Griffith *et al.* (2008) studied calcium isotopic fractionation during precipitation of barite, where isomorphic substitution of Ba by Ca ions in the barite crystal structure appears (see section 3.7). Experimental results indicate that ^{44}Ca is discriminated somewhat more strongly *vs.* ^{40}Ca during barite formation compared to calcite formation ($\Delta^{44/40}Ca_{Ca(barite)\text{-}aq} = -3.4$ to -2.4‰ *vs.* $\Delta^{44/40}Ca_{calcite\text{-}aq}$ down to -2‰ (Gussone *et al.*, 2005; Tang *et al.*, 2008b)). Barite is a highly stable mineral and the related Ca-isotope signatures may be used as an archive for palaeo-environment reconstruction, *e.g.* in marine environments. Accordingly, Ca isotope fractionation into sedimentary barite has been used to estimate changes in marine palaeoproductivity. In analogy to the results from calcium carbonate formation, kinetic fractionation processes probably control the apparent calcium isotopic fractionation for barite. The most sensitive parameters are temperature, crystal growth rates, and the chemical composition of the solution.

As a new isotope tool, the distribution of stable Ba isotopes in Ba-bearing sulphate and carbonate minerals has recently been introduced (von Allmen *et al.*, 2010), and this is under current, further investigation.

6. Biomineralization

As mentioned above, biomineral phases often have shape, crystallinity, trace and isotopic signatures different from their inorganic counterparts. The variability can be caused by interaction with organic macromolecules, internal pools of components, and an amorphous precursor phase. Deciphering individual physiological effects from environmental signals is a challenge for further scientific research (*e.g.* Weiner & Dove, 2003).

The effect of element partitioning in carbonates on internal pools of dissolved cations was mentioned by Elderfield *et al.* (1996) for Sr and Ba incorporation into foraminifers. Calcium carbonate formation may occur at a site which is isolated from the bulk solution but supplied by extraneous ions for mineral precipitation. In the case of foraminifera, an internal pool provides aqueous carbonate as well as Ca, Ba and Sr for calcification, which is not necessarily identical to sea water. The main reasons for the variability of internal pool composition are divalent ion fractionation due to crossing the cell membrane fraction through channels and potential Rayleigh distillation behaviour for solutions trapped within the organisms during calcification (see equation 3). In this case, the environmental conditions of solid-phase formation are referred to the internal pool condition rather than to the external bulk solution like sea water.

The formation of precursor phases such as thermodynamically metastable amorphous calcium carbonate (ACC), observed, *e.g.* by Griesshaber *et al.* (2009) for brachiopods, may affect element partitioning in organic induced carbonates. Precipitation experiments by Wang *et al.* (2009) showed that the Mg content in ACC can be regulated by macromolecules. In that study, simple organic molecules push Mg/Ca levels of the precipitated ACC into the ranges of high-magnesium carbonates and even dolomite. These high Mg/Ca levels in the precipitated carbonate were obtained in solutions with Mg/Ca ratios between 2 and 5 (modern sea water) without considering an extreme Mg enrichment in the past ocean. It is suggested that extreme Mg-rich carbonates in biogenic and sedimentary environments may be formed *via* a transient phase such as ACC, where carboxylated molecules control the Mg content of the ACC during calcification. Subsequent transformation of amorphous calcium carbonate into crystallized calcium carbonates can be regulated by proteins and inorganic ions (Xu *et al.*, 2008). However, as an amorphous calcium carbonate phase is doped with foreign ions such as Mg^{2+}, these compounds may function as growth and crystal-structure modifiers of $CaCO_3$ (Lam *et al.*, 2007; Han & Aizenberg, 2008). On another point of view, such foreign components are incorporated into the crystalline $CaCO_3$ during the transformation of amorphous calcium carbonate. Thus, incorporation of trace elements into the final calcium carbonate skeleton is related to the incorporation behaviour of amorphous calcium carbonate rather than to that of the final (Mg) calcite or aragonite structure.

7. Outlook

The discussion here is focused on binary carbonate and sulphate SS–AS systems at low temperatures. In previous decades, extensive studies have been carried out for a number of important systems. The intense discussion on low-temperature SS–AS systems was stimulated by the pioneering works of Friedrich Lippmann, Niel Plummer and co-workers more than three decades ago (Thorstenson & Plummer, 1977; Lippmann, 1977, 1980; Garrels & Wollast, 1978; Lafon, 1978), focusing on the SS–AS equilibrium characterization, and continued to attract scientists' interest through the 1980s and 90s (*e.g.* Königsberger & Gamsjäger, 1987; Glynn & Reardon, 1990; Glynn *et al.*, 1992; Gamsjäger *et al.*, 2000). Although individual mechanisms are mostly not fully understood in all details, large steps in understanding have been undertaken in the past 30 years. It should be noted, however, that natural carbonate and sulphate minerals are often complex mixtures of more than two end members (*e.g.* Lippmann, 1973; Morse & Mackenzie, 1990; Church, 1979; Romanek *et al.*, 2009), and it has been found that single trace-element partition coefficients may be influenced by the presence and co-incorporation of further foreign ions (*e.g.* Pingitore, 1986). With the present lack of experimental data this remains a mostly open question for future experimental, field and theoretical work. Another mechanistic question which remains to be answered is the impact of microbial processes on mineral (trans)formation and its role in the development of geochemical and isotope signatures. Current discussion on the impact of climate change on processes in the global low-temperature biogeochemical element cycles and the likely need for human intervention (*e.g.* carbon dioxide sequestration) requires a deeper understanding of factors influencing quantitatively the interactions between solid solutions and aqueous solutions. The mechanistic base for metal-ion partitioning between solids and aqueous fluids must be understood quantitatively before predictions can be made with confidence. With the recent development of new concepts in aqueous solution–solid surface boundary processes (*e.g.* Wolthers *et al.*, 2008) and the need for a proper quantitative evaluation of low-temperature biogeochemical element cycles involving the carbonate and sulphate systems, SS–AS will continue to be the focus of low-temperature biogeochemical research. Metal-ion partitioning upon solid-solution (trans)formations influences the kinetics of dissolution, precipitation and reaction paths in open and closed systems. They are nowadays included in reaction-transport modelling approaches to understand both the thermodynamics of SS–AS systems (Glynn & Parkhurst, 1992; Kulik *et al.*, 2000; Kulik, 2006; Prieto, 2009) but in particular also the inherent kinetic aspects quantitatively (Lichtner & Carey, 2006; Noguera *et al.*, 2009). A quantitative understanding of reaction kinetics, however, is not possible without a proper thermodynamics characterization of the investigated system. Further developments are needed and expected in the future. This will include experimental and modelling approaches and their application to field situations. The combined evaluation of different metal ions and/or metal isotope systems ('multi-proxy approach') promises, in particular, to open exciting new routes for the interpretation of palaeosystems and separating abiotic from direct or indirect biological factors in SS–AS systems.

Acknowledgements

The authors thank Manuel Prieto and Heather Stoll for the invitation to contribute to the EMU 2010 summer school, and their continuous encouragement and help during the preparation of the manuscript. The research on ODP Leg 207 barites used samples and data provided by the Ocean Drilling Program (ODP). ODP is sponsored by the U.S. National Science Foundation (NSF) and participating countries under the management of Joint Oceanographic Institutions (JOI), Inc. This research was funded by Deutsche Forschungsgemeinschaft (grants BO 1584/2 and BR 775/17), Max Planck Society, and Leibniz Institute for Baltic Sea Research. Thanks are due to Hans Brumsack for providing the photograph of the Leg 207 barite. Experimental work on Mn-Ca carbonate solid solution was supported by Deutsche Forschungsgemeinschaft during the priority program 'Intraformational Ore Deposits'. The part on carbonate solid-solution reactivity was supported by the German Federal Ministry of Education and Research (BMBF, FKZ 03F0608 O) during the Verbundprojekt "BIOACID". Experimental studies on isotope fractionation and trace element incorporation during carbonate crystallization were supported by the European Science Foundation (ESF-04-ECLIM-FP08) and the Austrian Science Fund (FWF-I34-B06). Field studies focused on respective carbonate formation were funded by FFG-Austrian Science Fund within the Competence Centre WaterPool (WP-431) and Leibniz IOW. The authors are further indebted to Hans-Eberhard Usdowski for initially stimulating their interest for different aspects of the carbonate system. The authors wish to thank a number of colleagues for inspiring discussions within the past twenty years on low-temperature carbonates, isotope and element partitioning, and SS–AS systems, especially Tom Church, Wolfgang Dreybrodt, Anton Eisenhauer, Jonathan Erez, Pierre Glynn, Wolfgang Krumbein, Friedrich Lippmann, Jack Middelburg, John Morse, Niel Plummer, and Manuel Prieto.

This publication is dedicated to John W. Morse, who recently passed away, for his long-term encouragement and inspiration.

References

Aller, R.C. & Rude, P.D. (1988) Complete oxidation of solid phase sulfides by manganese and bacteria in anoxic sediments. *Geochimica Cosmochimica Acta*, **52**, 751–765.

Arndt, S., Brumsack, H.J. & Wirtz, K. (2006) Cretaceous black shales as active bioreactors: A biogeochemical model for the deep biosphere encountered during ODP Leg 207 (Demerara Rise). *Geochimica et Cosmochimica Acta*, **70**, 408–425.

Astilleros, J.M., Pina, C.M., Fernandez-Diaz, L. & Putnis A. (2002) Molecular-scale surface processes during the growth of calcite in the presence of manganese. *Geochimica et Cosmochimica Acta*, **66**, 3177–3189.

Averyt, K.B. & Paytan, A. (2003) Empirical partition coefficients for Sr and Ca in marine barite: Implications for reconstructing seawater Sr and Ca concentrations. *Geochemistry Geophysics Geosystems*, **4**, 14.

Averyt, K.B. & Paytan, A. (2007) Reply to comment by D. Wang *et al.* on "Empirical partition coefficients for Sr and Ca in marine barite: Implications for reconstructing seawater Sr and Ca concentrations". *Geochemistry Geophysics Geosystems*, **8**, 3.

Banner, J.L. & Hanson, G.N. (1990) Calculation of simultaneous isotopic and trace element variations during water-rock interaction with application to carbonate diagenesis. *Geochimica et Cosmochimica Acta*, **54**, 3123–3137.

Barone, J.P., Nancollas, G.H. & Yoshikawa, Y. (1983) Crystal growth as a function of the surface area. *Journal of Crystal Growth*, **63**, 91–96.

Beck, J.W., Edwards, R.L., Ito, E., Taylor, F.W., Recy, J., Rougerie, F., Joannot, P. & Henin, C. (1992) Sea-surface temperature from coral skeletal strontium/calcium ratio. *Science*, **257**, 644–647.

Bertram, M.A., Mackenzie, F.T., Bishop, F.C. & Bischoff, W.D. (1991) Influence of temperature on the stability of magnesian calcite. *American Mineralogist*, **76**, 1889–1896.

Bisset, A., de Beer, D., Schoon, R., Shiraishi, F., Reimer, A., Reitner, J. & Arp, G. (2008) Microbial mediation of stromatolite formation in karst-water creeks. *Limnology & Oceanography*, **53**, 1159–1168.

Böttcher, M.E. (1993) *The experimental investigation of ore-deposit relevant metal enrichment reactions from aqueous solutions with special regards to the formation of rhodochrosite* ($MnCO_3$) (in German). PhD thesis, University of Göttingen, Germany, 237 pp.

Böttcher, M.E. (1995) The formation of rhodochrosite-smithsonite ($MnCO_3$-$ZnCO_3$) solid-solutions at 5°C. *Mineralogical Magazine*, **59**, 481–488.

Böttcher, M.E. (1996) $^{18}O/^{16}O$ and $^{13}C/^{12}C$ fractionation during the reaction of carbonates with phosphoric acid: Effects of cationic substitution and reaction temperature. *Isotopes in Environmental and Health Studies*, **32**, 299–305.

Böttcher, M.E. (1997a) The transformation of aragonite to $Mn_XCa_{(1-X)}CO_3$ solid-solutions at 20°C: An experimental study. *Marine Chemistry*, **57**, 97–106.

Böttcher, M.E. (1997b) Experimental dissolution of $CaCO_3$–$MnCO_3$ solid-solutions in CO_2–H_2O solutions at 20°C. I. Synthetic low-temperature carbonates. *Solid State Ionics*, **101/103**, 1263–1266.

Böttcher, M.E. (1997c) Comment on "Solid solution partitioning of Sr^{2+} and Ba^{2+}, and Cd^{2+} to calcite" by A.J. Tesoriero and J.F. Pankow. *Geochimica et Cosmochimica Acta*, **61**, 661–662.

Böttcher, M.E. (1998) Manganese(II) partitioning during experimental precipitation of rhodochrosite-calcite solid-solutions from aqueous solutions. *Marine Chemistry*, **62**, 287–297.

Böttcher, M.E. (2010) Manganese (sedimentary carbonates and sulfides). In: *Encyclopaedia of Geobiology* (J. Reitner & V. Thiel, editors). Springer Science, Berlin (in press).

Böttcher, M.E. & Parafiniuk, J. (1998) Methane-derived carbonates in a native sulfur deposit: Stable isotope and trace element discriminations related to the transformation of aragonite to calcite. *Isotopes in Environmental and Health Studies*, **34**, 177–190.

Böttcher, M.E., Gehlken, P.L. & Steele, D.F. (1997) Characterization of inorganic and biogenic magnesian calcites by Fourier Transform infrared spectroscopy. *Solid State Ionics*, **101–103**, 1379–1385.

Boudreau, B.P. (1997) *Diagenetic Models and their Implementation - Modelling Transport and Reactions in Aquatic Sediments*. Springer, Heidelberg, Germany, 414 pp.

Buhl, D., Immenhauser, A., Smeulders, G., Kabiri, L. & Richter, D.K. (2007) Time series δ^{26}Mg analysis in speleothem calcite: Kinetic versus equilibrium fractionation, comparison with other proxies and implications for palaeoclimate research. *Chemical Geology*, **244**, 715–729.

Burgess, J. (1978) *Metal Ions in Solution*. Ellis Horwood, Chichester, UK & Wiley, New York, 481 pp.

Burton, E.A. & Walter, L.M. (1991) The effects of P_{CO_2} and temperature on magnesium incorporation in calcite in seawater and $MgCl_2$-$CaCl_2$ solutions. *Geochimica et Cosmochimica Acta*, **55**, 777–785.

Busenberg, E. & Plummer, L.N. (1982) The kinetics of dissolution of dolomite in CO_2–H_2O solutions at 1.5 to 65°C and 0 to 1 atm P_{CO_2}. *American Journal of Science*, **282**, 45–78.

Busenberg, E. & Plummer, L.N. (1989) Thermodynamics of magnesian calcite solid-solutions at 25°C and 1 atm total pressure. *Geochimica et Cosmochimica Acta*, **53**, 1189–1208.

Chou, L., Garrels, R.M. & Wollast, R. (1989) Comparative study of the kinetics and mechanisms of dissolution of carbonate minerals. *Chemical Geology*, **78**, 269–282.

Church, T.M. (1979) Marine barite. In: *Marine Minerals* (R.G. Burns, editor). Reviews in Mineralogy, **6**, 175–209. Mineralogical Society of America, Washington, D.C.

Cubillas, P., Köhler, S., Prieto, M., Causserand, C. & Oelkers, E.H. (2005) How do mineral coatings affect dissolution rates? An experimental study of coupled $CaCO_3$ dissolution-$CdCO_3$ precipitation. *Geochimica et Cosmochimica Acta*, **69**, 5459–5476.

Curtis, C.D., Coleman, M.L. & Love, L.G. (1986) Pore water evolution during sediment burial from isotopic and mineral chemistry of calcite, dolomite and siderite concretions. *Geochimica et Cosmochimica Acta*, **50**, 2321–2334.

Davis, J.A. & Kent, D.B. (1990) Surface complexation modeling in aqueous geochemistry. In: *Mineral-water Interface Geochemistry* (M.F. Hochella & A.F. White, editors). Reviews in Mineralogy & Geochemistry, **23**, 177–260. Mineralogical Society of America, Washington, D.C.

Davis, K.J., Dove, P.M. & De Yoreo, J.J. (2000) The role of Mg^{2+} as an impurity in calcite growth. *Science*, **290**, 1134–1137.

Davis, K.J., Dove, P.M., Wasylenki, L.E. & De Yoreo, J.J. (2004) Morphological consequences of differential Mg^{2+} incorporation at structurally distinct steps on calcite. *American Mineralogist*, **89**, 714–720.

De Villiers, S., Shen, G.T. & Nelson, B.K. (1994) The Sr/Ca-temperature relationship in coralline aragonite: Influence of variability in $(Sr/Ca)_{seawater}$ and skeletal growth parameters. *Geochimica et Cosmochimica Acta*, **58**, 197–208.

Denis, J. & Michard, G. (1983) Dissolution of a solid solution: A theoretical and experimental study. *Bulletin de Mineralogie*, **106**, 309–319.

Devery, D.M. & Ehlmann, A.J. (1981) Morphological changes in a series of synthetic Mg-calcites. *American Mineralogist*, **66**, 592–595.

Dietzel, M. (2010) Carbonates. In: *Encyclopaedia of Geobiology* (J. Reitner & V. Thiel, editors). Springer Science, Berlin (in press).

Dietzel, M., Gussone, N. & Eisenhauer, A. (2004) Co-precipitation of Sr^{2+} and Ba^{2+} with aragonite by membrane diffusion of CO_2 between 10 and 50°C. *Chemical Geology*, **203**, 139–151.

Dietzel, M., Rinder, T., Leis, A., Reichl, P., Sellner, P., Draschitz, C., Plank, G., Klammer, D. & Schöfer, H. (2008) Koralm tunnel as a case study for sinter formation in drainage systems – precipitation mechanisms and retaliatory action. *Geomechanik und Tunnelbau*, **1**, 271–278.

Dissard, D., Nehrke, G., Reichart, G.J. & Bijma, J. (2010) The impact of salinity on the Mg/Ca and Sr/Ca ratio in the benthic foraminifera Ammonia tepida: Results from culture experiments. *Geochimica et Cosmochimica Acta*, **74**, 928–940.

Dove, P.M. & Hochella, M.F. (1993) Calcite precipitation mechanism and inhibition by orthophosphate: In situ observations by Scanning force microscopy. *Geochimica et Cosmochimica Acta*, **57**, 705–714.

Dove, P.M., DeYoreo, J.J. & Weiner, S. (editors) (2003) *Biomineralization.* Reviews in Mineralogy and Geochemistry, **54**. Mineralogical Society of America, Washington, D.C., 381 pp.

Dromgoole, E.L. & Walter, L.M. (1990a) Iron and manganese incorporation into calcite: Effects of growth kinetics, temperature and solution chemistry. *Chemical Geology*, **81**, 311–336.

Dromgoole, E.L. & Walter, L.M. (1990b) Inhibition of calcite growth rates by Mn^{2+} in $CaCl_2$ solutions at 10, 25, and 50°C. *Geochimica et Cosmochimica Acta*, **54**, 2991–3000.

Elderfield, H., Bertram, C.J. & Erez, J. (1996) A biomineralization model for the incorporation of trace elements into foraminiferal calcium carbonate. *Earth and Planetary Science Letters*, **142**, 409–423.

Elderfield, H. & Ganssen, G. (2000) Past temperature and $\delta^{18}O$ of surface ocean waters inferred from foraminiferal Mg/Ca ratios. *Nature*, **405**, 442–445.

Elderfield, H., Yu, J., Anand, P., Kiefer, T. & Nyland, B. (2006) Calibrations for benthic foraminiferal Mg/Ca paleothermometry and the carbonate ion hypothesis. *Earth and Planetary Science Letters*, **250**, 633–649.

Fairchild, I.J. & Treble, P.C. (2009) Trace elements in speleothems as recorders of environmental change. *Quaternary Science Reviews*, **28**, 449–468.

Fantle, M.S. & DePaolo, D.J. (2007) Ca isotopes in carbonate sediment and pore fluid from ODP Site 807A: The $Ca^{2+}_{(aq)}$-calcite equilibrium fractionation factor and calcite recrystallization rates in Pleistocene sediments. *Geochimica et Cosmochimica Acta*, **71**, 2524–2546.

Felmy, A.R., Rai, D. & Moore, D.A. (1993) The solubility of $(Ba,Sr)SO_4$ precipitates: Thermodynamic equilibrium and reaction path analysis. *Geochimica et Cosmochimica Acta*, **57,** 4345–4363.

Franke, W. & Hofer, A. (1986) The range of metastable formation and the morphology of the crystals in the system $CaCO_3$-$MnCO_3$. *Crystal Research & Technology*, **21**, 1333–1337.

Franklin, M.L. & Morse, J.W. (1983) The interaction of manganese(II) with the surface of calcite in dilute solutions and seawater. *Marine Chemistry*, **12**, 241–254.

Freij, S. J., Putnis, A. & Astilleros, J.M. (2004) Nanoscale observations of the effect of cobalt on calcite growth and dissolution. *Journal of Crystal Growth*, **267**, 288–300.

Friedl, G., Wehrli, B. & Manceau, A. (1993) Solid phases in the cycling of manganese in eutrophic lakes: New insights from EXAFS spectroscopy. *Geochimica et Cosmochimica Acta*, **61**, 275–290.

Gaetani, G.A. & Cohen, A.L. (2006) Element partitioning during precipitation of aragonite from seawater: A framework for understanding paleoproxies. *Geochimica et Cosmochimica Acta*, **70**, 4617–4634.

Gagan, M.K., Ayliffe, L.K., Hoplry, D., Cali, J.A., Mortimer, G.E., Chappell, J., McCulloch, M.T. & Head, M.J. (1998) Temperature and surface ocean water balance of the mid-Holocene tropical western Pacific. *Science*, **279**, 1014–1018.

Galinier, C., Dandurand, J.L., Souissi, F. & Schott, J. (1989) On the non-ideal character of the $(Ba,Sr)SO_4$ solid solutions: Evidence from the determination of thermodynamic parameters by dissolution approaches at 25°C. *Comptes Rendus de l'Academie des Sciences Séries 2, Mécanique, Physique, Chimie, Sciences de l'Universe, Sciences de la Terre*, **308**, 1363–1368.

Galy, A., Bar-Matthews, M., Halicz, L. & O'Nions, R.K. (2002) Mg isotopic composition of carbonate: Insight from speleothem formation. *Earth and Planetary Science Letters*, **201**, 105–115.

Gamsjäger, H., Königsberger, E. & Preis, W. (2000) Lippmann diagrams: Theory and application to carbonate systems. In: *The geochemistry of carbonates and solid-solution aqueous-solution systems at low temperatures: A special issue honoring Friedrich Lippmann* (M.E. Böttcher, editor). Aquatic Geochemistry, **6**, 119–132.

Ganeshram, R.S., François, R., Commeau, J. & Brown-Leger, S.L. (2003) An experimental investigation of barite formation in seawater. *Geochimica et Cosmochimica Acta*, **67**, 2599–2605.

Garrels, R.M. & Wollast, R. (1978) Discussion of: "Equilibrium criteria for two-component solids reacting with fixed composition in an aqueous phase. Example: The magnesian calcites". *American Journal of Science*, **278**, 1469–1474.

Gilg, G.A., Struck, U., Vennemann, T. & Bon, M. (2003) Phosphoric acid fractionation factors for smithsonite and cerussite between 25 and 72°C. *Geochimica et Cosmochimica Acta*, **67**, 4049–4055.

Gingele, F.X. & Dahmke, A. (1994) Discrete barite particles as tracers of laleoproductivity in South Atlantic sediments. *Paleoceanography*, **91**, 151–168.

Gingele, F.X. & Kasten, S. (1994) Solid-phase manganese in Southeast Atlantic sediments: Implications for the paleoenvironment. *Marine Geology*, **121**, 317–332.

Glynn, P.D. (1992) Effect of impurities in gypsum on contaminant transport at Pinal Creek, Arizona. *US Geological Survey Water-Resources Investigations Report*, **91–4034**, 466–474.

Glynn, P.D. (2000) Solid-solution solubilities and thermodynamics: Sulfates, carbonates and halides. In: *Crystallography, Geochemistry and Environmental Significance* (C.N. Alpers, J.L. Jambor & D.K. Nordstrom, editors). Reviews in Mineralogy and Geochemistry, **40**, 481–511. Mineralogical Society of America, Washington, D.C.

Glynn, P.D. & Parkhurst, D.L. (1992) Modeling non-ideal solid-solution aqueous-solulion reactions in mass transfer computer codes. In: *Water-Rock Interaction* (Y.K. Kharaka & A.S. Maest, editors), **7**, 175–179. Balkema Press. Rotterdam.

Glynn, P.D. & Reardon, E.J. (1990) Solid-solution aqueous-solution equilibria: Thermodynamic theory and representation. *American Journal of Science*, **290**, 164–201.

Glynn, P.D., Reardon, E.J., Plummer, L.N. & Busenberg, E. (1990) Reaction paths and equilibrium end-points in solid solution aqueous solution systems. *Geochimica et Cosmochimica Acta*, **54**, 267–282.

Goldsmith, J.R. (1983) Phase relations of rhombohedral carbonates. In: *Carbonates: Mineralogy and Chemistry* (R.J. Reeder, editor). Reviews in Mineralogy, **11**, 49–76. Mineralogical Society of America, Washington, D.C.

Gonzales-Munoz, M.T., Fernández-Luque, B., Martínez-Ruiz, F., Chekroun, K.B., Arias, J.M., Rodríguez-Gallego, M., Martínez-Canamero, M., de Linares, C. & Paytan, A. (2003) Precipitation of barite by *Myxococcus Xanthus*: Possible implications for the biogeochemical cycle of barium. *Applied Environmental Microbiology*, **60**, 5722–5725.

Griesshaber, E., Kelm, K., Sehrbrock, A., Mader, W., Mutterlose, J., Brand, U. & Schmahl, W.W. (2009) Amorphous calcium carbonate in the shell material of the brachiopod Megerlia truncata. *European Journal of Mineralogy*, **21**, 715–723.

Griffith, E.M., Schauble, E.A., Bullen, T.D. & Paytan, A. (2008) Characterization of calcium isotopes in natural and synthetic barite. *Geochimica et Cosmochimica Acta*, **72**, 5641–5658.

Guo, W., Mosenfelder, J., Goddard III, W.A. & Eiler, J.M. (2009) Isotopic fractionations associated with phosphoric acid digestion of carbonate minerals: Insights from first-principles theoretical modeling and clumped isotope measurements. *Geochimica et Cosmochimica Acta*, **73**, 7203–7225.

Gussone, N., Eisenhauer, A., Heuser, A., Dietzel, M., Bock, B., Böhm, F., Spero, H.J., Lea, D.W., Bijma, J. & Nägler, T.F. (2003) Model for kinetic effects on calcium isotope fractionation (δ^{44}Ca) in inorganic aragonite and cultured planktonic foraminifera. *Geochimica et Cosmochimica Acta*, **67**, 1375–1382.

Gussone, N., Böhm, F., Eisenhauer, A., Dietzel, M., Heuser, A., Teichert, B.M., Reitner, J., Wörheide, G. & Dullo, W.C. (2005) Calcium isotope fractionation in calcite and aragonite. *Geochimica et Cosmochimica Acta*, **69**, 4485–4494.

Han, T.Y.J. & Aizenberg, J. (2008) Calcium carbonate storage in amorphous form and its template-induced crystallization. *Chemistry of Materials*, **20**, 1064–1068.

Hanor, J.S. (2000) Barite–celestine geochemistry and environments of formation. In: *Sulfate Minerals: Crystallography, Geochemistry and Environmental Significance* (C.N. Alpers, J.L. Jambor & D.K. Nordstrom, editors). Reviews in Mineralogy and Geochemistry, **40**, 193–275. Mineralogical Society of America, Washington, D.C. and the Geochemical Society, St. Louis, Missouri, USA.

Hanor, J.S. (2004) A model for the origin of large carbonate- and evaporate-hosted celestine ($SrSO_4$) deposits. *Journal of Sedimentary Research*, **74**, 168–175.

Hennessy, A.J.B. & Graham, G.M. (2002) The effect of additives on the co-crystallisation of calcium with barium sulphate. *Journal of Crystal Growth*, **237–239**, 2153–2159.

Hippler, D., Buhl, D., Witbaard, R., Richter, D.K. & Immenhauser, A. (2009) Towards a better understanding of magnesium-isotope ratios from marine skeletal carbonates. *Geochimica et Cosmochimica Acta*, **73**, 6134–6146.

Holland, H.D., Borcsik, M., Munoz, J. & Oxburgh, U.M. (1963) The co-precipitation of Sr^{2+} with aragonite and of Ca^{2+} with strontianite between 90 and 100°C. *Geochimica et Cosmochimica Acta*, **27**, 957–977.

Holland, H.D., Holland, H.J. & Munoz, J.L. (1964) The coprecipitation of cations with $CaCO_3$ – II. The coprecipitation of Sr^{+2} with calcite between 90° and 100°C. *Geochimica et Cosmochimica Acta*, **28**, 1287–1301.

Huang, Y. & Fairchild, I.J. (2001) Partitioning of Sr^{2+} and Mg^{2+} into calcite under karst-analogue experimental conditions. *Geochimica et Cosmochimica Acta*, **65**, 47–62.

Jakobsen, R. & Postma, D. (1989) Formation and solid solution behaviour of Ca-rhodochrosites in marine muds of the Baltic deeps. *Geochimica et Cosmochimica Acta*, **53**, 2639–2648.

Johnson, C.J., Brian, L.B. & Albarede, F. (editors) (2004) *Geochemistry of Non-traditional Stable Isotopes*. Reviews in Mineralogy and Geochemistry, **55**. Mineralogical Society of America, Washington, D.C. and the Geochemical Society, St. Louis, Missouri, USA, 454 pp.

Johnson, J.W., Oelkers, E. & Helgeson, H.C. (1992) SUPCRT92: A software package for calculating the standard molal thermodynamic properties of minerals, gases, aqueous species, and reactions from 1 to 5000 bars and 0°C to 1000°C. *Computers and Geosiences*, **18**, 899–947.

Joschi, M.S. (1960) *Precipitation of Carbonates from Sea Water*. Progress Report of the U.S. AEC Contract No. AT(30-1)-2266.

Katsikopoulos, D., Fernández-González, Á., Prieto, A.C. & Prieto, M. (2008) Co-crystallization of Co(II) with calcite: Implications for the mobility of cobalt in aqueous environments. *Chemical Geology*, **254**, 87–100.

Katsikopoulos, D., Fernández-González, Á. & Prieto M. (2009) Precipitation and mixing properties of the 'disordered' (Mn, Ca)CO_3 solid solution. *Geochimica et Cosmochimica Acta*, **73**, 6147–6161.

Katz, A., Sass, E., Starinsky, A. & Holland, H.D. (1972) Strontium behavior in the aragonite-calcite transformation: An experimental study at 40–98°C. *Geochimica et Cosmochimica Acta*, **36**, 481–496.

Kinsman, D.J.J. & Holland, H.D. (1969) The coprecipitation of cations with $CaCO_3$ – IV. The coprecipitation of Sr^{2+} with aragonite between 16° and 96°. *Geochimica et Cosmochimica Acta*, **33**, 1–17.

Kitano, Y., Kanamori, N. & Oomori, T. (1971) Measurements of distribution coefficients of strontium and barium carbonates between carbonate precipitate and solution – Abnormally high values of distribution coefficients measured at early stages of carbonate formation. *Geochemical Journal*, **4**, 183–206.

Köhler, S.J., Cubillas, P., Rodríguez-Blanco, J.D., Bauer, C. & Prieto, M. (2007) Removal of cadmium from wastewaters by aragonite shells and the influence of other divalent cations. *Environmental Science and Technology*, **41**, 112–118.

Königsberger, E. & Gamsjäger, H. (1987) Solid-solute equilibria in aqueous solution. 1. Solubility constant and free enthalpy of formation of huntite. *Berichte Bunsengesellschaft für Physikalische Chemie*, **91**, 785–790.

Kornicker, W.A., Presta, P.A., Paige, C., Johnson, D., Hileman, O. & Snodgrass, W. (1991) The aqueous dissolution kinetics of the barium/lead sulfate solid solution series at 25°C and 60°C. *Geochimica et Cosmochimica Acta*, **55**, 3531–3541.

Kulik, D.A. (2006) Dual-thermodynamic estimation of stoichiometry and stability of solid solution end members in aqueous-solid solution systems. *Chemical Geology*, **225**, 189–212.

Kulik, D.A., Kersten, M., Heiser, U. & Neumann, T. (2000) Application of Gibbs energy minimization to model early diagenetic solid-solution aqueous-solution equilibria involving authigenic rhodochrosites in anoxic Baltic sea sediments. *Aquatic Geochemistry*, **6**, 147–149.

Kushnir, J. (1980) The coprecipitation of strontium, magnesium, sodium, potassium and chloride ions with gypsum. An experimental study. *Geochimica et Cosmochimica Acta*, **44**, 1471–1482.

Kushnir, J. (1982) The partitioning of seawater cations during the transformation of gypsum to anhydrite. *Geochimica et Cosmochimica Acta*, **46**, 433–446.

Lafon, G.M. (1978) Discussion of "Equilibrium criteria for two-component solids reacting with fixed composition in aqueous phase. Example: the magnesian calcites". *American Journal of Science*, **274**, 1455–1468.

Lahann, R.W. & Siebert, R.M. (1982) A kinetic model for distribution coefficients and application to Mg-calcites. *Geochimica et Cosmochimica Acta*, **46**, 2229–2237.

Lam, R.S.K., Charnock, J.M., Lennie, A. & Meldrum, F.C. (2007) Synthesis-dependant structural variations in amorphous calcium carbonate. *Crystal Engineering Communications*, **9**, 1226–1236.

Lea, A.S., Hurt, T.T., El-Azab, A., Amonette, J.E. & Baer, D.R. (2003) Heteroepitaxial growth of a manganese carbonate secondary nano-phase on the (1014) surface of calcite in solution. *Surface Science*, **524**, 63–77.

Lea, D.W. (2004) Elemental and isotopic proxies of the past ocean temperature. In: *Treatise on Geochemistry* (H. Elderfield, editor), **6**, 365–390. Elsevier, Amsterdam.

Lea, D.W., Mashiotta, T.A. & Spero, H.J. (1999) Controls on magnesium and strontium uptake in planktonic foraminifera determined by live culturing. *Geochimica et Cosmochimica Acta*, **63**, 2369–2379.

Lemarchand, D., Wasserburg, G.J. & Papanastassiou, D.A. (2004) Rate-controlled calcium isotope fractionation in synthetic calcite. *Geochimica et Cosmochimica Acta*, **68**, 4665–4678.

Lichtner, P.C. & Carey, J.W. (2006) Incorporating solid solutions in reactive transport equations using a kinetic discrete-composition approach. *Geochimica et Cosmochimica Acta*, **70**, 1356–1378.

Lind, C.J. & Hem, J.D. (1993) Manganese minerals and associated fine particulates in the streambed of Pinal Creel, Arizona, U.S.A.: A mining-related acid-drainage problem. *Applied Geochemistry*, **8**, 67–80.

Lippmann, F. (1973) *Sedimentary Carbonate Minerals*. Springer-Verlag, Berlin, 228 pp.

Lippmann, F. (1977) The solubility product of complex minerals, mixed crystals and three-layer clay minerals. *Neues Jahrbuch für Mineralogie Abhandlungen*, **130**, 243–263.

Lippmann, F. (1980) Phase diagrams depicting aqueous solubility of binary mineral systems. *Neues Jahrbuch für Mineralogie Abhandlungen*, **139**, 1–25.

Lopez, O., Zuddas, P. & Faivre, D. (2009) The influence of temperature and seawater composition on calcite crystal growth mechanisms and kinetics: Implications for Mg incorporation in calcite lattice. *Geochimica et Cosmochimica Acta*, **73**, 337–347.

Lorens, R.B. (1981) Sr, Cd, Mn and Co distribution coefficients in calcite as a function of precipitation rate. *Geochimica et Cosmochimica Acta*, **45**, 553–561.

Lu, F.H., Meyers, W.J. & Schoonen, M.A.A. (1997) Minor and trace element analyses on gypsum: An experimental study. *Chemical Geology*, **142**, 1–10.

Mackenzie, F., Bischoff, W.D., Bishop, F.C., Loijens, M., Schoonmaker, J. & Wollast, R. (1983) Magnesian calcites: Low-temperature occurence, solubility and solid-solution behavior. In: *Carbonates: Mineralogy and Chemistry* (R.J. Reeder, editor). Reviews in Mineralogy, **11**, 97–144. Mineralogical Society of America, Washington, D.C.

Malone, M.J. & Baker, P.A. (1999) Temperature dependence of the strontium distribution coefficient in calcite: An experimental study from 40 to 200°C and application to natural diagenetic calcites. *Journal of Sedimentary Research*, **69**, 216–223.

Marriott, C.S., Henderson, G.M., Belshaw, N.S. & Tudhope, A.W. (2004) Temperature dependence of δ^7Li, δ^{44}Ca and Li/Ca during growth of calcium carbonate. *Earth and Planetary Science Letters*, **222**, 615–624.

McIntire, W.L. (1963) Trace element partition coefficients — A review and application to geology. *Geochimica et Cosmochimica Acta*, **27,** 1209–1264.

McCrea, J.M. (1950) On the isotopic chemistry of carbonates and a paleotemperature scale. *Journal of Chemical Physics*, **18**, 849–857.

Middelburg, J.J., de Lange, G. & van der Weijden, C.H. (1987) Manganese solubility control in marine pore waters. *Geochimica et Cosmochimica Acta*, **51**, 759–763.

Mikutta, C., Wiederhold, J.G., Cirpka, O.A., Hofstetter, T.B., Bourdon, B. & Gunten, U.V. (2009) Iron isotope fractionation and atom exchange during sorption of ferrous iron to mineral surfaces. *Geochimica et Cosmochimica Acta*, **73**, 1795–1812.

Mitchell, A.C. & Ferris, F.G. (2005) The incorporation of Sr into calcite precipitates induced by bacterial ureolysis in artificial groundwater: Temperature and kinetic dependence. *Geochimica et Cosmochimica Acta*, **69,** 4199–4210.

Monnin, C. (2006) The saturation state of the world's ocean with respect to (Ba,Sr)SO_4 solid solutions. *Geochimica et Cosmochimica Acta*, **70**, 3290–3298.

Monnin, C. & Hoareau, G. (2010) Chemical equilibrium between aqueous fluids and minerals in the marine environment. In: *Ion Partitioning in Ambient-Temperature Aqueous Systems* (M. Prieto, editor). EMU

Notes in Mineralogy, **10**, 227–258. European Mineralogical Union and the Mineralogical Society of Great Britain & Ireland, London.

Morse, J.W. & Bender, M.L. (1990) Partition coefficients in calcite: Examination of factors influencing the validity of experimental results and their application to natural systems. *Chemical Geology*, **82**, 265–277.

Morse, J.W. & Mackenzie, F.T. (1990) *Geochemistry of Sedimentary Carbonates*. Developments in Sedimentology, **48**, Elsevier, Amsterdam, 707 pp.

Morse, J.W., Arvidson, R.S. & Lüttge, A. (2007) Calcium carbonate formation and dissolution. *Chemical Reviews*, **107**, 342–381.

Mucci, A. (1987) Influence of temperature on the composition of magnesian calcite overgrowths precipitated from seawater. *Geochimica et Cosmochimica Acta*, **51**, 1977–1984.

Mucci, A. (1988) Manganese uptake during calcite precipitation from seawater: Conditions leading to the formation of pseudo-kutnahorite. *Geochimica et Cosmochimica Acta*, **52**, 1859–1868.

Mucci, A. (2004) The behaviour of mixed Ca-Mn carbonates in water and seawater: Controls of manganese concentrations in marine pore waters. *Aquatic Geochemistry*, **10**, 113–121.

Mucci, A. & Morse, J.W. (1983) The incorporation of Mg^{2+} and Sr^{2+} into calcite overgrowths: influences of growth rate and solution composition. *Geochimica et Cosmochimica Acta*, **47**, 217–233.

Mucci, A., Canuel, R. & Zhong, S. (1989) The solubility of calcite and aragonite in sulfate-free seawater and the seeded growth kinetics and composition of the precipitates at 25°C. *Chemical Geology*, **74**, 309–320.

Nehrke, G., Reichart, G.J., van Cappellen, P., Meile, C. & Bijma, J. (2007) Dependence of calcite growth rate and Sr partitioning on solution stoichiometry: Non-Kossel crystal growth. *Geochimica et Cosmochimica Acta*, **71**, 2240–2249.

Nielsen, A.E. (1984) Electrolyte crystal growth mechanisms. *Journal of Crystal Growth*, **67**, 289–310.

Noguera, C., Fritz, B., Clément, A., & Amal Y. (2009) Simulation of the nucleation and growth of binary solid solutions in aqueous solutions. *Chemical Geology*, **269**, 89–100.

Oomori, T., Kaneshima, H., Maezato, Y. & Kitano, Y. (1987) Distribution coefficient of Mg^{2+} ions between calcite and solution at 10–50°C. *Marine Chemistry*, **20**, 327–336.

Paquette, J. & Reeder, R.J. (1995) Relationship between surface structure, growth mechanism and trace element incorporation in calcite. *Geochimica et Cosmochimica Acta*, **59**, 735–749.

Paytan, A. & Griffith, E.M. (2007) Marine barite: Recorder of variations in ocean export productivity. *Deep-Sea Research: Part II, Topical Studies in Oceanography*, **54**, 687–705.

Paytan, A., Kastner, M., Campbell, D. & Thiemens, M.H. (1998) Sulfur isotopic composition of Cenozoic seawater sulphate. *Science*, **282**, 1459–1462.

Peacor, D.R., Essene, E.J. & Gaines, A.M. (1987) Petrologic and crystal-chemical implications of the cation order-disorder in kutnahorite [$CaMn(CO_3)_2$]. *American Mineralogist*, **72**, 319–328.

Pingitore, N.E. (1978) The behavior of Zn^{2+} and Mn^{2+} during carbonate diagenesis: Theory and applications. *Journal of Sedimentary Petrology*, **48**, 799–814.

Pingitore, N.E. (1986) Modes of coprecipitation of Sr^{2+} and Ba^{2+} with calcite. *American Chemical Society Symposium Series*, **323**, 574–586.

Pingitore, N.E. & Eastman, M.P. (1985) Barium partitioning during the transformation of corals from aragonite to calcite. *Chemical Geology*, **48**, 183–187.

Pingitore, N.E., Eastman, M.P., Sandige, M., Oden, K. & Freiha, B. (1988) The coprecipitation of manganese(II) with calcite: an experimental study. *Marine Chemistry*, **25**, 107–120.

Pingitore, N.E., Lytle, F.W., Davies, B.M., Eastman, M.P., Eller, P.G. & Larson, E.M. (1992) Mode of incorporation of Sr^{2+} into calcite: Determination by X-ray absorption spectroscopy. *Geochimica et Cosmochimica Acta*, **56**, 1531–1538.

Playà, E. & Rosell, L. (2005) The celestite problem in gypsum Sr geochemistry: An evaluation of purifying methods of gypsiferous samples. *Chemical Geology*, **221**, 102–116.

Plummer, L.N. & Busenberg, E. (1987) Thermodyamics of aragonite-strontianite solid solutions: Results from stoichiometric solubility at 25 and 76°C. *Geochimica et Cosmochimica Acta*, **51**, 1393–1411.

Plummer, L.N., Busenberg, E., Glynn, P.D. & Blum, A.E. (1992) Dissolution of aragonite-strontianite solid-solutions in non-stoichiometric $Sr(HCO_3)$-$Ca(HCO_3)$-CO_2-H_2O solutions. *Geochimica et Cosmochimica Acta*, **56**, 3045–3072.

Pokrovsky, O.S. (1998) Precipitation of calcium and magnesium carbonates from homogeneous supersaturated solutions. *Journal of Crystal Growth*, **186**, 233–239.

Prieto, M. (2009) Thermodynamics of solid-solution aqueous-solution systems. In: *Thermodynamics and Kinetics of Water-rock Interaction* (E.H. Oelkers & J. Schott, editors). Reviews in Mineralogy and Geochemistry, **70**, 47–85. Mineralogical Society of America, Washington, D.C. and the Geochemical Society, St. Louis, Missouri, USA.

Prieto, M., Fernández-González, A., Putnis, A. & Fernández-Díaz, L. (1997) Nucleation, growth and zoning phenomena in crystallizing (Ba,Sr)CO_3, Ba(SO_4, CrO_4), (Ba,Sr)SO_4, and (Cd,Ca)CO_3 solid solutions from aqueous solutions. *Geochimica et Cosmochimica Acta*, **61**, 3383–3397.

Prieto, M., Fernández-González, A., Becker, U. & Putnis, A. (2000) Computing Lippmann Diagrams from Direct Calculation of Mixing Properties of Solid Solutions: Application to the Barite-Celestite System. *Aquatic Geochemistry*, **6,** 133–146.

Purkayastha, B.C. & Chatterjee, A. (1966) On the study of the uptake of strontium tracer by different forms of calcium sulphate. *Journal of the Indian Chemical Society*, **43**, 687–693.

Purton, L.M.A., Shields, G.A., Brasier, M.D. & Grime, G.W. (1999) Metabolism controls Sr/Ca ratios in fossil aragonitic mollusks. *Geology*, **27**, 1083–1086.

Putnis, A., Fernández-Díaz, L. & Prieto, M. (1992) Experimentally produced oscillatory zoning in the (Ba,Sr)SO_4 solid solution. *Nature*, **358,** 743–745.

Raiswell, R. & Brimblecombe, P. (1977) The partition of manganese into aragonite between 30 and 60°C. *Chemical Geology*, **19**, 145–151.

Rimstidt, J.D., Balog, A. & Webb, J. (1998) Distribution of trace elements between carbonate minerals and aqueous solutions. *Geochimica et Cosmochimica Acta*, **62**, 1851–1863.

Romanek, C.S., Jiménez-López, C., Navarro, A.R., Sánchez-Román, A., Sahai, N. & Coleman, M. (2009) Inorganic synthesis of Fe–Ca–Mg carbonates at low temperature. *Geochimica et Cosmochimica Acta*, **73**, 5361–5376.

Rüggeberg, A., Fietzke, J., Liebetrau, V., Eisenhauer, A., Dullo, W.C., & Freiwald, A. (2008) Stable strontium isotopes ($\delta^{88/86}Sr$) in cold-water corals – A new proxy for reconstruction of intermediate ocean water temperatures. *Earth and Planetary Science Letters*, **269**, 569–574.

Schauble, E.A., Rossman, G.R. & Taylor, H.P. (2001) Theoretical estimates of equilibrium Fe-isotope fractionations from vibrational spectroscopy. *Geochimica et Cosmochimica Acta*, **65**, 2487–2497.

Sharma, S.K. & Sharma, T. (1969) Intramolecular kinetic isotope effect in the acid decomposition of carbonates. *International Journal of Mass Spectrometry and Ion Physics*, **2,** 485–493.

Smith, S.V., Buddemeier, R.W., Redalje, R.C. & Houck, J.E. (1979) Strontium-calcium thermometry in coral skeletons. *Science*, **204**, 404–407.

Speer, J.A. (1983) Crystal chemistry and phase relations of orthorhombic carbonates. In: *Carbonates: Mineralogy and Chemistry* (R.J. Reeder, editor). Reviews in Mineralogy, **11**, 145–225. Mineralogical Society of America, Washington, D.C.

Steefel, C.I. & van Cappellen, P. (1990) A new kinetic approach to modelling water-rock interactions: The role of nucleation, precursors, and Ostwald ripening. *Geochimica et Cosmochimica Acta*, **54**, 2657–2677.

Stoessell, R.K., Klimentidis, R.E. & Prezbindowski, D.R. (1987) Dedolomitization in Na-Ca-Cl brines from 100 to 200°C at 300 bars. *Geochimica et Cosmochimica Acta*, **51**, 847–855.

Stoll, H.M., Rosenthal, Y. & Falkowski, P. (2002a) Climate proxies from Sr/Ca of coccolith calcite: Calibrations from continuous culture of *Emiliania huxleyi*. *Geochimica et Cosmochimica Acta*, **66**, 927–936.

Stoll, H.M., Klaas, C.M., Probert, I., Encinar, J.R. & García-Alonso, J.I. (2002b) Calcification rate and temperature effects on Sr partitioning in coccoliths of multiple species of coccolithophorids in culture. *Global and Planetary Change*, **34**, 153–171.

Sverjensky, D.A. (1984) Prediction of free energies of calcite-type carbonates and the equilibrium distribution of trace elements between carbonates and aqueous solutions. *Geochimica et Cosmochimica Acta*, **48**, 1127–1134.

Sverjensky, D.A. & Molling, P.A. (1992) A linear free energy relationship for crystalline solids and aqueous ions. *Nature*, **356**, 231–234.

Tang, J., Köhler, S.J. & Dietzel, M. (2008a) Sr^{2+}/Ca^{2+} and $^{44}Ca/^{40}Ca$ fractionation during inorganic calcite formation: I. Sr incorporation. *Geochimica et Cosmochimica Acta*, **72**, 3718–3732.

Tang J., Dietzel, M., Böhm, F., Köhler, S.J. & Eisenhauer, A. (2008b) Sr^{2+}/Ca^{2+} and $^{44}Ca/^{40}Ca$ fractionation during inorganic calcite formation: II. Ca isotopes. *Geochimica et Cosmochimica Acta*, **72**, 3733–3745.

Tarutani, T., Clayton, R.N., & Mayeda, T.K. (1969) The effect of polymorphism and magnesium substitution on oxygen isotope fractionation between calcium carbonate and water. *Geochimica et Cosmochimica Acta*, **33,** 987–996.

Tesoriero, A.J. & Pankow, J. F. (1996) Solid solution partitioning of Sr^{2+}, Ba^{2+}, and Cd^{2+} to calcite. *Geochimica et Cosmochimica Acta*, **60**, 1053–1063.

Thornber, M.R. & Nickel, E.H. (1976) Supergene alterations of sulphides. III. The composition of associated carbonates. *Chemical Geology*, **17**, 45–72.

Thorstenson, D.C. & Plummer, N.L. (1977) Equilibrium criteria for two-component solids reacting with fixed composition in an aqueous phase – Example: The magnesian calcites. *American Journal of Science*, **277**, 1203–1223.

Torres, M.E., Brumsack, H.J., Bohrmann, G. & Emeis, K.C. (1996) Barite fronts in continental margin sediments: A new look at barium remobilization in the zone of sulfate reduction and formation of heavy barites in authigenic fronts. *Chemical Geology*, **127**, 125–139.

Usdowski, H.E. (1973) The geochemical behaviour of strontium upon genesis and diagenesis of Ca carbonate and Ca sulfate minerals. *Contributions to Mineralogy & Petrology* **38**, 177–195.

Usdowski, H.E. (1975) *Fractionation of Trace Elements Upon Crystallization.* Hochschultext, Springer-Verlag, Berlin, 104 pp. (in German).

Veizer, J. (1983) Trace elements and isotopes in sedimentary carbonates. In: *Carbonates: Mineralogy and Chemistry* (R.J. Reeder, editor). Reviews in Mineralogy, **11**, 263–299. Mineralogical Society of America, Washington, D.C.

von Allmen, K., Böttcher, M.E., Samankassou, E. & Nägler, T.F. (2010) Barium isotope fractionation in the global barium cycle: First evidence from barium minerals and precipitation experiments. *Chemical Geology* (in press).

Wang, D., Lu, X. & Xu, S. (2007) Comment on "Empirical partition coefficients for Sr and Ca in marine barite: Implications for reconstructing seawater Sr and Ca concentrations" by Kristen B. Averyt and Adina Paytan. *Geochemistry Geophysics Geosystems*, **8**, doi:10.1029/2006GC001456.

Wang, D., Wallace, A.F., DeYoreo, J.J. & Dove, P.M. (2009) Carboxylated molecules regulate magnesium content of amorphous calcium carbonates during calcification. *Proceedings of the National Academy of Sciences*, **106**, 21511–21516.

Wang, Y. & Xu, H. (2001) Prediction of trace metal partitioning between minerals and aqueous solutions: a linear free energy correlation approach. *Geochimica et Cosmochimica Acta*, **65**, 1529–1543.

Wasylenki, L.E., Dove, P.M., Wilson, D.S. & De Yoreo, J.J. (2005a) Nanoscale effects of strontium on calcite growth: An in situ AFM study in the absence of vital effects. *Geochimica et Cosmochimica Acta*, **69**, 3017–3027.

Wasylenki, L.E., Dove, P.M. & De Yoreo, J.J. (2005b) Effects of temperature and transport conditions on calcite growth in the presence of Mg^{2+}: Implications for paleothermometry. *Geochimica et Cosmochimica Acta*, **69**, 4227–4236.

Watson, E.B. & Liang, L. (1995) A simple model for sector zoning in slowly grown crystals: implications for growth rate and lattice diffusion, with emphasis on accessory minerals in crustal rocks. *American Mineralogist*, **80**, 1179–1187.

Weiner, S. & Dove, P.M. (2003) An overview of biomineralization processes and the problem of the vital effect. In: *Biomineralization* (P.M. Dove, J.J. De Yoreo & S. Weiner, editors). Reviews in Mineralogy & Geochemistry, **54**, 1–29. Mineralogical Society of America, Washington, D.C.

Welch, S.A., Beard, B.L., Johnson, C.M. & Braterman, P.S. (2003) Kinetic and equilibrium Fe isotope fractionation between aqueous Fe(II) and Fe(III). *Geochimica et Cosmochimica Acta*, **67**, 4231–4250.

Wersin, P., Charlet L., Karthein, R. & Stumm, W. (1989) From adsorption to precipitation: Sorption of Mn^{2+} on $FeCO_{3(s)}$. *Geochimica et Cosmochimica Acta*, **53**, 2787–2796.

Wiesli, R.A., Beard, B.L. & Johnson, C.M. (2004) Experimental determination of Fe isotope fractionation between aqueous Fe(II), siderite and "green rust" in abiotic systems. *Chemical Geology*, **211**, 343–362.

Wolthers, M., Charlet, L. & Van Cappellen, P.V. (2008) The surface chemistry of divalent metal carbonate minerals: A critical assessment of surface charge and potential data using the charge distribution multi-site ion complexation model. *American Journal of Science*, **308**, 905–941.

Xu, X.R., Cai, A.H., Liu, R., Pan, H.H. & Tang, R.K. (2008) Amorphous calcium carbonate in biomineralization. *Progress in Chemistry*, **20**, 54–59.

Yoshida, Y., Nakazawa, T., Yoshikawa, H. & Nakanishi, T. (2009) Partition coefficient of Ra in gypsum. *Journal of Radioanalytical and Nuclear Chemistry*, **280**, 541–545.

Young, E.D. & Galy, A. (2004) The isotope geochemistry and cosmochemistry of magnesium. In: *Geochemistry of Non-traditional Stable Isotopes* (C.M. Johnson, B.L. Beard & F. Albarède, editors). Reviews in Mineralogy & Geochemistry, **55**, 197–230. Mineralogical Society of America, Washington, D.C. and the Geochemical Society, St Louis, Missour, USA.

Zhong, S. & Mucci, A. (1989) Calcite and aragonite precipitation from seawater solutions of various salinities: Precipitation rates and overgrowth compositions. *Chemical Geology*, **78**, 283–299.

Zhu, C. (2004) Coprecipitation in the barite isostructural family: 1. Binary mixing properties. *Geochimica et Cosmochimica Acta*, **68**, 3327–3337.

Zielinski, R.A. & Budahn, J.R. (2007) Mode of occurrence and environmental mobility of oil-field radioactive material at the US Geological Survey Research Suite B, Osaga-Skiatook Project, northeastern Oklahoma. *Applied Geochemistry*, **22**, 2125–2137.

EMU Notes in Mineralogy, Vol. 10 (2010), Chapter 5, 189–226

Ion partitioning and element mobilization during mineral replacement reactions in natural and experimental systems

Christine V. PUTNIS[1,*] and Lurdes FERNÁNDEZ DÍAZ[2]

[1]*Institute of Mineralogy, University of Münster, Corrensstrasse 24, 48149 Münster, Germany, e-mail: putnisc@uni-muenster.de*
[2]*Department of Crystallography and Mineralogy, Complutense University of Madrid, C/ José Antonio Novais, 2, 28040 Madrid, Spain*
**Corresponding author*

Mineral replacement is a common phenomenon in a wide range of geological environments. Metasomatism, metamorphism, weathering, diagenesis and fossilization are examples of processes that can involve the replacement of one or more minerals associated with extensive chemical change. Such replacements are very often pseudomorphic. At relatively low temperatures such as those at the surface or within the upper part of the Earth's crust, where solid-state diffusion can be considered negligible, mineral replacement is the result of interface-coupled dissolution-precipitation reactions, driven by the interplay between the degrees of saturation of natural fluids with respect to different mineral phases. These reactions can play an important role in the mobilization and partitioning of elements. Mineral replacement is accompanied by the generation of porosity, which provides a pathway for the penetration of the fluid within the original parent mineral and facilitates mass transport. This chapter highlights the importance of dissolution-precipitation mineral replacements for element mobilization and ion partitioning in the Earth by presenting and discussing both experimental models and examples of natural processes, development of which can involve significant chemical change, such as during serpentinization and bone fossilization.

1. Introduction

Whenever minerals are in contact with aqueous fluids, reactions take place in an attempt to reach a new equilibrium. Aqueous fluids are ubiquitous in the crust of the Earth and so minerals are constantly re-equilibrating. There are several possibilities for this equilibration:

(1) The mineral may dissolve into the fluid phase. If the resultant fluid is undersaturated with respect to all possible mineral phases, the mineral will continue to dissolve, *e.g.* this occurs extensively during limestone weathering to result in holes and caves in the rock.

(2) However, if the fluid becomes supersaturated with respect to another mineral phase, then this mineral may precipitate, *e.g.* the infilling of veins with phases, such as quartz, calcite, zeolites, different from the parent rock.

DOI: 10.1180/EMU-notes.10.5

(3) If the rate of dissolution of the parent mineral phase is equal to the rate of precipitation of the product phase, *i.e.* if there is an interface coupling between dissolution and precipitation, a pseudomorphic replacement may occur. In this case the original mineral is replaced by another chemically different mineral, the external dimensions and, when structural similarities exist, also crystallographic characteristics, of which, are inherited from the original parent mineral (Putnis, 2002, 2009). This is the re-equilibration mechanism referred to as interface-coupled dissolution-precipitation. The factors determining how and where one mineral may be replaced by another are complex. However the fluid composition will be the controlling factor for the dissolution and precipitation processes. A new mineral can only precipitate or grow when a fluid is supersaturated with respect to that particular phase. During a pseudomorphic replacement reaction the composition of the fluid at the mineral interface places an important control on the progress of the reaction. The initial interface is the mineral surface or grain boundary. As the interface-coupled dissolution-precipitation reaction proceeds, the interface moves into the mineral producing a reaction rim. The penetration of the fluid within an originally solid mineral depends on the production of porosity, which then provides the pathway for mass element transport. The formation of porosity as a result of the replacement reaction depends not only on molar volume differences between parent mineral and product but also on the solubility differences between the parent and the product. This will be discussed below.

Interface-coupled dissolution-precipitation replacement reactions occur continuously within the Earth, during such processes as metasomatism, metamorphism, weathering, diagenesis and fossilization. In fact, whenever minerals are in contact with aqueous fluids, pseudomorphic replacement reactions may take place. Furthermore, these reactions may occur at relatively low temperatures consistent with temperatures found at the surface or within the upper part of the Earth's crust and certainly at temperatures where solid-state diffusion can only play a very minor role in element partitioning or re-equilibration.

This chapter presents examples of such pseudomorphic or interface-coupled dissolution-precipitation mineral replacements in order to indicate the significance of this replacement mechanism for element mobilization and ion partitioning in the Earth.

2. Replacement in the KCl-KBr-H_2O solid solution-aqueous solution system

Natural mineral replacement reactions occurring in the Earth often take place at elevated temperatures and pressures and in unknown fluid compositions. In order to establish the potential conditions for pseudomorphic replacement reactions, the first example of replacement is from a simple salt system model, the KBr-KCl-H_2O solid solution-aqueous solution system, where *in situ* observations of the evolution of both the solid and fluid compositions at the interface during the replacement reaction give insights into the reaction mechanism (Fig. 1, reproduced from Putnis *et al.*, 2005). Under ambient conditions of temperature and pressure and free from imposed stress, a single crystal of KBr can be replaced by a single crystal of KCl. The pseudomorphism is initiated by the epitaxial growth of a new product phase at the fluid-mineral interface, when the dissolution of the parent phase results in an interfacial fluid layer, which is supersaturated with respect to a different solid composition. The thermodynamics of this system has been thoroughly investigated (Lippmann, 1980; Glynn *et al.*, 1990). The subsequent evolution of the coupled dissolution and growth can be related to local equilibrium defined by a Lippmann diagram (Fig. 2).

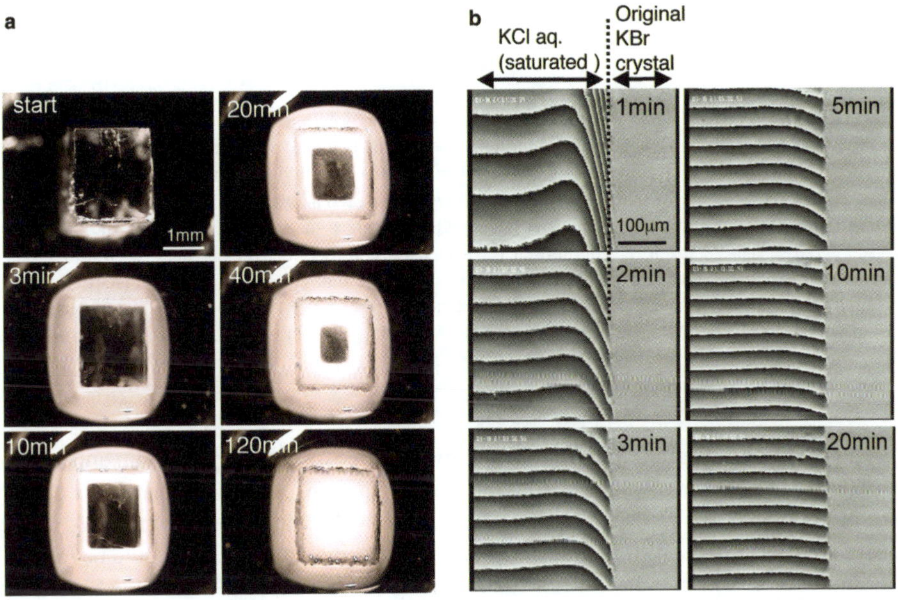

Fig. 1. (***a***) A time sequence of optical micrographs showing the migration of the replacement interface when a KBr parent crystal (2 mm × 2 mm × 1 mm) was in contact with a saturated KCl solution. The initial KBr crystal was optically clear, but as the reaction proceeded, the reaction rim, composed of the K(Br,Cl) product, appeared cloudy, due to the development of a fine porosity. In order to view the advancement of the reaction interface, the KCl solution was added up to but not covering the top of the KBr crystal. The resulting meniscus is seen as the white ring surrounding the crystal. After 120 min the KBr crystal was replaced pseudomorphically by KCl. (***b***) A sequence of images from real-time phase-shift interferometry of the same experiment as (*a*). The fringes on the left represent the characteristic refractive index (RI) values of the fluid. Changes in RI with time indicate compositional changes in the fluid. The curved interference fringes near the surface of the KBr crystal show an initial steep compositional gradient at the crystal surface as a result of the initial replacement reaction. The fringes only show fluid changes outside the crystal being replaced. As the replacement proceeded, the reaction interface moved within the crystal, but the original external surface was maintained in the pseudomorphic replacement. Real-time interferometry showed that the composition of the fluid boundary layer at the crystal interface is very different from further away from the crystal (reproduced with the permission of the Mineralogical Society of America from Putnis *et al.* (2005) in *American Mineralogist*, fig. 2).

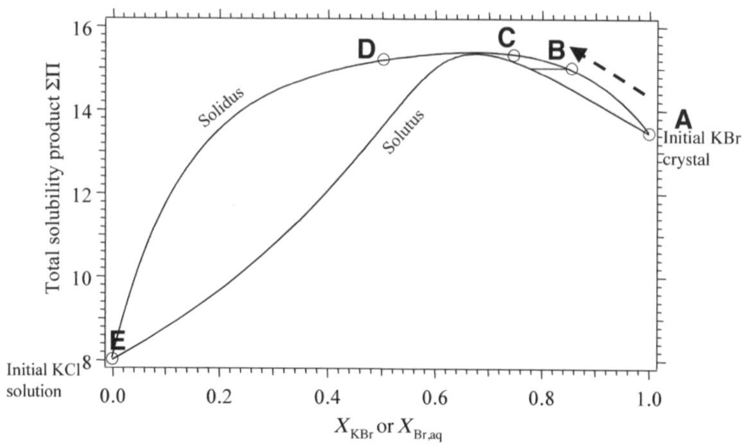

Fig. 2. Lippmann phase diagram for the solid solution–aqueous solution system KCl-KBr-H_2O. Horizontal tie lines connect solid and aqueous compositions at thermodynamic equilibrium. The total solubility product, $\Sigma\Pi$, is defined as $\{K^+\}(\{Br^-\} + \{Cl^-\})$. The x axis is a double scale, the activity fraction of the aqueous phase, together with the mole fraction of the solid phase. The arrow indicates the direction of the K(Br,Cl) replacement evolution from the original KBr crystal at the reaction interface, starting at the crystal surface. We assume local equilibrium (or near equilibrium) at the replacement reaction interface. The points (A–E) represent KBr molar fractions after successive time intervals of reaction (start, 3, 10, 60, 120 min). After 120 min the original KBr crystal was totally replaced by KCl (point E). (Reproduced with the permission of the Mineralogical Society of America from Putnis *et al.* (2005) in *American Mineralogist*, fig. 4).

The Lippmann diagram relates fluid and solid compositions at equilibrium and so represents a solid solution-aqueous solution phase diagram. Assuming that local equilibrium applies at the replacement interface, even when the system as a whole is far out of equilibrium, the initial replacement rim composition after 1 min would be in equilibrium with a fluid of composition connected by a horizontal tie line on the Lippmann diagram. This is consistent with the data from real-time phase shift interferometry (Tsukamoto, 1993; Onuma *et al.*, 1994; Putnis *et al.*, 2005), which indicate a high KBr concentration at the interface after 1 min (Fig. 1b). The evolution of the rim composition with time records successive local interface equilibrium states.

In the replacement reaction all elements of the parent KBr crystal are replaced by the K^+ and Cl^- in the fluid phase, as shown by the tracing of a ^{40}K-enriched KCl solution, using inductively coupled plasma mass spectrometry, ICP-MS (Putnis & Mezger, 2004). The only way that both Cl and K from the fluid phase can form part of the new product structure is if the original KBr structure is broken down and a new structure precipitates.

The development of porosity in the new solid phase (Fig. 3), is the important textural change during this replacement reaction, allowing fluid to penetrate within the new product phase up to the reaction interface where the replacement reaction takes place. This interface can then move further within the mineral crystal.

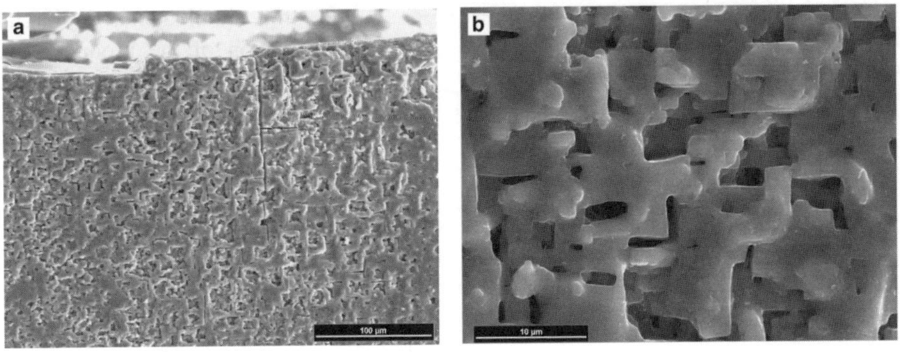

Fig. 3. (***a***) SEM image of the surface of a replaced KBr crystal, showing the high porosity in the K(Br,Cl) product. (***b***) Close up from the centre of (a) showing that the porosity is crystallographically controlled. The amount of porosity generated in this replacement reaction is $\gg$ the molar volume differences between the KBr and KCl end-members and so an explanation for the constant-volume reaction must also depend on the relative solubilities of solid phases in the fluid at the reaction interface (reprinted from *Geochimica et Cosmochimica Acta*, **68** (13), C.V. Putnis and K. Mezger: A mechanism of mineral replacement: Isotope tracing in the model system KCl-KBr-H_2O. Copyright (2004) with permission from Elsevier.)

The amount of porosity generated depends not only on the molar volumes of the parent and product crystals, but also on their relative solubilities in the fluid at the reaction interface (Pollok *et al.*, 2002). Considering only molar volume changes, the expected molar volume decrease in the replacement of KBr by KCl would be ~13.2%. However, considering the solubility differences, as well as molar volume changes, we would expect a total porosity of ~30%, which is in accordance with the observed porosity development, estimated from scanning electron microscopy (SEM) image analysis (Fig. 3). This interpretation uses both the KBr-KCl solubility diagram and the Lippmann diagram in which the equilibrium compositions indicate that the amount of solid dissolved must be more than the amount of product precipitated, *i.e.* a volume deficit reaction. To express this in another way, the parent phase is more soluble than the precipitating product in the fluid phase at the interface. Therefore a volume deficit results as seen in the porosity containing the fluid phase. So the interfacial control, which dictates that the reaction interface is the original crystal surface, inevitably leads to the requirement of porosity in the product.

The evolution of the reaction is limited by diffusion of elements through the fluid phase from the outside of the crystal or the grain boundary up to the reaction interface. This is confirmed from the plot of the reaction kinetics (Fig. 4). The rate of migration of the reaction interface through the parent crystal is consistent with a diffusion-controlled mechanism with a value of $D \sim 10^{-6}\ cm^2\ s^{-1}$, typical of self-diffusion coefficients in an aqueous solution and approximately 10 orders of magnitude larger than solid-state diffusion coefficients in alkali halides (Putnis & Mezger, 2004).

The experimental results indicate that the replacement mechanism is controlled by the initial composition of a fluid boundary layer formed by dissolution of the parent crystal,

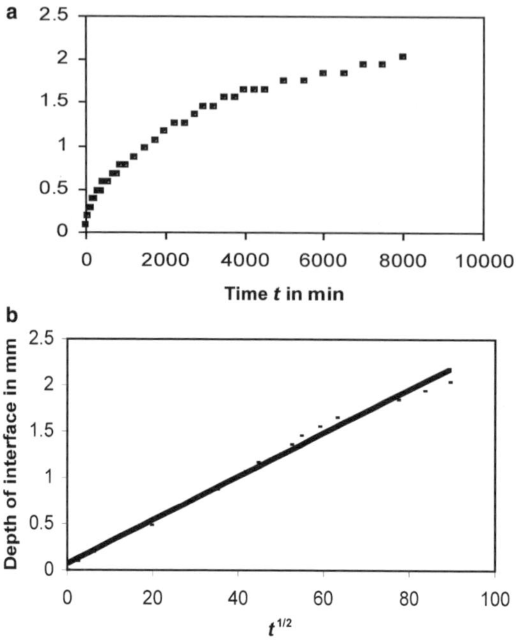

Fig. 4. (***a***) Rate of advance of the replacement interface in depth (mm) plotted *vs.* time *t* (min). (***b***) Depth of the replacement interface plotted *vs.* $t^{1/2}$. The good approximation to a straight line indicates that the reaction rate is probably controlled by diffusion of ions through the fluid phase to and from the reaction interface.

and the simultaneous epitaxial precipitation of the product. The fluid-solid equilibrium is only maintained within this boundary layer the thickness of which depends on the relative dissolution rate of the parent crystal, the epitaxial nucleation of the product crystal and the mass transport rate through the fluid, both to the crystal surface and within the porous replacement up to the reaction interface. This mechanism of pseudomorphism does not require stress to preserve the external dimensions of the crystal. The transfer of crystallographic information from parent to product depends on the degree of lattice mismatch allowing for possible epitaxial precipitation, but in solid solution–aqueous solution systems, where continuous changes in lattice parameters accompany the incremental changes in equilibrium compositions, a single crystal of the parent can be completely replaced by a single crystal of the product. This is also relevant for many rock-forming mineral systems that are solid solution systems. When the KBr crystal is completely replaced, its composition has reached the final bulk equilibrium located at point E in Figure 2.

$$\mathrm{KBr} + \mathrm{Cl}^-_{(aq)} \Longrightarrow \mathrm{KCl} + \mathrm{Br}^-_{(aq)}$$

The reaction features of this replacement, share many characteristics of replacement reactions in nature, such as: (1) a reaction rim, that is compositionally and texturally different from the parent crystal; (2) the retention of the morphology and crystal structure of the parent mineral; and (3) the presence of porosity, which may be transient (Putnis *et al.*, 2005), in the new product mineral. These textural characteristics of natural replacements suggest a general coupled dissolution-precipitation mechanism for the re-equilibration of minerals in solid-fluid interactions (Putnis & Putnis, 2007).

It is the development of porosity that is essential for an interface-coupled dissolution-precipitation reaction to result in a pseudomorphic replacement, because a porosity provides a pathway for fluid transport and hence for element transfer, ultimately resulting in ion partitioning and element mobilization.

3. Replacements in common rock-forming mineral systems

3.1. Feldspars in natural rocks

Feldspars are common rock-forming minerals that mostly appear as cloudy or turbid mineral grains. This textural characteristic of feldspars results from the presence of a pervasive porosity within the mineral grain. This porosity is commonly on a nanoscale and the presence of nanocrystals of hematite is responsible for the pink to red colouration commonly seen in the feldspars in rocks such as granites (Putnis *et al.*, 2007a). The presence of porosity in feldspars indicates that these minerals have re-equilibrated in the presence of a fluid phase, either sub-solidus reactions within a magma body or as a later re-equilibration reaction when the rock may have come into contact with aqueous fluids, during a metasomatic or metamorphic event.

Feldspars have been studied extensively over many decades. It has been recognized that the chemical re-equilibration of a crystalline solid in an aqueous fluid often results in the replacement of one crystal by another while retaining the external form and crystallographic details of the parent (O'Neil & Taylor, 1967; Lee & Parsons, 1997; Fiebig & Hoefs, 2002; Labotka *et al.*, 2004; Putnis *et al.*, 2007a; Parsons & Lee, 2009a, 2009b).

Parsons & Lee (2009a) emphasize that intracrystalline microtextures formed by a process of mutual replacement in alkali feldspars record fluid-rock reactions that have affected large volumes of the Earth's crust. They present data from alkali feldspars occurring in layered syenites from the Klokken intrusion in South Greenland. The original or parent feldspar was a sodic sanidine, which unmixed and ordered by volume diffusion during cooling into the microcline field temperatures of ~450°C, to result into fully coherent 'braid' crytoperthite. At temperatures below 450°C the crystals reacted with a circulating post-magmatic aqueous fluid. The braid perthite acted like a single reactant phase, which was replaced by two product phases, incoherent subgrains of albite and microcline, a 'patch' perthite, with micropores at their boundaries (Fig. 5). The bulk-element composition of the parent remained unchanged and the replacement was pseudomorphic, maintaining the euhedral crystal shape of the original perthite. The driving force for the reaction was coherency strain energy at the reaction interface. However, most mineral replacement reactions involve a chemical change and so we need to consider a thermodynamic driving force for the reaction.

Major- and trace-element analyses for the above examples of replaced perthite have also been made in order to investigate the trace element partitioning and geothermometry (Parsons & Lee, 2009b). Laser ablation ICP-MS was used to confirm the element change during the replacement reaction, including any element partitioning between the two Na- and K-rich solid product phases. In this case, while the major-element bulk composition of the starting crystals was retained, many trace elements, including *REE*s, were lost to the aqueous phase during the reaction. This loss was particularly large for Fe and Ti. Other trace elements also partitioned between the Na and K product phases, some on the basis of ionic radii. Parsons & Lee (2009b, and references therein) give an excellent account of the trace-element partitioning during replacement

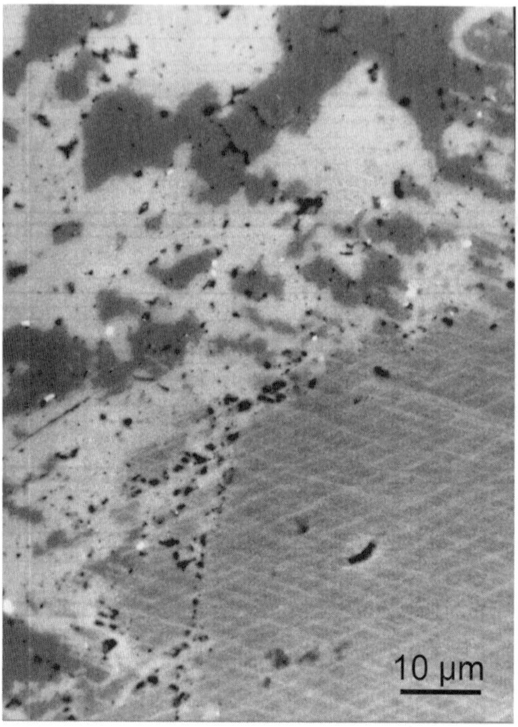

Fig. 5. SEM backscatter image of the replacement of a braid perthite by patch perthite. The parent perthite acted like a single reactant phase, which was replaced by two product phases, a patch perthite, composed of incoherent subgrains of albite ($NaAlSi_3O_8$), and microcline, ($KAlSi_3O_8$). The bulk-element composition of both parent and product perthites remained unchanged and the replacement was pseudomorphic, maintaining the euhedral crystal shape of the original perthite. The replacement was driven by the reduction in strain energy at the reaction interface. Note the porosity in the product patch perthite (image courtesy of I. Parsons).

in the Klokken syenites and show how element partitioning has the potential to provide a new type of feldspar geothermometer.

Albitization is a very common process occurring in many rocks but especially those associated with large-scale mineralization. Engvik *et al.* (2008) investigated the albitization of the rocks in the Bamble Sector of southern Norway. Here, where Na-bearing fluids have had enhanced access to the rock, through fractures, the contact areas of the rock along the fractures show extensive albitization up to an interface where presumably the fluid was no longer available (Fig. 6a). Observations and analyses using SEM, electron microprobe analysis (EMPA) and transition electron microscopy (TEM) provide information about the nature of the replacement:

(1) The reaction interface is sharp on a nanoscale, with little evidence of any significant solid-state diffusion between parent and product phases. The sharpness of the interface is evident from both back-scatter SEM images, as well as at higher magnification in TEM images (Fig. 6b,c).

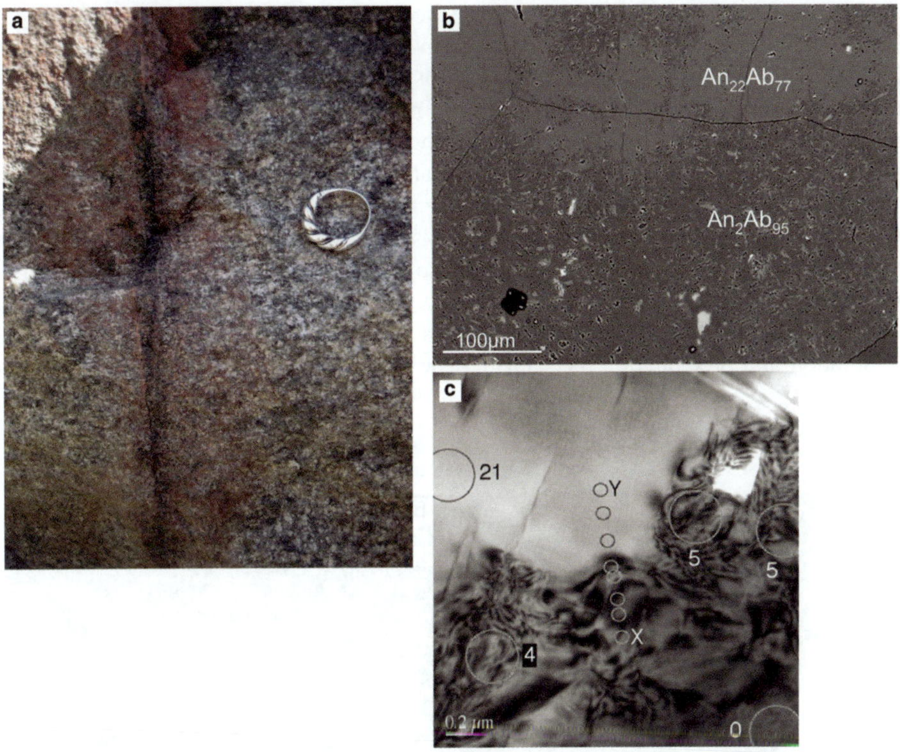

Fig. 6. (*a*) Albitization of granitoid rock in the Bamble Sector, Norway. The fracture in the rock is surrounded by a zone of alteration, marked by a pink colouration. In this zone a fluid-induced chemical re-equilibration of feldspars has occurred, as indicated in (*b*) which shows a backscatter electron (SEM) image of the same rock at the replacement interface. The plagioclase (oligoclase, $An_{22}Ab_{77}$, lighter area) is replaced by almost pure albite (An_2Ab_{95}, darker area). The numbers represent the anorthite:albite % in the plagioclase. Note the porosity development in the new albite phase. (*c*) TEM image of the sharp albitization interface. The albite and oligoclase have been shown to be crystallographically continuous across the reaction interface and the albite phase has nanoscale pores (Engvik *et al.*, 2008). The circles are the positions of microprobe analyses, confirming the sharpness of the interface.

(2) The product is porous. The porosity is the volume deficit between parent and product and is occupied by the fluid phase.

(3) The original external volume occupied by the plagioclase crystals in the rock as well as their crystallographic orientation are preserved during the albitization process.

(4) Other phases may co-precipitate during the albitization reaction. In many albitization reactions in nature, the precipitation of very fine (nanoscale) hematite is associated with the albitization and causes the red/pink colouration in the rock, a feature which has also been associated with K-metasomatism in granitic rocks (Putnis *et al.*, 2007a, and references therein).

Another example of albitization is taken from the albitized rocks of the Proterozoic Willyama Supergroup in the Olary Domain of the Curnamona Province of South Australia and Western New South Wales, Australia (Fig. 7). This area includes extensive mineralization of Pb, Zn and Ag in the massive Broken Hill deposits as well as many other deposits of ore minerals containing metals such as Cu, Au and U (Bierlein *et al.*, 1995). It is clear from looking at these albitized rocks that they have experienced several large-scale fluid events. The albitized areas can be seen to follow fractures within the rocks. These fractures provide easy pathways for fluid access. However, there has to be a mechanism for the fluid to reach within all mineral grains of the rock. The grain boundaries can provide fluid access to all grains but the mechanism of albitization within the grains themselves can be easily achieved by a coupled dissolution and precipitation mechanism. The albite mineral grains are highly porous

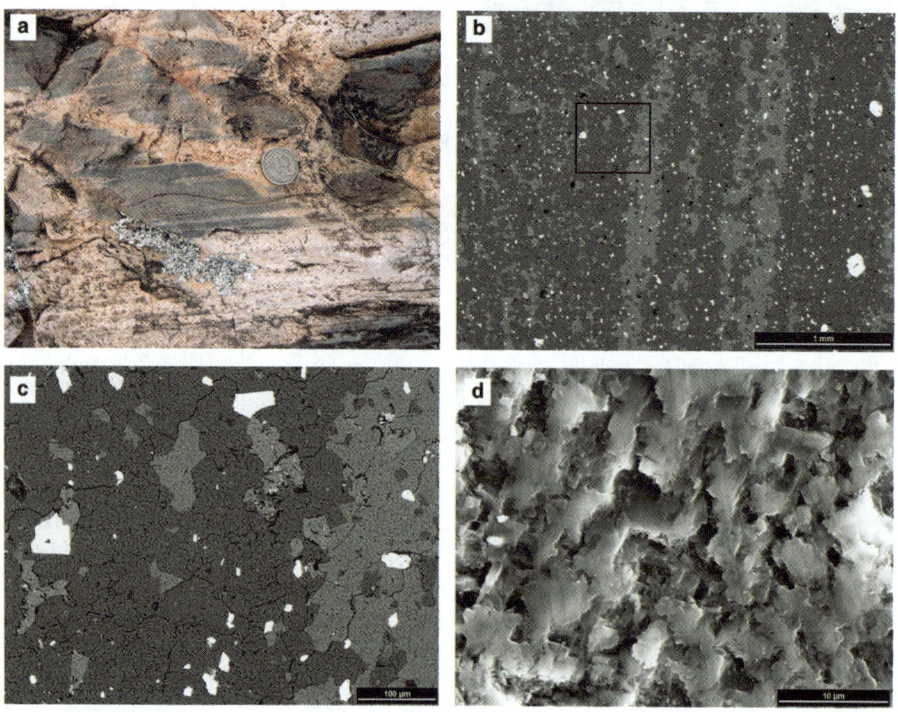

Fig. 7. (***a***) Albitization, marked by a pink colouration in metamorphosed rocks in the Curnamona Province, South Australia. The albitization starts along fractures and then affects the whole rock by a process of infiltration through porosity, which provides a pathway for mass transport, developed within the minerals. (***b***) a SEM backscatter image across the grey, predominantly K feldspar to the pink, albite interface: very bright spots are Fe or Fe,Ti oxides, light areas K feldspar and darker areas albite. (***c***) Close-up image from (*b*), indicated by the square, showing the porous texture of the feldspars. (***d***) SEM image of the porosity in the albite.

(Fig.7d). The pervasive fluids in this case must have been Na-rich, which is highly likely given the availability of Na fluids within the crust of the Earth.

The development of shear zones is known to be synchronous with widespread fluid flow resulting in albitization and calc-silicate alteration. This is discussed by Clark *et al.* (2005) and Schmidt-Mumm *et al.* (2006), who emphasize how shear zones play an important role in providing major pathways for large volumes of fluid flow during deformation, and therefore has a significant role in the genesis of hydrothermal ore deposits. The variations in chemical and isotopic compositions of sheared and altered rocks relative to the unaltered protolith can be used to quantify sources of the fluids involved.

The iron oxide-copper-gold deposits in the Mount Isa Block, Queensland, Australia have been studied extensively by Oliver *et al.* (2004) and constitute another example of mineralization related to sodic alteration and mineral replacement. Those authors recognize that regional-scale albitization can be a mechanism for 'stripping' large quantities of metals from the affected rock. The resulting 'charged' fluids are enriched in the metals stripped from the albitized rocks and will leave a characteristic alteration trend in their pathway. Ultimately these fluids have great potential to become the mineralizing fluids of ore deposits.

Albitization involves the complete destruction and replacement of existing mineral phases in a rock. Obviously this transformation has severe implications for the physical and chemical properties of the rocks affected and it has been shown that during albitization, a given rock may entirely lose its chemical (*i.e.* major, trace and rare earth element) characteristics as well as its isotopic signatures (O'Neil & Taylor, 1967; Oliver *et al.*, 2004). The process also has major implications for the fluids involved as they change composition in a similarly dramatic way. The 'charged fluids' resulting from albitization are enriched in a range of major elements, trace metals as well as rare earth elements. Thus potentially they are powerful, metal-bearing, mineralizing fluids.

These examples suggest that large-scale fluid transport has allowed element mobilization to occur. Although fluids will move within the Earth's crust as a response to gravity or temperature or pressure gradients, there must be a mechanism for element exchange, *i.e.* for stripping metals from original minerals. Interface-coupled dissolution-reprecipitation of minerals reactive in the aqueous fluids at their interfaces and consequent porosity generation provides an important mechanism of element mobilization, at temperatures well below the solidus for these rock-forming minerals and at temperatures where solid-state diffusion would be extremely slow. However to test this hypothesis, the experimental replacement of one feldspar by another chemically different feldspar can shed insights into the validity of the proposed mechanism.

3.2. Feldspar replacement – experimental examples

Experiments carried out by Hövelmann *et al.* (2010) follow directly an albitization pathway. Starting with oligioclase, (Ca,Na)$Al_2Si_3O_8$ ($Ab_{73}An_{23}$ + minor K_2O and FeO) and labradorite, (Ca,Na)$Al_2Si_3O_8$ ($Ab_{39}An_{60}$ + trace K_20), the crystals + an aqueous solution containing NaCl + Si were sealed in hydrothermal autoclaves at 500°C and

2 kbar for 14 days. Analyses by SEM of cross sections of the reacted crystals revealed reaction rims composed of almost pure albite (Fig. 8). Using a FIB (focused ion beam) cut section, TEM samples were imaged and showed a sharp reaction interface, on a nanoscale (Fig. 8b). The albite rim was highly porous. Quantitative chemical analyses (microprobe and laser ablation ICP-MS) demonstrated that the albitization was accompanied by the mobilization of major, minor and trace elements, also including elements such as Al and Ti, which are commonly regarded as immobile during hydrothermal alteration. The re-equilibration of the plagioclase in an ^{18}O-enriched Na and Si-bearing solution resulted in the incorporation of the ^{18}O in the product albite alumino-silicate framework structure (analysed using Raman spectrometry) and indicate a complete breakdown of the original crystal framework and precipitation of the new albite phase

$$\begin{aligned} &CaAl_2Si_3{}^{16}O_8 + Na^+ + Si^{4+} + H_2{}^{18}O \\ &\quad \Longleftrightarrow NaAlSi_3{}^{18}O_8 + Ca^{2+} + Al^{3+} + H_2{}^{16}O. \end{aligned}$$

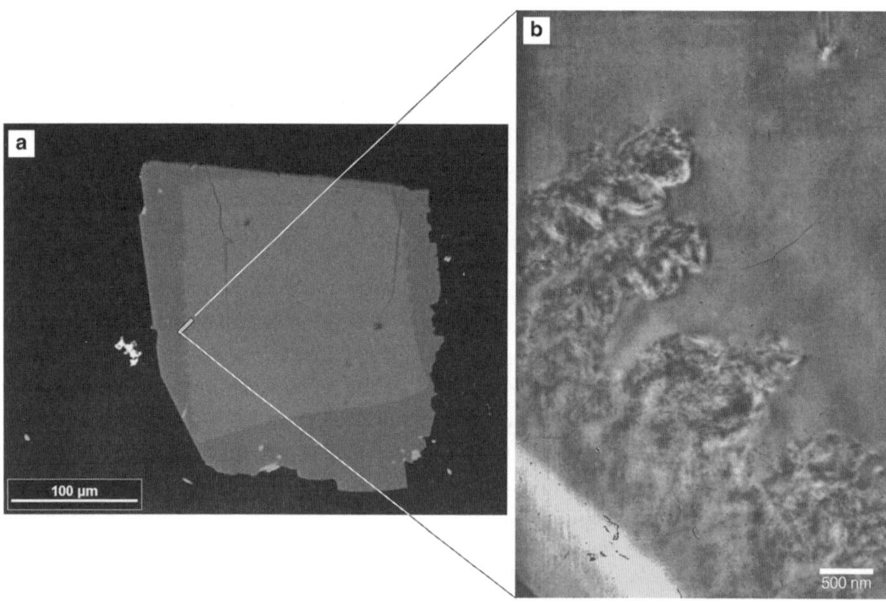

Fig. 8. (***a***) SEM backscatter image of a section through a partly replaced oligoclase plagioclase. The lighter centre is the original oligoclase, which was a single pure transparent crystal, $Ab_{74}An_{22}Or_4$, the external dimensions of which were retained during the replacement process. It was reacted in an aqueous solution containing Na and Si at 600°C, 2 kbar for 17 days. The darker replacement rim is pure albite. (***b***) TEM image from a FIB cut across the reaction interface shown in (*a*), indicating a very sharp interface and a turbid, porous albite replacement product (Hövelmann *et al.*, 2010).

These experiments demonstrated how the replacement mechanism was responsible for the partitioning of major, minor and trace elements between the solid (both parent and product) and fluid phases. The textural and compositional characteristics of the reaction products were similar to naturally occurring albitized plagioclase and are indicative of an interface-coupled dissolution-precipitation mechanism.

Experiments by Niedermeier *et al.* (2009) to investigate the equilibration of albite in the presence of K-rich solutions have also shown that crystalline albite can be similarly pseudomorphically replaced at a moving interface by K-feldspar (Fig. 9). Experiments made with ^{18}O-enriched KCl solutions and analysed with Raman spectrometry also revealed the incorporation of ^{18}O from the water of the aqueous phase into the product feldspar alumino-silicate framework structure.

$$NaAlSi_3{}^{16}O_8 + K^+ + H_2{}^{18}O \Longleftrightarrow KAlSi_3{}^{18}O_8 + Na^+ + H_2{}^{16}O$$

Tranmission electron microscopy images confirmed a sharp interface and diffraction patterns showed the crystallographic continuity across the replacement interface. Again this reaction took place within a few days, a reaction rate that is not consistent with solid-state diffusion.

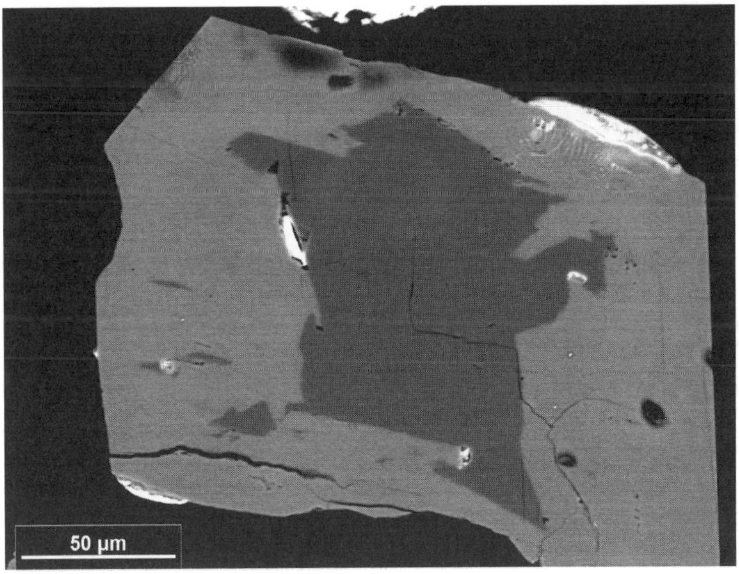

Fig. 9. SEM backscatter image of the pseudomorphic replacement of a single crystal of albite by K-feldspar. The image is from a cross section of the interface between pristine albite (darker grey) in the core and reacted K-feldspar (lighter grey) in the rim. The experiment time was 5 days at 600°C and 2 kbar. Note the sharp interface and the occurrence of numerous pores at the interface (Niedermeier *et al.*, 2009). Note also the similarities with the replacement shown in Figure 8. (Reproduced, with the permission of Springer, from a paper by Niedermeier *et al.*, 2009 in *Contributions to Mineralogy and Petrology*).

These experiments suggest that the composition of the fluid is a controlling factor for the replacement process. Such a process is often referred to as ion exchange, implying the simple substitution of one ion by another. However, to achieve this effective ion exchange, isotope tracing experiments confirm the breakdown of the parent phase and the recrystallization of the product, *i.e.* coupled dissolution-precipitation.

3.3. The experimental replacement of leucite by analcime

The replacement of leucite $KAlSi_2O_6$ by analcime $NaAlSi_2O_6.H_2O$ in a NaCl solution similar to seawater concentration (35%) at 100–150°C also proceeds by a dissolution-precipitation mechanism (Putnis *et al.*, 2007b; Fig. 10), although these are zeolitic minerals in which the channels within the alumino-silicate framework are large enough that a simple ion-exchange mechanism might be predicted.

The reaction is pseudomorphic and a porosity develops in the analcime product despite a 10% molar volume increase from parent to product. A sharp reaction interface, as well as the fact that ^{18}O in the NaCl aqueous solution is incorporated into the analcime alumino-silicate framework, confirms that a dissolution-precipitation, rather than a solid-state diffusion mechanism is involved

$$KAlSi_2{}^{16}O_6 + Na^+ + H_2{}^{18}O \Longrightarrow NaAlSi_2{}^{18}O_6.H_2{}^{18}O + K^+ + H_2{}^{16}O$$

To further support the former mechanism, this reaction begins at low temperatures and is completed at 150°C in a day or two, representing temperatures and times not possible with solid-state diffusion. In nature the fast replacement of leucite by analcime is confirmed by the fact that leucite does not survive in surface rocks and in general leucite is never found in rocks older than Tertiary.

This example of element mobilization during a replacement reaction emphasizes that the formation of porosity is not solely dependent on molar volume changes between parent and product but also on the relative solubilities of the solids in the aqueous phase at the reaction interface (Pollok *et al.*, 2002). This means that it is possible to generate porosity even when the molar volume of the product is greater than that of the parent solid. The aqueous phase provides the new elements and accepts elements originally present in the parent phase, *i.e.* the aqueous phase facilitates an effective exchange of elements while also actively being involved in the replacement process. This is also confirmed from the replacement of ^{18}O in enriched solutions with ^{16}O in the parent solid. Figure 11a shows Raman spectra of the Raman shift towards smaller wavenumbers when ^{18}O is incorporated within the analcime silicate framework structure. Time of flight secondary ion mass spectrometry (ToF-SIMS) has been used to confirm the simultaneous ^{18}O and cation exchange (Fig. 11b).

The replacement of leucite by analcime has also recently been investigated by Xia *et al.* (2009), who were able to further constrain the replacement process, in terms of the fluid control at the reaction interface. Furthermore, Xia *et al.* (2009) have shown how functionally designed materials can be made using the principles of coupled dissolution and precipitation. A material with certain desired properties, such as porosity or an

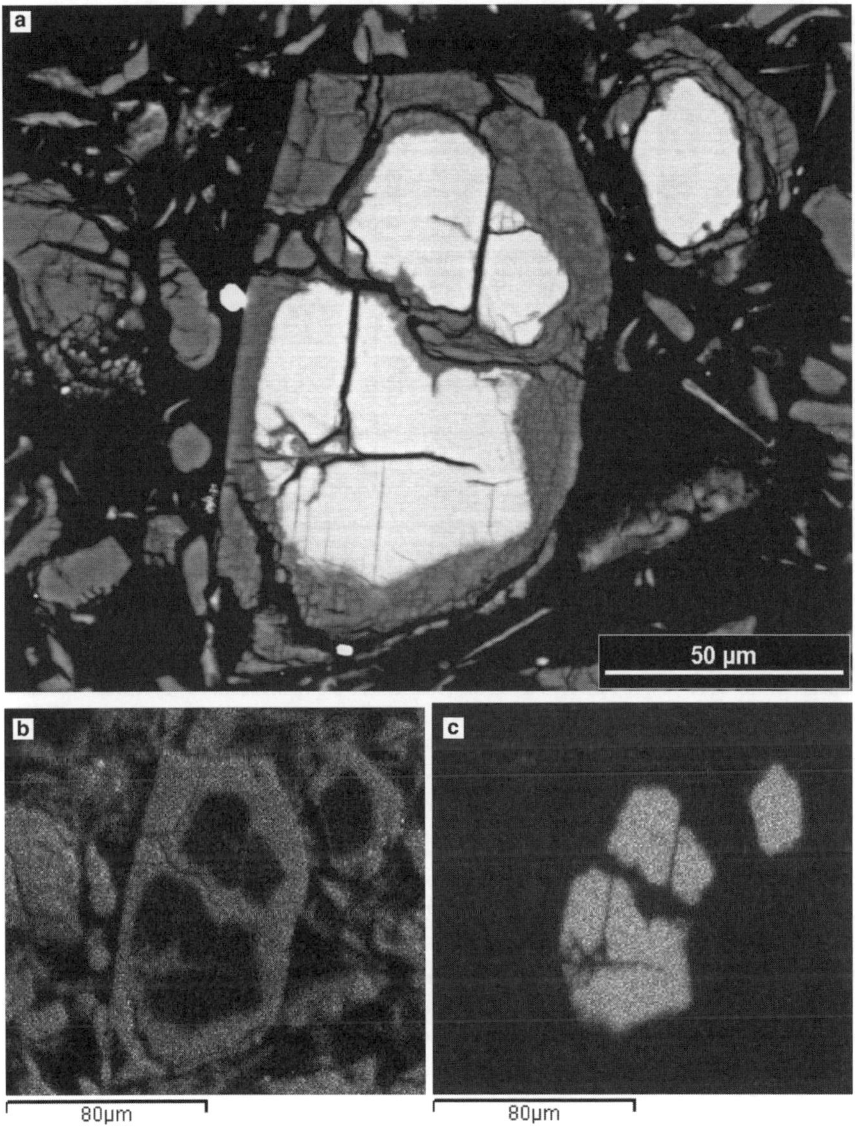

Fig. 10. The replacement of leucite ($KAlSi_2O_6$) by analcime ($NaAlSi_2O_6 \cdot H_2O$). (***a***) Backscatter SEM image of a typical leucite grain after partial replacement by analcime (with permission of the Mineralogical Society of America). The dark areas are analcime and the lighter centre is the original leucite. The replacement took place within 3 days at 150°C and proceeded from the outside of the leucite crystal inwards along a moving interface (Putnis *et al.*, 2007b). This interface is characterized by a sharp compositional front. The reaction also proceeded along small cracks, which then helped to break the crystal apart (Jamtveit *et al.*, 2009). Note the porosity in the analcime phase. Parts (***b***) and (***c***) are EDX elemental distribution maps of Na and K, respectively, showing the sharp compositional interface. (Reproduced with the permission of the Mineralogical Society of America from Putnis *et al.* (2007) in *American Mineralogist*, fig. 2).

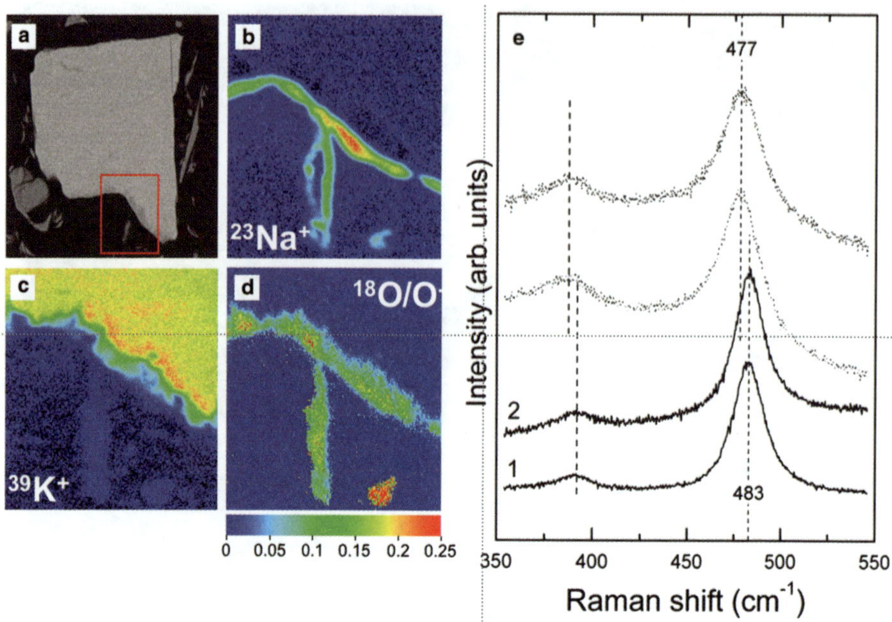

Fig. 11. (*a–d*) Images from time of flight secondary ion mass spectrometry (ToF-SIMS) used to confirm the simultaneous ^{18}O and cation K-Na exchange in the replacement shown in Figure 10. (*e*) Raman spectra of the Raman shift towards lower wavenumbers when ^{18}O is incorporated within the analcime silicate framework structure (Putnis *et al.*, 2007b). (Reproduced with permission of the Mineralogical Society of America from a paper by Putnis *et al.*, 2007b).

incorporated trace element content, may be produced from a precursor material through a dissolution-precipitation mechanism. In the case of the leucite-anacime replacement, the morphology of the product analcime nanocrystals is 'tunable' by simply changing solution conditions, such as *pH*. The pseudomorphic replacement route has the potential to synthesize other ordered arrays of functional nanocrystals with controlled shape, size, and crystallographic orientation, as well as a range of designed physical and/or chemical properties.

3.4. Serpentinization

Serpentinization is a low-temperature metamorphic/metasomatic process occurring on a large scale at the seafloor, at plate margins, *i.e.* mid-ocean spreading ridges and subduction zones. During serpentinization, low-silica mafic and ultramafic oceanic crustal rocks (peridotites) and sediments are replaced predominantly by serpentine, brucite and magnetite. Serpentinization is a hydration process whereby serpentine is formed from olivine *via* several possible reactions, some of which are complementary. Olivine is a solid solution between the magnesium endmember, forsterite and the iron end member fayalite. Serpentinite reactions (1) and (2), below, exchange silica between forsterite and fayalite to form serpentine-group minerals and magnetite. These are highly exothermic reactions.

(1) Fayalite + water ⇒ Magnetite + aqueous silica + hydrogen

$$3Fe_2SiO_4 + 2H_2O \Longrightarrow 2Fe_3O_4 + 3SiO_2 + 2H_2$$

(2) Forsterite + aqueous silica ⇒ Serpentine

$$3Mg_2SiO_4 + SiO_2 + 4H_2O \Longrightarrow 2Mg_3Si_2O_5(OH)_4$$

Other possibilities include the reaction of olivine plus water to give serpentine plus $Mg(OH)_2$ (brucite), and reactions in which magnesium and silica are transported out of the serpentinizing volume.

In the presence of carbon dioxide, however, serpentinitization may form either magnesite ($MgCO_3$) or generate methane (CH_4). It is thought that some hydrocarbon gases may be produced by serpentinite reactions within the oceanic crust, and the serpentinite reaction is a key argument for the theory of an abiogenic origin of hydrocarbon gases (Berndt *et al.*, 1996) and petroleum.

(3) Olivine + Water + Carbonic acid ⇒ Serpentine + Magnetite + Methane

$$9Mg_2SiO_4 + 3Fe_2SiO_4 + 13H_2O + 0.5CO_2$$
$$\Longrightarrow 6Mg_3Si_2O_5(OH)_4 + 2Fe_3O_4 + 0.5CH_4$$

(4) Olivine + Water + Carbonic acid ⇒ Serpentine + Magnetite + Magnesite + Silica

$$2Mg_2SiO_4 + 3Fe_2SiO_4 + 2H_2O + CO_2$$
$$\Longrightarrow Mg_3Si_2O_5(OH)_4 + 2Fe_3O_4 + MgCO_3 + 3SiO_2$$

The degree to which a mass of ultramafic rock undergoes serpentinization depends on both the starting rock and fluid compositions and on whether fluids transport Ca, Mg and other elements away during the process. If an olivine composition contains sufficient Fe, then olivine plus water can be completely replaced by serpentine and magnetite in a closed system. In most ultramafic rocks formed in the Earth's mantle, however, the olivine is ~90% forsterite, Mg end member, and for that olivine to react completely to serpentine, Mg must be transported out of the reacting volume.

Serpentinitization of a mass of peridotite often destroys all previous textural evidence because the serpentine minerals are weak and behave in a very ductile fashion. However, serpentinite may be less severely deformed, and in this case the rock retains textures of the former peridotite, as indicators of the replacement process. There have been many studies on the varied and interesting textures developed during serpentinization (such as in Wicks & Whittaker, 1977). An example of a replacement texture can be seen in Figure 12 showing a typical serpentine mesh texture, where

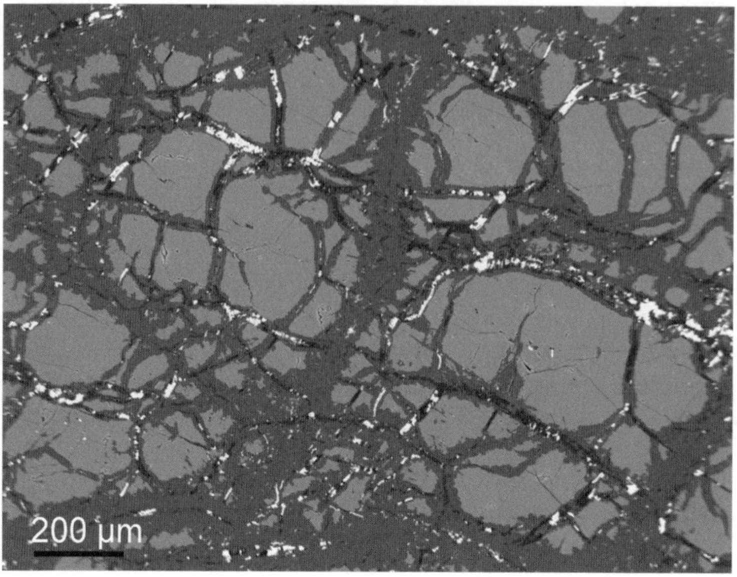

Fig. 12. SEM backscatter image of a typical mesh texture in serpentinized olivine. The lighter grey areas are the original olivine, which is intersected by fractures along which a reacting fluid could penetrate. The olivine is replaced by serpentine (darker grey areas) along the fractures. The blacker mineral in the fractures is magnesite and the very bright mineral is hematite (image courtesty of H. Austrheim).

the original parent olivine is intersected by cross-cutting veins of serpentine, indicating that a possible dissolution-replacement reaction occurred where fluid was able to access the mineral grains through fractures. There has been a long-running debate regarding possible volume changes during serpentinization. The formation of serpentine after olivine involves a large molar volume increase (30–40%) and so researchers have been divided between those arguing for a physical volume increase and those suggesting a constant-volume reaction (Thayer, 1966). O'Hanley (1992) gives an overview of this debate. However, reaction-induced fracturing has been reported for the pseudomorphic replacement of leucite by analcime, which also involves a molar volume increase (Fig. 10; Jamtveit *et al.*, 2009). Therefore a constant volume reaction is not unlikely. For such a scenario to occur, significant amounts of MgO and SiO_2 must be mobilized and removed. As seen in Figure 12, the serpentine is highly porous and the reaction interface is sharp. The veins of serpentine also contain magnesite, $MgCO_3$, and Fe oxide, indicating that the fluid phase contained Mg, Fe and carbonate, and implying that these elements were mobilized by the fluid phase. All these features are consistent with a dissolution-precipitation mechanism.

Clearly, for serpentinization to occur on a large scale, fluids have been available to bring in elements as well as to remove elements from the reaction site. Furthermore, sepentinization of deep subducting oceanic crustal rocks may play a significant role in pervasive hydration of the deep crust as well as the sub-Moho mantle (Ranero *et al.*, 2003). The implication is that an understanding of serpentinization is important for an understanding of the Earth's global water (Rüpke *et al.*, 2004) and carbon cycles.

4. Replacements in other mineral systems

4.1. The replacement of apatite

Apatite, has the general formula $Ca_5(PO_4)_3(OH,F,Cl)$, where the OH, F, and Cl ions can mutually replace each other to form the almost pure end-members. Apatite has the ability to incorporate trace elements and REEs into its structure. This may be by the substitution for Ca, the most common of which are trace amounts of Sr and REEs, where the Y group generally predominates. During replacement reactions, many if not most of these trace elements can be lost to the fluid phase. Harlov & Förster (2003) and Harlov *et al.* (2002, 2005) provide evidence that a combination of observation of natural samples as well as hydrothermal experiments, give important information about the nature of aqueous fluids and the mechanism of reactions. Y + *REE*s may be present in apatites in solid solution, which can be released as *REE*-bearing phosphate phases, such as monazite and xenotime, occurring as inclusions when the apatite reacts with fluids. This can be seen from chemical analyses and textural observations of the partially metasomatized apatites from the Bamble Sector, SE Norway, in which Cl-rich apatites have been replaced by OH and F apatites (Engvik *et al.*, 2009).

$$Ca_5(PO_4)_3Cl + OH^- + F^- \Longleftrightarrow Ca_5(PO_4)_3(OH,F) + Cl^-$$

The product phases are nanoporous but retain the crystallographic characteristics of the parent apatite and contain numerous inclusions of monazite and xenotime. Harlov and co-workers have been able to experimentally reproduce textures similar to those recorded from natural apatites. They found that the extent of metasomatism depended on temperature, but the textures and product compositions depended on initial fluid composition. In pure H_2O the replaced rims were porous, depleted in *REE*s compared with the parent apatite and contained monazite and xenotime inclusions. However, when the reacting fluid contained CaF_2 the altered parts were F-rich but retained their Y and *REE* composition and so had very few inclusions.

The conclusions from these observations and experimental results were that the presence of *REE*-rich inclusions in apatite resulted from a metasomatic replacement process controlled by the composition of the infiltrating aqueous fluid and to a much lesser extent temperature and pressure. Thus the results present another example of fluid-induced phase separation or ion partitioning by a coupled dissolution-precipitation mechanism.

4.2. The experimental replacement of aragonite by apatite

Recent hydrothermal experiments by Kasioptas *et al.* (2008, 2010) have shown that biogenic aragonite, $CaCO_3$, from a cuttlebone of *Sepia officinalis*, can very easily

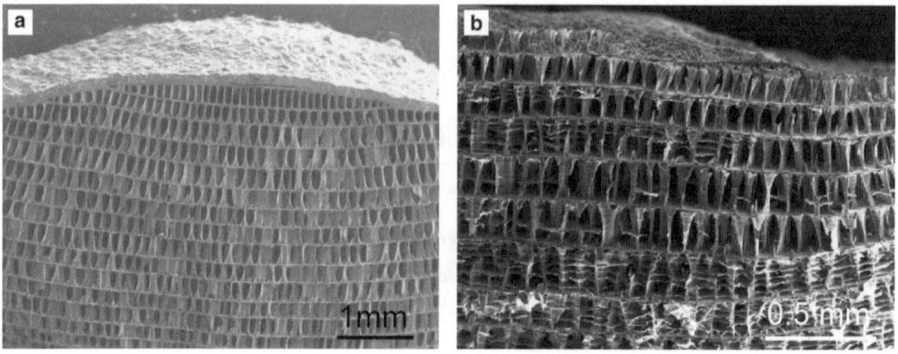

Fig. 13. SEM backscatter images of (*a*) the cuttlebone, composed of aragonite, $CaCO_3$, before replacement, and (*b*) after treatment in $(NH_4)_2PO_4$ solution at 50°C for 2 days. Note the retention of the fine internal structure of the cuttlebone, which has been replaced pseudomorphically by hydroxyapatite (image courtesty of A. Kasioptas).

be replaced by hydroxyapatite, $Ca_5(PO_4)_3(OH,CO_3)$ (also containing a small amount of carbonate, probably substituting for PO_4), at low temperatures (50–150°C), within a day or two, by reaction with a solution of ammonium hydrogen phosphate, $(NH_4)_2HPO_4$ in hydrothermal experimental reactors. The reaction is fast and results in a beautiful pseudomorphic replacement of the delicate chambered columnar internal aragonite structure (Fig. 13)

$$5CaCO_3 + 3(NH_4)_2HPO_4 \Longrightarrow Ca_5(PO_4)_3(OH,CO_3) + 3(NH_4)_2CO_3.$$

In the specific case of apatite formation, there are several hypotheses for the origin of marine phosphorites (sediments composed mainly of carbonate-fluorapatite), including diagenetic replacement of pre-existing carbonate deposits by carbonate-bearing apatite (Knudsen & Gunter, 2002). In medical sciences the pseudomorphic replacement of coral ($CaCO_3$) by coralline hydroxyapatite, preserving the unique microstructure, has resulted in material which can be used as a bone implant. As a biomaterial this offers biocompatibility, bioactivity and osteoconductivity, meaning that it can bond directly with bones and can slowly be replaced by the host bone after implantation (Roy & Linnehan, 1974).

4.3. *REE* and other trace-element enrichment during bone fossilization

Bone is calcified tissue which can be described as a composite of an organic matrix (mainly collagen) and an inorganic component (the nanocrystalline biomineral, carbonated hydroxylapatite, $Ca_{10}(CO_3,PO_4)_6(OH)_2$). In living organisms, as bone matures, the crystallinity and the Ca/P ratio of the inorganic component increase, whereas its content of carbonate decreases (Boskey & Mendelsohn, 2005; Glimcher, 1998; Rodríguez-Navarro *et al.*, 2006). Due to their specific shape and reduced size, hydroxylapatite crystals in bones have a very large surface area in comparison to those in other biomaterials such as enamel (Weiner & Price, 1986). Such a characteristic makes bone especially reactive (Nriagu, 1983). During burial and fossilization, the bone collagen is destroyed as a result of the hydrolysis of gelatine. Consequently, a very important volume of porosity is generated. Such porosity facilitates the interaction between bone hydroxylapatite crystals and groundwaters and pore waters in sediments. Since bone hydroxylapatite crystals are highly reactive, they also are susceptible to modification of their crystallinity, textural relations and composition during fossilization. Moreover, there is a general agreement that bone fossilization involves a recrystallization process through which the size of the hydroxylapatite crystals increases. Such a recrystallization process is supposed to occur very rapidly. As a result, the porosity volume generated after the lost of the bone collagen would become closed, to a large extent, shortly after burial.

Although not frequently discussed, fossilization is accompanied by bone apatite recrystallization. However, the actual characteristics of such recrystallization are controversial. To date, two models have been proposed to explain the process, the most significant features of which relate to the reduction of the porosity left after the loss of collagen and the increase in the crystallinity, and the original size of bone apatite crystals. One of the models considers that fossilization occurs *via* dissolution-reprecipitation of apatite. Therefore, in this model it is implicitly assumed that the formation of fossil bones is a process of pseudomorphism (Kolodny *et al.*, 1996), with similar characteristics to others described in the previous sections in this chapter. The other model assumes that the recrystallization process exclusively involves the growth of the previously existing apatite crystals, which act as 'seeds'. The increase in the size of bone apatite crystals resulting from their growth would explain the drastic reduction in the porosity generated after the lost of collagen (Hubert *et al.*, 1996). The controversy between the two models proposed to explain fossilization arises from the fact that most fossil bones retain exquisite patterns originally present in fresh bones (Zocco & Schwartz, 1994; Schweitzer *et al.*, 1997; Truemann & Tuross, 2002), which according to most authors would be impossible if a dissolution-precipitation mechanism were involved in bone fossilization. However, such a statement needs to be reconsidered in the light of recent experimental results. As has been explained in the previous section, Katsiopas *et al.* (2010) demonstrated that dissolution-precipitation processes can preserve tiny and delicate internal patterns if certain conditions are fulfilled.

The changes in size and crystallinity experienced by bone apatite crystals during fossilization correlate with changes in composition. These changes are facilitated by the ability of the apatite structure to incorporate a wide range of elements. The main factors influencing the modification of bone composition during fossilization are the original concentration of the different elements in the fresh bone and the chemistry of pore waters in the hosting sediment (Trueman & Tuross, 2002). Generally, some trace elements tend to be incorporated into bones during fossilization, increasing its concentration in comparison to concentrations found in fresh bones, while others tend to be partially removed. Although compositional changes during fossilization involve many trace metals, up to now most research have been focused on the variation of Sr, Ba, Pb, U, Th and the *REE* contents. It has been stated that Sr, a common component in fresh bones, can substitute extensively for Ca in fossil bones (Tuross *et al.*, 1989).

Pb can also accumulate in fossil bones. Furthermore, the concentration of elements such as U, Th and the *REE*s in fossil bones is markedly greater than in fresh bones.

The concentration of radiogenic and *REE* elements within fossil bones are often not homogeneous. On the contrary, concentration profiles show the existence of steep gradients between the surface and the interior of the bone. Moreover, it is worth noting that the greatest concentrations of U, Th and *REE* are found at bone margins, while their concentrations decrease inwards within the bone cortex (Williams, 1988; Williams & Marlow, 1987; Elliott & Grime, 1993; Janssens *et al.*, 1999). This distribution has been interpreted as resulting from the absorption of these elements onto the surface of bone apatite crystallites *via* cation exchange with Ca^{2+}. Such an incorporation mechanism would be facilitated by the high reactivity of the defective and poorly crystalline apatite crystals of fresh bones. It has been interpreted that the development of radiogenic and *REE* concentration profiles would reflect a diffusion controlled-transport through the bone. Finally, the heterogeneous distribution of these elements would become irreversible as the bone porosity closes up as a result of the recystallization process, which accompanies fossilization.

Despite overwhelming evidence for a correlation between trace-elements incorporation and crystallinity and size changes in fossil-bone apatite crystals, attempts to conciliate the characteristics of the recrystallization mechanism which could control fossilization and the enrichment of fossil bones in U, Th and the *REE*, as well as in other trace elements, and their distribution profiles within the fossil bones do not exist, to the knowledge of the authors. There are several facts that should be considered in order to reach a conclusion on which recrystallization is more likely to predominate in bone fossilization. For example, if radiogenic and *REE* incorporation into fossil bones were the result of absorption on the surface of apatite crystals prior to the closure of porosity, one would expect that the distribution pattern of these elements in the bone would be disrupted by the distribution of the original porosity of fresh bones after collagen loss. However, such a disruption is not observed. Moreover, the characteristics of the distribution profiles of U, Th and *REE* described above are inconsistent with bone fossilization occurring as a result of the re-growth of the original apatite crystals, acting as 'seeds'. If this were the mechanism, a concentric distribution of radiogenic and *REE* elements around each small apatite crystal would be expected. With reference to the whole bone this could be considered equivalent to an homogeneous distribution. Therefore, the steep gradients shown by radiogenic and *REE* concentration profiles, with the composition decreasing inwards, points to dissolution-precipitation as the most likely mechanism operating during bone fossilization. Such a process could be described as a pseudomorphic replacement with similar characteristics to other replacement examples described above.

4.4. Formation of diagenetic carbonates after replacement of evaporitic calcium sulphates

Soil waters generally have *pH* values as low as 4 or 5 but after percolation through carbonate rocks, meteoric waters previously equilibrated with oxygen from the atmosphere can dissolve significant amounts of carbonate ions and increase their *pH*. The interaction of carbonate-bearing waters with different rock-forming minerals can lead to the development of carbonation processes, which proceed *via* dissolution-precipitation and frequently occur with the preservation of the original external shape of the parent phase, *i.e.* with the formation of pseudomorphs. One of the most extended and geologically relevant mineral carbonation processes is that which leads to the replacement of evaporitic calcium sulphates (anhydrite and gypsum) by calcium carbonate. This process has been pointed out by numerous authors as responsible for the formation of large volumes of diagenetic carbonates (Rouchy *et al.*, 1994; Cañaveras *et al.*, 1996, 1998; Rouchy *et al.*, 1998; Sanz-Rubio *et al.*, 2001). It is considered that the replacement of sedimentary gypsum by $CaCO_3$ can involve two different processes, both of which involve element mobilization, including ionic and isotopic partitioning. Anadón *et al.* (1992) proposed a mechanism which would be promoted by sulphate-reducing bacteria activity in sediments with a large amount of organic matter. In this case, the replacement of gypsum or anhydrite by carbonate would be accompanied by the formation of native sulphur accumulations, which would be the result of the very significant remobilization of this element. Other authors (Wigley, 1973; Pierre & Rouchy, 1988) consider that the replacement of gypsum by carbonate is the direct result of sulphate dissolution by meteoric carbonate-rich groundwaters according to the following reaction (Sanz-Rubio *et al.*, 2001):

$$\begin{aligned} &CaSO_4 \cdot 2H_2O_{(gypsum)} + 2HCO_3^- \\ &\quad \Longrightarrow CaCO_{3(calcite)} + CO_2 + H_2O + SO_4^{2-}, \end{aligned}$$

where bicarbonate ions are produced by the oxidation of lacustrine or terrestrial organic matter after interaction with waters charged with CO_2. This mechanism is supported by the results of stable isotope analyses in a number of sedimentary basins (Sanz-Rubio *et al.*, 2001), which also indicate the importance of the changes in isotopic signatures resulting from replacement processes.

4.5. Pseudomorphic carbonation of gypsum. Experimental models

Flörke & Flörke (1961) reported for the first time the transformation of gypsum into vaterite/calcite under laboratory-controlled conditions. Fernández-Diaz *et al.* (2009) studied the development of carbonation of gypsum experimentally by immersing single crystals in aqueous solutions with different concentrations of dissolved carbonate. The results obtained by those authors show that gypsum single crystals are totally or partially replaced by calcite after a very short time (<1 day) in contact with carbonate-rich solutions. The replacement progresses inwards in the gypsum crystal and occurs with the preservation of its external shape. Moreover, when the replacement is partial, the rimming of the gypsum relict by an aggregate of calcite crystals is clearly distinguished (Fig. 14a). The advance of the replacement front is facilitated by the development of a significant volume of microporosity (Fig. 14b). As explained in previous sections, the generation of porosity is a requirement to guarantee the preservation of the external

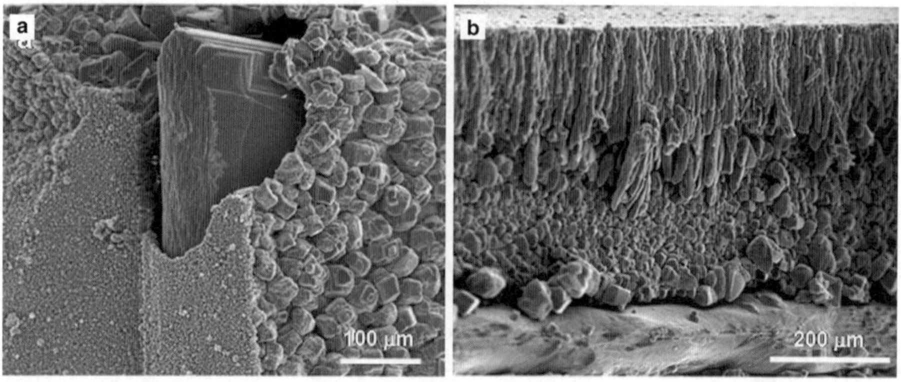

Fig. 14. SEM images of gypsum partially replaced by calcite. (***a***) Note the rimming of the gypsum relict by an aggregate of calcite crystals. The gap between the gypsum relict and the calcite rim can be clearly observed. Also note the very different scale of the pseudomorphism on different surfaces of the original gypsum crystal. (***b***) The replaced layer is porous and highly structured, with calcite columns penetrating towards the interface.

shape during mineral replacement reactions and arises from differences between both the solubilities and the molar volumes of the phases involved in the reaction (Yanagisawa *et al.*, 1999; Pollok *et al.*, 2002; Glikin *et al.*, 1994; Putnis & Mezger, 2004; Putnis & Putnis, 2007; Suárez-Orduña *et al.*, 2004; Rendón-Ángeles *et al.*, 2006; Pöml *et al.*, 2007). Gypsum is much more soluble than calcite ($K_{sp(\text{gypsum})} = 10^{-4.58}$; Parkhurst & Appelo, 1999) ($K_{sp(\text{calcite})} = 10^{-8.48}$; Plummer & Busenberg, 1982) and has a much larger molar volume (gypsum molar volume = 74.5 cm^3/mol; calcite molar volume = 36.40 cm^3/mol). Both factors determine that the transformation of gypsum into calcite is accompanied by a significant volume reduction. Such a volume reduction needs to be balanced to guarantee the preservation of the external shape of gypsum crystals. Moreover, in this particular case the generation of microporosity is insufficient to reach such a balance and a gap between the inward penetrating gypsum reaction interface and the replaced layer arises (Fig. 14a). The development of a similar gap during the replacement of pyrochlore as a result of hydrothermal alteration has been reported by Pöml *et al.* (2007). The microporosity facilitates the communication between the carbonate-rich aqueous phase and the interfacial liquid, which fills the gap, ensuring the progress of the replacement as long as the supply of carbonate is guaranteed by a high concentration of reactant in the bulk solution.

Unlike other mineral-replacement processes described in this chapter, the solvent-mediated transformation of gypsum into calcite does not involve the epitactic nucleation of the new phase. Therefore, no crystallographic relationships can be observed between gypsum and calcite. Moreover, the carbonation of gypsum occurs through a sequence of dissolution-crystallization reactions that involve the formation of a number of $CaCO_3$ metastable polymorphs (amorphous calcium carbonate (ACC), vaterite, aragonite). All these polymorphs subsequently transform into calcite *via* the solvent. The reaction pathway is controlled by the initial concentration of dissolved carbonate in the bulk solution. Since the ability of different $CaCO_3$ polymorphs to accommodate minor elements in their structure is variable, ion partitioning is likely to occur along sequences of polymorphic solvent-mediated transformations.

The initial concentration of dissolved carbonate in the bulk solution also has an influence on the scale of the pseudomorphism (Xia *et al.*, 2009) with greater concentrations of dissolved carbonate leading to a more faithful reproduction of the surface characteristics of the gypsum crystals. The lack of crystallographic relationships between the gypsum substrate and the product phases could also favour both the reproduction of fine morphological details of the original surface, as pointed out by Pöml *et al.* (2007), and the progress of the replacement since; as Prieto *et al.* (2003) demonstrated, the formation of epitaxial layers can prevent the development of a connected porosity net by protecting the dissolving substrate from further contact with the reactant solution.

4.6. Origin of the compositional bimodality of the $(Ba,Sr)CO_3$ ideal solid solution in natural occurrences

Carbonation processes have been considered relevant in many different contexts and can explain the genesis of both minerals that constitute large volumes of rocks, such as calcite in limestone, and minerals that appear as minor phases. This is the case for the end members of the $(Ba,Sr)CO_3$ solid solution, which is considered to be ideal and, therefore, complete at all temperatures and pressures (Chang, 1971; Brice & Chang, 1973). However, intermediate compositions between strontianite ($SrCO_3$) and witherite ($BaCO_3$) are absent in nature (Speer, 1983). The most Sr-rich witherite reported contains up to 11 mol.% $SrCO_3$, while the Ba contents of natural strontianites are minor (Baldasari & Speer, 1979). Such a bimodal composition of $(Ba,Sr)CO_3$ minerals introduced some doubts regarding the thermodynamic properties of this solid solution, also supported by the large difference between the cationic sizes of Sr^{2+} and Ba^{2+} (Chang, 1971; Speer & Hensley-Dunn, 1976). However, intermediate members of this solid solution have been synthesized in the laboratory. Fernández-González (1996) obtained $(Ba,Sr)CO_3$ spherulites showing many Ba/Sr ratios (Böttcher *et al.*, 1997) (Fig. 15), by using a counter-diffusion experimental set-up through silica gel. Similar results were obtained by Sánchez-Pastor (2007). In order to explain the lack of natural intermediate compositions, two possible reasons have been argued: (1) the non-occurrence of natural fluids from which $(Ba,Sr)CO_3$ with intermediate compositions could precipitate; and (2) a secondary origin for strontianite and witherite, which would form after $(Ba,Sr)SO_4$ precursors from which the Ba/Sr ratio could be inherited (Baldasari & Speer, 1979; Prieto *et al.*, 1997). The first argument can be ruled out as Speer (1983) stated that, in the geological environments in which strontianite and witherite are found, there is no lack of any of the substituting cations. Moreover, the results of Prieto *et al.* (1997) demonstrated that in the $(Ba,Sr)CO_3$-H_2O solid solution–aqueous solution system the composition of $(Ba,Sr)CO_3$ crystals is close to that of the aqueous phase from which they form even when growth occurs near equilibrium. This is a direct consequence of the similarity of the solubility products of the two

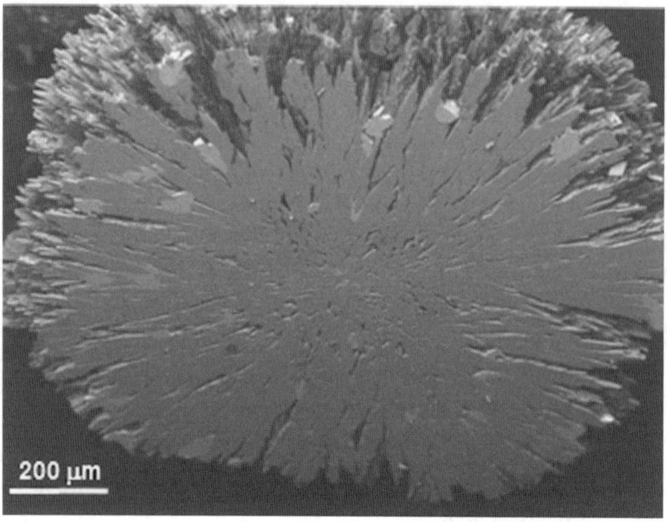

Fig. 15. SEM backscatter image of a $(Ba,Sr)CO_3$ spherulite grown in a silica gel. The composition of the spherulite is homogeneous and intermediate (image courtesty of N. Sánchez-Pastor).

end-members ($K_{sp(\text{strontianite})} = 10^{-9.27}$, Busenberg *et al.*, 1984; $K_{sp(\text{witherite})} = 10^{-8.56}$, Busenberg & Plummer, 1986), which determines that the solutus and solidus curves in the Lippmann diagram plot near each other (Fig. 16a).

The formation of strontianite and witherite as a result of the replacement of $(Ba,Sr)SO_4$ precursors gives a reasonable explanation for the compositional bimodality of a $(Ba,Sr)CO_3$ solid solution in natural occurrences. $(Ba,Sr)SO_4$ solid solution crystals show a strong compositional bimodality both in nature and laboratory experiments (Prieto *et al.*, 1993, 1997). Such a compositional bimodality is a direct consequence of the large difference in solubility between the two end-members (at 25°C $K_{sp(\text{barite})} = 10^{-9.98}$, Blount, 1977; $K_{sp(\text{celestite})} = 10^{-6.63}$; Reardon & Armstrong, 1987) of the solid solution. The strong preferential ion partitioning during crystallization in the $(Ba,Sr)SO_4$-H_2O system can be illustrated through its Lippmann diagram. In this case, the solutus and solidus curves plot far from each other (Fig. 16b). As a result, almost pure barite forms from a wide range of aqueous solutions, including very Sr-rich solutions, whereas Sr-rich solids only precipitate when the aqueous phase is extremely poor in Ba^{2+}. The range of aqueous-phase compositions that can precipitate intermediate members of the solid solution is extremely narrow.

The formation of witherite and strontianite *via* replacement of precursor sulphates is consistent with the fact that most natural occurrences of these minerals correspond to low or ambient temperatures and pressures, up to 130°C and up to 275 bar, respectively (Speer, 1983). Most strontianites appear in sedimentary rocks, mainly limestones, while witherite commonly occurs in hydrothermal vein deposits and rarely in sedimentary rocks (Baldasari & Speer, 1979; Speer & Hensley-Dunn, 1976).

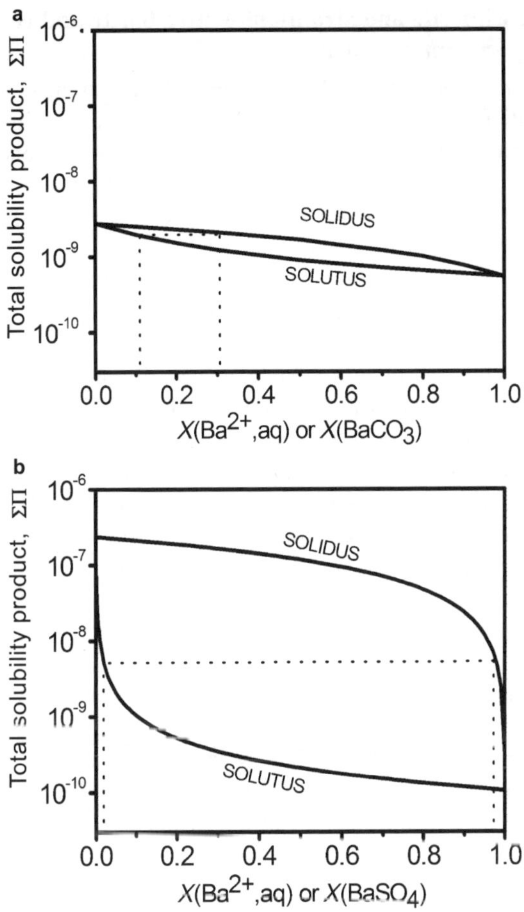

Fig. 16. Lippmann diagrams for (***a***) (Ba,Sr)CO_3-H_2O and (***b***) (Ba,Sr)SO_4-H_2O systems. In both plots, the dashed lines connect aqueous, x(aq), and solid phase, x(s) compositions at thermodynamic equilibrium. Note the contrast between the relatively close compositions of the aqueous and solid phases at equilibrium in the (Ba,Sr)CO_3-H_2O system and the widely different compositions of both phases in the (Ba,Sr)SO_4-H_2O system.

However, both strontianite and witherite are secondary minerals in these occurrences and its origin has been related to the interaction of carbonate-rich waters with celestite or strontian anhydrite, in the case of strontianite, and barite, in the case of witherite (Speer, 1983; Chang *et al.*, 1998).

The strong preferential ion partitioning during crystallization in the (Ba,Sr)SO_4-H_2O system has also been considered by Hanor (2004) as a strong argument to support the hypothesis that large celestite deposits hosted in sedimentary rocks (coastal carbonates and evaporites) form as a result of the epigenetic replacement of calcium sulphates (gypsum and anhydrite).

4.7. Formation of witherite and strontianite after barite and celestite replacement – experimental models

Suárez-Orduña *et al.* (2004) studied experimentally the transformation of celestite ($SrSO_4$) into strontianite ($SrCO_3$) as a result of interaction with alkaline aqueous phases under hydrothermal conditions. The experiments were conducted in a closed system at different temperatures in the range 150–250°C, using solutions with CO_3^{2-}/SO_4^{2-} molar ratios between 1.5 and 10. The complete transformation of square plates of celestite (10 mm wide and 3 mm thick) into strontianite occurred within 96 h when solutions with $CO_3^{2-}/SO_4^{2-} = 10$ were used. The replacement occurred *via* coupled dissolution-crystallization, proceeding without severe changes in the original shape or dimensions of the celestite crystals. Strontianite pseudomorphs after celestite show a characteristic internal texture, which consists of randomly distributed channels and holes. This texture can be considered a relict of the porosity that controlled the mass transport during the replacement and is evidence for a coupled dissolution-crystallization mechanism (Suárez-Orduña *et al.*, 2007). As in other examples explained above, this porosity originates as a result of a combination between the differences in solubility and in molar volume between the two phases involved in the solvent mediated transformation. Moreover, Suárez-Orduña *et al.* (2004) reported that the celestite-strontianite transformation is topotactic. Therefore, an epitaxial nucleation of strontianite on the surface of the dissolving celestite crystal can be considered to be the initial step for the pseudomorphic replacement.

A similar experimental study of the replacement of barite ($BaSO_4$) by witherite ($BaCO_3$) under hydrothermal conditions was carried out by Rendón-Ángeles *et al.* (2008). These authors reported that barite crystals in contact with alkaline carbonate-rich aqueous solutions in the temperature range between 150 and 250°C are replaced by witherite without significant changes in the original shape and dimensions of the crystals. Also, in this case, the pseudomorphic replacement has a topotactic character and, as it proceeds, a specific microstructure develops, which consists of an arrangement of elongated channels and pores, with similar characteristics to that described for the replacement of celestite by strontianite.

The coupling between dissolution and crystallization during the interaction of celestite (001) surfaces with carbonate bearing solutions was observed at the nanoscale by Sánchez-Pastor *et al.* (2007a,b). Atomic force microscopy (AFM) experiments showed that islands of a new phase form rapidly on celestite (001) surfaces in contact with carbonate-rich aqueous solutions (Fig. 17). The growth of these islands correlates with the dissolution of the celestite surface, which is enhanced in comparison to dissolution in pure water. On the basis of the height of monolayers of the newly formed phase and the orientation of the islands on the celestite substrate, Sanchez-Pastor *et al.* (2007a) concluded that the islands were strontianite and their growth was epitaxial. Moreover, those authors proposed the parallelism between (001) planes $b_{strontianite} \parallel a_{celestite}$, and $a_{strontianite} \parallel b_{celestite}$, as the most probable epitaxial relationships, considering *Pmcn* and *Pnma* space group set-ups for strontianite and celestite, respectively. Similar results were obtained for the interaction between carbonate-bearing solutions and barite crystals (Sanchez-Pastor *et al.*, 2007b).

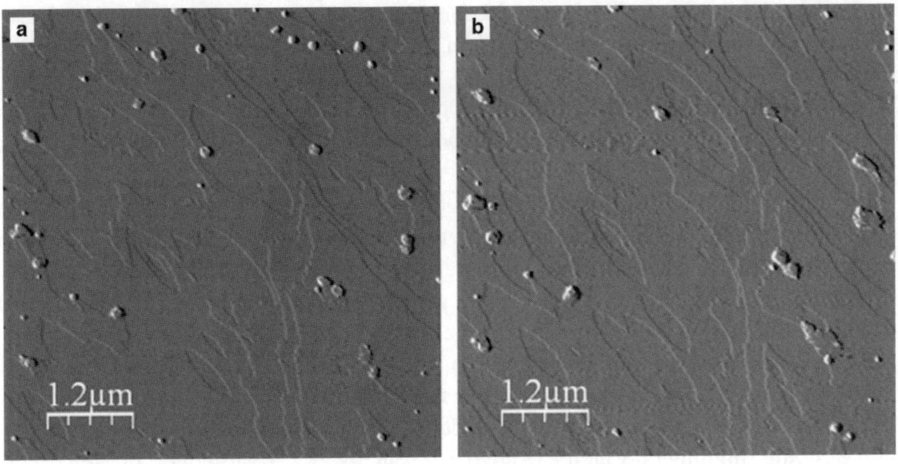

Fig. 17. AFM images of a celestite (001) surface dissolving in contact with a carbonate-bearing aqueous solution. Note the simultaneous growth of oriented islands. The time elapsed between images (***a***) and (***b***) was 200 s. Sánchez-Pastor *et al.* (2007a) concluded that the islands are strontianite on the basis of the almost identical values of $\frac{1}{2}\ d_{001}$ in strontianite (3.015 Å) and the height of growing monolayers (2.9 ± 0.2 Å) (images courtesy of N. Sánchez-Pastor).

It can be concluded from the experimental results obtained in the $Sr\text{-}SO_4^{2-}\text{-}CO_3^{2-}\text{-}H_2O$ and $Ba\text{-}SO_4^{2-}\text{-}CO_3^{2-}\text{-}H_2O$ systems that the replacement of sulphate precursors through a coupled dissolution-precipitation mechanism is a likely process to occur, which could satisfactorily explain the bimodal composition shown by the ideal $(Ba,Sr)CO_3$ (witherite-strontianite) solid solution in natural occurrences.

4.8. Replacement processes involving environmentally hazardous products: The transformation of phosgenite into cerussite

Chemical weathering of the upper part of numerous sulphide deposits commonly produces local enrichments in metals. Such enrichment is promoted by the interaction between acidic groundwaters and primary sulphide minerals. The breakdown of sulphide minerals *via* oxidation reactions leads to the mobilization of the metal ions into the aqueous phase (Chávez, 2000; Gilg *et al.*, 2008). Dissolved metals can subsequently reprecipitate where physicochemical conditions become more reducing as secondary minerals, mainly sulphates and carbonates like anglesite, cerussite, malachite, siderite, *etc.* (Robb, 2005). Similar processes are also common in slag heaps and tailings around sulphide mine sites. The evolution of the secondary minerals formed as a result of sulphide alteration processes is gaining relevance since these minerals are exposed to weathering and pose the potential threat of environmental contamination (Walder & Chavez, 1995). Many secondary minerals bear high concentrations of environmentally hazardous metals like lead or copper. In particular, the interest in the behaviour of lead secondary minerals, as well as in the aqueous geochemistry of lead, has increased considerably in recent decades due to the importance of this metal as a pollutant of soils and groundwaters (Hem, 1976; Saether *et al.*, 1988; Frost *et al.*, 2003; Porter *et al.*, 2004).

Phosgenite ($Pb_2C1_2CO_3$) and cerussite ($PbCO_3$) are secondary minerals that can form as a result of interaction of carbonate-rich waters with lead minerals, both primary, *e.g.* galena, and secondary, *e.g.* anglesite, in supergene deposits, slag heaps or as degradation products of art works (Bersani *et al.*, 2008). The transformation of phosgenite into cerussite is a relatively common phenomenon in nature, which can readily occur in the presence of an aqueous phase. The relative stability of both phases is strongly *pH*-dependent (Humphreys *et al.*, 1980; Abdul-Samad *et al.*, 1981; Abdul-Samad *et al.*, 1982). Phosgenite grows from acidic waters with large chloride anion contents (Bersani *et al.*, 2008). An increase in the *pH* above 4.5 can trigger the replacement of phosgenite crystals by cersussite *via* dissolution-precipitation. Evidence for this process has been found in numerous lead deposits where both minerals are associated, with cerussite rimming phosgenite (Lattanzi & Tanelli, 1978). Moreover, in such occurrences, epitatic relationships between phosgenite and cerussite have been observed (Von Vultée, 1950). Physicochemical conditions similar to those that induce the transformation of phosgenite into cerussite were reproduced by Pina *et al.* (1996) using the silica-gel crystal growth method. The experiments conducted by those authors showed that the transformation of phosgenite by cerussite was triggered by the heterogeneous nucleation of cerussite on specific sites of phosgenite crystal surfaces, such as corners and edges. The transformation proceeded, showing typical characteristics of mineral-replacement processes: a sharp boundary between the parent phase, phosgenite and the newly formed phase, cerussite, could be clearly distinguished (Fig. 18a); the

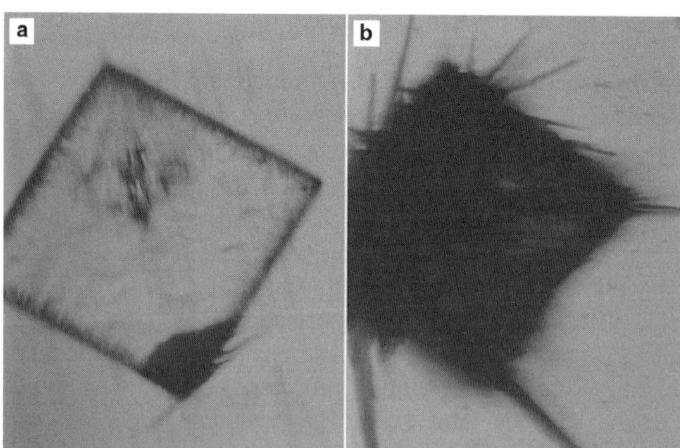

Fig. 18. Optical microscope images of phosgenite crystals (***a***) partially and (***b***) totally transformed into cerussite within a silica gel matrix. In (*a*) note the sharp boundary between the replaced region and the parent phase. The blurred edges of the phosgenite crystal are a sign of incipient dissolution. In (*b*) the replacement is complete. The overgrowth of some cerussite crystals on the original surface can be observed.

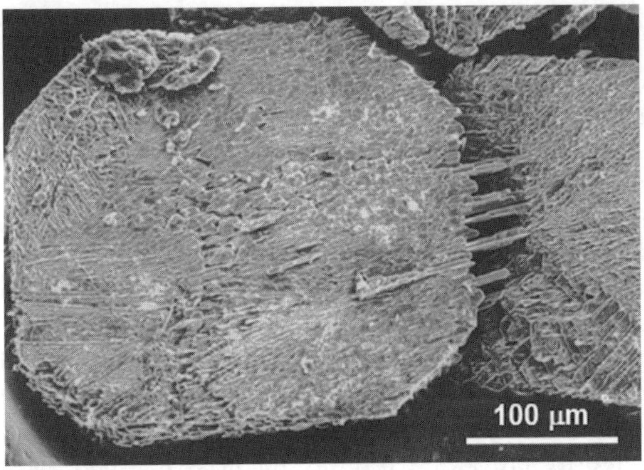

Fig. 19. SEM image of cerussite pseudomorphs after phosgenite. The pseudomorphs consist of an aggregate of cerussite crystals that have grown oriented with respect to the original phosgenite surface.

transformation front moved into the phosgenite crystal as the transformation proceeded; and during the process the external shape of the parent phase was fairly well preserved (Fig. 18b). Considering that this transformation occurred under ambient conditions, the only likely operating mechanism is a solvent-mediated transformation, which involves the dissolution of phosgenite and precipitation of cerussite and is mediated by the presence of a thin aqueous film at the interface between both minerals. Observations by AFM confirmed that this solvent-mediated transformation fulfils another characteristic of mineral-replacement processes (Pina *et al.*, 2000): the coupling between the dissolution of the parental phase and the growth of the product phase. Such a coupling is proven by a significant enhancement of the dissolution rate of phosgenite once cerussite nuclei form on its surface, subsequently followed by the achievement of a coupling between the rates of phosgenite dissolution and cerussite precipitation.

The coupling between the dissolution of phosgenite and the growth of cerussite guarantees the fairly good preservation of the original shape of phosgenite crystal after the replacement is completed. Figure 19 shows a cerussite pseudomorph after phosgenite, suggesting strong structural control of this replacement. The nucleation of cerussite on the phosgenite original surface is epitactic. This is seen in the microstructure of the pseudomorph, which consists of an aggregate of parallel cerussite crystals elongated along their *c* axis.

5. Conclusions

The examples above have been presented to indicate how element mobilization and ion partitioning may occur both in natural Earth environments, as well as in experimental samples of minerals, by a mechanism involving dissolution of the solid in an aqueous fluid, coupled with precipitation of a new solid phase. The main features of such a coupled dissolution-precipitation mechanism are:

(1) the dissolution of the parent solid in a fluid layer at the mineral interface, such that this boundary fluid layer becomes saturated with respect to another phase (or phases), which can then immediately/simultaneously precipitate as the new product solid mineral phase;

(2) the generation of porosity in the product phase which is less soluble in the fluid at the interface than the parent phase;

(3) the movement of the fluid phase through the generated porosity within the product up to the interface between parent and product minerals.

The replacement reaction takes place at an interface that moves continuously within the mineral. The rate of replacement depends on the rates of dissolution and precipitation as well as the diffusion of elements within the aqueous fluid to and from the reaction interface. Porosity provides a pathway for mass transport and hence, element mobilization.

The main observation is that by a careful analysis of textural and compositional characteristics, valuable information about the past history of a mineral can be collected, as well as elucidating possible element exchanges between solid mineral and aqueous fluid phases. As soon as minerals form or crystallize in a given environment of temperature, pressure and chemistry, they begin to undergo changes or re-equilibrate in response to changes in the physical and chemical surrounding conditions. The examples illustrated here show that it is possible to retain external form and crystallographic information, while the chemical composition is completely changed, that is, a pseudomorphic replacement. During this process elements may be mobilized into solution while others are precipitated from solution into a new mineral replacement. Therefore coupled dissolution-precipitation is a mechanism whereby element mobilization effectively partitions elements into new mineral phases. This mechanism is likely to be relevant in the Earth whenever one mineral is pseudomorphically replaced by another.

Acknowledgements

The authors are grateful for funding received from various sources, enabling much of the research reported here to be carried out. CVP acknowledges support from Marie Curie Initial Training networks, MIR (Contract No. MEST-CT-2005-021120), MinGro (Contract No. MRTN-CT-2006-035488) and DeltaMin (Contract No. PITN-GA-2008-215360) and the German Research Council (DFG), as well as the support from the technical staff at the Institut für Mineralogie, Münster. LFD acknowledges support from the Spanish Ministry of Science and Innovation (Project CGL2007-65523-C02-01) and the Comunidad de Madrid (Project CAM-2009 910148), as well as the assistance and support from technical staff at the Central Services (Microscopy Centre and Diffraction Centre) of the Universidad Complutense de Madrid.

References

Abdul-Samad, F.A., Humphreys, D.A., Thomas, J.H. & Williams, P.A. (1981) Chemical studies on the stabilities of boileite and pseudoboileite. *Mineralogical Magazine*, **44**, 101–104.

Abdul-Samad, F., Thomas, J.H., Williams, P.A., Bideaux, R. & Symes, R.F. (1982) Mode of formation of some rare copper (II) and lead (II) minerals from aqueous solution, with particular reference to deposits at Tiger, Arizona. *Transition Metal Chemistry*, **7**, 32–37.

Anadón, P., Rosell, L. & Talbot, M.R. (1992) Carbonate replacement of lacustrine gypsum deposits in two Neogene continental basins, eastern Spain. *Sedimentary Geology*, **78**, 201–216.

Baldasari, A. & Speer, J.A. (1979) Witherite composition physical properties, and genesis. *American Mineralogist*, **64**, 742–747.

Berndt, M.E., Allen, D.E. & Seyfried Jr., W.E. (1996) Reduction of CO_2 during serpentinization of olivine at 300°C and 500bar. *Geology*, **24**, 351–354.

Bersani, D., Campani, E., Casoli, A., Lotici, P.P. & Marino, I.G. (2008) Spectroscopic studies of the degradation products in the holy water fonts in Santa Maria della Steccata Church in Parma (Italy) *Analytica Chimica Acta*, **610**, 74–79.

Bierlein, F.P., Ashley, P.M., & Plimer, I.R. (1995) Sulphide mineralisation in the Olary Block, South Australia: evidence for syn-tectonic to late-stage mobilisation. *Mineralium Deposita*, **30**, 424–438.

Blount, C.W. (1977) Barite solubilities and thermodynamics quantities up to 300°C and 1400 bars. *American Mineralogist*, **62**, 942–957.

Boskey, A. & Mendelsohn, R. (2005) Infrared analysis of bone in health and disease. *Journal of Biomedicine. Optics*, **10**, 031102-1–031102-9.

Böttcher, M.E., Gehleken, P.L., Fernández-González, A. & Prieto, M. (1997) Characterization of synthetic $BaCO_3$–$SrCO_3$ (witherite–strontianite) solid-solution by Fourier transform infrared spectroscopy. *European Journal of Mineralogy*, **9**, 519–528.

Brice, W.R. & Chang, L.L.Y. (1973) Subsolidus phase relations in aragonite-type carbonates. III. The systems $MgCO_3$–$CaCO_3$–$BaCO_3$, $MgCO_3$–$CaCO_3$–$SrCO_3$, and $MgCO_3$–$SrCO_3$–$BaCO_3$. *American Mineralogist*, **58**, 970–985.

Busenberg, E. & Plummer, L.N. (1986) The solubility of $BaCO_3$ (cr) (witherite) in CO_2-H_2O solutions between 0 and 90°C, evaluation of the association constants of $BaHCO_3^+$(aq) and $BaCO_3^0$(aq) between 5 and 80°C, and a preliminary evaluation of the thermodynamic properties of Ba^{2+} (aq). *Geochimica et Cosmochimica Acta*, **50**, 2225–2233.

Busenberg, E., Plummer, L.N. & Parker, V.B. (1984) The solubility of strontianite ($SrCO_3$) in CO_2–H_2O solutions between 2 and 91°C, the association constants of $SrHCO_3^+$ and $SrCO_3$ (aq) between 5 and 80°C, and an evaluation of the thermodynamic properties of Sr^{2+}(aq) and $SrCO_3$(cr) at 25°C and 1 atm total pressure. *Geochimica et Cosmochimica Acta*, **48**, 2021–2035.

Cañaveras, J.C., Sánchez-Moral, S., Calvo, J.P., Hoyos, M. & Ordoñez, S. (1996) Dedolomites associated with karstification, an example of early dedolomitización in lacustrine sequences from the Tertiary of the Madrid Bassin, Central Spain. *Carbonates and Evaporites*, **11,** 85–103.

Cañaveras, J.C., Sánchez-Moral, S., Sanz-Rubio, E. & Hoyos, M. (1998) Meteoric calcitization of magnesite in Miocene lacustrine deposits (Calatayud Basi, NE Spain). *Sedimentary Geology*, **119,** 183–194.

Chang, L.L.Y. (1971) Subsolidus phase relations in aragonite-type carbonates. 1. System $CaCO_3$–$SrCO_3$–$BaCO_3$, *American Mineralogist*, **56**, 1660–1673.

Chang, L.L.Y., Howie, R.A. & Zussman, J. (1998) *Non-silicates: Sulphates, Carbonates, Phosphates, Halides.* Rock-Forming Minerals, vol. 5B. The Geological Society London, London, 383 pp.

Chávez, W.X. (2000) Supergene oxidation of copper deposits: zoning and distribution of copper oxide minerals. *Society of Economic Geologists Newsletter*, **41**, 10–21.

Clark, C., Schmidt Mumm, A. & Faure, K. (2005) Timing and nature of fluid flow and alteration during Mesoproterozoic shear zone formation, Olary Domain, South Australia. *Journal of Metamorphic Geology*, **23**, 147–164.

Elliott, T.A. & Grime, G.W. (1993) Examining the diagenetic alteration of human bone material from a range of archaeological burial sites using nuclear microscopy. *Nuclear Instruments and Methods in Physics Research Section B*, **77**, 537–547.

Engvik, A.K., Putnis, A., FitzGerald, J.D. & Austrheim, H. (2008) Albitisation of granitic rocks: the mechanism of replacement of oligoclase by albite. *The Canadian Mineralogist*, **46**, 1401–1415.

Engvik, A.K., Golla-Schindler, U., Berndt-Gerdes, J., Austrheim, H. & Putnis, A. (2009) Intragranular replacement of chlorapatite by hydroxyl-fluor–apatite during metasomatism. *Lithos*, **112**, 236–246.

Fernández-Díaz, L., Pina, C.M., Astilleros, J.M. & Sánchez-Pastor, N. (2009) The carbonatation of gypsum: Pathways and formation of pseudomorphs. *American Mineralogist*, **94**, 1223–1234.

Fernández-González, A. (1996) *Cristalización de soluciones sólidas completas en sistemas acuosos diffusion-reacción.* PhD Thesis, University of Oviedo, Oviedo, Spain, 241 pp. (in Spanish).

Fiebig, J. & Hoefs, J. (2002) Hydrothermal alteration of biotite and plagioclase as inferred from intragranular oxygen isotope- and cation-distribution patterns. *European Journal of Mineralogy*, **14**, 49–60.

Flörke, W. & Flörke, O.W. (1961) Vateritbildung aus Gips in Sodalösung. Neues *Jahrbuch fürMineralogie Monatshaften*, 179–181 (in German).

Frost, R.L., Martens, W., Kloprogge, T. & Ding, Z. (2003) Raman spectroscopy of selected lead minerals of environmental significance. *Spectrochimica Acta*, **A59**, 2705–2711.

Gilg, H.A., Boni, M., Hochleitner, R. & Struck, U. (2008) Stable isotope geochemistry of carbonate minerals in supergene oxidation zones of Zn-Pb deposits. *Ore Geology Reviews*, **33**, 117–133.

Glikin, A.E., Leontieva, O.A., & Sinai, M.Y. (1994) Mechanisms of exchange of isomorphous components between crystal and solution and macro-defectiveness in secondary crystals. *Journal of Structural Chemistry*, **35**, 642–646.

Glimcher, M. J. (1998) The nature of the mineral phase in bone: Biological and clinical implications. In: *Metabolic Bone Disease and Clinically Related Disorders* (Alvioli, L.V. & Krane, S. M., editors), 23–50. Academic Press, San Diego, California, USA.

Glynn, P.D., Reardon, E.J., Plummer, L.N., & Busenberg, E. (1990) Reaction paths and equilibrium end-points in solid solution-aqueous solution systems. *Geochimica et Cosmochimica Acta*, **54**, 267–282.

Hanor, J.S. (2004) A model for the origin of large carbonate- and evaporite-hosted Celestine ($SrSO_4$) deposits. *Journal of Sedimentary Research*, **74**, 168–175.

Harlov, D.E. & Förster, H-J. (2003) Fluid-induced nucleation of (Y+ REE)-phosphate minerals within apatite: Nature and experiment. Part II. Fluorapatite. *American Mineralogist*, **88**, 1209–1229.

Harlov, D.E., Förster, H-J. & Nijland, T.G. (2002) Fluid-induced nucleation of (Y+ REE)-phosphate minerals within apatite: Nature and experiment. Part I. Chlorapatite. *American Mineralogist*, **87**, 245–261.

Harlov, D.E., Wirth, R. & Förster, H-J. (2005) An experimental study of dissolution-reprecipitation in fluorapatite: Fluid infiltration and the formation of monazite. *Contributions to Mineralogy and Petrology*, **150**, 268–286.

Hem, J.D. (1976) Geochemical controls on lead concentrations in stream water and sediments. *Geochimica et Cosmochimica Acta*, **40**, 599–609.

Hövelmann, J., Putnis, A., Geisler, T., Schmidt, B.C. & Golla-Schindler, U. (2010) The replacement of plagioclase feldspar by albite: observations from hydrothermal experiments. *Contributions to Mineralogy and Petrology*, **159**, 43–59.

Hubert, J.F., Panish, P.T., Chure, D.J. & Prostak, K.S. (1996) Chemistry, microstructure, petrology, and diagenetic model of Jurassic dinosaur bones, Dinosaur National Monument, Utah. *Journal of Sedimentary Research*, **66**, 531–547.

Humphreys, D.A., Thomas, J.H., Williams, P.A. & Symes, R.F. (1980) The chemical-stability of mendipite, diabolite, chloroxiphite, cumengeite, and their relationships to other secondary lead(II) minerals. *Mineralogical Magazine*, **43**, 901–904.

Jamtveit, B., Putnis, C.V. & Malthe-Sorenssen, M. (2009) Reaction induced fracturing during replacement processes. *Contributions to Mineralogy and Petrology*, **157**, 127–133.

Janssens, K., Vincze, L., Vekemans, B., Williams, C.T., Radtke, M., Haller, M. & Knöchel, A. (1999) The non-destructive determination of REE in fossilized bone using synchrotron radiation induced K-line X-raymicrofluorescence analysis. *Fresenius' Journal of Analytical Chemistry*, **363**, 413–420.

Kasioptas, A., Perdikouri, C., Putnis, C.V. & Putnis, A. (2008) Pseudomorphic replacement of single calcium carbonate crystals by polycrystalline apatite. *Mineralogical Magazine*, **72**, 77–80.

Kasioptas, A., Geisler, T., Putnis, C.V., Perdikouri, C. & Putnis, A. (2010) Crystal growth of apatite by replacement of an aragonite precursor. *Journal of Crystal Growth*, doi: 10.1016/j.jcrysgro.2010.05.014

Knudsen, A.C. & Gunter, M.E. (2002) Sedimentary Phosphorites - An Example: Phosphoria Formation, Southeastern Idaho, U.S.A. In: *Phosphates – Geochemical, Geobiological and Materials Importance* (M.L. Kohn, J. Rakovan & J.M. Hughes, editors). Reviews in Mineralogy and Geochemistry, **48**, 363–389. Washington, D.C.

Kolodny, Y., Luz, B., Sander, M. & Clemens, W.A. (1996) Dinosaur bones: fossils or pseudomorphs? The pitfalls of physiology reconstruction from apatitic fossils. *Palaeogeography, Palaeoclimatology, Palaeoecology*, **126**, 161–171.

Labotka, T.C., Cole, D.R., Fayek, M., Riciputi, L.R. & Stadermann, F.J. (2004) Coupled cation and oxygen-isotope exchange between alkali feldspar and aqueous chloride solution. *American Mineralogist*, **89**, 1822–1825.

Lattanzi, P. & Tanelli, G. (1978) Coniderazioni genetiche sulla cotunnite del giacimento a pirite di Niccioleta (Toscana). *Rendiconti della Società Italiana di Mineralogia e Petrologia*, **34**, 37–44 (in Italian).

Lee, M.R. & Parsons, I. (1997) Dislocation formation and albitization in alkali feldspars from the Shap granite. *American Mineralogist*, **82**, 557–570.

Lippmann, F. (1980) Phase diagrams depicting the aqueous solubility of binary mineral systems. *Neues Jahrbuch für Mineralogie Abhandlungen*, **139**, 1–25.

Niedermeier, D.R.D., Putnis, A., Geisler, T., Golla-Schindler, U. & Putnis, C.V., (2009) The mechanism of cation and oxygen isotope exchange in alkali feldspars under hydrothermal conditions. *Contributions to Mineralogy and Petrology*, **157**, 65–76.

Nriagu, J.O. (1983) Rapid decomposition of fish bones in Lake Erie sediments. *Hydrobiologica*, **106**, 217–222.

O'Hanley, D.S. (1992) Solution to the volume problem in serpentinization. *Geology*, **20**, 705–708.

Oliver, N.H.S., Cleverley, J.S., Mark, G., Pollard, P.J., Fu, B., Marshall, L.J., Rubenach, M.J., Williams, P.J. & Baker, T. (2004) Modeling the role of sodic alteration in the genesis of iron oxide-copper-gold deposits, Eastern Mount Isa Block, Australia. *Economic Geology*, **99**, 1145–1176.

O'Neil, J.R. & Taylor, H.P. (1967) The oxygen isotope and cation exchange chemistry of feldspars. *American Mineralogist*, **52**, 1414–1437.

Onuma, K., Kameyama, T. & Tsukamoto, K. (1994) In situ study of surface phenomena by real time phase shift interferometry. *Journal of Crystal Growth*, **137**, 610–622.

Parkhurst, D.L. & Appelo, C.A.J. (1999) *User's guide to PHREEQC (version 2) A computer program for speciation, batch-reaction, one-dimensional transport, and inverse geochemical calculations.* U.S. Geological Survey Water Resources Investigations Report **99–4259**, U.S. Geological Survey, Denver, 312 pp.

Parsons, I. & Lee, M.R. (2009a) Mutual replacement reactions in alkali feldspars I: microtextures and mechanisms. *Contributions to Mineralogy and Petrology*, **157**, 641–661.

Parsons, I. & Lee, M.R. (2009b) Mutual replacement reactions in alkali feldspars II: trace element partitioning and geothermometry. *Contributions to Mineralogy and Petrology*, **157**, 663–687.

Pierre, C. & Rouchy, J.M. (1988) Carbonate replacements after sulphate evaporites in the Middle Miocene of Egypt. *Journal of Sedimentary Petrology*, **58**, 446–456.

Pina, C.M., Fernández-Díaz, L. & Prieto, M. (1996) Topotaxy relationships in the transformation phosgenite-cerussite. *Journal of Crystal Growth*, **158**, 340–345.

Pina, C.M., Fernández-Díaz, L., Prieto, M. & Putnis, A. (2000) In situ atomic force microscope observations of a dissolution-crystallisation reaction: The phosgenite- cerussite transformation. *Geochimica et Cosmochimica Acta*, **64**, 215–221.

Plummer, L.N. & Busenberg, E. (1982) The solubilities of calcite, aragonite and vaterite in CO_2-H_2O solutions between 0 and 90°C, and an evaluation of the aqueous model for the system $CaCO_3$-CO_2-H_2O. *Geochimica et Cosmochimica Acta*, **46**, 1011–1040.

Pollok, K., Pina, C.M., Putnis, C.V., Gilkin, A.E., & Putnis, A. (2002) Replacement reactions in solid solution-aqueous solution systems using the KBr-KCl-H_2O system as a model: theoretical considerations on volume changes and kinetics. In *Program and Abstracts of the 18th General Meeting IMA*, MO9, 111.

Pöml, P., Menneken, M., Stephan, T., Niedermeier, D.R.D., Geisler, T., & Putnis, A. (2007) Mechanism of hydrothermal alteration of natural self-irradiated and synthetic crystalline titanate-based pyrochlore. *Geochimica et Cosmochimica Acta*, **71**, 3311–3322.

Porter, S.K., Scheckel, K.G., Impellitteri, C.A. & Ryan, J.A. (2004) Toxic metals in the envorinment: Thermodynamic considerations for possible immobilization strategies for Pb, Cd, As, and Hg. *Critical Reviews in Environmental Science and Technology*, **34**, 495–604.

Prieto, M., Putnis, A. & Fernández-Díaz, L. (1993) Crystallization of solid-solutions from aqueous-solutions in a porous-medium: Zoning in $(Ba,Sr)SO_4$. *Geological Magazine*, **130**, 289–299.

Prieto, M., Fernández-González, A. & Fernández-Díaz, L. (1997) Nucleation, growth, and zoning phenomena in crystallizing $(Ba,Sr)CO_3$, $Ba(SO_4,CrO_4)$, $(Ba,Sr)SO_4$, and $(Cd,Ca)CO_3$ solid solutions from aqueous solutions. *Geochimica et Cosmochimica Acta*, **61**, 3383–3397.

Prieto, M., Cubillas, P., & Fernández-González, A. (2003) Uptake of dissolved Cd by biogenic aragonite: a comparison with sorption onto calcite. *Geochimica et Cosmochimica Acta*, **67**, 3859–3869.

Putnis, A. (2002) Mineral replacement reactions: from macroscopic observations to microscopic mechanisms. *Mineralogical Magazine*, **66**, 689–708.

Putnis, A. (2009) Mineral replacement reactions. In: *Thermodynamics and Kinetics of Water-Rock Interaction* (E.H. Oelkers & J. Schott, editors). Reviews in Mineralogy and Geochemistry, **70**, 87–124. Mineralogical Society of America, Washington, D.C.

Putnis, C.V. & Mezger, K. (2004) A mechanism of mineral replacement: Isotope tracing in the model system KCl-KBr-H_2O. *Geochimica et Cosmochimica Acta*, **68**, 2839–2848.

Putnis, A. & Putnis, C.V. (2007) The mechanism of equilibration of solids in the presence of a fluid phase. *Journal of Solid State Chemistry*, **180**, 1783–1786.

Putnis, C.V., Tsukamoto, K. & Nishimura, Y. (2005) Direct observations of pseudomorphism: compositional and textural evolution at a fluid-solid interface. *American Mineralogist*, **90**, 1909–1912.

Putnis, A., Hinrichs, R., Putnis, C.V., Golla-Schindler, U. & Collins, L.G. (2007a) Hematite in porous red-clouded feldspars: Evidence of large-scale crustal fluid-rock interaction. *Lithos*, **95**, 10–18.

Putnis, C.V., Geisler, T., Schmid-Beurmann, P., Stephan, T. & Giampaolo, C. (2007b) An experimental study of the replacement of leucite by analcime. *American Mineralogist*, **92**, 19–26.

Ranero, C.R., Phipps Morgan, J., McIntosh, K. & Reichert, C. (2003) Bending-related faulting and mantle serpentinization at the Middle America trench. *Nature*, **425**, 367–373.

Reardon, E.J. & Armstrong, D.K. (1987) Celestite ($SrSO_4$(s)) solubility in water, sea water and NaCl solution. *Geochimica et Cosmochimica Acta*, **51**, 63–72.

Rendón-Ángeles, J.C., Pech-Canul, M.I., López-Cuevas, J., Matamoros-Veloza, Z., & Yanagisawa, K. (2006) Differences on the conversion of celestite in solutions bearing monovalent ions under hydrothermal conditions. *Journal of Solid State Chemistry*, **179**, 3645–3652.

Rendón-Ángeles, J.C., Matamoros-Veloza, Z., López-Cuevas J., Pech-Canul, M. I. & Yanagisawa, K. (2008) Stability and direct conversion of mineral barite crystals in carbonated hydrothermal fluids. *Journal of Material Science*, **43**, 2189–2197.

Robb, L. (2005) Introduction to ore-forming processes. Blackwell Publishing, Oxford, 373 pp.

Rodríguez-Navarro, A.B., Romanek, C.S., Pérez-Lloret, P. & Gaines, K.F. (2006) Effect of in ovo exposure to PBCs and Hg on Clapper rail bone mineral chemistry from a contaminated salt marsh in coastal Georgia. *Environmental Science and Technology*, **40**, 4936–4942.

Rouchy, J.M., Bernet-Rolland, M.C. & Maurin, A. (1994) Descriptive petrography of evaporates: applications in the field, subsurface and the laboratory. In: *Evaporite Sequences in Petroleum Exploration.* Technip, Paris, pp. 70–123.

Rouchy, J.M., Taberner, C., Blanc-Valleron, M.M., Sprovieri, R., Russell, M., Pierre, C., Di Stefano, E., Pueyo, J.J., Caruso, A., Dinarés, J., Gomis-Coll, E., Cespuglio, G., Wolff, G., Ditchfield, P., Santisteban, C., Pestrea, S., Comboureu-Nebout, N., Santisteban, S. & Grimalt, J.O. (1998) Sedimentary and diagenetic markers of the restriction in a marine basin: the Lorca basin (SE Spain) during the Messinian. *Sedimentary Geology*, **121**, 23–55.

Roy, D.M. & Linnehan, S.K. (1974) Hydroxyapatite formed from coral skeletal carbonate by hydrothermal exchange. *Nature*, **247**, 220–222.

Rüpke, L.H., Phipps Morgan, J., Hort, M. & Connolly, J.A.D. (2004) Serpentine and the subduction zone water cycle. *Earth and Planetary Science Letters*, **223**, 17–34.

Saether, O.M., Bølviken, B., Låg, J. & Steinnes, E. (1988) Concentration and chemical form of lead during natural transportation in groundwater. *Chemical Geology*, **69**, 309–319.

Sánchez-Pastor, N. (2007) *Cristalización secuencial de sulfatos y carbonatos de bario y estroncio a partir de soluciones acuosas: Relaciones entre morfología cristalina, composición química y fenómenos de superficie a escala molecular.* PhD Thesis, Universidad Complutense de Madrid, Spain, 178 pp. (in Spanish).

Sánchez-Pastor, N., Pina, C.M. & Fernández-Díaz, L. (2007a) A combined in situ AFM and SEM study of the interaction between celestite (001) surfaces and carbonate-bearing aqueous solutions. *Surface Science*, **601**, 2973–2982.

Sánchez-Pastor, N., Pina, C.M. & Fernández-Díaz, L. (2007b) Interaction between barite and celestite cleavage surfaces and carbonate bearing aqueous solutions. *Geochimica et Cosmochimica Acta*, **71**, A872.

Sanz-Rubio, E., Sánchez-Moral, S., Cañaveras, J.C., Calvo, J.P. & Rouchy, J.M. (2001) Calcitization of Mg-Ca carbonate and Ca sulphate deposits in a continental Tertiary basin (Calatayud Basin, NE Spain). *Sedimentary Geology*, **140**, 123–142.

Schmidt-Mumm, A., Clark, C. & Skorrow, R. (2006) Mineral exploration under cover: characterizing mineralizing fluid systems. *Journal of Geochemical Exploration*, **89**, 359–362.

Schweitzer, M.H., Johnson, C., Zocco, T.G., Horner, J.R. & Starkey, J.R. (1997) Preservation of biomolecules in cancellous bone of *Tyrannosaurus rex*. *Journal of Vertebrate Paleontology*, **17**, 349–359.

Speer, J.A. (1983) Crystal chemistry and phase relations of orthorhombic carbonates. In: *Crystal Chemistry and Face Relations of Orthorhombic Carbonates*. (R. Reeder, editor). Reviews in Mineralogy, **11**, 145–190. Mineralogical Society of America, Washington, D.C.

Speer, J.A. & Hensley-Dunn, M. (1976) Strontianite composition and physical properties. *American Mineralogist*, **61**, 1001–1004.

Suárez-Orduña, R. Rendón-Angeles, J.C., López-Cuevas, J. & Yanagisawa, K. (2004) The conversion of mineral celestite to strontianite under alkaline hydrothermal conditions. *Journal of Physics: Condensed Matter*, **16**, S1331–S1344.

Suárez-Orduña, R., Rendón-Angeles, J.C. & Yanagisawa, K. (2007) Kinetic study of the conversion of mineral celestite to strontianite under alkaline hydrothermal conditions. *International Journal of Mineral Processing*, **83**, 12–18.

Thayer, T.P. (1966) Serpentinization considered as a constant-volume metasomatic process. *American Mineralogist*, **51**, 685–710.

Truemann, C.N. & Tuross, N. (2002) Trace Elements in Recent and Fossil Bone Apatite. In: *Phosphates: Geochemical, Geobiological, and Materials Importance* (M.L. Kohn, J. Rakovan & J.M. Hughes, editors). Reviews in Mineralogy and Geochemistry, **48**, 489–521. Mineralogical Society of America, Washington, D.C. and the Geochemical Society, St. Louis, Missouri, USA.

Tsukamoto, K. (1993) In situ observations of crystal growth from solution. *Faraday Discussions of the Chemical Society*. **95**, 183–189.

Tuross, N., Behrensmeyer, A.K. & Eanes, E.D. (1989) Strontium increases and crystallinity changes in taphonomic and archaeological bone. *Journal of Archaeological Science*, **16**, 661–672.

Von Vultée J. (1950) Die orientierten verwachsungen der Mineralien. *Fortschrift für Mineralogie*, **29/30**, 297–378 (in German).

Walder, I.F. & Chavez, W.X. (1995) Mineralogical and geochemical behavior of mill tailing material produced from lead-zinc skarn mineralization, Hanover, Grant County, New Mexico, USA. *Environmental Geology*, **26**, 1–18.

Weiner, S. & Price, P.A. (1986) Disaggregation of bone into crystals. *Calcified Tissue International*, **39**, 365–375.

Wicks, F.J. & Whittaker, E.J.W. (1977) Serpentine textures and serpentinization, *The Canadian Mineralogist*, **15**, 459–488.

Wigley, T.M.L. (1973) Chemical evolution of the system calcite-gypsum-water. *Canadian Journal of Earth Sciences*, **10,** 306–315.

Williams, C.T. (1988) Alteration of chemical composition of fossil bones by soil processes and groundwater. In: *Trace Elements in Environmental History* (G. Grupe & B. Herrmann, editors), 27–40. Springer-Verlag, Berlin.

Williams, C.T. & Marlow, C.A. (1987) Uranium and Thorium distributions in fossil bones from Olduvai Gorge,Tanzania and Kanam, Kenya. *Journal of Archaeological Science*, **14**, 297–309.

Xia, F. Zhou, J., Brugger, J., Ngothai, Y., O'Neill, B., Chen, G. & Pring, A. (2008) Novel route to synthesise complex metal sulphides: Hydrothermal coupled dissolution-reprecipiation reactions. *Chemistry of Materials*, **20**, 2808–2817.

Xia, F. Brugger, J., Ngothai, Y., O'Neill, B., Chen, G. & Pring, A. (2009) Three-dimensional ordered arrays of zeolite nanocrystals with uniform size and orientation by a pseudomorphic coupled dissolution-reprecipitation replacement route. *Crystal Growth and Design*, **9**, 4902–4906.

Yanagisawa, K., Rendón-Ángeles, J.C., Ishiwara, N. & Oishi, S. (1999) Topotaxial replacement of chlorapatite by hydroxyapatite during hydrothermal ion exchange. *American Mineralogist*, **84**, 1861–1869.

Zocco, T.G. & Schwartz, H.L. (1994) Microstructural analysis of bone of the sauropod dinosaur *Seismosaurus* by transmission electron microscopy. *Palaeontology* **37**, 493–503.

EMU Notes in Mineralogy, Vol. 10 (2010), Chapter 6, 227–258

Chemical equilibrium between aqueous fluids and minerals in the marine environment

CHRISTOPHE MONNIN[1,2,*] and GUILHEM HOAREAU[1]

[1]*Université de Toulouse, UPS (OMP), LMTG, 14, Av. Edouard Belin, F-31400 Toulouse, France, e-mail: monnin@lmtg.obs-mip.fr*
[2] *CNRS, LMTG, F-31400 Toulouse, France*
**Corresponding author*

The purpose of this chapter is to give an account of the accuracy required to characterize chemical equilibria between aqueous fluids and minerals in the marine environment, and the answer brought to this challenge by the recent improvements in the calculation of the thermodynamic properties of multicomponent aqueous electrolyte solution models, namely Pitzer's ion interaction approach, and mineral solubility models based on this formalism. We focus on the general outline leading to the construction of a chemical model for seawater type solutions in the Na-K-Ca-Mg-Sr-Ba-Cl-SO_4-H_2O system, and to the solubility calculations it allows for a few minerals (barite, celestine, gypsum) as a function of solution composition, temperature and pressure. This model is then applied to the analysis of the saturation state of ocean waters and marine sediment pore waters, as examples of what can be expected from the use of such models to address the general question of chemical equilibrium between minerals and aqueous fluids in natural settings from the case of the marine environments.

1. Temperature, pressure and composition of ocean waters and of marine sediment interstitial waters

The marine environment, for the purposes of this exercise, refers to all that is related to seawater, oceans and to marine sediments. Our analysis focuses on fluid composition so it will be restricted to environments where these fluids can be and have been sampled. This pertains of course to the water column, but also to interstitial waters extracted from sediment cores. More precisely we will take the example of those retrieved by drilling during the successive international scientific drilling programs: Deep Sea Drilling Program (DSDP), Ocean Drilling Program (ODP) and Integrated Ocean Drilling Program (IODP), which extensively sampled the first kilometre of sediments in a wide variety of geological settings all over the globe. As such, the temperature and pressure conditions will be bound by those found on one hand in the water column, *i.e.* roughly between $-2°C$ and $40°C$ and up to 1 kbar, and on the other hand those encountered in marine sediments that have been drilled. If we exclude the peculiar high-temperature hydrothermal environments found at Middle Valley on the Juan de Fuca ridge (North East Pacific), the highest temperature found at the base of a

DOI: 10.1180/EMU-notes.10.6

DSDP/ODP/IODP borehole is 150°C in the Nankai Through (Leg 190 – site 1174). To sum up, the temperatures that we have to take into account range from −2°C to 150°C while pressure spans from 1 bar to 630 bars for the deepest boreholes.

Although in modern seawater the concentrations of major elements are roughly constant all over the globe, it is not the case when seawater is trapped during sediment burial where it undergoes very important changes due to what is called chemical diagenesis.

1.1. The composition of seawater – Model seawater and seawater models

The total solute load of normal seawater is 35 g/l. This means that when 1 litre of seawater is evaporated to dryness, the residue weighs 35 g. Although this definition relates to a process that seems straightforward, there are considerable technical difficulties in measuring the salinity in this way (Millero, 2006). The standard procedure is to heat seawater at 480°C to speed up evaporation. At such temperatures chloride can be lost as volatile HCl and carbonates decompose to oxides. It has then been proposed to relate salinity to the measurement of the chloride content of the solution which gave rise to a variety of relationships between salinity and chlorinity as a way to characterize seawater samples.

These 35 g of dissolved solids are largely dominated by sodium and chloride which constitute 95% of the solute load (Table 1). The other major elements in order of decreasing concentration are Mg, SO_4, Ca and K. The concentrations of these major elements are 'reasonably' constant in all ocean waters because of their conservative behaviour (meaning that their concentrations are not modified by any chemical reaction). It seems strange to use such a vague wording in a context of quantification, but this is indeed the reflection of the variability of natural seawater composition on one hand, and on the other hand of the illustration of the development of the concepts and measurements of salinity, chlorinity and seawater composition and the long way to the definition of a standard seawater composition (see Millero, 2006 for a very detailed presentation of these topics). This to say that the notion of constant seawater composition remains vague if the chemical system (*i.e.* the ensemble of the independent constituents used to describe the system, to use a thermodynamic definition) is not defined. The major element concentrations can change in peculiar locations like closed basins (the Red Sea, the Mediterranean Sea), in estuaries or close to hydrothermal vents.

Following these major elements are what can be called minor components, such as Br, Sr, carbonate, bicarbonate, *etc.*, the concentration of which is generally in the millimole range. The remaining 1% of the solute load is composed of elements and compounds, termed minor or trace, the concentrations of which can vary in extended ranges and can be used as tracers of various processes: water mass mixing, biochemical reactions, tracers of hydrothermal or aeolian inputs, *etc.* These tracers also include isotopes. Of particular interest is the alkalinity, which is defined as the quantity that can be titrated by a strong acid, and as such encompasses the carbonate ions, boron, phosphate, *etc.* (Zeebe & Wolf-Gladrow, 2001; Millero, 2006).

Table 1. The composition of standard IAPSO seawater and selected pore waters from the ODP. Pore waters were selected for their highly altered composition compared to seawater. Greatest and smallest concentration values (among examples) are in bold.

	Standard seawater (IAPSO)[1]	**Japan Sea – Kita-Yamato through**	**Mariana arc – South Chamorro seamount**	**Great Bahama Bank**	**Vanuatu island arc**	**Great Australian Bight**	**Mediterranean Sea**
Leg-site-hole-core-section		128-799-A-45-3	195-1200-D-8-2	166-1006-A-50-2	134-833-B-46-3	182-1126-B-8-4	160-973-A-16-1
Depth (m.b.s.f.)		408.3	27.5	453	494.5	69.4	142.8
Temperature (°C)	25	39.5	2	25	35.8	4.4	16.5
Pressure (bar)	1	248	296	111	312.3	85.2	383.8
Salinity (g/kg)	35.16	**31.6**	35.7[(2)]	46	88.3	100	**403**[(2)]
pH	8.1	**4.95**	**12.4**	7.2	8.6	6.6	5.8
Titration alkalinity (meq/L)	2.388	19.3	**62.3**	7.5	**0.5**	2.66	17.4
Na (mmol/l)	469	ND	617	652	**197**	**1453.5**	433
K (mmol/l)	10.2	6	18.6	7.6	**0.7**	29	**112.9**
Ca (mmol/l)	10.3	4.8	**0.2**	14.6	**548.5**	45.3	15
Mg (mmol/l)	52.8	26.6	**0**	31.1	**0**	**127.2**	**0**
Ba (μmol/l)	0.102	**2310**	1400	206	ND	ND	ND
Sr (μmol/l)	94	281	**5.2**	**7005**	132.7	218	ND
Cl (mmol/l)	566	**482**	534.5	795	1241	1630	**5504**
SO_4 (mmol/l)	29.3	**0**	25.7	**0**	12	**70.1**	37.8

[1] Concentration data in the molal scale, from Millero *et al.* (2008), except Ba from Summerhayes and Thorpe (1996).
[2] Data in g/l, calculated with the VOPO program Monnin (1994).

The constant salinity of ocean waters in the major oceans and seas leads naturally to the concept of constant ionic medium and to the idea of constant ionic strength. This has led to the development of specific models based on measurements of mineral solubilities either directly in seawater or in aqueous electrolyte solutions of the same salinity or ionic strength. On the other hand seawater is a multicomponent electrolyte solution. As such its properties can be calculated from those of the building blocks, *i.e.* binary and ternary with common ion solutions of the constituent aqueous electrolytes, and adequate mixing rules. This latter approach requires that properties of aqueous solutions of NaCl, KCl, $MgCl_2$, Na_2SO_4, *etc.* are known over the temperature, pressure and concentration ranges of oceanic interest. This is the subject of the classical thermodynamic study of aqueous fluids, a part of physical chemistry. It aims at elaborating a thermodynamic model of a multicomponent system within which seawater is only a single point in the whole composition space. When the properties of seawater are the primary goal, this method is much less straightforward than direct specific measurements using seawater itself, but it can take into account composition variations at constant salinity, at the difference with the constant ionic medium approach. Taking the example of a physical property, the density of seawater has been measured as a function of temperature, pressure and salinity by a large number of authors. These measurements have been represented by empirical polynomials that are used as an international standard (Millero & Poisson, 1981; Safarov *et al.*, 2009) that allow seawater density to be calculated with an accuracy better than 0.00001 g/cm^3 at oceanic conditions (*i.e.* temperatures from -2 to 40°C, pressures to 1 kbar and salinities up to 40‰). This high accuracy is required by models of ocean circulation where density-driven flow is a key process. On the other hand, a model allowing the calculation of the volumetric properties (solute partial molal volumes and solution density) of the Na-K-Ca-Mg-Cl-SO_4-CO_3-HCO_3-H_2O system at 25°C from the composition of the solution has been proposed by Monnin (1989). The density of seawater calculated this way has an uncertainty of 0.0001 g/cm^3 (Monnin, 1994) which does not match that of the international equation of state, but the model allows the calculation of densities of waters of any composition up to very high concentrations (up to halite saturation). It has been used by Hoareau (2009) who calculated the density at 25°C, of ~14,000 marine pore-water samples from their composition. The results have been represented with an accuracy between 0.001 and 0.0001 g/cm^3 by the following expression in which the density is in g/cm^3, which can be used up to a Total Dissolved Solid (TDS) load of 150 g/l:

$$d = 0.99761585 + 0.0006910582 \cdot TDS \quad (1)$$

Such an expression applies to the vast majority of marine sediment pore fluids, but there may be exceptions of peculiar compositions very far away from those of NaCl-dominated waters where the density must be calculated for example with the VOPO code (Monnin, 1994) for an optimum accuracy, instead of using equation 1.

1.2. Composition changes at constant salinity

The composition of seawater is the result of large-scale processes at play during plate tectonics. These processes lead to mass and energy (heat) transfers between the main geochemical reservoirs, namely the oceanic crust and sediments, the atmosphere and the continents (Holland, 2003, 2005). The composition of seawater is the result of the budget between the inputs to and outputs from these reservoirs. Modelling these coupled processes is the subject of palaeo-environmental modelling which is an area of very active research due to the question of the climatic history of the earth, in echo to the present and highly debated focus on climatic change (Hardie, 1996; Horita *et al.*, 2002; Stanley & Hardie, 1998; Tyrrell & Zeebe, 2004). If the composition of modern seawater is constant all over the globe, it has varied during geological time (Holland, 2005). For example, the ratio of the Mg over Ca concentration has varied considerably during the Phanerozoic, from the modern seawater value of 5.2 (in mole per mole) to values as low as 1, although the salinity has admittedly remained constant (Berner, 2004; Brennan & Lowenstein, 2002; Lowenstein *et al.*, 2005). These studies also point to the role of the sulphate/calcium ratio as well as the alkalinity content of the oceans, with far-reaching conclusions on the sedimentary record.

It then appears that there are several ways to characterize seawater. Oceanographers use CTD data as the signature of water masses, CTD standing for Conductivity-Temperature-Density, conductivity being related to salinity. Chemical oceanography goes into the details of seawater composition in order to use minor and trace elements (including isotopes) to trace oceanic processes in a constant major-element-composition background. There exist international standards for seawater (Bacon *et al.*, 2007), the main use of which is the calibration of analytical chemical methods. On the other hand, a seawater model can be defined as a chemical model containing dissolved elements relevant to the composition of seawater. This can range from very simple models such as a binary NaCl solution of the same ionic strength (*i.e.* 0.75 M), a six-ion system such as the Na-K-Ca-Mg-Cl-SO_4-H_2O used, for example, by Eugster *et al.* (1980) and by Harvie & Weare (1980) to calculate the mineral sequences forming during the evaporation of seawater, to the extended databases of the computer codes devoted to geochemical modelling (Oelkers *et al.*, 2009). Several concentration scales are used to characterize the quantities of matter in seawater. It is very common to find in the oceanographic literature concentrations expressed in moles or grams per kilo. This is the definition of molality if kilo means one kilogram of the solvent, *i.e.* water, but most of the times in the authors' mind it refers to one kilogram of seawater. The conversion is straightforward in this case and does not need any information other than the molar masses of the solutes. But the conversion of molarity (moles or grams per litre) to any other scale requires knowledge of the solution density. At 25°C, a density value of 1.0232 can be used for such a conversion without introducing any significant error (Monnin, 1994). The methods and computer codes proposed by Monnin (1994) and more recently by Marion (2007) provide easy ways to calculate the density of seawater-like solutions and convert various concentration scales.

To summarize, we see that seawater nowadays has an almost constant salinity and major-ion concentrations in all oceans apart from marginal basins, but this has not been the case over geological time.

1.3. The composition of marine sediment interstitial waters

1.3.1. Processes modifying the composition of pore waters

Marine sediment pore waters display widely varying compositions, up to the point where they bear only a minor resemblance to the seawater from which they were derived. These significant compositional changes are the result of chemical reactions taking place in the sediment column that describe briefly here. Table 1 gives examples of the difference between the composition of normal seawater and of sediment interstitial waters. All the information that the Ocean Drilling Program has produced during ODP Legs 100 to 210 can be found at http://www-odp.tamu.edu/database/, including, in particular, all kinds of data, the cruise Initial Reports and the Scientific Results volumes. The Integrated Ocean Drilling Program (IODP) website is at http://www.iodp.org/ (see Hoareau, 2009, for a critical compilation of the ODP/IODP pore-water compositions).

The processes responsible for the change in the composition of seawater to those of interstitial waters are grouped under the term diagenesis, which refers in its widest definition to all the physical and chemical transformations of the sedimentary material after its deposition. This covers temperature and pressure conditions up to ~200°C and 1 kbar, after which the domain known as metamorphism begins, although the limit is a matter of debate. During diagenesis, unconsolidated marine sediments, which have a large water content at the start of deposition, transform progressively into sedimentary rocks during burial. From the seafloor to a depth of a few hundred meters in the sediment, chemical diagenesis is driven by microbially mediated redox reactions. These reactions take place in distinct sediment layers, defining successive stratified biogeochemical zones characterized by a dominant reaction. Once the most easily available oxidant has been consumed, living organisms use what they can graze upon for their survival (see Claypool & Kaplan, 1974; Gieskes, 1975; Sayles & Manheim, 1975; Froelich *et al.*, 1979; Berner, 1980; Gieskes, 1983; Curtis, 1995; among others for the description of early diagenesis).

At the very top of the sediment pile (*i.e.* in the first centimetres), oxygen is rapidly consumed by aerobic bacteria. Anoxic reactions occurring beneath this zone involve a series of minor elements such as N, Fe, Mn or H. One of the main reactions affecting the composition of a major element is the sulphate-reduction reaction accompanying the anaerobic oxidation of organic matter. Using CH_2O as a very simplified representation of organic matter, this reaction can be written as

$$2CH_2O + SO_4^{2-}(aq) \rightleftharpoons HS^-(aq) + H^+(aq) + 2HCO_3^-(aq) \qquad (2)$$

It produces, on one hand, reduced sulphur species that can react with iron leading to the formation of iron sulphide. Indeed pyrite is a common mineral in marine sediments

(Goldhaber, 2002). On the other hand, it produces carbonate species that increase the alkalinity of the pore waters. This reaction can lead to the total consumption of dissolved sulphate. In this case, for example, biogenic barite produced in the water column and scavenged into the sediments may dissolve releasing large quantities of barium to the pore waters, the so-called barium remobilization (Torres *et al.*, 1996; McManus *et al.*, 1998; Dickens, 2001).

When dissolved sulphate is totally depleted, organic matter may continue to be degraded by a reaction producing methane:

$$2CH_2O + H_2O(l) \rightleftharpoons CH_4(aq) + H^+(aq) + HCO_3^-(aq) \qquad (3)$$

The reaction in equation 3 is referred to as methanogenesis. Again it produces carbonate alkalinity that may react with dissolved calcium to produce peculiar solid carbonate formations (*e.g.* Mazzullo, 2000).

Another important diagenetic reaction is the recrystallization of calcium carbonates. They are deposited onto the seafloor as biogenic Sr-rich aragonite and Sr-rich calcite produced by living organisms in the water column. Biogenic aragonite and calcite are metastable with respect to inorganic calcite in the marine environment at low temperature. They tend to dissolve, allowing inorganic calcite to precipitate (Sayles & Manheim, 1975; Hawkesworth & Elderfield, 1978). The Sr content of inorganic calcite being less than that of biogenic carbonate Sr is released to the pore waters, up to the point where strontium sulphate saturation may be reached (Baker & Bloomer, 1988).

Marine sediments may contain volcanic material either as ashes or basaltic intrusions (sills). This material is highly reactive and it also leads to strong modifications of the pore-water compositions. Also, it has long been recognized that in deep marine sediments, the pore-water Ca concentration often increases progressively with depth to reach large values just above the basaltic basement, while the Mg concentration decreases to 0 near the basement (*e.g.* McDuff & Gieskes, 1976; McDuff, 1981). This reflects the alteration of the basaltic basement below sediments (McDuff & Gieskes, 1976; Gieskes, 1981; McDuff, 1981). In this case, diffusive fluxes must be invoked to explain the changes in pore-water composition above the basement-sediment interface, but the physical process is the same: the alteration of volcanic material either below the sediment column or within the sediment itself. This process leads to a Ca-Mg exchange in the fluid with one mole of Mg lost for each mole of Ca gained (*e.g.* Mottl & Wheat, 1994), but this stoichiometry is not followed all the time. Hardie (1983) indicates that the concentrations of aqueous Na, Mg and SO_4 decrease while those of Ca and K increase. Also, secondary reactions such as calcium carbonate precipitation may lead to a different stoichiometry (Schacht *et al.*, 2008).

In addition to these chemical reactions linked to diagenesis and seawater-basalt interaction, the composition of pore waters can be influenced by fluid circulation in the sediments and in the basaltic basement (Elderfield *et al.*, 1999; Mottl & Wheat, 1994; Wheat & Mottl, 2000). Fluid circulation in marine sediments and the oceanic crust is now recognized as a major process affecting almost the entire seafloor and not just the ridge axis and subduction zones (Davis & Elderfield, 2004).

1.3.2. The composition of marine sediment pore waters from the ODP/IODP database

Most of the 14,400 pore waters reported in the ODP/IODP data record have salinities close to that of normal seawater, *i.e.* ~35 g/l. Pore waters more saline than seawater (brines) remain the exception with only 6% of the pore waters having a salinity greater than 40 g/l. About half of such very saline fluids have been collected in peculiar environments such as the Mediterranean Sea (ODP Legs 107, 160 and 161), characterized by evaporitic sediments deposited during the Messinian Salinity Event ~5.5 My ago, and the Great Australian Bight (ODP Leg 182), where brines derived from evaporitic coastal deposits are currently circulating within the sediments. Salinities reach values as high as 400 g/l in the evaporitic sediments of the Eastern Mediterranean basin (ODP Leg 160 site 973), and 106 g/l in the Great Australian Bight. On the other hand, salinities significantly lower than normal seawater are rare and generally found in continental margins. They can also result from the recovery of sediments containing gas hydrates, as an artefact linked to the fusion of these gas hydrates during core recovery.

The pore-water compositions can be so different from seawater that Na and Cl may no longer be the major dissolved elements. For example, the diagenetic alteration of volcanogenic material in sediments can lead to the substitution of Na and Mg by Ca which in turn can become the dominant cation (Egeberg *et al.*, 1990). The largest Ca concentration found in the ODP/IODP database (0.5 mol/l, 50 times greater than normal seawater) is found in pore waters sampled in volcanic-rich sediments of the Vanuatu island arc (ODP Leg 134 Site 833).

Various dissolved elements can become major species in marine sediment pore waters, especially in highly concentrated brines where their concentrations can reach extreme values. For example, the greatest Na, Cl, Mg and K pore-water concentrations found in the entire ODP/IODP database are 5500, 4100, 850 and 110 mmol/l, respectively. They have been measured in the brines collected during ODP Leg 160 in the Eastern Mediterranean basin. Similarly, SO_4 concentrations reach 70 mmol/l (2.5 times the seawater value) in the high-salinity pore waters of the Great Australian Bight. But for some elements, very large concentrations are reached in pore waters of seawater-like salinity. For example, a Sr concentration of 7 mmol/l (70 times the seawater value) was found in pore waters retrieved at ODP Site 1006 (Leg 166, Bahamas). Pore-water Ba concentrations are even more variable, from the average seawater value of 100 nmol/l to 2.3 mmol/l (ODP Leg 128 Site 799), which represents a four order of magnitude range.

Diagenetic alteration also has a strong impact on the pore-water alkalinity and *pH*. Intense bacterial oxidation of organic matter (mainly sulphate reduction) explains the extremely large alkalinity values (up to ~120 meq/l, 60 times the seawater value) measured in pore waters of organic-rich sediments of the Peru margin (ODP Leg 112) and of the Blake Ridge, off the coast of Florida (ODP Leg 164). Pore-water *pH* values cover a very large range, from 5 in the sediments of the Japan Sea (ODP Leg 128 Site 799) to 12.5 in the mud volcanoes of the Mariana forearc (ODP Leg 125 Site 780). This latter value is one of the largest observed in the natural environment. It is due to the hydrothermal serpentinization of peridotites.

These very large modifications of the composition of seawater either during its evolution in geological times or during its burial in marine sediments call for advanced thermodynamic models of water-rock interactions in order to characterize chemical equilibrium in aqueous solutions exhibiting large compositional changes. It is necessary to go beyond models based on seawater-like solutions of constant ionic strength.

2. Some generalities on solubility models based on Pitzer's ion-interaction formalism

Geoscientists have rapidly seen the solutions that Pitzer's ion-interaction approach to the calculation of the thermodynamic properties of aqueous electrolyte solutions could bring to the many problems involving aqueous fluids in natural environments. After the publication in the 1970s, by Pitzer and coworkers of a series of papers presenting the fundamentals of this approach (Pitzer, 1973; Pitzer & Mayorga, 1973, 1974; Pitzer & Kim, 1974), the landmark paper of Harvie & Weare (1980) describing a model allowing the calculation of mineral solubilities in the Na-K-Ca-Mg-Cl-SO_4-H_2O system at 25°C was followed by the calculation of mineral sequences forming during the evaporation of seawater (Eugster *et al.*, 1980; Harvie *et al.*, 1980), therefore solving a problem which had persisted since the early work of Vant'Hoff at the end of the nineteenth century. This success opened the way to the development of a series of models for increasingly complex systems. Pitzer's ion-interaction model and the development of solubility models of interest in Earth sciences and beyond have been the subject of a large number of reviews, books and monographs that present the formalism in great detail, besides the original papers in the primary literature. The interested reader can refer to Weare (1987), Pitzer (1991a,b), Anderson & Crerar (1993), Pitzer (1995), Millero (2001) and Marion & Kargel (2008). It is nevertheless necessary to recall a few points that are always a source of misunderstanding, although they have been already underlined by Pitzer & Mayorga (1974), Pitzer (1987), Weare (1987) and Ptacek & Blowes (2000), among others.

First of all it must be recalled that Pitzer's formalism is an aqueous-solution model that, in its final form, is merely an extension of the former Debye-Hückel and Guggenheim models (Pitzer, 1977). Its master equation for the activity coefficient of an aqueous *MX* electrolyte in a binary solution is a Debye-Hückel term followed by a polynomial series function of the solute molality:

$$\ln \gamma_{MX} = |z_M z_X| f_{DH} + \left(2\frac{\nu_M \nu_X}{\nu}\right) m B_{MX} + \left(2\frac{(\nu_M \nu_X)^{3/2}}{\nu}\right) m^2 C_{MX} \qquad (4)$$

expression in which m is the molality, z is the charge of the designated species and ν its stoichiometric coefficient in the nominal formula of the *MX* salt. The Debye-Hückel term f_{DH} is:

$$f_{DB} = -A_\Phi\left[\frac{\sqrt{I}}{1 + b\sqrt{I}} + \frac{2}{b}\ln\left(1 + b\sqrt{I}\right)\right] \qquad (5)$$

where I is the ionic strength, b is a constant equal to 1.2 and A_Φ is the Debye-Hückel slope. B_{MX} and C_{MX} are the second and third Virial coefficients.

$$B_{MX} = \beta_{MX}^{(0)} + \beta_{MX}^{(1)} \cdot g(\alpha_1 \sqrt{I}) + \beta_{MX}^{(2)} \cdot g(\alpha_2 \sqrt{I}) \quad (6)$$

$\beta_{MX}^{(0)}$, $\beta_{MX}^{(1)}$, $\beta_{MX}^{(2)}$ and C_{MX} are adjustable constants specific of each aqueous electrolyte *MX*. They can be calculated from a simple least-squares fit to the data. For 2-2 electrolytes, α_1 is equal to 1.4 and α_2 to 12. For electrolytes of lesser valence type α_1 is equal to 2 and α_2 to 0. Equation 4 resembles the well known Virial expansion equation of state for real gases. So far the first success of the method lies in the fact that it allows the mathematical representation of the experimental data over a large concentration range within experimental uncertainty (Pitzer, 1973, 1977). The concentration range within which the data can be represented this way depends on the electrolyte under consideration, but it usually extends to the solubility limit.

The statistical mechanical theory of real gases shows that the Virial coefficients can be expressed in terms of ionic interactions, the second Virial coefficient taking into account pairwise interactions, the third coefficient triplets and so forth (Hill, 1960; MacQuarrie, 1973). Strong pairwise interactions can lead to the formation of a stable Van der Waals species called an ion pair or an aqueous complex. It then follows that any Debye-Hückel-type aqueous solution model to which a polynomial function of the solution composition is added implicitly takes into account ion association in solution through a second Virial coefficient.

Pitzer's model extends the notion of strong electrolytes: the thermodynamic properties of a large number of aqueous electrolytes can be represented on the basis of total dissociation. The physical reality of cation-anion interaction in solution is taken into account by the second Virial coefficient. This has been very elegantly illustrated by Harvie *et al.* (1984) who showed that the activity coefficient of an aqueous electrolyte can be equally well represented by a Virial-type expression or an ion association model, and that the stronger the interactions in solutions (*i.e.* the more stable the ion pair and therefore the larger the association constant), the more negative the second Virial coefficient. For example Millero & Thurmond (1983) have calculated a $\beta_{MX}^{(2)}$ parameter for the $MgCO_3$ aqueous electrolyte of about -200, while the same data could be fitted with the same accuracy to an ion pairing model with a $pK = 3$ for the $MgCO_3^0$ ion pair dissociation constant. When the association in solution becomes large, it is more efficient to explicitly include an ion pair in Pitzer's model (Pitzer & Silvester, 1976; Pitzer, 1987; Weare, 1987). This confirms that the limit of the strong electrolyte model (total dissociation) is reached. Pitzer (1987) and Weare (1987) set this limit at a pK of about 2 for the ion-pair dissociation constant. Models including ion pairs in Pitzer's formalism have been termed extended or hybrid (Whitfield, 1975; Pitzer, 1987). At the risk of repeating, it must be stressed that in this case, Pitzer's parameters and the association constants for aqueous complexes are not independent. In fact, saying that calculations have been done using Pitzer's model does not provide any useful information. A proper wording should be "Pitzer's formalism along with the parameter set for the system under study proposed by authors X and Y".

Changing a single parameter in an internally consistent parameter set amounts to proposing a new model and as such should be justified following the same criteria (performance, approximations, validation) used in the construction of the original model. Parameter consistency is always critical in thermodynamic calculations. This becomes crucial in the case of calculations using Pitzer's model. Inconsistencies in the parameter set can deteriorate the results to the point that it becomes useless to revert to the Pitzer's formalism involved. For example, Felmy *et al.* (1989) have shown how the calculation of the solubility of plutonium hydroxide deteriorates when inadequate interaction parameters are included, to the point where Pitzer's formalism is no longer more efficient than a Debye-Hückel-type universal function of the ionic strength such as the Davies equation.

Pitzer's formalism allows the calculation of the thermodynamic properties of concentrated aqueous solutions, up to the solubility limit. As such, the solubility products of salts can be calculated from their solubilities. This gave rise to a number of solubility models such as the HMW model (Harvie *et al.*, 1984). In such models, the solid solubility products are generally calculated at the same time as Pitzer's mixing parameters, θ_{ij} and ψ_{ijk}, where i and j designate ions of the same type (cations or anions) and k an ion of a type different than i and j. Such parameters can be best calculated from activity or osmotic coefficient data for common-ion ternary mixtures. Their evaluation from solubility data often at the same time as the mineral solubility products put the emphasis on solubility calculations, which is of course of primary importance in water-rock interaction modelling, at the risk of sometimes reducing the model performance in the lower concentration range (Weare, 1987).

To illustrate the procedure, one can take the example of a ternary with a common ion system where the solubility of a *MX* salt, say halite, has been measured in solutions of various concentrations of the *NX* aqueous electrolyte (say in KCl solutions). Pitzer's expression for the activity coefficient of aqueous sodium chloride can be schematically expressed as:

$$\ln \gamma_{\mathrm{NaCl},aq} = \ln \gamma_{\mathrm{NaCl},aq}(\theta = \psi = 0) + m_{\mathrm{K}}\theta_{\mathrm{NaK}} + \frac{1}{2} m_{\mathrm{K}} m_{\mathrm{Cl}} \psi_{\mathrm{NaKCl}}, \qquad (7)$$

expression in which $\ln \gamma_{\mathrm{NaCl},aq}(\theta = \psi = 0)$ is the contribution to the activity coefficient of all the terms involving parameters for the binary interactions between ions of opposite sign, *i.e.* the Na-Cl and K-Cl interactions. Because the aqueous fluid is at equilibrium with NaCl(*s*), one can write:

$$\ln K_{sp}(\mathrm{NaCl},\, s) = \ln (m_{\mathrm{Na}} m_{\mathrm{Cl}}) + 2 \ln \gamma_{\mathrm{NaCl},aq} \qquad (8)$$

Combining equations 7 and 8 and rearranging, one gets:

$$\ln K_{sp}(\mathrm{NaCl},\, s) - m_{\mathrm{K}}\theta_{\mathrm{NaK}} - m_{\mathrm{K}} m_{\mathrm{Cl}} \psi_{\mathrm{NaKCl}} = \ln (m_{\mathrm{Na}} m_{\mathrm{Cl}}) + 2 \ln \gamma_{\mathrm{NaCl},aq}(\theta_{\mathrm{NaK}} = \psi_{\mathrm{NaKCl}} = 0) \qquad (9)$$

The terms on the right hand side are all known either from measurements (the Na and Cl molalities are the experimental halite solubility) or calculated from known parameters. The unknowns on the left hand side are the halite solubility product and the Pitzer's mixing parameters. A simple, linear, least-squares fit of the solubility data for the common ion ternary system provides an adequate representation of the experimental data and a consistent set of values of these parameters. In the many situations where data for simple ternary systems are not available, one may use more complex systems (see the examples of Møller, 1988; Greenberg & Møller, 1989; Monnin, 1999; among others). There may be different strategies for different authors as to the choice of data to use or the primary thermodynamic quantities to rely on, but the principle is the same. For example Pitzer and his coworkers (Pabalan & Pitzer, 1987; Pitzer, 1991a) rely on the NBS values of the thermodynamic properties of solid phases while Møller, Weare and coworkers (*e.g.* Harvie *et al.*, 1984; Greenberg & Møller, 1989) have chosen to adjust them to the solubility data.

Again this stresses the need for consistency, this time between the thermodynamic parameters of the aqueous solution and those of the solids. Solubility products calculated from solubility data should, in principle, agree with those calculated from calorimetric data, *i.e.* from the standard enthalpy and entropy of the dissolution reaction. An example between so many others is given by Monnin (1995) who calculated the thermodynamic properties of solid barium chlorides and showed how well these compared with the NBS values. Another example is the work of Pabalan & Pitzer (1987) for the Na-K-Mg-SO_4-OH-H_2O system to high temperature. Quite often, the use of the solubility products derived from calorimetric data can lead to discrepancies between calculated and measured solubilities large enough to preclude any conclusion to be drawn on the characterization of chemical equilibrium in marine environments.

In echo of the above discussion, the User Guides for computer codes often contain warnings about the consistency of the databases. In the PHREEQC user guide, Parkhurst & Appelo (1999) write that "the database files provided with the program should be considered to be preliminary. Careful selection of aqueous species and thermodynamic data is left to the users of the program." PHREEQC version 2 incorporates Pitzer's ion-interaction model and as such the previous PHREEQPITZ model (Plummer *et al.*, 1998) is now obsolete. The final report of the Joint European Thermodynamic Database for Environmental Modelling Project (http://www.fzd.de/projects/JETDEM/) provides an interesting overview of the capacities and limitations of a variety of widely distributed geochemical codes. Ideally, the databases of the computer codes should be clearly identified by users in any water-rock interaction modelling result.

The work of Helgeson and coworkers (Helgeson & Kirkham, 1974a, 1974b; Helgeson *et al.*, 1981) is certainly the most extensive and powerful method for estimating the thermodynamic properties of aqueous solutes and minerals in extended temperature and pressure ranges. These authors clearly stated that the uncertainties in the thermodynamic parameters generated by their model were not critical in comparison to those inherent in the variability of nature and to the errors of field observations and measurements. For example, at metamorphic conditions of 700°C, an error of 50°C on the equilibrium temperature of a mineral assemblage is certainly acceptable. This is certainly not the case for low-temperature geochemistry. So Weare (1987) indicated that Pitzer's formalism indeed allows the generation of solubility models of "sufficient accuracy". What is sufficient? Following the current usage, we here define the saturation state of a mineral as the ratio of the ionic product to the solubility product at a given T and P. Some authors may call it the saturation index and they define the saturation state as the log of the saturation index, and *vice versa*, which has absolutely no importance as "thermodynamics is not difficult as long as you keep track of what you are talking about" (cited in Anderson & Crerar, 1993). Equilibrium is indicated when the saturation state is equal to 1, which never happens numerically. In practice one must define a saturation state range within which equilibrium is assumed. Ideally again, this range should not be left at the user's discretion, but should be the result of the error-propagation calculation taking into account the uncertainties on all the parameters of the model.

An important step in the development of accurate solubility models is the validation of the model's predictions against experimental data not used during the construction of the model. This is indeed a way to evaluate model performances and an indirect way to estimate how well solubilities are calculated. For example, Monnin (1999) considers that barite and celestine solubilities can be calculated within 10% in the Na-K-Ca-Mg-Ba-Sr-Cl-SO_4-H_2O system. This means that any mineral saturation state calculated outside the range within which equilibrium is assumed will be an indication of the dynamics of the system (out of equilibrium states, *i.e.* supersaturation or undersaturation) and not the result of the limitation of the thermodynamic model or the inconsistencies in the database.

An important distinction must be made between the chemical system used to describe the natural environment, and the chemical model used to calculate the thermodynamic properties of this chemical system. For example, the composition of the geological object, seawater, can be fully determined due to the advances in modern analytical chemistry. Indeed Millero (2006) reports the concentrations of almost all the elements of the periodic table in seawater. As such, an 'exact' seawater model should take into account all the components, but due to the limitations of thermodynamic models, not all the interactions involving these components can be taken into account in any model aiming at the calculation of the properties of seawater. As such this 'exact' chemical system will be described by an approximate chemical model. Conversely, reducing the chemical system to an approximate (simplified) system will allows an 'exact' chemical model to be used. What 'exact' means is that the model results are within the accuracy of measured values. For example, we describe below an accurate chemical model for the Na-K-Ca-Mg-Ba-Sr-Cl-SO_4-H_2O system, which is a model system for seawater-type solutions. The chemical model is certainly able to accurately calculate mineral solubilities, but the chemical system is definitely approximate for seawater in the sense that it does not take into account important ligands such as the carbonate ions. The present remarks are only an echo of the long-standing discussion in the physical sciences on the trade-off between the exact solution of an approximate model and the approximate solution of an exact model.

3. Example: a model for the Na-K-Ca-Ba-Sr-Cl-SO_4-H_2O system to 200°C and 1 kbar

Monnin (1999) has elaborated a model of the Na-K-Ca-Mg-Ba-Sr-Cl-SO_4-H_2O system that allows the thermodynamic properties of aqueous solutions of any composition to be calculated to high concentrations up to 200°C and 1 kbar. We discuss here some of the main features of the way it has been built. First it takes into account a variety of previous models for subsystems such as the high-temperature models for the Na-K-Ca-Cl-SO_4-H_2O system of Greenberg & Møller (1989), the $MgCl_2$ model of Pabalan & Pitzer (1987), the $SrCl_2$ model of Phutela *et al.* (1987) and the $BaCl_2$ model of Monnin (1995). Interactions between aqueous barium, strontium, calcium and sulphate are explicitly taken into account by chemical equilibria leading to the formation of the $BaSO_4^0(aq)$, $SrSO_4^0(aq)$ and $CaSO_4^0(aq)$ ion pairs, while all other ionic interactions in solution are described by adequate parameters in Pitzer's formalism.

Pressure effects on the thermodynamic properties of aqueous electrolytes are evaluated through the calculation of their partial molal volumes. Pitzer's parameters for the volumetric properties (*i.e.* the first and second derivatives with respect to pressure, of Pitzer's ion-interaction parameters for the partial excess free energy) are obtained from regression of density and compressibility data for binary aqueous solutions (Monnin, 1989, 1990).

This aqueous-solution model is then combined with the solubility products of a few minerals (barite, celestine, gypsum, anhydrite among others) to calculate their solubilities in the composition, temperature and pressure ranges of interest for the marine environment. Predicted solubilities of barite and celestine have been tested against data not used in the model construction, which led to an uncertainty estimate of $\pm 10\%$ in the calculated solubilities and from there the somewhat stringent criterion retained for the characterization of chemical equilibrium in the natural environment using this model.

As recalled by Oelkers *et al.* (2009), such models rely upon a lot of experimental data. As such they are as good as the data they are built upon. For example, there are large discrepancies between authors on the gypsum-anhydrite transition in pure water, from 40°C to 60°C, mainly due to the sluggish kinetics of anhydrite precipitation below 70°C (Freyer & Voigt, 2003). The Greenberg-Møller model (and therefore Monnin's model) predicts a transition temperature of 49°C. Anhydrite precipitates when seawater is heated during its downward percolation in the oceanic crust leading to porosity/permeability reduction. A 20°C difference in the anhydrite precipitation temperature translates into a significant depth difference in the basaltic basement (namely 500 m for an average thermal gradient of 35°C/km) and a change in the circulation patterns at ridge crests or in ridge flanks (Lowell & Yao, 2002; Lowell *et al.*, 2003). The anhydrite precipitation temperature also depends on pressure (Monnin *et al.*, 2003).

To summarize, it must be kept in mind that, as good as they are, the existing solubility models based on Pitzer's ion-interaction approach have intrinsic limits that can be bypassed only by new experimental data (Freyer & Voigt, 2004).

4. Case studies: the characterization of chemical equilibrium in the marine environment

4.1. The saturation state of the oceans with respect to barium sulphate

The marine chemistry of barium has received and continues to receive attention. The dissolved barium concentration is correlated with the alkalinity and the silica contents of seawater. This correlation is characteristic of each oceanic basin (Jeandel *et al.*, 1996). Furthermore, the main solid phase carrying barium in the water column is barite, although other solids found in surficial sediments may also contain significant amounts of barium (Gonneea & Paytan, 2006). When deposited on the seafloor, insoluble barite is supposed to resist dissolution unlike silica and carbonate phases which are more sensitive to the effects of diagenesis. As such, barite may provide a good record of oceanic conditions. Indeed it has been used extensively, *e.g.* by Paytan and coworkers, to monitor variations in the strontium (Paytan *et al.*, 1993; Paytan & Griffith, 2007) and sulphur (Paytan *et al.*, 2004) isotopic compositions of seawater during geological times.

One ongoing question is the mechanism of barite formation in the water columns. There is no doubt that it is linked to the biological activity where it may form in micro-environments of decaying organic matter (Dehairs *et al.*, 1980). On the other hand, there is a class of living organism, *Acantharia*, which build their skeleton out of strontium sulphate, a process during which barium is incorporated into the celestine structure (Bernstein *et al.*, 1992, 1998; Bernstein & Byrne, 2004). Because seawater is largely undersaturated with respect to celestine, the *Acantharia* skeletons dissolve rapidly when these organisms die thereby releasing barium to the nearby environment with barite formation as a consequence. Whatever the mechanism of barite formation in the water column, the question of the saturation state of the waters is central. It has been addressed by Monnin *et al.* (1999) who used the chemical model of the Na-K-Ca-Mg-Ba-Sr-Cl-SO_4-H_2O system of Monnin (1999) to compute the barite saturation state from the water-composition data collected throughout the oceans by the GEOSECS program (Ostlund *et al.*, 1987). This program (GEOchemical SECtionS of the oceans), which operated between 1970 and 1972, has allowed the construction of a major database for the characterization of the water column throughout the globe from data collected by numerous cruises. This database is still a major reference in modern oceanography.

Reported in the GEOSECS database for each seawater sample are the depth at which it was collected, the potential temperature (*i.e.* the temperature measured *in situ* and corrected for the effects of decompression; Millero, 2006), the salinity and the barium concentration. For thermodynamic calculations, depth is simply converted to pressure by assuming that 10 m equals 1 bar. The seawater composition in the Na-K-Ca-Mg-Ba-Sr-Cl-SO_4-H_2O system can be calculated from the composition of standard seawater of a 35 g/l salinity (Millero, 2006) using:

$$m_i(S) = m_i(35) \times \frac{S}{35} \qquad (10)$$

S is the salinity (in g/l), $m_i(S)$ is the concentration of the i element in the sample and $m_i(35)$ its concentration in standard seawater. The sodium concentration is calculated from the electroneutrality condition. Molarities are converted to molalities using VOPO (Monnin, 1994). The final input data for the calculation of the barite saturation state is then temperature, pressure and the molalities of Na, K, Ca, Mg, Sr, Ba, Cl and SO_4. The barium concentration in the ocean varies from 40 nmol/l to 150 nmol/l, while all the other concentrations are constant (within the variations linked to the salinity changes, which are very minor).

The results are depicted in Figure 1 for the Atlantic, Indian, Pacific and Southern oceans. The Atlantic Ocean is undersaturated everywhere, but equilibrium (depicted by a saturation state between 0.9 and 1.1) is reached in a number of places: intermediate

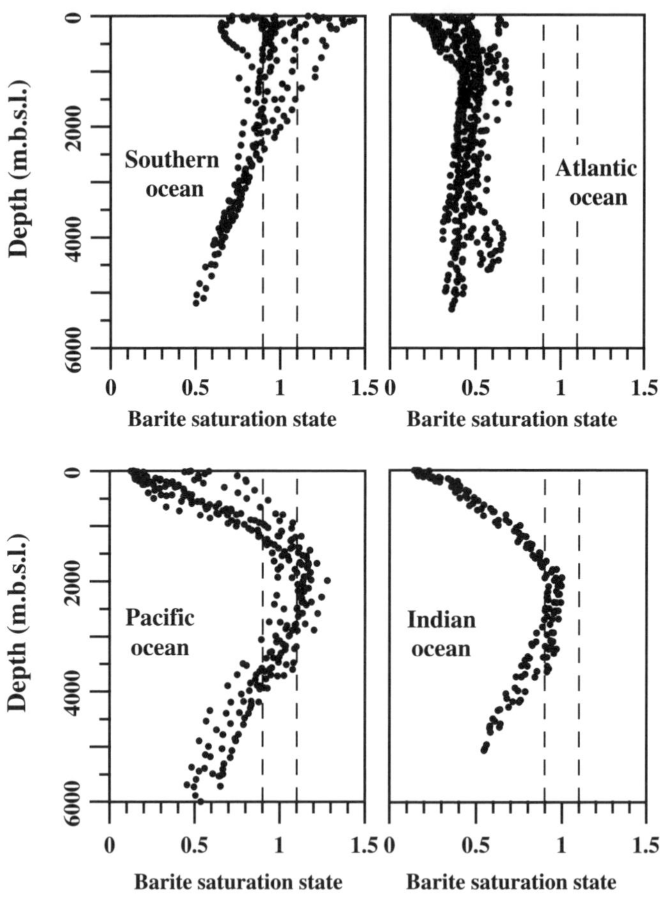

Fig. 1. The barite saturation state (the ratio of the ionic product to the solubility product) in the major world oceans for the GEOSECS stations (from Monnin *et al.*, 1999). The area between the two vertical dashed lines delineates the zone of equilibrium.

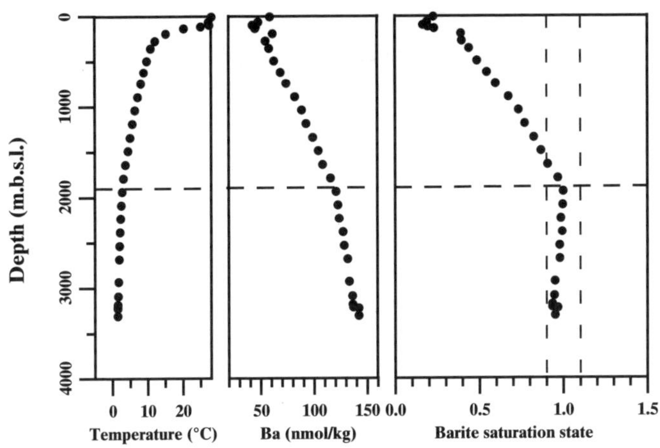

Fig. 2. The temperature, Ba concentration and barite saturation state *vs.* the depth in the water column at GEOSECS Station 446 in the Gulf of Bengal (after Monnin *et al.*, 1999). The horizontal dashed line indicates the depth at which saturation is reached, which coincides with a change in the slope of the Ba concentration and temperature variations. The vertical dashed lines delimitate the saturation-state range within which equilibrium is assumed.

waters of the Indian and Pacific oceans and surface waters of the Southern ocean. Of particular interest is GEOSECS station 446 in the Bay of Bengal where it can be seen in Figure 2 that the slight change in the slope of the Ba content variation with depth at a depth of 2000 m below sea level can be explained by the fact that equilibrium (saturation) is reached.

Other examples of the calculation of the barite saturation state in ocean-type waters are given by Schenau *et al.* (2001), Dehairs *et al.* (2008) and Jacquet *et al.* (2008). Using the same GEOSECS database, Monnin & Cividini (2006) investigated the role that Sr substitution in barium sulphate may play on the stability of barite in the ocean, basing their analysis of the aqueous solution–solid solution system on the methods described in detail by Prieto (2009) among others. They concluded that the change is very minor and cannot explain the general undersaturation, although barite is commonly found in suspended matter in ocean waters.

4.2. The stability of celestine in marine sediments

Celestine is the Sr-rich end member of the solid-solution series $(Ba,Sr)SO_4$. Due to its solubility being greater than that of barite, its stability conditions are quite different from those of barite (Hanor, 2000). Celestine can control the Sr content of marine sediment pore waters, an important element that can be used as a tracer of diagenetic reactions (in particular of carbonate recrystallization). As such, the conditions of its stability in marine sediments must be known. At the difference with diagenetic barite the precipitation of which is favoured in organic-rich marine sediments (Torres *et al.*, 1996), celestine is a common minor diagenetic mineral in carbonate-rich marine sediments or in evaporitic environments (Hanor, 2000). As we have seen, biogenic celestine produced in the ocean-water column by *Acantharia* cannot be deposited onto the seafloor because it dissolves in the water column. Therefore, the Sr concentration must increase for celestine equilibrium to be reached in sediments. This can be achieved by calcium carbonate recrystallization during early diagenesis (Baker & Bloomer, 1988). Moreover, many studies based on ancient marine environments (*e.g.* Scholle *et al.*, 1990; Taberner *et al.*, 2002) propose that the sulphate necessary for celestine to precipitate is supplied by brines from evaporite formations, although in certain cases sulphate from seawater can also lead to celestine saturation (Baker & Bloomer 1988).

The stability of celestine in marine sediments cored during the DSDP and ODP programs has been thoroughly investigated by Baker & Bloomer (1988) and by Hampt & Delaney (1997) through thermodynamic calculations, which emphasized the role of carbonate recrystallization in celestine precipitation. These latter authors, as well as Richter (1996), who evaluated the role of celestine equilibrium in the sediment Sr budget, used the celestine solubility model of Reardon & Armstrong (1987). Some other authors have also tentatively characterized the saturation of ODP pore waters with respect to celestine, using more or less empirical estimates of the thermodynamic properties of the system (*e.g.* Swart & Guzikowski, 1988; Swart & Burns, 1990; Kramer *et al.*, 2000). More recently, Hoareau (2009) has carried out a systematic study of celestine stability in marine sediments through the calculation of its saturation state for the entire ODP/IODP database (14416 samples), using the model of Monnin (1999) for the Na-K-Ca-Mg-Sr-Ba-Cl-SO_4-H_2O system. The pore water *in situ* pressure was obtained from the water-column depth and the depth in the sediment at which the sample is collected. The *in situ* sample temperatures in a given borehole have been calculated from the bottom seawater temperature and from the reported thermal gradient, and, in the many cases where this thermal gradient has not been directly measured, either using a default value of 35°C/km or assuming the same value of the T gradient as that of a closely located site within the same geological environment. Molarities have then been converted to molalities through the calculation of the pore-water densities at 25°C with the VOPO model of Monnin (1994). All the details are given in Hoareau (2009).

The results show that celestine saturation of pore waters is not uncommon, as it is reached in a number of sites much larger than previously indicated by the seldom reported presence of this mineral in marine sediments cored during DSDP/ODP/IODP. Starting from ocean bottom waters largely undersaturated with respect to celestine, pore-water saturation is reached in 83 boreholes (*i.e.* ~10% of all boreholes) drilled during 22 ODP/IODP Legs in various geological environments. When saturation is reached, equilibrium is the rule. Only very slight supersaturation values (with a maximum saturation state of 1.4) have rarely been found. Two different and non-exclusive mechanisms leading to celestine formation have been found. The first, already determined by previous studies (Baker & Bloomer, 1988) is linked to carbonate recrystallization. The second is an increase in the Sr and sulphate pore-water concentrations following the salinity increase due to the presence of brines, in accordance with numerous previous observations of celestine in evaporitic settings.

Sediments at most of the sites exhibiting celestine equilibrium share common characteristics such as a carbonate-rich lithology (typically >80 wt.%) and a small organic carbon content (generally <1 wt.%), which favours an important Sr release through carbonate recrystallization and a low bacterial sulphate reduction.

We illustrate these results by a brief description of three representative cases.

4.2.1. Celestine equilibrium controlled by carbonate recrystallization: example of ODP Leg 130 site 806 (Ontong Java Plateau)

Site 806 was drilled during ODP Leg 130 in a foraminifer nanofossil ooze of the Ontong Java Plateau, with the aim of recovering tropical assemblages of pelagic and benthic microfossils for palaeo-oceanographic and biostratigraphic studies of the Neogene. This site provides an illustration of equilibrium with respect to celestine reached as a consequence of carbonate recrystallization, in pore waters of seawater-like salinity (Fig. 3). The high carbonate content of sediments (~92 wt.%) favours an important release of Sr, the concentration of which reaches 1.2 mmol/l in the pore waters during carbonate recrystallization, while their small organic-matter content (~0.1 wt.%) limits bacterial-sulphate reduction. The thermodynamic model demonstrates that celestine equilibrium is reached 150 m below the water-sediment interface and is maintained up to the bottom of the borehole. This also shows that celestine equilibrium controls the pore-water Sr concentration and explains why it stops increasing when equilibrium is reached.

4.2.2. Celestine equilibrium in brines: Eastern Mediterranean Basin (ODP Leg 160 site 968)

ODP Leg 160 site 968 has been drilled to investigate the sedimentary and tectonic evolution of the Cyprus margin. It demonstrates that equilibrium with respect to celestine can be reached in highly saline sulphate-rich pore waters (Fig. 4). The composition of the sediments cored at this site is more complex than the preceding example as they contain nanofossil ooze, sapropels, clay-rich and authigenic gypsum layers, among others. The presence of evaporitic deposits linked to the Messinian Salinity Event in and below the cored sediment interval explains the rapid downhole increase in salinity (up to 90‰), in the strontium concentration (up to 1.1 mmol/l), and in dissolved sulphate (up to 43 mmol/l), leading to celestine saturation.

4.2.3. Example of a complex saturation state variation: ODP Leg 133 site 815 (northeast Australian Margin)

ODP Leg 133 Site 815 was drilled at the margin of a carbonate platform in order to investigate the record of the evolution of the north-eastern Australian passive margin. This example illustrates the complexity of the saturation-state profile which can result from carbonate recrystallization in sulphate-rich brines (Fig. 5). Sediments at this site are composed mainly of nanofossil foraminifer ooze and chalk with variable amounts of clays, with mean carbonate and organic-matter contents of 70 wt.% and 0.25 wt.%, respectively. Celestine equilibrium is reached at two distinct depths in the sediments section, owing to the complex interplay between the Sr release due to carbonate

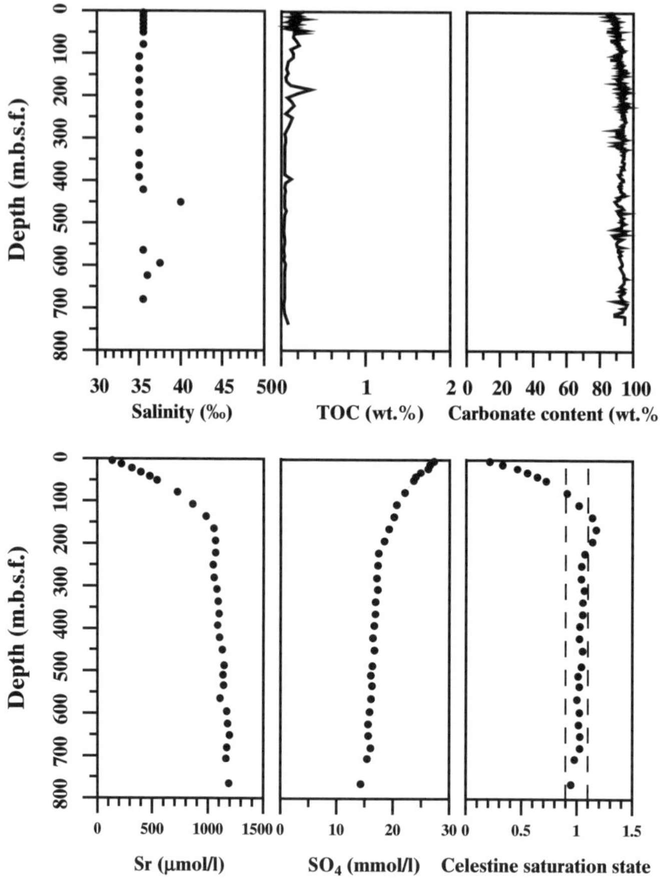

Fig. 3. The pore-water salinity, the Total Organic Carbon (TOC) content of the sediments, the carbonate content, the pore-water Sr and SO_4 concentrations and the celestine saturation state *vs.* depth (in metres below the seafloor) of a borehole drilled in the sediments of the Ontong Java plateau (ODP Leg 130 Site 806) (modified from Hoareau, 2009). Note the limitation of the pore-water Sr content by celestine equilibrium.

recrystallization and the progressive increase in salinity with depth. Again note that equilibrium is fully depicted by the model, with a celestine saturation state between 0.9 and 1.1 for very different pore-water compositions, temperature and pressure.

4.3. Authigenic gypsum in marine sediments

Gypsum is usually found in evaporitic environments (Chang *et al.*, 1998) and more rarely in deep marine sediments. The example of ODP Leg 160 Site 968 (Eastern Mediterranean Basin) shows that authigenic gypsum can be found in high-salinity sediment pore waters, but it is also reported, for example, in sediments of the Japan sea (ODP Leg 126) were pore waters have a salinity close to the seawater value (Tazaki & Fyfe, 1992).

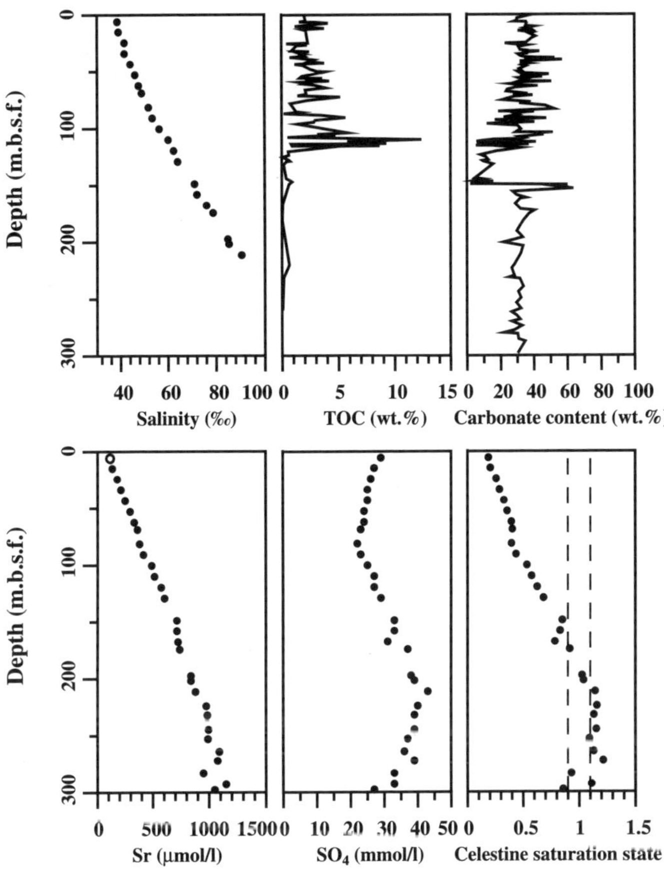

Fig. 4. The pore-water salinity, the Total Organic Carbon (TOC) content of the sediments, the carbonate content, the pore-water Sr and SO_4 concentrations and the celestine saturation state *vs.* depth (in metres below the seafloor) of a borehole drilled in the sediments of the Eastern Mediterranean Basin (ODP Leg 160 Site 968) (modified from Hoareau, 2009). The regular pore-water salinity increase is leading to a regular Sr concentration increase and to celestine saturation at the base of the borehole.

As for celestine, we have calculated the gypsum saturation state using the entire ODP pore-water composition database and the model of Monnin (1999) (Hoareau, 2009). Again, starting from largely undersaturated bottom ocean waters (with a gypsum saturation state of 0.12 at 2°C and 300 bars), the pore-water concentration of calcium and/or sulphate must increase to reach gypsum saturation. Calculations indicate that saturation is obtained in sediment pore waters of 26 boreholes drilled at 23 different sites, during 12 ODP/IODP Legs (Hoareau, 2009). As for celestine, saturation is mainly characterized by equilibrium (no significant supersaturation). Two main mechanisms lead to saturation. First, equilibrium is reached in submarine evaporitic formations (gypsum layers in sediments and brines with elevated major dissolved species concentrations, including

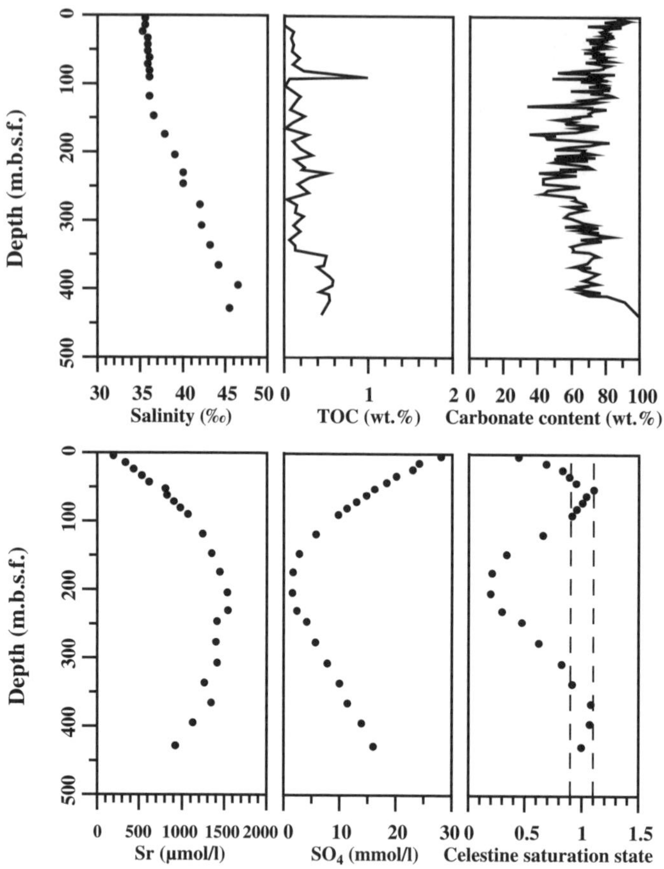

Fig. 5. The pore-water salinity, the Total Organic Carbon (TOC) content of the sediments, the carbonate content, the pore-water Sr and SO_4 concentrations and the celestine saturation state *vs.* depth (in metres below the seafloor) of a borehole drilled in the sediments of the Eastern Mediterranean Basin (ODP Leg 133 Site 815) (modified from Hoareau, 2009). The complex interplay between the pore-water Sr and SO_4 concentrations is leading to two distinct zones of celestine equilibrium.

Ca and SO_4). Secondly, the diagenetic alteration of volcanogenic material present in the sediments releases very large amounts of Ca to the pore waters, leading to saturation. This latter case often pertains to pore waters with salinity close to that of normal seawater. As for celestine, we describe below examples of sites which illustrate these two mechanisms, leading to gypsum equilibrium.

4.3.1. Saturation in brines in sediments without evaporates: ODP Leg 182 (Great Australian Bight)

An unexpected observation of ODP Leg 182, primarily intended to study the thick 'cold carbonate' Australian margin, was to find at most sites a sulphate-rich brine extending

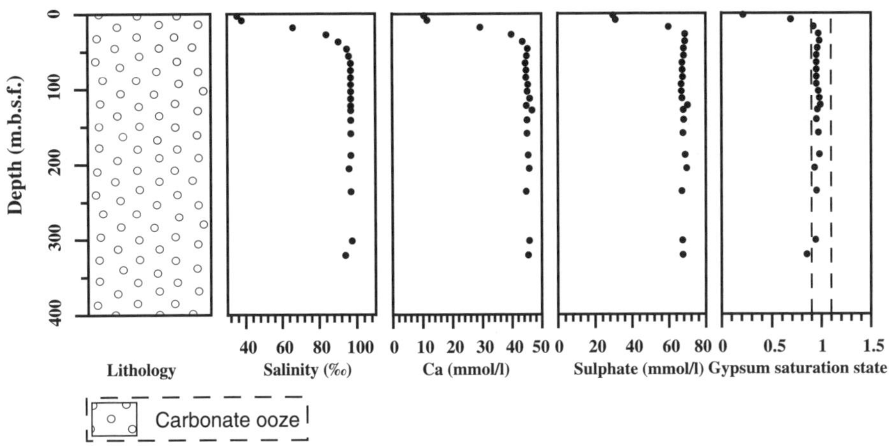

Fig. 6. The simplified lithology (left column), variation in pore-water salinity, Ca and SO_4 concentrations, and the gypsum and anhydrite saturation state *vs.* depth for ODP Leg 182 sites 1134 (modified from Hoareau, 2009). Note the perfect gypsum equilibrium in almost the entire borehole.

sub-horizontally on a regional scale. Indeed, pore waters collected at boreholes drilled during ODP Leg 182 have salinities up to 106‰ (three times that of seawater). Gypsum equilibrium is reached at five sites (1126, 1129, 1130, 1132 and 1134), although the mineral is not reported in the sediment cores. At these five sites, the salinity increase leads to an increase in the concentration of nearly all the dissolved species with depth, including Ca and SO_4. Our results show that the limitations of the Ca concentration indicated by an almost constant value below 50 m.b.s.f. (sites 1126 and 1134) and below ~230 m.b.s.f. (sites 1130 and 1132) are due to the equilibrium with respect to gypsum (Fig. 6).

4.3.2. Saturation due to the alteration of volcanogenic material in sediments: example of ODP Leg 126 (Izu-Boni Arc-Trench System)

The equilibrium between gypsum and pore waters at the sites of ODP Leg 126 in the Izu-Bonin forearc basin was demonstrated by Egeberg (1992) using the SOLMINEQ computer code (Kharaka *et al.*, 1988). The three sites drilled during this Leg are characterized by abundant volcanogenic material throughout the sedimentary pile at ODP site 787 and in Oligocene deposits at Sites 792 and 793. At these two latter sites, secondary gypsum was reported, in ash-rich units, by Tazaki & Fyfe (1992). At all sites of ODP Leg 126, the Ca concentration increases with depth to values up to 310 mmol/l (30 times that of normal seawater) while the SO_4 concentrations decrease down to the bottom of the holes (Fig. 7). Our calculations, in accordance with those of Egeberg (1992), show that saturation with respect to gypsum is reached at ODP Sites 787, 792 and 793 in the Oligocene ash-rich units only.

Again, these examples show how accurate thermodynamic calculations can point to chemical equilibrium between minerals and pore waters in marine sediments.

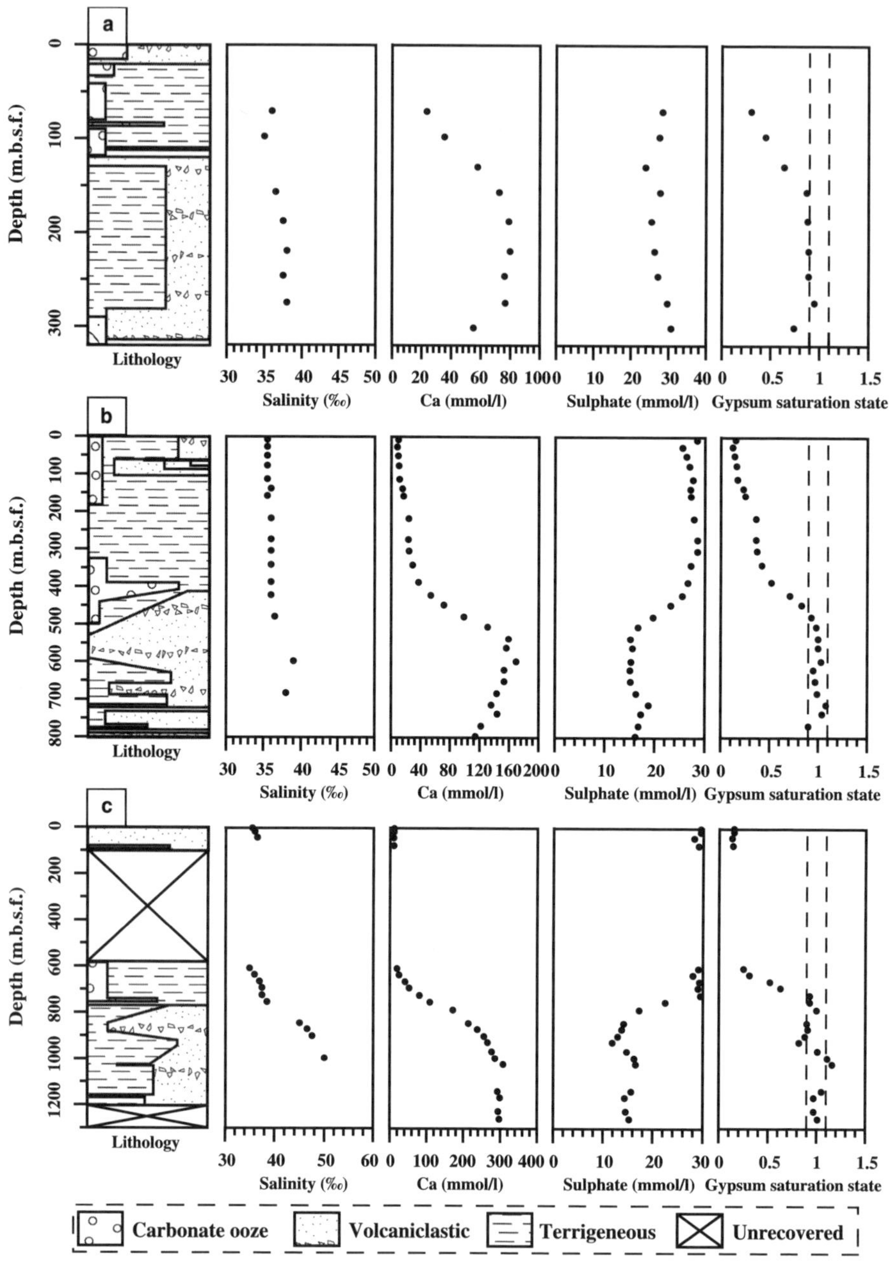

Fig. 7. The simplified lithology (left column), variation of pore-water salinity, Ca and SO_4 concentrations, and the gypsum and anhydrite saturation state *vs.* depth for ODP Leg 126 sites 787 (**a**), 792 (**b**) and 793 (**c**) (modified from Hoareau, 2009). Note that at the three sites gypsum equilibrium is found in the ash-rich units only.

5. The carbonate system in the marine environment: room for improvement

Needless to say, the carbonate system in seawater is of paramount importance. Methods for the calculation of its thermodynamic properties and of the solubility of calcium carbonate minerals based on the use of apparent thermodynamic constants have been and continue to be developed (see Millero, 2007 for a detailed review). Schematically, these constants use concentrations instead of activities and as such they depend on the solution composition as well as on temperature and pressure. They have been measured in seawater by many workers. Critically compiled data is generally fitted to empirical polynomial functions of T, P and the salinity or the ionic strength (see Zeebe & Wolf-Gladrow, 2001; Millero, 2006, 2007). A variety of such models has conveniently been embodied in a computer code, CO_2SYS (Lewis & Wallace, 1998; Pierrot *et al.*, 2006), which allows the calculation of the carbonate speciation and the saturation state of calcium-carbonate minerals in seawater. This program has been used extensively to recalculate the calcite and aragonite compensation depths in the various oceans and their evolution with respect to atmospheric CO_2 increase and the subsequent ocean acidification (Jutterstrom & Anderson, 2005; Friis *et al.*, 2006; Tanhua *et al.*, 2007; see Millero, 2007, for an up-to-date account of this topic). Because of computational performance and ease of use, reactive transport models often rely on such apparent thermodynamic constants for the carbonate system in seawater to study the dynamics of the marine environment in modern sediments and in palaeo-environmental reconstructions (*e.g.* Tribble & Mackenzie, 1998; Tyrrell & Zeebe, 2004; Arndt *et al.*, 2006), although sediment pore waters are markedly different from standard seawater, as we have seen above.

The behaviour of the carbonate system in aqueous solutions can be extracted from several combinations of two out of four measurable parameters: *pH*, alkalinity, dissolved inorganic carbon, CO_2 fugacity or partial pressure (Zeebe & Wolf-Gladrow, 2001). The two parameters relevant to the carbonate system routinely measured during the various ocean drilling programs DSDP, ODP and IODP are *pH* and alkalinity. Unfortunately, these two parameters are both the subject of large uncertainties, sampling artefacts, ambiguous definitions varying with authors, *etc.* (Zeebe & Wolf-Gladrow, 2001). In their 1986 ODP technical note, Gieskes & Peretsman (1986) wrote that "during sample retrieval on board ship, often substantial gas losses can occur which would severely affect the *pH*, but not the alkalinity. At very small CO_2 contents, on the other hand, CO_2 uptake into the sample may occur, and this, again, will affect the *pH* and not the alkalinity. For these reasons the use of *pH* for equilibrium calculations is limited." To these sampling artefacts that may eventually taint the measurements of the parameters relevant to the aqueous carbonate system, one can add the uncertainties introduced by the various definitions of *pH*, the different *pH* scales and electrode-calibration procedures (see Zeebe & Wolf-Gladrow, 2001 for a thorough discussion for seawater). The NBS *pH* scale has been in use for a long time aboard the *Joides Resolution* in spite of the recommendations of Gieskes & Peretsman (1986) and Gieskes *et al.* (1991) to revert to the Bis-Tris *pH* scale to avoid liquid-junction potential effects due to the difference in salinity between the dilute standard solutions used for the NBS buffers and the sediment pore waters. Zeebe & Wolf-Gladrow (2001) indicate that, starting from the same electrical potential difference between the measurement and the reference electrodes, a change in *pH* scale can lead to a *pH* change of ~0.1 *pH* unit.

The alkalinity concept also has its own ambiguities and uncertainties, starting by its definition (Zeebe & Wolf-Gladrow, 2001). The total alkalinity is the sum of all bases (weak and strong) that can be titrated by a strong acid. The carbonate alkalinity must then be extracted from the total alkalinity by a series of corrections of the boron, silica, phosphate, hydroxide, *etc.* contributions, most of which are based on apparent stability constants measured in standard seawater (Zeebe & Wolf-Gladrow, 2001; Millero, 2007).

While models based on apparent thermodynamic constants are very accurate for seawater, they cannot achieve the same accuracy for marine sediment pore waters (see the attempt of Monnin *et al.*, 2000). Unfortunately no carbonate model of an accuracy equivalent to those based on Pitzer's ion-interaction approach for sulphates is currently available. For example, the extension of the 25°C Harvie-Møller-Weare model (Harvie *et al.*, 1984) for the Na-K-Ca-Mg-Cl-SO_4-HCO_3-CO_3-CO_2-H_2O to high temperature and pressure is still incomplete in spite of noticeable advances in the description of various subsystems (see Millero, 2001, 2007), simply because of the lack of experimental data needed for the parameterization. Of noticeable interest is the model for the solubility of $CaCO_3$ minerals in the Na-K-Ca-Mg-Cl-SO_4-H_2O system developed by He & Morse (1993). It has been used by Malone *et al.* (2001) who investigated calcite and aragonite stabilities at ODP Sites 1008 and 1009 (Great Bahamas Bank). Those authors report calcite equilibrium and aragonite undersaturation at ODP Site 1008 and aragonite equilibrium and a slight calcite supersaturation at ODP Site 1009, despite large alkalinity differences in the pore waters of the two boreholes.

Finally, the example of sulphate minerals illustrates the accuracy with which chemical equilibrium can be depicted at the conditions of the ocean-water column but also at those encountered in marine waters with compositions markedly different from seawater. It also points to the requirements of the construction of accurate solubility models and of their extension to more and more complex models for increasingly realistic systems.

References

Anderson, G.M. & Crerar, D.A. (1993) *Thermodynamics in Geochemistry – The Equilibrium Model.* Oxford University Press, Oxford, UK, 588 pp.

Arndt, S., Brumsack, H.J. & Wirtz, K.W. (2006) Cretaceous black shales as active bioreactors: A biogeochemical model for the deep biosphere encountered during ODP Leg 207 (Demerara Rise). *Geochimica et Cosmochimica Acta*, **70**, 408–425.

Bacon, S., Culkin, F., Higgs, N. & Ridout, P. (2007) IAPSO standard seawater: Definition of the uncertainty in the calibration procedure, and stability of recent batches. *Journal of Atmospheric and Oceanic Technology*, **24**, 1785–1799.

Baker, P.A. & Bloomer, S.H. (1988) The origin of celestite in deep-sea carbonate sediments. *Geochimica et Cosmochimica Acta*, **52**, 335–339.

Berner, R.A. (1980) *Early Diagenesis: A Theoretical Approach*. Princeton University Press, Princeton, New Jersey, USA, 241 pp.

Berner, R.A. (2004) A model for calcium, magnesium and sulfate in seawater over Phanerozoic time. *American Journal of Science*, **304**, 438–453.

Bernstein, R.E. & Byrne, R.H. (2004) Acantharians and marine barite. *Marine Chemistry*, **86**, 45–50.

Bernstein, R.E., Byrne, R.H., Betzer, P.R. & Greco, A.M. (1992) Morphologies and transformations of celestite in seawater: The role of acantharians in strontium and barium geochemistry. *Geochimica et Cosmochimica Acta*, **56**, 3273–3279.

Bernstein, R.E., Byrne, R.H. & Schijf, J. (1998) Acantharians: a missing link in the oceanic biogeochemistry of barium. *Deep-Sea Research Part I – Oceanographic Research Papers*, **45**, 491–505.

Brennan, S.T. & Lowenstein, T.K. (2002) The major-ion composition of silurian seawater. *Geochimica et Cosmochimica Acta*, **66**, 2683–2700.

Chang, L.L.Y., Howie, R.A. & Zussman, J. (1998) *Rock-Forming Minerals. Volume 5B, Non-silicates: sulphates, carbonates, phosphates, halides (2nd edition)*. The Geological Society of London, London, 383 pp.

Claypool, L.E. & Kaplan, I.R. (1974) The origin and distribution of methane in marine sediments. In: *Natural Gases in Marine Sediments* (I.R. Kaplan, editor), Plenum Press, New York, pp. 99–139.

Curtis, C.D. (1995) Post-depositional evolution of mudstones l: early days and parental influences. *Journal of the Geological Society, London*, **152**, 577–586.

Davis, E.E. & Elderfield, H. (2004) *Hydrogeology of the Oceanic Lithosphere*. Cambridge University Press, Cambridge, UK, 706 pp.

Dehairs, F., Chesselet, R. & Jedwab, J. (1980) Discrete suspended particles of barite and the barium cycle in the open ocean. *Earth and Planetary Science Letters*, **49**, 528–550.

Dehairs, F., Jacquet, S., Savoye, N., Van Mooy, B.A.S., Buesseler, K.O., Bishop, J.K.B., Lamborg, C.H., Elskens, M., Baeyens, W., Boyd, P.W., Casciotti, K.L & Monnin, C. (2008) Barium in twilight zone suspended matter as a potential proxy for particulate organic carbon remineralization: Results for the North Pacific. *Deep-Sea Research Part Ii-Topical Studies in Oceanography*, **55**, 1673–1683.

Dickens, G.R. (2001) Sulfate profiles and barium fronts in sediment on the Blake Ridge: present and past methane fluxes through a large gas hydrate reservoir, *Geochimica et Cosmochimica Acta*, **65**, 529–543.

Egeberg, P.K. (1992) Thermodynamic aspects of Leg 126 interstitial waters. In: *Proceedings of the Ocean Drilling Program, Scientific Results, 126* (B. Taylor, K. Fujioka *et al.*, editors). Ocean Drilling Program, College Station, Texas, USA, pp. 519–529.

Egeberg, P.K. & Leg 126 Shipboard Scientific Party (1990) Unusual composition of pore waters found in the Izu-Bonin fore-arc sedimentary basin. *Nature*, **344**, 215–218.

Elderfield, H., Wheat, C.G., Mottl, M.J., Monnin, C. & Spiro, B. (1999) Fluid and geochemical transport through oceanic crust: a transect across the eastern flank of the Juan de Fuca Ridge. *Earth and Planetary Science Letters*, **172**, 151–165.

Eugster, H.P., Harvie, C.E. & Weare, J.H. (1980) Mineral equilibria in a 6-component seawater system, Na-K-Mg-Ca-SO_4-Cl-H_2O, at 25°C. *Geochimica et Cosmochimica Acta*, **44**, 1335–1347.

Felmy, A.R., Rai, D., Schramke, J.A. & Ryan, J.L. (1989) The solubility of plutonium hydroxide in dilute-solution and in high-ionic-strength chloride brines. *Radiochimica Acta*, **48**, 29–35.

Freyer, D. & Voigt, W. (2003) Cystallisation and phase stability of $CaSO_4$ and $CaSO_4$-based salts. *Monatshefte für Chemie*, **134**, 693–719.

Freyer, D. & Voigt, W. (2004) The measurement of sulfate mineral solubilities in the Na-K-Ca-Cl-SO_4-H_2O system at temperatures of 100, 150 and 200 degrees C. *Geochimica et Cosmochimica Acta*, **68**, 307–318.

Friis, K., Najjar, R.G., Follows, M.J., Dutkiewicz, S., Körtzinger, A. & Johnson, K.M. (2006) Dissolution of calcium carbonate: observations and model results in the North Atlantic. *Biogeoscience Discussions*, **3**, 1715–1738.

Froelich, P.N., Klinkhammer, G.P., Bender, M.L., Luedtke, N.A. Heath, G.R., Cullen, D., Dauphin, P., Hammond, D., Hartman, B. & Maynard, V. (1979) Early oxidation of organic matter in pelagic sediments of the equatorial Atlantic: suboxic diagenesis. *Geochimica et Cosmochimica Acta*, **43**, 1075–1090.

Gieskes, J.M. (1975) Chemistry of interstitial waters of marine sediments. *Annual Reviews of Earth and Planetary Science*, **3**, 433–453.

Gieskes, J.M. (1981) Deep-sea drilling interstitial water studies: implications for chemical alteration of the oceanic crust, Layers I and II. *Society of Economic Palaeontologists and Mineralogists, Special Publication*, **32**, 149–167.

Gieskes, J.M. (1983) The chemistry of interstitial waters of deep-sea sediments: interpretation of Deep Sea drilling data. In: *Chemical Oceanography* (J.P. Riley & R. Chester, editors), **8**, 221–269. Academic Press, London.

Gieskes, J.M. & Peretsman, G. (1986) Water chemistry procedures aboard JOIDES 35 Resolution – some comments. *ODP Technical Note 5*. Ocean Drilling Program, College Station, Texas, USA.

Gieskes, J.M., Gamo, T. & Brumsack, H. (1991) Chemical methods for interstitial water analysis aboard JOIDES Resolution. *ODP Technical Note 15*. Ocean Drilling Program, College Station, Texas, USA.

Goldhaber, M.B. (2002) Sulfur-rich sediments. In: *Treatise on Geochemistry* (F.T. MacKenzie, editor), **7**, 257–288. Elsevier, New York.

Gonneea, M.E. & Paytan, A. (2006) Phase associations of barium in marine sediments. *Marine Chemistry*, **100**, 124–135.

Greenberg, J.P. & Møller, N. (1989) The prediction of mineral solubilities in natural waters: A chemical equilibrium model for the Na-K-Ca-Cl-SO_4-H_2O system to high concentration from 0 to 250°C. *Geochimica et Cosmochimica Acta*, **53**, 2503–2518.

Hampt, G. & Delaney, M.L. (1997) Influences on calcite Sr/Ca records from Ceara Rise and other regions: distinguishing ocean history and calcite recrystallisation. In: *Proceedings of the Ocean Drilling Project, Scientific Results* (N.J. Shackleton, W.B. Curry, C. Richter & T.J. Bralower, editors), **154**, 491–500. Ocean Drilling Program, College Station, Texas, USA.

Hanor, J. (2000) Baryte-celestine geochemistry and environments of formation. In: *Sulfate Minerals* (C. Alpers, J. Jambor & K. Norstrom, editors). Reviews in Mineralogy and Geochemistry, **40**, 193–275. Mineralogical Society of America, Washington, D.C.

Hardie, L.A. (1983) Origin of $CaCl_2$ Brines by Basalt – Seawater Interaction: Insights Provided by Some Simple Mass Balance Calculations. *Contributions to Mineralogy and Petrology*, **82**, 205–213.

Hardie, L.A. (1996) Secular variation in seawater chemistry: an explanation for the coupled secular variation in the mineralogies of marine limestones and potash evaporites over the past 600 m.y. *Geology*, **24**, 279–283.

Harvie, C.E. & Weare, J.H. (1980) The prediction of mineral solubilities in natural waters – the Na-K-Mg-Ca-Cl-SO_4-H_2O system from zero to high concentration at 25 degrees C. *Geochimica et Cosmochimica Acta*, **44**, 981–997.

Harvie, C.E., Møller, N. & Weare, J.H. (1984) The prediction of mineral solubilities in natural waters: The Na-K-Mg-Ca-H-Cl-SO_4-OH-HCO_3-CO_3-CO_2-H_2O system to high ionic strengths at 25°C. *Geochimica et Cosmochimica Acta*, **48**, 723–751.

Harvie, C.E., Weare, J.H., Hardie, L.A. & Eugster, H.P. (1980) Evaporation of seawater – calculated mineral sequences. *Science*, **208**, 498–500.

Hawkesworth, C.J. & Elderfield, H. (1978) The strontium isotopic composition of interstitial waters from Sites 245 and 336 of the Deep Sea Drilling Project. *Earth and Planetary Science Letters*, **40**, 423–432.

He, S. & Morse, J.W. (1993) The carbonic acid system and calcite solubility in aqueous Na-K-Ca-Mg-Cl-SO_4 solutions from 0 to 90°C. *Geochimica et Cosmochimica Acta*, **57**, 3533–3554.

Helgeson, H.C. & Kirkham, D.H. (1974a) Theoretical prediction of the thermodynamic behavior of aqueous electrolytes at high pressures and temperatures: I. Summary of the thermodynamic/electrostatic properties of the solvent. *American Journal of Science*, **274**, 1089–1198.

Helgeson, H.C. & Kirkham, D.H. (1974b) Theoretical prediction of the thermodynamic behavior of aqueous electrolytes at high pressures and temperatures: II. Debye-Huckel parameters for activity coefficients and relative partial molal properties. *American Journal of Science*, **274,** 1199–1261.

Helgeson, H.C., Kirkham, D.H. & Flowers, G.C. (1981) Theoretical prediction of the thermodynamic behavior of aqueous electrolytes by high pressures and temperatures; IV, Calculation of activity coefficients, osmotic coefficients, and apparent molal and standard and relative partial molal properties to 600 degrees C and 5kb. *American Journal of Science*, **281**, 1249–1516.

Hill, T.L. (1960) *An Introduction to Statistical Thermodynamics.* Addison Wesley, Reading, UK, 265 pp.

Hoareau, G. (2009) *Sulfate/carbonate relationships during the diagenesis of marine sediments : sedimentological study of the Eocene Sobrarbe delta (Spain) and thermodynamic modelling of ODP/IODP data.* PhD thesis, University of Toulouse, France.

Holland, H.D. (2003) The Geologic History of Seawater. In: *Treatise on Geochemistry* (H.D. Holland & K. K. Turekian, editors), **6**, 1–46. Elsevier, New York.

Holland, H.D. (2005) Sea level, sediments and the composition of seawater. *American Journal of Science*, **305**, 220–239.

Horita, J., Zimmermann, H. & Holland, H.D. (2002) Chemical evolution of seawater during the Phanerozoic: implications from the record of marine evaporites. *Geochimica et Cosmochimica Acta*, **66**, 3733–3756.

Jacquet, S.H.M., Dehairs, F., Savoye, N., Obernosterer, I., Christaki, U., Monnin, C. & Cardinal, D. (2008) Mesopelagic organic carbon remineralization in the Kerguelen Plateau region tracked by biogenic particulate Ba. *Deep-Sea Research Part I – Topical Studies in Oceanography*, **55**, 868–879.

Jeandel, C., Dupre, B., Lebaron, G., Monnin, C. & Minster, J.-F. (1996) Longitudinal distributions of dissolved barium, silica and alkalinity in the western and southern Indian *Ocean. Deep Sea Research Part I: Oceanographic Research Papers*, **43**, 1–31.

Jutterstrom, S. & Anderson, L.G. (2005) The saturation of calcite and aragonite in the Arctic Ocean. *Marine Chemistry*, **94**, 101–110.

Kharaka, Y.K., Gunter, W.D., Aggarwall, P.K., Perkins, E.H. & DeBraal, J.D. (1988) *SOLMINEQ.88: a computer program for geochemical modelling of water rock interactions.* U.S. Geological Survey Water Resources Invest. Rep., **88-4277**.

Kramer, P.A., Swart, P.K., De Carlo, E.H. & Schovsbo, N.H. (2000) 17. Overview of interstitial fluid and sediment geochemistry, sites 1003–1007 (Bahamas transect). In: *Proceedings of the Ocean Drilling Project, Scientific Results 166* (P.K. Swart, G.P. Eberli, et al., editors), Ocean Drilling Program, College Station, Texas, USA, pp. 179–195.

Lewis, E. & Wallace, D.W.R. (1998) *Program Developed for CO_2 System Calculations. ORNL/CDIAC-105.* Carbon Dioxide Information Analysis Center, Oak Ridge National Laboratory, U.S. Department of Energy, Oak Ridge, Tennessee, USA.

Lowell, R.P. & Yao, Y.F. (2002) Anhydrite precipitation and the extent of hydrothermal recharge zones at ocean ridge crests. *Journal of Geophysical Research – Solid Earth*, **107**, 2424–2430.

Lowell, R.P., Yao, Y. & Germanovich, L.N. (2003) Anhydrite precipitation and the relationship between focused and diffuse flow in seafloor hydrothermal systems. *Journal of Geophysical Research – Solid Earth*, **108**, 2424, doi:10.1029/2002JB002371.

Lowenstein, T.K., Timofeeff, M.N., Kovalevych, V.M. & Horita, J. (2005) The major-ion composition of Permian seawater. *Geochimica et Cosmochimica Acta*, **69**, 1701–1719.

MacQuarrie, D. (1973) *Statistical Thermodynamics.* Harper and Row, New York, 273 pp.

Malone, M.J., Slowey, N.C. & Henderson, G.M. (2001) Early diagenesis of shallow water, periplatform carbonate sediments, leeward margin, western Great Bahama Bank (ODP Leg 166). *Geological Society of America Bulletin*, **113**, 881–894.

Marion, G.M. (2007) Adapting molar data (without density) for molal models. *Computers & Geosciences*, **33**, 829–834.

Marion, G.M. & Kargel, J.S. (2008) *Cold Aqueous Planetary Geochemistry with FREZCHEM: From Modelling to the Search for Life at the Limits.* Springer, Berlin, 251 pp.

Mazzullo, S.J. (2000) Organogenic dolomitization in peritidal to deep-sea sediments. *Journal of Sedimentary Research*, **70**, 10–23.

McDuff, R.E. (1981) Major cation gradients in DSDP interstitial waters: the role of diffusive exchange between seawater and the upper oceanic crust. *Geochimica et Cosmochimica Acta*, **45**, 1705–1713.

McDuff, R.E. & Gieskes, J.M. (1976) Calcium and magnesium in DSDP interstitial waters: diffusion or reaction? *Earth and Planetary Science Letters*, **33**, 1–10.

McManus, J., Berelson, W.M., Klinkhammer, G.P., Johnson, K.S., Coale, K.H., Anderson, R.F., Kumar, N., Burdige, D.J., Hammond, D.E., Brumsack, H.J., McCorkle, D.C. & Rhusdi, A. (1998). Geochemistry of barium in marine sediments: Implications for its use as a paleoproxy. *Geochimica et Cosmochimica Acta*, **62**, 3453–3473.

Millero, F.J. (2001) *Physical Chemistry of Natural Waters*. Wiley Interscience, New York, 654 pp.

Millero, F.J. (2006) *Chemical Oceanography (3rd edition)*. CRC Taylor and Frances Group, Boca Raton, Florida, USA, 496 pp.

Millero, F.J. (2007) The marine inorganic carbon cycle. *Chemical Reviews*, **107**, 308–341.

Millero, F.J. & Poisson, A. (1981) International One–Atmosphere Equation of State of Seawater. *Deep-Sea Research Part A – Oceanographic Research Papers*, **28**, 625–629.

Millero, F.J. & Thurmond, V. (1983) The ionization of carbonic acid in Na-Mg-Cl solutions at 25°C. *Journal of Solution Chemistry*, **6**, 401–412.

Millero, F.J., Feistel, R., Wright, D.G. & McDougall, T.J. (2008) The composition of Standard Seawater and the definition of the Reference-Composition Salinity Scale, *Deep-Sea Research I*, **55**, 50–72.

Møller, N. (1988) The prediction of mineral solubilities in natural waters: a chemical equilibrium model for the Na-Ca-Cl-SO_4-H_2O system to high temperature and concentration. *Geochimica et Cosmochimica Acta*, **52**, 821–837.

Mottl, M.J. & Wheat, C.G. (1994) Hydrothermal circulation through mid-ocean ridge flanks: fluxes of heat and magnesium. *Geochimica et Cosmochimica Acta*, **58**, 2225–2237.

Monnin, C. (1989) An ion interaction model for the volumetric properties of natural waters: Density of the solution and partial molal volumes of electrolytes to high concentrations at 25°C. *Geochimica et Cosmochimica Acta*, **53**, 1177–1188.

Monnin, C. (1990) The influence of pressure on the activity coefficients of the solutes and on the solubility of minerals in the system Na-Ca-Cl-SO_4-H_2O to 200°C and 1 kbar, and to high NaCl concentration, *Geochimica et Cosmochimica Acta*, **54**, 3265–3282.

Monnin, C. (1994) Density calculation and concentration scale conversions for natural waters. *Computers & Geosciences*, **20**, 1435–1445.

Monnin, C. (1995) Thermodynamic properties of the Na-K-Ca-Ba-Cl-H_2O system to 473.15 K and solubility of barium-chloride hydrates. *Journal of Chemical and Engineering Data*, **40**, 828–832.

Monnin, C. (1999) A thermodynamic model for the solubility of barite and celestite in electrolyte solutions and seawater to 200°C and to 1 kbar. *Chemical Geology*, **153**, 187–209.

Monnin, C. & Cividini, D. (2006) The saturation state of the world's ocean with respect to (Ba,Sr)SO_4 solid solutions. *Geochimica et Cosmochimica Acta*, **70**, 3290–3298.

Monnin, C., Jeandel, C., Cattaldo, T. & Dehairs, F. (1999) The marine barite saturation state of the world's oceans. *Marine Chemistry*, **65**, 253–261.

Monnin, C., Karpoff, A.-M. & Buatier, M., (2000) Calcium carbonate stability in the sediments of the eastern flank of the Juan de Fuca ridge. In: *Proceedings of the Ocean Drilling Project, Scientific Results*, (A. Fisher, E.E. Davis & C. Escutia, editors), **168**, 95–103. Ocean Drilling Program, College Station, Texas, USA.

Monnin, C., Balleur, S. & Goffe, B. (2003) A thermodynamic investigation of barium and calcium sulfate stability in sediments at an oceanic ridge axis (Juan de Fuca, ODP legs 139 and 169). *Geochimica et Cosmochimica Acta*, **67**, 2965–2976.

Mottl, M.J. & Wheat, C.G. (1994) Hydrothermal circulation through mid-ocean ridge flanks: fluxes of heat and magnesium. *Geochimica et Cosmochimica Acta*, **58**, 2225–2237.

Oelkers, E.H., Benezeth, P. & Pokrovski, G.S. (2009) Thermodynamic databases for water-rock interaction. In: *Thermodynamics and Kinetics of Water-Rock Interaction* (E.H. Oelkers & J. Schott, editors). Reviews in Mineralogy and Geochemistry, **70**, 1–46. Mineralogical Society of America, Washington, D.C. and the Geochemical Society, St. Louis, Missouri, USA.

Ostlund, H., Craig, H., Broecker, W. & Spencer, D. (1987) *Geosecs Atlantic, Pacific and Indians Ocean Expeditions. Shorebased Data and Graphics*. National Science Foundation, USA.

Pabalan, R.T. & Pitzer, K.S. (1987) Thermodynamics of concentrated electrolyte mixtures and the prediction of mineral solubilities to high-temperatures for mixtures in the system Na-K-Mg-Cl-SO_4-OH-H_2O. *Geochimica et Cosmochimica Acta*, **51**, 2429–2443.

Parkhurst, D.L. & Appelo, C.A.J. (1999) *User's Guide to PHREEQC (Version 2) – A Computer Program for Speciation, Batch-Reaction, One-Dimensional Transport, and Inverse Geochemical Calculations*. U.S. Geological Survey Water Resources Investigations Report **99-4259**, U.S. Geological Survey, Denver, Colorado, USA, 312 pp.

Paytan, A. & Griffith, E.M. (2007) Marine barite: Recorder of variations in ocean export productivity. *Deep-Sea Research Part I – Topical Studies in Oceanography*, **54**, 687–705.

Paytan, A., Kastner, M., Martin, E.E., Macdougall, J.D. & Herbert, T. (1993) Marine barite as a monitor of seawater strontium isotope composition. *Nature*, **366**, 445–449.

Paytan, A., Kastner, M., Campbell, D. & Thiemens, M.H. (2004) Seawater sulfur isotope fluctuations in the cretaceous. *Science*, **304**, 1663–1665.

Phutela, R.C., Pitzer, K.S. & Saluja, P.P.S. (1987) Thermodynamics of aqueous magnesium chloride, calcium chloride and strontium chloride at elevated temperatures. *Journal of Chemical and Engineering Data*, **32**, 76–80.

Pierrot, D., Lewis, E. & Wallace, D.W.R. (2006) *MS Excel Program Developed for CO_2 System Calculations*. Carbon Dioxide Information Analysis Center, Oak Ridge National Laboratory, U.S. Department of Energy, Oak Ridge, Tennessee, USA.

Pitzer, K.S. (1973) Thermodynamics of electrolytes. 1. Theoretical basis and general equations. *Journal of Physical Chemistry*, **77**, 268–277.

Pitzer, K.S. (1977) Electrolyte Theory – Improvements since Debye and Huckel. *Accounts of Chemical Research*, **10**, 371–377.

Pitzer, K.S. (1987) Thermodynamic model for aqueous solutions of liquid-like density. In: *Thermodynamic Modeling of Geological Materials: Minerals, Fluids and Melts* (I.S.E. Carmichael & H.P. Eugster, editors). Reviews in Mineralogy, **17**, 97–142. Mineralogical Society of America, Washington, D.C.

Pitzer, K.S. (1991a) *Activity Coefficients in Electrolyte Solutions*. CRC Press, Boca Raton, Florida, USA, 542 pp.

Pitzer, K.S. (1991b) Ion interaction approach: Theory and data correlation. In: *Activity Coefficients in Electrolyte Solutions* (K.S. Pitzer, editor). CRC Press, Boca Raton, Florida, USA, pp. 75–153.

Pitzer, K.S. (1995) *Thermodynamics (3rd edition)*. McGraw Hill, New York, 626 pp.

Pitzer, K.S. & Kim, J.J. (1974) Thermodynamics of electrolytes. IV. Activity and osmotic coefficients for mixed electrolytes. *Journal of the American Chemical Society*, **96**, 5701–5707.

Pitzer, K.S. & Mayorga, G. (1973) Thermodynamics of electrolytes. II. Activity and osmotic coefficients for strong electrolytes with one or both ions univalent. *Journal of Physical Chemistry*, **77**, 2300–2308.

Pitzer, K.S. & Mayorga, G. (1974) Thermodynamics of Electrolytes. III. Activity and osmotic coefficients for 2-2 electrolytes. *Journal of Solution Chemistry*, **3**, 539–546.

Pitzer, K.S. & Silvester, L.F. (1976) Thermodynamics of electrolytes. 6. Weak electrolytes including H_3PO_4. *Journal of Solution Chemistry*, **5**, 269–278.

Plummer, L.N., Parkhurst, D.L., Fleming, G.W. & Dunkle, S.A. (1988) *A computer program incorporating Pitzer's equations for calculation of geochemical reactions in brines*. U.S. Geological Survey Water-Resources Investigation Report **88-4153**.

Prieto, M. (2009) Thermodynamics of solid solution-aqueous solution Systems. In: *Thermodynamics and Kinetics of Water-Rock Interaction* (E.H. Oelkers & J. Schott, editors). Reviews in Mineralogy and Geochemistry, **70**, 47–85. Mineralogical Society of America, Washington, D.C. and the Geochemical Society, St. Louis, Missouri, USA.

Ptacek, C. & Blowes, D. (2000) Predicting sulfate-mineral solubility in concentrated waters. In: *Sulfate Minerals* (C.N. Alpers, J.L. Jambor & D.K. Nordstrom, editors). Reviews in Mineralogy and Geochemistry, **40**, 511–540. Mineralogical Society of America, Washington, D.C.

Reardon, E.J. & Armstrong, D.K. (1987) Celestine ($SrSO_4$(s)) solubility in water, seawater and NaCl solution. *Geochimica et Cosmochimica Acta*, **51**, 63–72.

Richter, F.M. (1996) Models for the coupled Sr and sulfate budget in deep-sea carbonates. *Earth and Planetary Science Letters*, **141**, 199–211.

Safarov, J., Millero, F., Feistel, R., Heintz, A. & Hassel, E. (2009) Thermodynamic properties of standard seawater: extensions to high temperatures and pressures. *Ocean Science*, **5**, 235–246.

Sayles, F.L. & Manheim, F.T. (1975) Interstitial solutions and diagenesis in deeply buried marine sediments – Results from the Deep Sea Drilling Project. *Geochimica et Cosmochimica Acta*, **39**, 103–127.

Schacht, U., Wallmann, K., Kutterolf, S. & Schmidt, M., (2008) Volcanogenic sediment-seawater interactions and the geochemistry of pore waters. *Chemical Geology*, **249**, 321–338.

Schenau, S.J., Prins, M.A., De Lange, G.J. & Monnin, C. (2001) Barium accumulation in the Arabian Sea: controls on barite preservation in marine sediments. *Geochimica et Cosmochimica Acta*, **65**, 1545–1556.

Scholle, P.A., Stemmerik, L. & Harpoth, O. (1990) Origin of major karst-associated celestine mineralization in Karstrygen, central East Greenland. *Journal of Sedimentary Petrology*, **60**, 397–410.

Stanley, S.M. & Hardie, L.A. (1998) Secular oscillations in the carbonate mineralogy of reef-building and sediment-producing organisms driven by tectonically forced shifts in seawater chemistry. *Palaeogeography, Palaeoclimatology, Palaeoecology*, **144**, 3–19.

Summerhayes, C.P. & Thorpe, S.A. (1996) *Oceanography – An Illustrated Guide.* Manson Publishing, London, 352 pp.

Swart, P.K. & Burns, S.J. (1990) Pore-water chemistry and carbonate diagenesis in sediments from Leg 115: Indian Ocean. In: *Proceedings of the Ocean Drilling Project, Scientific Results* (R.A. Duncan, J. Backman *et al.*, editors), **115**, 629–645. Ocean Drilling Program, College Station, Texas, USA.

Swart, P.K. & Guzikowski, M. (1988) Interstitial-water chemistry and diagenesis of periplatform sediments from the Bahamas, ODP Leg 101. In: *Proceedings of the Ocean Drilling Project, Scientific Results* (J.A. Austin Jr., W. Schlager *et al.*, editors), **101**, 363–380. Ocean Drilling Program, College Station, Texas, USA.

Taberner, C., Marshall, J.D., Hendry, J.P., Pierre, C. & Thirlwall, M.F. (2002) Celestine formation, bacterial sulphate reduction and carbonate cementation of Eocene reefs and basinal sediments (Igualada, NE Spain). *Sedimentology*, **49**, 171–190.

Tanhua, T., Körtzinger, A., Friis, K., Waugh, D.W. & Wallace, D.W.R. (2007) An estimate of anthropogenic CO_2 inventory from decadal changes in oceanic carbon content. *Proceedings of the National Academy of Science*, **104**, 3037–3042.

Tazaki, K. & Fyfe, W.S. (1992) Diagenetic and hydrothermal mineral alteration observed in Izu-Bonin deep-sea sediments. In: *Proceedings of the Ocean Drilling Project, Scientific Results* (B. Taylor, K. Fujioka *et al.*, editors), **126**, 101–112. Ocean Drilling Program, College Station, Texas, USA.

Torres, M.E., Brumsack, H.J., Bohrmann, G. & Emeis, K.C. (1996) Barite fronts in continental margin sediments: a new look at barium remobilization in the zone of sulfate reduction and formation of heavy barites in diagenetic fronts. *Chemical Geology*, **127**, 125–139.

Tribble, J.S. & Mackenzie, F.T. (1998) Recrystallization of magnesian calcite overgrowths on calcite seeds suspended in seawater. *Aquatic Geochemistry*, **4**, 337–360.

Tyrrell, T. & Zeebe R.E. (2004) History of carbonate ion concentration over the last 100 million years. *Geochimica et Cosmochimica Acta*, **68**, 3521–3530.

Weare, J.H. (1987) Models of mineral solubility in concentrated brines with application to field observations. In: *Thermodynamic Modeling of Geological Materials: Minerals, Fluids and Melts* (I. S. E. Carmichael & H. P. Eugster, editors). Reviews in Mineralogy, **17**, 143–176. Mineralogical Society of America, Washington, D.C.

Wheat, C.G. & Mottl, M.M. (2000) Composition of pore and spring waters from Baby Bare: global implications of geochemical fluxes from a ridge flank hydrothermal system. *Geochimica et Cosmochimica Acta*, **64**, 629–642.

Whitfield, M. (1975) The extension of chemical models for sea water to include trace components at 25°C and 1 atm pressure. *Geochimica et Cosmochimica Acta*, **39**, 1545–1557.

Zeebe, R.E. & Wolf-Gladrow, D.A. (2001) *CO_2 in Seawater: Equilibrium, Kinetics, Isotopes*. Elsevier, Amsterdam, 346 pp.

EMU Notes in Mineralogy, Vol. 10 (2010), Chapter 7, 259–287

Trace element variations in stalagmites: controls by climate and by karst system processes

IAN J. FAIRCHILD* and ADAM HARTLAND

School of Geography, Earth and Environmental Sciences, University of Birmingham, Edgbaston, Birmingham B15 2TT, UK, e-mail: i.j.fairchild@bham.ac.uk
**Corresponding author*

Here, we seek to address four issues. Firstly, we use strontium isotope ratios and sulphate concentrations and isotope ratios to illustrate how the changing elemental composition of the atmosphere (reflecting windiness or pollution *e.g.*) can be recorded in speleothems, although in the case of sulphur there can be significant lag and storage effects in the soil-ecosystem. Secondly, we illustrate how the differential dissolution rates of calcite (faster) and dolomite (slower) influences alkaline earth element chemistry in caves and speleothems, and how the occurrence of prior calcite precipitation in seasonally dry caves can be a useful signature of aridity. Thirdly, the impact of drip hydrology on speleothems is illustrated with two examples. Enrichments in alkaline earth elements tend to correspond with drier conditions. Many elements are transported preferentially in colloidal complexes, as is particularly clearly illustrated in hyperalkaline environments where the colloids are stabilized. Seasonal increases in colloidal abundance tend to be associated with enrichments in stalagmites in associated elements. Fourthly, the annual temperature cycle is difficult to recognize directly, but is seen indirectly in terms of its impact on cave air circulation, and in turn on P_{CO_2}, *pH* and hence sulphate concentrations in speleothems. Multiproxy approaches prove to be particularly valuable.

1. Introduction

Speleothems are precipitates forming in caves caused by drips, thin films, or pools of water. The most common forms to be studied are stalagmites which are generated at the sites of impact of dripping water. Speleothems are well placed to record information about past environments and climates because of their location in protected underground landscapes (Fig. 1a) and the variety of environmental proxies that they contain (Fairchild *et al.*, 2006a). The mostly commonly studied characteristics are growth rate (Baker *et al.*, 1998, 2008), oxygen (Lachniet, 2009) and carbon isotopes (McDermott *et al.*, 2005), and trace elements (Fairchild & Treble, 2009), of which the alkaline earth elements have been the most commonly studied to date. In recent years, there has been an increasing emphasis on the setting of speleothems as a controlling agent for their growth and Fairchild *et al.* (2006a) distinguished five domains in which the characteristic responses of speleothems to signals of external forcing are generated and modified (Fig. 1). Firstly, the atmosphere is the source of wet and dry deposition

DOI: 10.1180/EMU-notes.10.7

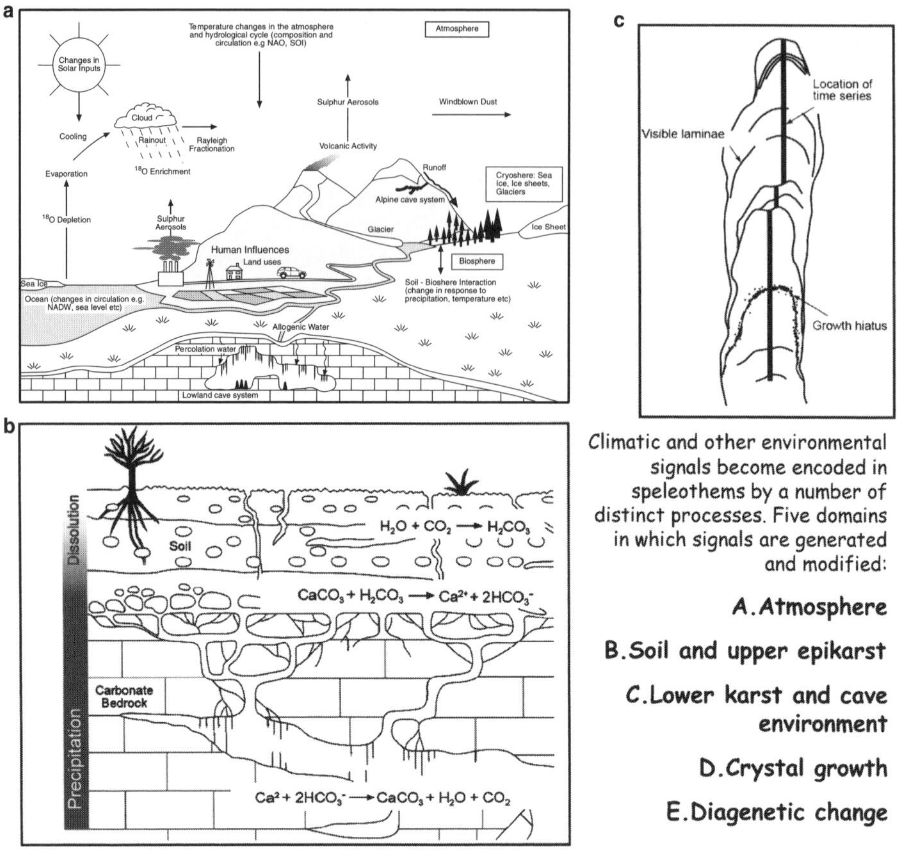

Fig. 1. (***a***) Cave environments in relation to surface environments and climates. (***b***) A schematic vertical profile illustrating the sequence of events that leads to speleothem formation. (***c***) The internal morphology of a candle-shaped speleothem illustrating growth layers and also hiatuses (after Fairchild *et al.* (2006a), with the permission of Elsevier).

(Fig. 2) from which are derived marine aerosol, pollutants, and water with a distinctive oxygen isotope composition. Second is the soil zone overlying the underlying epikarst which is a significantly porous and permeable region in the host rock created by long-term dissolution processes. Here dissolution of minerals, particularly carbonates, driven by high partial pressures of carbon dioxide, results in a solution with a high carbonate alkalinity, but near-neutral *pH*, and charged with a multitude of trace elements, both dissolved and attached to solid particles and smaller colloids (Fig. 3). The chemical composition of the fluid expelled downwards may depend on the intensity of the hydrological infiltration event and the time of year. Once the downward-percolating water encounters an air phase with a lower P_{CO_2} than it has previously experienced, degassing will tend to occur and we enter the third domain. The water that forms the

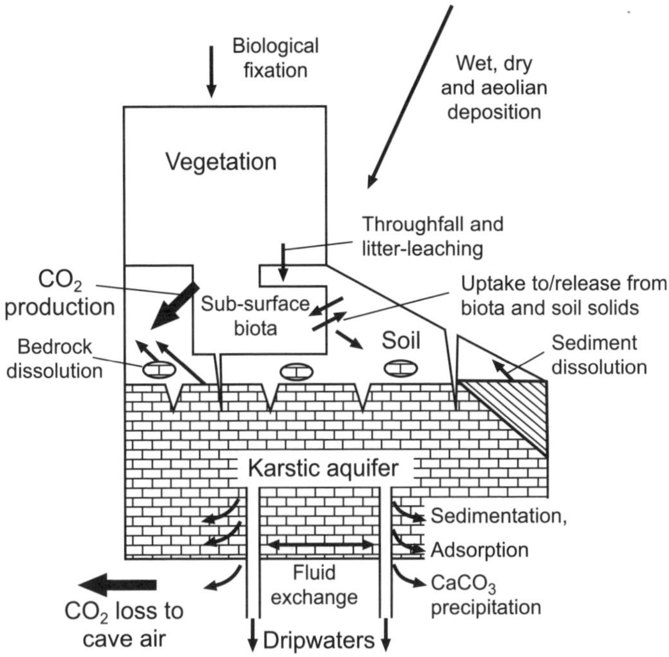

Fig. 2. Schematic flow chart to indicate sources of elements and processes involved in their transport and deposition in cave systems. Arrows denote element fluxes as particulates, colloids or solutes in aqueous solutions (from Fairchild & Treble (2009), with the permission of Elsevier).

stalagmite under study may already have precipitated calcite (prior calcite precipitation) which will have caused an evolution of its chemistry; colloidal species may also be lost by adsorption. The fourth domain is the stalagmite-growth surface and its covering by a thin water film from which a characteristic $CaCO_3$ crystal structure and chemistry emerges. Fifthly, there is the less well understood, post-depositional alteration of the calcareous deposit, principally by dissolution, but also including the conversion of aragonite to calcite.

Fairchild *et al.* (2006a) and Fairchild &Treble (2009) further distinguished several types of behaviour of the cave environment over time that could lead to distinctive time series of trace-element variation. These were stated to be: temperature-controlled patterns (direct mineralogical response to temperature), fluid-dominated patterns (changing elemental abundances that are reflected likewise in the speleothem) and crystal-dominated patterns. The latter are more complex, involving a chain of cause and effect, *e.g.* a change in temperature seasonally will tend to alter the cave-air circulation characteristics and hence the cave air P_{CO_2} which in turn leads to changes in *pH* in dripwater and solution saturation state. Finally such factors in turn can influence the style of growth and the incorporation of trace elements.

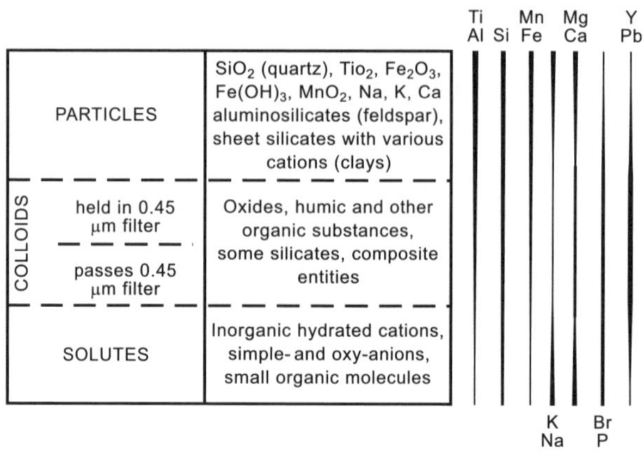

Fig. 3. Modes of transport of trace elements: particles, colloids, and solutes in karstic waters. Note that drip waters and hence speleothems are depleted in particles compared with general karstic waters. Colloids are defined as entities with at least one dimension in the size range of 1 nm to 1 µm (Lead & Wilkinson, 2006). The tendency of various elements to be transported in a given mode is shown schematically on the right: for example, Si is transported in abundance as insoluble particles, but is also present in colloids and in solution. For a more comprehensive discussion of colloid types and characterization methods see Lead & Wilkinson (2006) (from Fairchild & Treble (2009), with the permission of Elsevier).

2. How is the changing elemental geochemistry of the atmosphere reflected in speleothems?

In this section, we examine two examples where the atmosphere is the source of a distinctive variation in trace-element chemistry. Firstly we mention strontium, an element which has been studied extensively in carbonates (Morse & McKenzie, 1990). The controls on the absolute abundance of strontium are rather complex, as we discuss later, but the strontium isotope composition can serve as an effective indicator of the relative importance of different sources for the element. This task is made more straightforward because of the lack of Sr isotopic fractionation during $CaCO_3$ precipitation and the fact that Sr mainly substitutes for Ca (Pingitore & Eastman, 1986). Although carbonates tend to dissolve orders of magnitude more quickly than silicate phases, there may be strong differences in their Sr isotope composition that allows differing levels of contribution from silicate sources to be recognized. Also, the absolute magnitudes of Sr sourced from silicates may be relatively large, particularly if they are located in exchangeable sites. Sr studies have been most clearly successful in situations where they can be shown to either reflect changing weathering rates in relation to rainfall (*e.g.* Banner *et al.*, 1996), or be directly reflective of changing supply of aeolian dust, either of carbonate or silicate composition. These changes in dust deposition can reflect glacial-interglacial cycles, where major changes of source area in relation to sea level and climate occur (*e.g.* Goede *et al.*, 1998; Frumkin & Stein, 2004). Frumkin & Stein (2004) used a combination of Sr and U isotopes to characterize dust sources in Israel, and ongoing work in the region is in progress using Nd isotopes with the same objective.

The second example is that of sulphate and the potential for stalagmites to record histories of pollution or volcanic eruptions (Figs 4, 5). The first analyses of S in speleothems were by Frisia *et al.* (2005) who used synchrotron radiation to stimulate X-ray fluorescence in microscope domains whereby elements could be characterized in line profiles and maps. In a sample from a wooded Alpine site at Ernesto Cave (NE Italy) a distinct increase was seen during the second half of the 20th century, and X-ray absorption studies demonstrated that it is in the form of sulphate. Later a comparative study was carried out on trees growing in the region of the cave site both by dissolution of wood powders and micro-analysis using synchrotron radiation (Fairchild *et al.*, 2009). The increase was also seen, but the trees appeared to record it first (Fig. 4). Wynn *et al.* (2008) developed methods for extracting carbonate-associated sulphate (CAS) from small quantities of speleothems in order to derive time series of variation of stable isotopes of sulphur and oxygen in sulphate. The same techniques could not be used in the Alpine speleothems as the growth rates were too slow, but Wynn *et al.* (2010) have developed a micro-analytical method, using secondary ionization mass spectrometry (SIMS) to measure both the concentration and $\delta^{34}S$ composition of sulphate in the speleothem. The results show a distinct fall in $\delta^{34}S$ composition as the S concentration rises through the second half of the 20th century (Fig. 5b). This is what

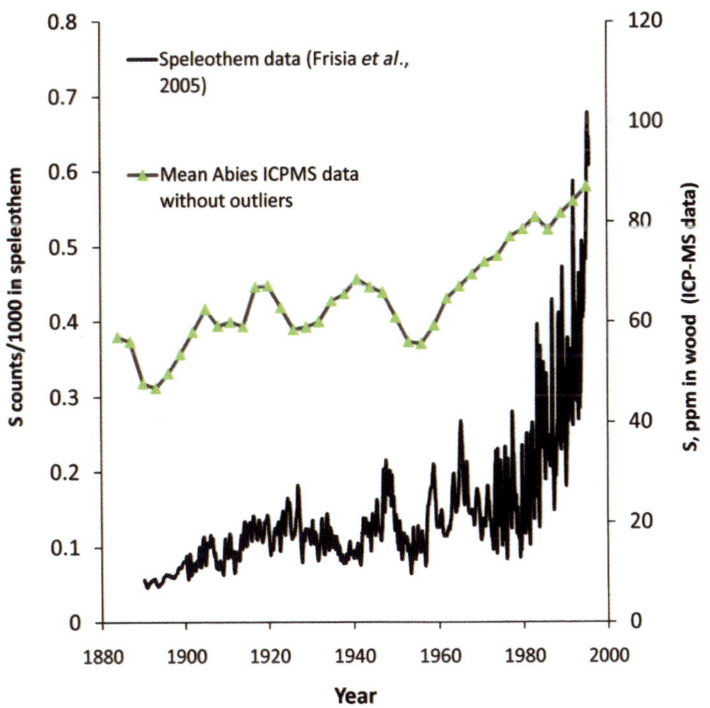

Fig. 4. Replotted from Fairchild *et al.* (2009). Comparison of 20th century S records from a speleothem and a tree (*Abies alba*) from the Ernesto cave area, NE Italy. The speleothem data were obtained using synchrotron μ-X-ray fluorescence and the wood data plotted are from ICP-MS analyses.

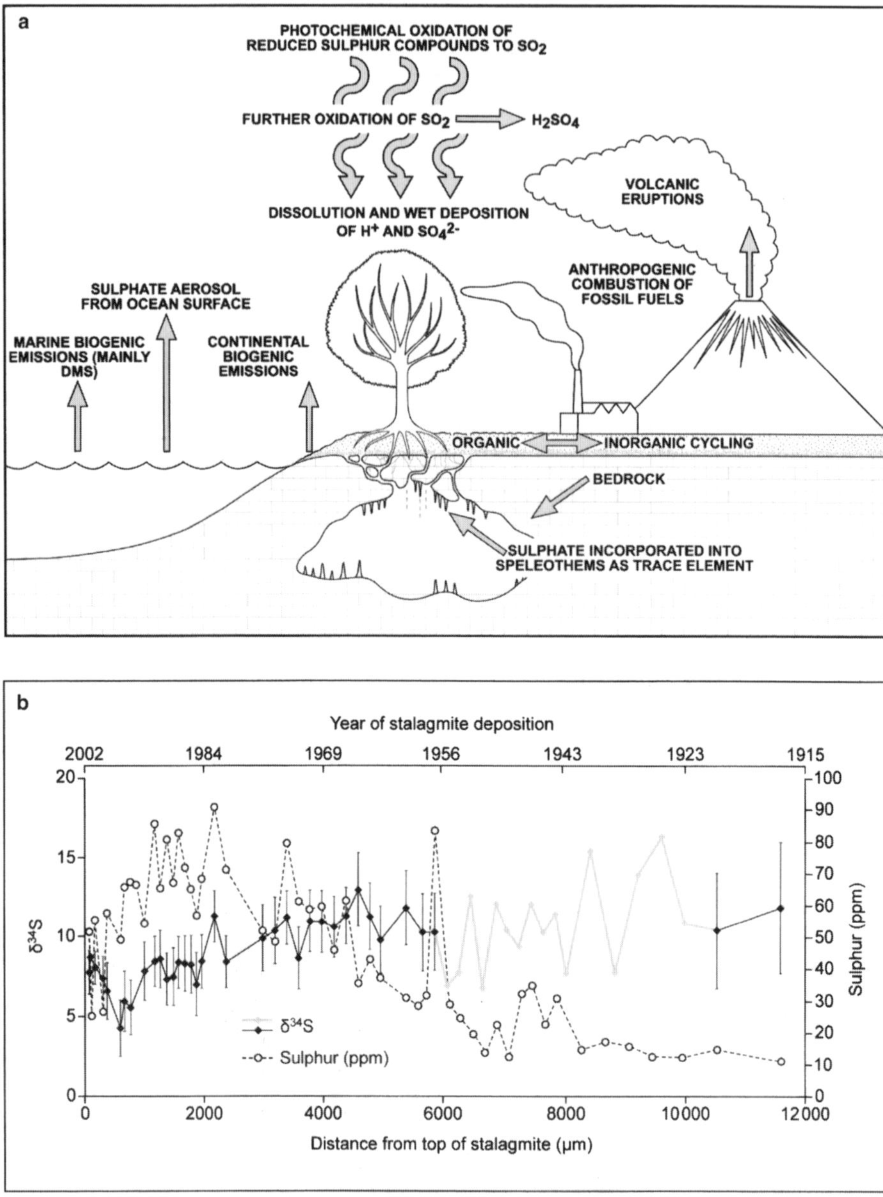

Fig. 5. (*a*) Sulphur sources with their isotopic composition (from Wynn *et al.* (2008), with the permission of Elsevier). (*b*) Microanalytical data demonstrating a late 20th century rise in sulphate content and decline in $\delta^{34}S$ composition in a stalagmite from Obir cave, SE Austria related to regional pollution. A mixing model between natural and pollution end members indicates that the pollution end member became heavier over time (Wynn *et al.*, 2010).

would be expected from the 20th century pollution history in western Europe, but an interesting point is that the S levels are still high in both trees and speleothems to the year 2000 yet the pollution levels peaked around 1980 (drip-water sulphate concentrations were seen to be declining by 2005). Unpublished $\delta^{18}O$ data in sulphate by P.M. Wynn demonstrates that the oxygen isotope composition has been re-set. This requires redox processes to have occurred as, otherwise, $\delta^{18}O$ in sulphate does not exchange. The range of S oxidation states found in local wood (Fairchild *et al.*, 2009) mirrors studies carried out elsewhere on humic substances, indicating that the soil-ecosystem partly reduces, mineralizes and stores S, later releasing it to solution.

Further back in time, pulses of S might be added by volcanism. They are more likely to be preserved in locations where there is a low background level of S, and one where the biomass in the overlying ecosystem is relatively small, hence avoiding too much smearing of the volcanic signal.

3. What relationship does speleothem geochemistry have to that of its enclosing karstic bedrocks?

In the broader field of aquatic chemistry there has been a lot of attention paid to sourcing chemical variations from dissolution of minerals (Appelo & Postma, 2005), but so far relatively little attention has been paid to determining how the source rock composition compares with that of cave drip waters and speleothems. A good reason for doing so is that it highlights processes occurring in the aquifer which could be significant in determining past environmental and climatic variations.

It is common to find a situation where both dolomite and calcite are available in the karstic aquifer. Whilst both are soluble, the dissolution kinetics of calcite is faster, but on the other hand the bulk of the Mg in the water is likely to be derived from dolomite. A check needs to be made that marine aerosols do not contribute a significant proportion of Mg, as this is sometimes the case (Baker *et al.*, 2000). Hence, it is better to work with water compositions which have been corrected for marine aerosol (Fairchild *et al.*, 2000).

Typical karstic solutions reach 95% of the equilibrium calcite solubility within a few hours (Appelo & Postma, 2005). Whilst calcite dissolution stops at saturation, dolomite dissolution can continue. The kinetics of dolomite dissolution are not well known close to its saturation, but it is likely to take months to years which could exceed the typical residence time of waters, in shallow caves in particular. Hence, the Mg/Ca ratio in drip waters in many shallow caves is likely to primarily reflect the relative proportions of dolomite and calcite that it has encountered during dissolution to calcite saturation. Figure 6 illustrates the relatively smaller Mg/Ca ratio expected in the drip waters compared with the bulk composition of rocks which it has encountered (Fairchild *et al.*, 2000).

If a second parameter is measured in the drip water and bedrocks, mixing lines can be constructed to illustrate the locus of bedrock composition in relation to drip waters. Figure 7 illustrates the results of such a study from two caves: Ernesto in Italy (Fig. 7a)

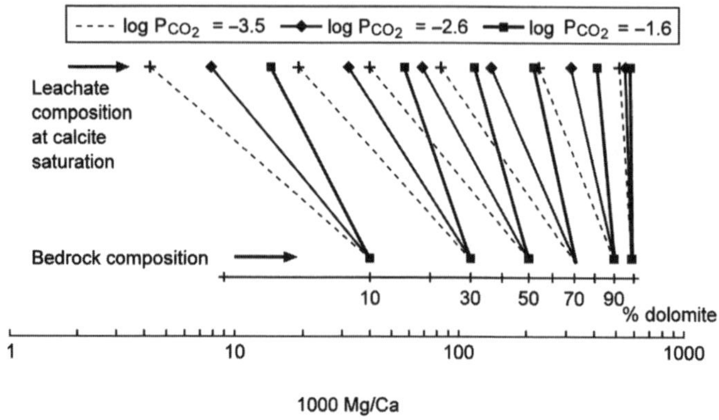

Fig. 6. Diagram to illustrate that drip waters tend to have smaller Mg/Ca ratios than mixed dolomite-limestone bedrock under conditions where simple dissolution (at various P_{CO_2} values) to calcite saturation occurs (after Fairchild *et al.* (2000), with the permission of Elsevier).

and Clamouse in southern France (Fig. 7b). In each there is a distinctive Sr composition of both dolomite and limestone end members that allows bedrock mixing lines to be constructed. In both cases, however, the water compositions deviate from this mixing line. Small pools in the cave have consistently larger Mg and Sr values compared with the drip waters that feed them. This is because of a loss of Ca from the water in calcite precipitation and so the Mg-Sr covariation that results is reflective of prior calcite precipitation.

Using Mg as an example, the composition of calcite that forms is given by:

$$Mg/Ca_{(calcite)} = Mg/Ca_{(water)} \cdot K_{Mg} \tag{1}$$

where K_{Mg} is the partition coefficient. Experimentally, the values of K_{Mg} and K_{Sr} are $\ll 1$, depending primarily on temperature for Mg and on growth kinetics for Sr (Huang & Fairchild, 2001). With increased PCP, Mg in calcite increases slightly more than Sr, since its partition coefficient is smaller.

Leaving aside the pool waters, there is also a Mg-Sr covariation shown in Figure 7a by the drip waters in the main chamber, but the puzzle here is that the Ca concentrations of water in this chamber are constant all year round. This demonstrates a constant P_{CO_2} in the epikarst and indicates an absence of prior calcite precipitation. This Mg/Ca-Sr/Ca covariation appears to be a form of selective trace-element leaching during surface weathering. This could be due to local salt precipitation and dissolution, but another universal affect arises whenever calcite grains are crushed to create fresh surfaces. The first stage of dissolution involves a selective leaching of Mg and Sr (McGillen & Fairchild, 2005) and can also be compared with the formation of a surface layer by exchange of protons with calcium as shown by Villegas-Jiménez *et al.* (2009). So this effect could be explained by the seasonal creation of fresh calcite surfaces by frost action as is indeed possible in this alpine environment.

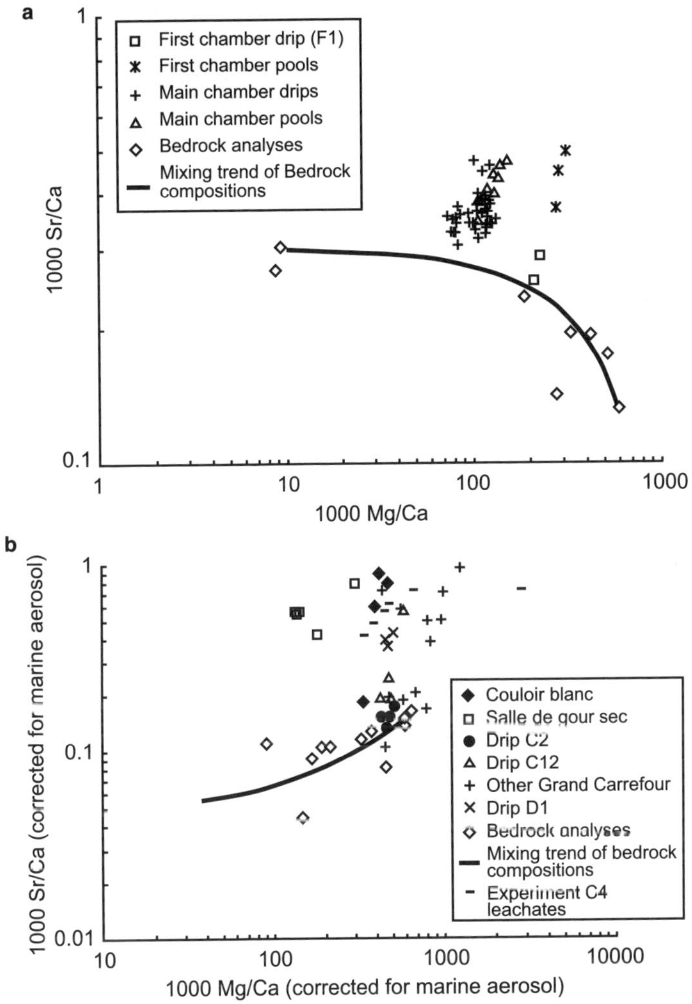

Fig. 7. Comparison of bedrock compositions and karstic waters. Data from (***a***) Ernesto and (***b***) Clamouse caves. The enrichment in element ratios compared with the bedrock mixing line shows that processes other than simple dissolution have occurred (after Fairchild *et al.* (2000), with the permission of Elsevier).

A second example is from the Clamouse Cave of southern France (Fig. 7b) where all the water is strongly displaced to higher Mg-Sr values compared with bedrocks. The aquifer here is a dedolomitized limestone which is porous and seems conducive to the operation of prior calcite precipitation leading to the displacement of the water values. In a high-resolution study of a speleothem from this site, McMillan *et al.* (2005) showed that enrichments in Mg and Sr occurred on an annual scale, but were also modulated over longer time periods. The evidence that prior calcite precipitation is the cause for these variations allows a confident interpretation of the longer-scale trends, which show the same type of variation, to be interpreted in the same way. A centennial-scale arid event, culminated around 1100 years B.P. with the nucleation of an aragonite layer, reflective of large Mg/Ca values and small drip rates (Frisia *et al.*, 2002). Cruz *et al.* (2007) demonstrated that such variations occurred on orbital scales in a speleothem from Botoverá Cave in Brazil.

4. How does drip hydrology impact on speleothems?

In shallow caves, the water discharge at many drip-water points display characteristic peaks associated with rainfall events followed by exponential decline superimposed on an annual-scale variation characteristic of the seasons. Slower drips tend to show much more muted variability without necessarily responding to individual events at all. Karst is known as a triple-porosity aquifer with major flow through conduits, a secondary flow along a complex network of two-dimensional fractures, and slow seepage flow through matrix. There can be distinctive impacts on hydrochemistry from both low flows (section 4.1) and high flows (section 4.2) which have different associated chemistry and mechanisms.

4.1. Alkaline earth elements: aquifer compartments and prior calcite precipitation

Studies in certain aquifers have demonstrated that drips display distinctive hydrogeochemical responses to hydrological changes. There is therefore the possibility that Mg/Ca ratios in drip waters and hence in stalagmites could reflect mean residence time. In reality this reflects different contributions of water that have followed different flow paths through the aquifer. This can be illustrated by means of a simple model of aquifer processes (Fig. 8a) that has been applied to drip water emerging in a limestone mine beneath a horizontally-bedded Mesozoic aquifer (Fairchild *et al.*, 2006b). The model was designed to explain certain observations such as the pattern of drip discharges in relation to rainfall, the observation that large discharges are associated with larger Ca than smaller discharges and the greatly enhanced Mg or Sr composition of certain drips at low flow. The model is configured so that water only infiltrates the ground above a threshold precipitation (more precise matching of model and data were found with the use of daily rainfall data rather than weekly cumulative totals). This water recharges the matrix, which is simulated as having a constant flow as long as water is present. Higher rates of infiltration result in recharge of fracture-flow, characterized by exponential decline of discharge following each event. Excessive recharge feeds a conduit (evidence for which is the constancy of the highest discharges at one drip site). Each flow compartment has a characteristic Mg/Ca and Sr/Ca ratio and water from each mixes and may precipitate $CaCO_3$ before the simulated drip water is sampled. This is the simplest type of model that can be used to simulate such an aquifer and it can be tuned for each drip to successfully account for the changing

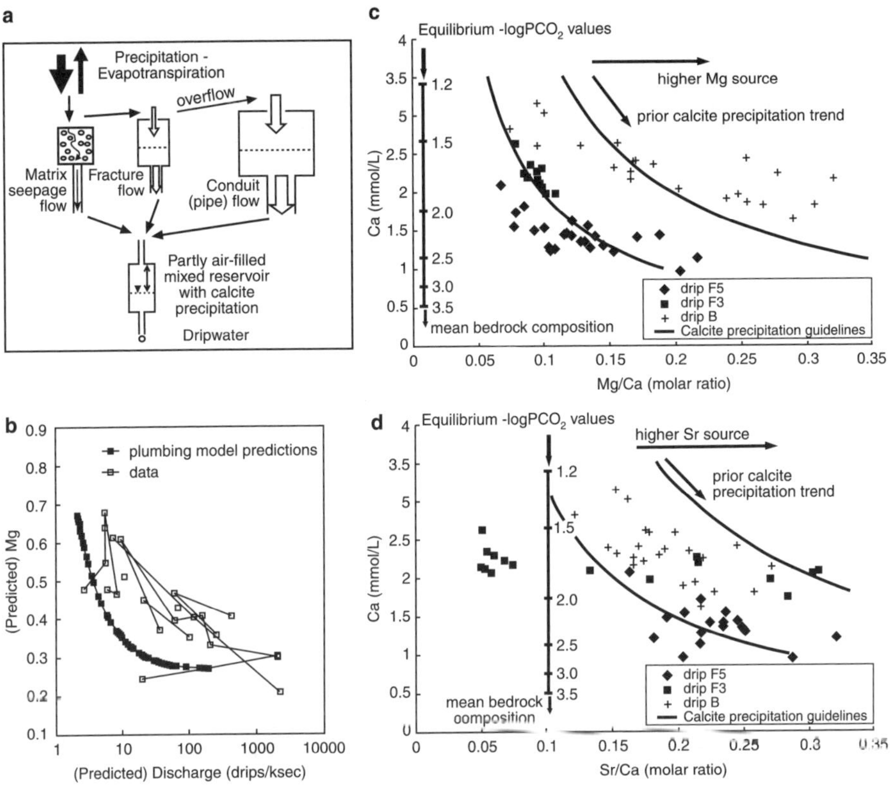

Fig. 8. Geochemical relationships at Brown's Folly Mine. (***a***) Hydrological model illustrating the role of processes such as fluid mixing and prior calcite precipitation (PCP). (***b***) Comparison of model predictions with data from a drip site; deviations are due to non-linear mixing effects. (***c***) and (***d***) Diagrams to determine the nature of processes causing chemical changes at low flow. Low Ca is found at low flows; in some cases this relationship can simply be explained by PCP (*e.g.* Mg/Ca data at site F3 and Sr/Ca data at sites B and F5). In the other cases, PCP must be augmented by enhanced trace-element composition of the aquifer segment being drawn on at low flow (after Fairchild *et al.* (2006b), with the permission of Elsevier).

discharge of many of them. Qualitatively, there is a clear explanation of the hydrogeochemical patterns noted above with data for Mg and/or Sr following the expected pattern for prior calcite precipitation, and in certain cases it can be inferred that there is a Mg- or Sr-rich aquifer compartment that is drawn on more at the slowest seepage flows (Fig. 8c,d).

Quantitatively however, there are deviations from the trends expected by linear mixing (Fig. 8b) indicating the role of effects such as dispersive exchange between matrix and fractures during high-flow episodes. More recently, variants of such linear systems models have been used to forward model oxygen-isotope variations, *e.g.* by allowing a Gaussian distribution of flow-through times for the seepage reservoir (Baker *et al.*, 2010).

The key point is whether trace-element enrichment results from prior calcite precipitation or from more trace element-rich slow-flow aquifer compartments; in both cases they are consistent with the effects of seasonal dryness in caves. This type of pattern has been demonstrated by monitoring and high-resolution speleothem studies at other cave sites in addition to those mentioned above (*e.g.* Verheyden *et al.*, 2000; Tooth & Fairchild, 2003; McDonald *et al.*, 2004; Johnson *et al.*, 2006; Karmann *et al.*, 2007; Mattey *et al.*, 2008, 2010). Such processes drive the distinctive trace-element responses in longer-term records such as those by McMillan *et al.* (2005) and Cruz *et al.* (2007) referred to above.

4.2. Colloidal transport of trace elements

Recently an association has been demonstrated between trace elements known to preferentially coordinate with oxygen-containing groups in organic ligands, and annual organic-rich laminae in speleothems (Borsato *et al.*, 2007). Attention is now turning to the transport processes of trace elements bound to organic-bearing colloids and some initial results are reported below. One relevant factor is that colloids are less prone to matrix diffusion (diffusion into micropores between fractures) than solutes and so may be especially effective in accelerating the migration of cationic species through fractured karstic aquifers (McKay *et al.*, 1993, 2000; McCarthy & McKay, 2004). We focus on water chemistry rather than speleothem chemistry since there is a lack of data on the relative partitioning of organics and trace elements from solution into speleothems. However, a first-order hypothesis can be that the binding of elements is sufficiently strong that the incorporation of such trace elements in the speleothem follows their relative abundance in water. Work is in progress to test this hypothesis.

4.2.1. Trace elements transported by colloids in karst waters: principles

Natural aquatic colloids are solids with a dimension between 1 μm and 1 nm, and encompass an array of materials and possible compositions. Colloids typically present as heterogeneous aggregates, generally consisting of organic matter (*e.g.* organic acids, proteinaceous materials, fibrilar polysaccharides) combined with crystalline solids such as aluminosilicates and oxides of Fe, Al and Mn (Lead & Wilkinson, 2006). Such aggregates, commonly referred to as organo-mineral complexes (OMC), are thought to form largely through bridging reactions between cementing cations (Ca^{2+}, Mg^{2+}) and exposed functional groups of fibrillar polysaccharides, organic acids and hydrolysed mineral surfaces (Buffle *et al.*, 1998; Davis *et al.*, 2001). Colloidal aggregates may also form through chemical precipitation from supersaturated solutions or through changes in redox potential (*e.g.* Mavrocordatos *et al.*, 2000).

Natural aquatic colloids possess large specific surface areas and, above neutral *pH*, typically have a negative surface charge. As a result, large proportions of trace metals may be bound to colloidal surfaces in fresh waters (Lead & Wilkinson, 2006). A suite of metal ions are known to preferentially coordinate with oxygen-containing functional groups in OMC. In particular, the transition metals (Cu, Ni, Zn, Cd) strongly coordinate with organic ligands (Xia *et al.*, 1997a, 1997b) and with hydroxyl groups at hydrous oxide surfaces (Collins *et al.*, 1999; Strawn & Sparks, 2000; Peacock, 2009) and layer silicate edges (Strawn & Sparks, 1999; Bourg *et al.*, 2007). Rare earth elements (REE^{3+}) also coordinate readily with organic acids (Plaschke *et al.*, 2004; Stern *et al.*, 2007; Marang *et al.*, 2009). These types of relationships have specifically been demonstrated in soil leachates (Tyler & Olsson, 2001).

Adsorption reactions between metal ions and surface functional groups essentially involve a single deprotonation and binding step:

$$\equiv \text{S–OH} + M^{Z+} \Longleftrightarrow \equiv \text{S–O}M^{Z-1} + \text{H}^{+} \qquad (2)$$

Where $\equiv$S represents the structural cation(s) to which a surface O atom is attached and M^{Z+} is a metal cation.

Adsorption reactions between transition metal (TM) ions and aquatic colloids typically correspond to inner-sphere complexation, whereby the metal is not readily exchangeable with bulk solution (Xia *et al.*, 1997a, 1997b). In contrast, reactions between colloidal surfaces and hydrolysing cations such as the alkaline earth elements and Pb^{2+} are non-specific, and occur *via* electrostatic interactions in which metal ions exchange readily between the colloid surface and bulk solution (Chen & Hayes, 1999; Zhang *et al.*, 2001). Outer-sphere complexes are insensitive to changes in *pH*, whereas inner-sphere complexation is enhanced at greater *pH* due to a combination of factors including progressive deprotonation, enhanced availability of functional groups with greater activation energies (*e.g.* phenolics) due to conformational changes (Tipping, 2001), and possible enhanced multi-denticity (Stern *et al.*, 2007). Multi-denticity is defined as the coordination of >1 functional group which is capable of bonding (donation of electron pair) to the central metal ion. Multi-denticity is exhibited by polydentate ligands, also known as chelating agents (Howard, 1998).

Free metal ions partition between the bulk phase and surface functional groups based on their reactivity (approximated by the ionic potential, z/r = charge/radius). The distribution of metals can thus be predicted based on their competitive affinity for available binding sites and successful models have been developed on this premise (*e.g.* NICA[non-ideal competitive adsorption]-Donnan model; Benedetti *et al.*, 1996; Kinniburgh *et al.*, 1999). Hence, in alkaline cave drip waters ($pH > 7$) the coordination of free metal ions at colloid surfaces is predicted to conform to a competitive hierarchy based on their selectivity for suitable and available binding sites.

The high-*pH* conditions found in cave drip waters may also have a significant effect on colloid stability through enhanced electrostatic repulsion between colloids (charge stabilization), especially where organic acids are present. Electrostatic repulsion between colloidal surfaces occurs because of the interaction between surface charge and surrounding electrolytes. In brief, a halo of counter-ions (electrostatic double layer) surrounds the negatively charged colloid surface, the concentration of which diminishes exponentially with distance (Stumm & Morgan, 1996). Electrostatic repulsion between colloids is exerted through this halo of counter ions. However, as ionic strength increases, the radius of the double layer becomes compressed. Colloids which come into close proximity may aggregate if the attractive London-van der Waals force exceeds the electrostatic repulsive force (Stumm & Morgan, 1996).

In karstic percolation waters, colloid stability and reactivity may, therefore, be constrained by the interplay between ionic strength and *pH*, as drip waters proceed towards equilibrium with respect to calcite. Indeed, colloidal abundance is known to be greater in groundwaters of lower ionic strength; in karstic systems, decreases in colloidal abundance have been attributed to increases in divalent cation concentrations associated with calcite dissolution (McCarthy & Shevenell, 1998; Atteia & Kozel, 1997; Atteia *et al.*, 1998).

The presence of colloids in groundwater implies stability: to remain in solution colloidal suspensions must be chemically stable (resistant to dissolution), physically stable (resistant to coagulation with other particulates), and have avoided filtration in immobile aquifer material (McCarthy & Shevenell, 1998). Colloidal composition is generally consistent with the local lithology (Atteia *et al.*, 1998) and Ca- and Mg-containing colloids are more abundant in karstic systems (McCarthy & Shevenell, 1998).

4.2.2. Trace elements transported by colloids in karst waters: example

Aside from studies conducted on samples from karstic springs and fast-flowing conduits, colloid-trace element interactions have not been examined widely in karstic systems and there have been no specific studies in cave drip waters. Ongoing research into trace element-colloid interactions in cave drip waters is being conducted at Pooles Cavern, Buxton, UK (Hartland *et al.*, 2010). Pooles Cavern (PC) is unusual because in some parts of the cave, instead of the normal *pH* (~8) of drip waters, a range from normal to hyperalkaline drip waters is found, with *pH* up to 13. This phenomenon results from rainwater percolating through lime waste beneath the forest soil. Limestone was quarried and burnt on the hillside above PC until 1820 A.D. when a deciduous forest was planted. Soils at the site consist of a leaf-litter layer, overlying a dark-brown, organic-rich top soil (brown ranker, Baker & Genty, 1999), beneath which are laterally varying proportions of lime waste and limestone fragments. The unusual geochemistry of PC provides a model system for understanding the effects of *pH* on colloidal stability and reactivity in cave drip waters. Current research at PC is aimed at understanding the interaction of trace metals and colloidal materials in order to better understand associated trace element variations in speleothems.

Colloidal materials in PC drips have been studied using Transmission Electron Microscopy-Energy Dispersive Spectroscopy (TEM-EDS). Samples were prepared onto grids using direct ultracentrifugation in order to obtain a representative sample of the native colloidal pool. Results from a typical karstic drip water (*pH* 8, EC 300–400) indicate that colloidal composition is consistent with other aquatic systems. Colloids are typically present in small concentrations and consist of aggregates of organic matter, aluminosilicates and metal oxides (Fig. 9).

In Figure 10, trace element and total organic carbon (TOC) data are presented from a high-*pH* site: here, a significant proportion of organic matter (and probably other entities as shown in Fig. 9) is present within the colloidal size range. Measured depletions in trace metals (Fig. 10) reflect the selective binding of Ni and Cu by colloids sized

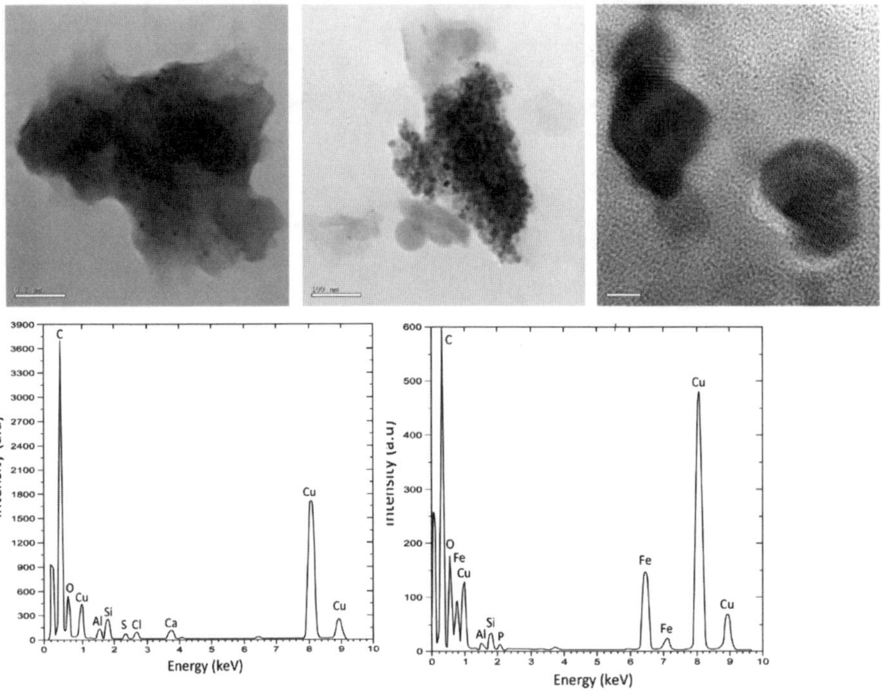

Fig. 9. TEM images of colloidal materials in Pooles Cavern drip water samples and corresponding EDS spectra. Left to right: aggregate (organic matter, aluminosilicate and iron oxide) with 200 nm scale bar; aggregate (organic matter and iron oxide) with 100 nm scale bar; enlargement of section of middle image showing iron oxide colloids, parallel lines are diffraction from atomic planes, 5 nm scale bar. The colloid composition was determined by energy-dispersive X ray analysis.

between 100 and 1 nm. Hence, the majority of available Cu and Ni ions may be assumed to be inner-spherically bound to colloidal surfaces. Reduction in the concentration of Br, primarily between 100 and 1 nm, probably relates to removal of organic matter, the organic-associated Br is thought to be present as a structural component of the organic matter (an organohalogen). Residual Br at <1 nm is a mixture of organo-Br in remaining organic matter and bromide. In Figure 10, Co reveals little association with colloidal materials, and a muted response to changes in *pH* (Fig. 11). Similarly, V is only slightly reduced in concentration by filtration. Indeed, a comparison of V data from all monitored sites at PC indicates that the *pH*-dependent binding characteristics of organic matter (OM) with V vary considerably between drip points (stars and crosses correspond to drips from different chambers in the cavern). The apparent divergence in binding affinity between different drip points may reflect qualitative differences in the composition of colloids, possibly resulting from hydrogeochemical forcing (*e.g.* ionic strength or *pH*). Alternatively, the divergence is likely to reflect differences in the availability of free V between dripwaters. Thus, as is the rule when seeking to understand the composition of speleothems, we must consider each drip

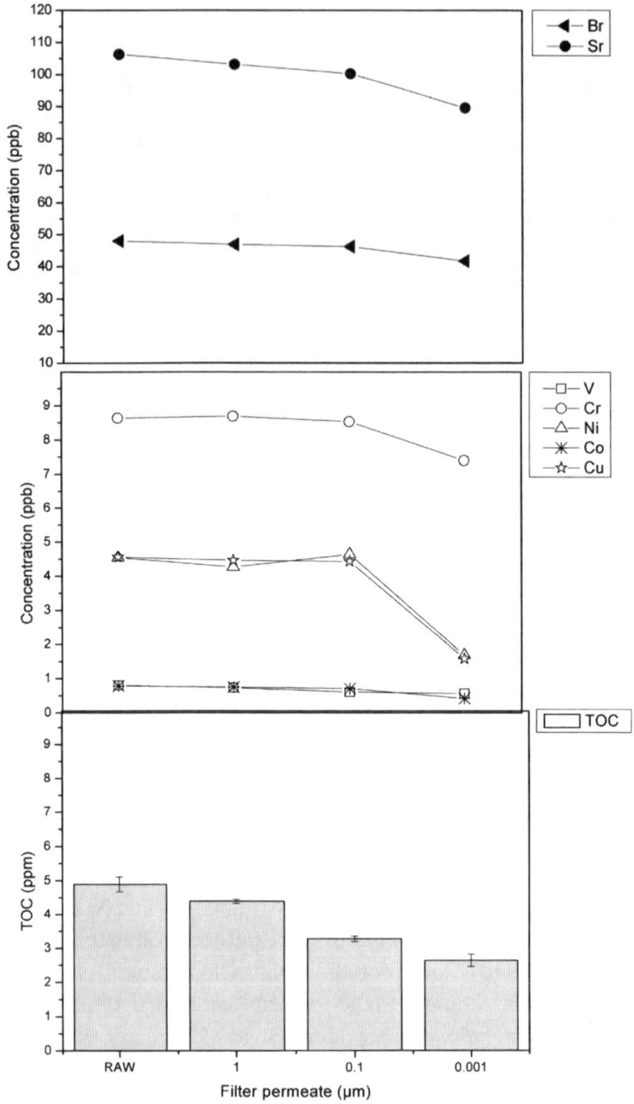

Fig. 10. Trace element and total organic carbon (TOC) concentrations in a filtered, hyper-alkaline cave dripwater sample from Pooles Cavern. TOC values are averaged, blank-subtracted measurements ($n = 4$); error bars represent one standard deviation of replicated analyses.

and each speleothem in isolation because of the inherent geochemical heterogeneity of cave and soil systems.

Of the monitored elements, the transition metals Cr, Ni and Cu displayed a strong association with OM in all monitored PC dripwaters. This is especially apparent at high *pH*, and probably reflects their ubiquity (*i.e.* binding was not limited by supply). Binding of trace elements specifically by colloidal materials (100 nm to 1 nm) was

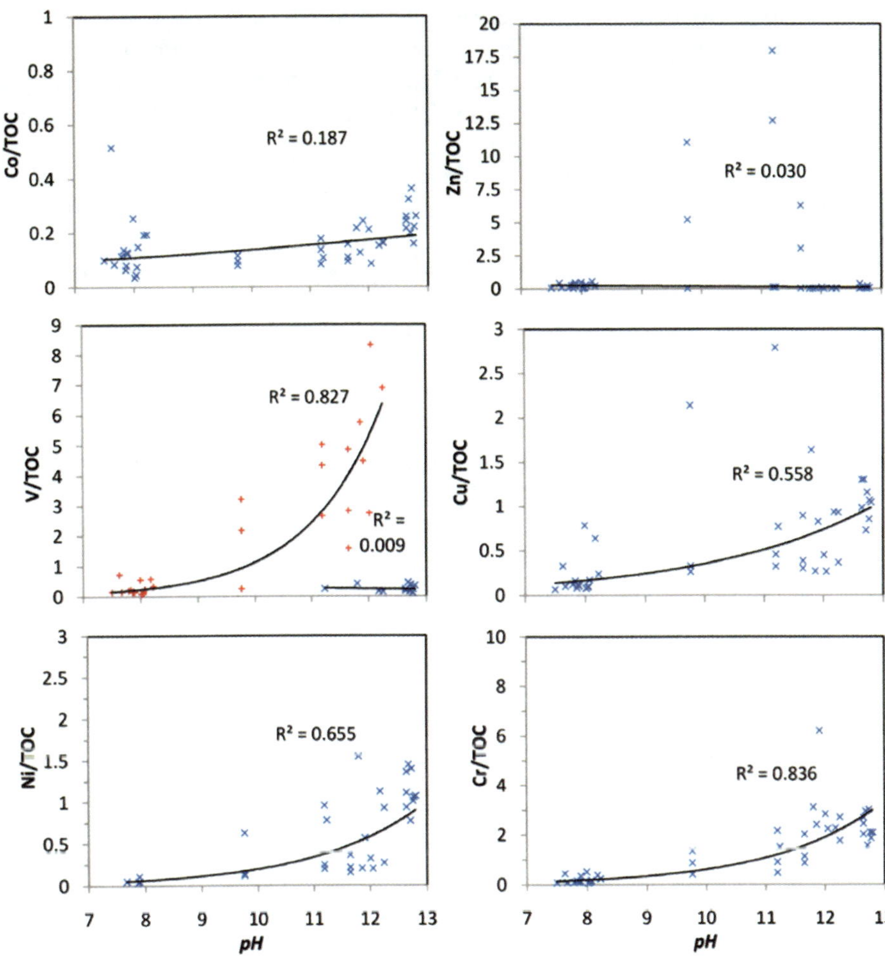

Fig. 11. Ratios of trace elements to total organic carbon (TOC) (μg trace element per mg organic carbon) in bulk Pooles Cavern drip-water samples plotted as a function of solution *pH*.

strongest for Ni and Cu. Zn may also display similar binding affinity to Cu and Ni, though issues with contamination from filters makes quantification of Zn fractionation impossible at trace concentrations using currently available materials. Also, the ratio of OM to Zn shows little *pH* sensitivity between drips in Pooles Cavern, possibly indicating that available Zn is limited and completely bound above *pH* 8, or that it is out-competed by other, more reactive species.

We can conclude from this initial study that the binding affinity of trace metals by karstic colloids is particularly sensitive to changes in *pH*, with an apparent hierarchy of binding strength between metal species Cr & V > Cu & Ni > Co > Zn. The particular *pH* dependency of Cr and V binding by OM probably reflects the increasing importance of multidenticity in the coordination of these ions. Other metals (*e.g.* Cd & Y & REE > Pb) are expected to show strong affinity for colloidal binding sites where present.

In contrast, Sr along with the other alkaline earth elements (Mg, Ca) adsorbs readily at available binding sites at colloid surfaces. This complexation corresponds to outer-sphere complexation, whereby the metal is readily exchangeable (labile) with bulk solution. Hence, Sr is reduced in each filter permeate indicating the non-specific formation of ion pairs with no apparent association with one size fraction.

4.2.3. High-flow enrichments in colloidally transported elements in speleothems

The coincidence of zonal enrichments of UV-fluorescent organic acids and trace elements provide circumstantial evidence supporting the hypothesis that organic colloids in karst drip waters act as vectors of trace element transport. Such zones were termed impurity pulse laminae (IPL) by Fairchild *et al.* (2007) based on their interpretation as reflecting short-lived flushes of organic colloids and associated trace elements to the karstic system. IPL may be visible by fluorescence, or in transmitted light in thin section, or both (Baker *et al.*, 1993; Borsato *et al.*, 2007; Fairchild *et al.*, 2001, 2010), or be primarily visible in this way. This is consistent with the results of analyses of trace elements in fractionated drip water sampled from PC and the corresponding speleothem, in which colloid-associated trace elements display a strong annual cyclicity, with a main autumnal peak coinciding with darker, organic-rich layers. In addition, these enrichments were found to be consistently mirrored by depletions in Sr and Ba.

Several authors have reported enrichments characteristic of IPL. The first report by Roberts *et al.* (1998) described annual chemical variation coincident with fluorescent laminae in stalagmite SU96-7 from Tartair Cave NW Scotland. The stalagmite displayed a consistent pattern of antipathetic variation of Mg with Sr and Ba coinciding with fluorescent laminations (Roberts *et al.*, 1998) and high P was later found to be the most distinctive characteristic of these and other examples (Fairchild *et al.*, 2001; Huang *et al.*, 2001).

Richter *et al.* (2004) reported periodic (mostly synchronous) enrichments in rare earth elements (REE^{3+}) and manganese (Mn^{2+}) within the laminae of eight late glacial German stalagmites, forming at times of inferred high rainfall. *REE* and Mn were generally concentrated in the more pigmented autumnal layers within the annual laminae (Richter *et al.*, 2004). Richter *et al.* (2004) attributed these peaks to relatively warm, humid climates within which elevated soil CO_2 concentrations in the summer, drove intense weathering of the host rock and allochthonous rock materials, but also referred to the possible role of transport by organic complexes. Indeed, Zhou *et al.* (2008) provided evidence of the co-abundance of Al and Mn in cave drip water and made the connection between zonal enrichments in Mn, P and *REE* with wetter climatic periods.

IPL in speleothems from Grotta d'Ernesto cave have been studied by several workers (Huang *et al.*, 2001; Fairchild *et al.*, 2001; Borsato *et al.*, 2007). Huang *et al.* (2001) showed that annual laminae from a stalagmite and a stalactite had different relations (one co-varying, one antipathetic) between Mg and Sr, but a consistent relationship was observed between enrichments in P and depletions in Sr. Comparison of the P and organic C contents indicated that the bulk of the P was inorganic and Mason *et al.* (2007) later found that P in an Ernesto speleothem was present both as phosphate

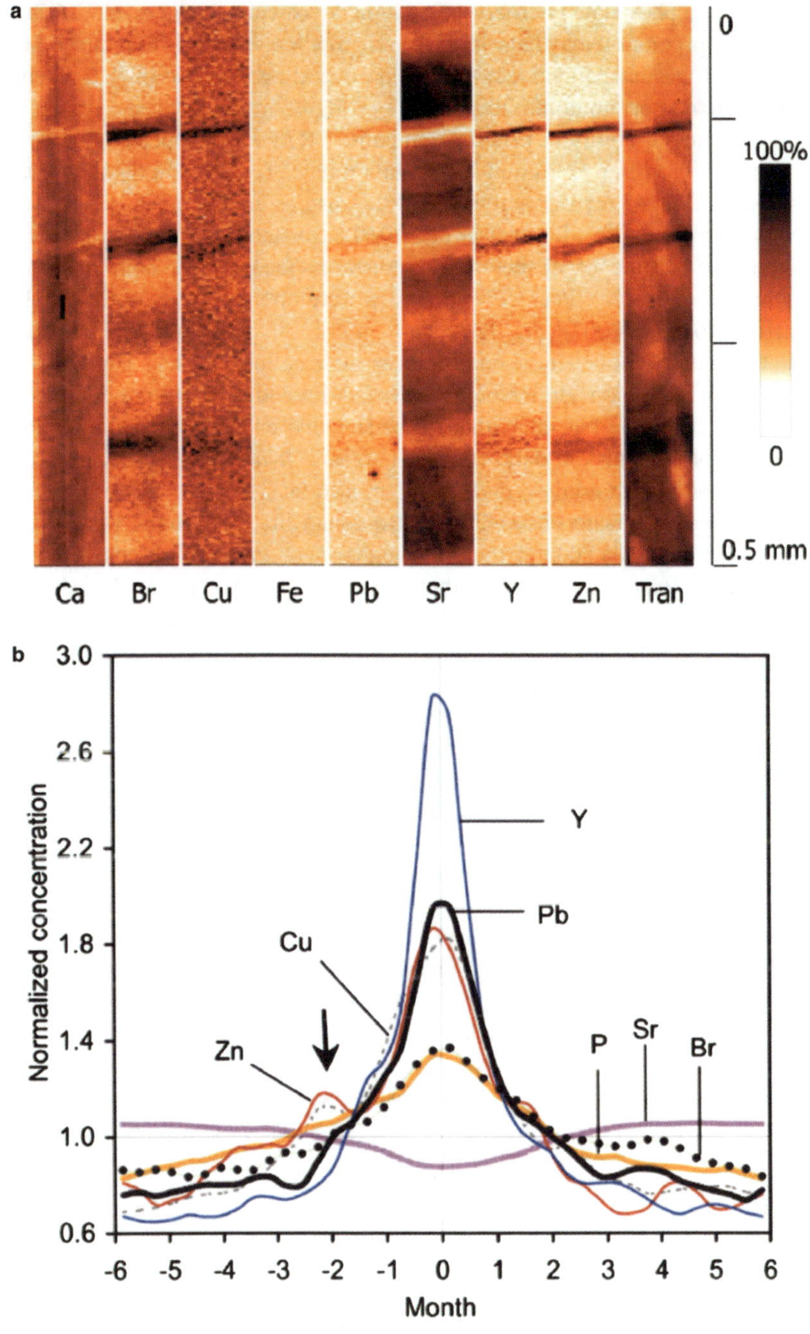

Fig. 12. (*a*) μ-XRF maps (30 μm × 500 μm) between 5.5 and 6.0 mm below the sample top showing Ca, Br, Cu, Fe, Pb, Sr, Y, Zn concentration and transmissivity (Tran) distribution for a 5-year interval (5 laminae). All linear scales are normalized to the minimum and maximum concentration for each element.

ions and calcium phosphate microcrystals within the calcite. Huang *et al.* (2001) speculated that P incorporation may have diminished growth rate leading to reduced Sr incorporation, but it was later found that there is no overall association of Sr concentration with growth rate (Borsato *et al.*, 2007).

Trace-element variability in Ernesto stalagmite ER78 was mapped by Borsato *et al.* (2007) using micrometer-resolution X-ray fluorescence (μXRF). In the autumn, a significant proportion of rapid fracture fed-flow ensures rapid transmissivity into the cave (Frisia *et al.*, 2000; Miorandi *et al.*, 2010). Borsato *et al.* (2007) expanded the observations to a suite of elements commonly associated with colloidal transport, centred on IPL (Fig. 12a). Peak concentrations were ordered Y > Zn, Cu and Pb > P and Br (Borsato *et al.*, 2007) (Fig. 12b). Those authors hypothesized that this hierarchy is due to the selectivity of element binding by organic colloids flushed from the soil zone during autumn infiltration. Enrichments in Y, Cu, Pb and Zn were not confined to the centre of dark laminae, but inhabited a broader zone encompassing the two months preceding and following the dark laminae (Borsato *et al.*, 2007). This may indicate that pronounced infiltration events during mid-to-late summer are represented by these enrichments. Borsato *et al.* (2007) also found that Sr troughs were associated with IPL. They speculated that the Sr troughs were induced by competition with other elements (Borsato *et al.*, 2007) that could arise due to increased availability of defect sites on rougher crystal surfaces. Indeed, seasonal changes in crystal morphology were also documented by Borsato *et al.* (2007). Crystals grown on glass slides before autumnal flushing were characterized by well developed rhombohedra, but those formed during autumn and early winter had developed complex surfaces with micrometer-sized holes between crystallites and a greater density of macrosteps, attributed to the presence of OM in drip waters (Fig. 13). This is consistent with the generation of more defect sites at which a wide variety of trace elements could be incorporated.

Borsato *et al.* (2007) noted that the peak concentration of all the colloidally transported elements was greater during a period around WWI which was known as a period of deforestation. Hence, changes in colloidally transported elements may be linked to changes in vegetative cover above the cave system. This idea was tested at an abandoned mine site, Browns Folly Mine, near Bath in SW England, whose drip hydrology was mentioned above. Baldini *et al.* (2005) presented a carbon and oxygen isotope study of the 20th century growth of a stalagmite from the mine. Growth rate increased significantly with time which was related to the revegetation of the site as it recovered from mining. Similarly, changes in $\delta^{13}C$ and $\delta^{18}O$ also occurred.

Fig. 12. (*Continued*) (***b***) Composite diagram of the average 'annual' signal for the seven elements which show an annual structure obtained by stacking ten successive cycles observed from 5.55 to 6.45 mm below the top, resampled over 81 points per cycle, and centred over the visible lamina, which marks 'month' 0. Each month is calculated as $1/12$ of the annual width from the top of one dark lamina layer to the next, representing the hydrological year. Three-point moving averages of primary data were normalized to the mean value. The arrow indicates secondary Cu and Zn peaks, 2–3 months prior to the major peak (from Borsato *et al.* (2007), used with the permission of Elsevier).

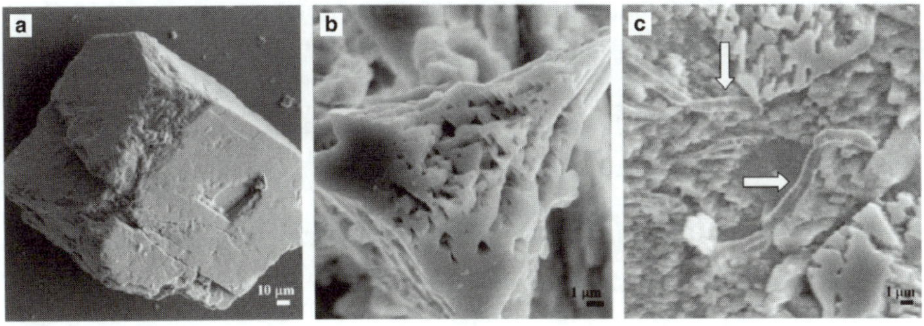

Fig. 13. SEM images of calcite crystals deposited on glass plates under ER78 drip. (***a***) Typical calcite crystal grown throughout one hydrological year and removed in September, prior to the autumnal discharge peak. The rhombohedral crystal shows macro steps, where calcite growth probably occurs. (***b***) Composite crystal removed in November, after autumnal discharge peak. The composite crystal consists of smaller crystallites of ~1 μm in size that do not completely coalesce, leaving voids between them. (***c***) Putative bacteria on the surface of calcite crystals removed after the autumnal discharge peak. No micro-pits or macro-steps on the calcite surface (from Borsato *et al.*, (2007), with the permission of Elsevier).

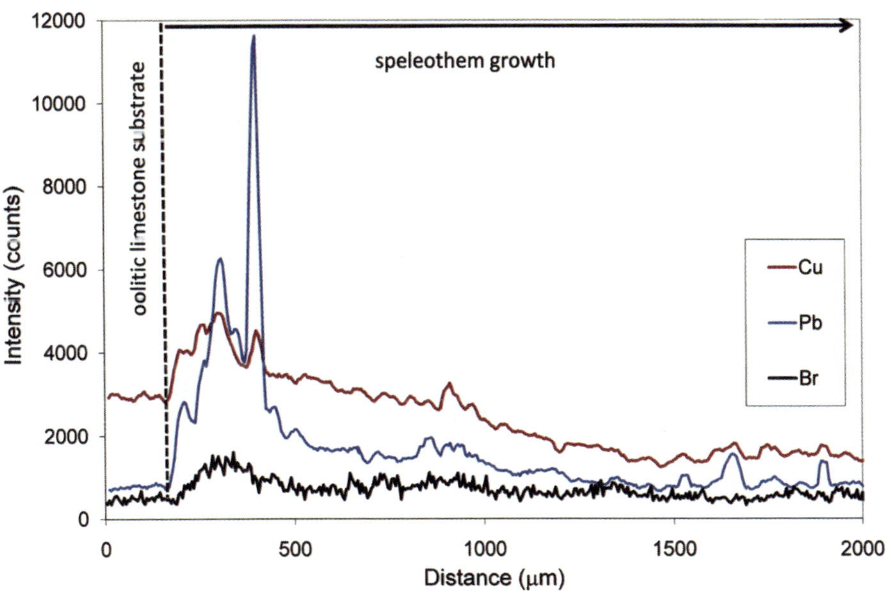

Fig. 14. Synchrotron μ-XRF scans of modern stalagmite BFM Boss from Browns Folly Mine, showing large abundances of trace elements (Cu [Zn is nearly identical to Cu], Pb, Br) in the early stages of growth in the early 20th century (starting around 1910/1915). The scan shows ~20 years of growth on the oolitic limestone substrate. Initially there was little vegetation at the site, leading to large trace-element contents, then decreasing over time. Five to six annual peaks are visible in the interval 1500–2000 μm from the start of the scan.

Figure 14 illustrates a strong trend in several trace elements that are known to be colloidally associated. They display much larger values than the host rock in the initial years of growth in the early 20th century, and these values then decline over time. This suggests that the use of such elements as proxies for palaeo-infiltration of rainfall is worth pursuing.

5. How is the annual temperature cycle reflected in speleothem composition?

The trace element known to relate directly to temperature changes under karst-analogue is Mg (Huang & Fairchild, 2001), and Mg variation in forams is important as a marine palaeotemperature proxy (Henderson, 2002). The effects of an annual 11°C change in stream temperature were also clearly shown in the Mg composition of tufas by Ihlenfeld *et al.* (2003). However, it should be noted that there is an inherent uncertainty of $\pm$15% in the experimental determination of Mg partition coefficients under identical conditions (Huang & Fairchild, 2001 and our unpublished data). This noise appears to be reflected by variability of natural speleothem calcites. An example is the speleothem SU96-7 from Tartair cave, Scotland from a cave chamber (Roberts *et al.*, 1998; Proctor *et al.*, 2000; Fairchild *et al.*, 2001) that is now known to vary seasonally by 4.5°C in temperature (Fuller *et al.*, 2008) and displays constant year-around Mg/Ca and Sr/Ca. Although Mg increases by ~5% per degree of temperature between 15 and 25°C, this dependency may be slightly less at lower temperatures (Huang & Fairchild, 2001). At any rate, an annual Mg cycle is not visible in the measurements (Fig. 15). One consistent feature in Figure 15, which applies to many speleothems (Fairchild & Treble, 2009) is a

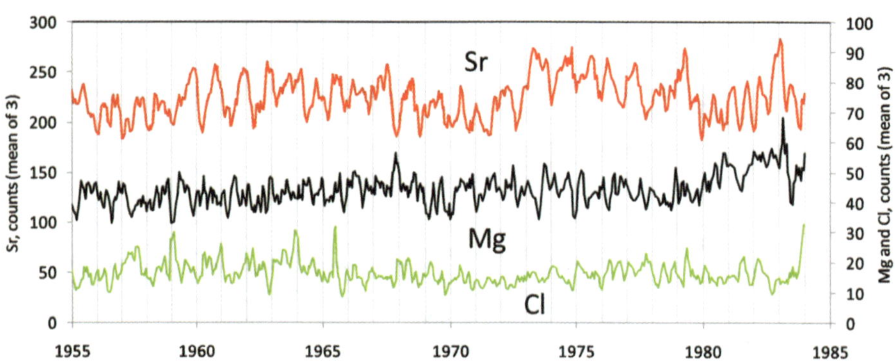

Fig. 15. Synchrotron µ-XRF scans across the top of sample SU96-7, Tartair cave, Sutherland. The time scale is based on lamina-thickness measurements from fluorescence images close to the profile location, but there is some lateral variation in thickness. Allowing for this, Sr displays a consistent pattern of annual variation, which can also be mapped (low Sr maps onto fluorescent layers). Low Sr was shown to coincide with high P by Fairchild *et al.* (2001). Mg and Cl do not display annual cycles. Data are three-point moving averages and their standard error is shown.

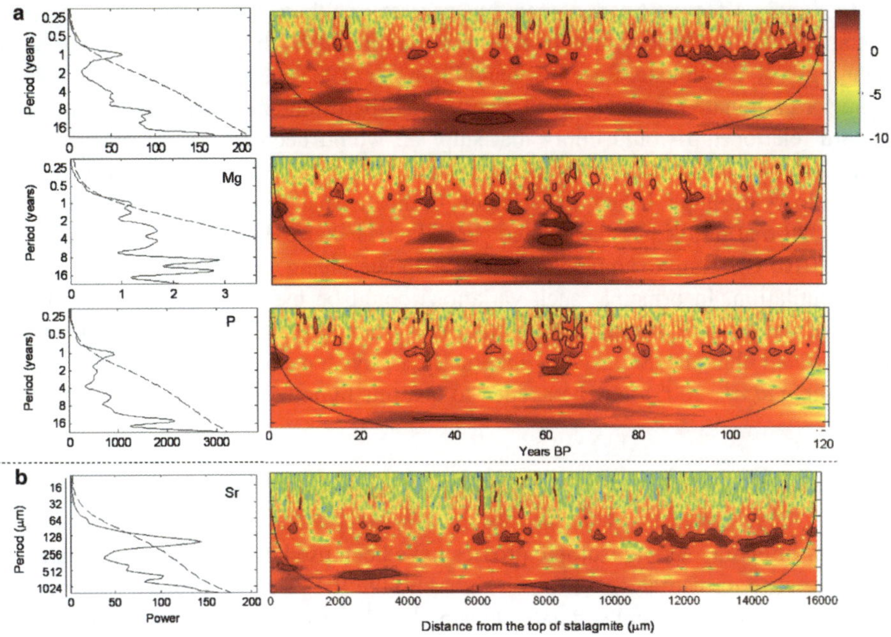

Fig. 16. (*a*) Spectral and wavelet plots for Sr, P and Mg from stalagmite Obi84, Obir cave Austria. Significance levels are calculated under the assumptions of a red noise background spectrum and c2 distributed power spectrum estimates. The dashed line in the spectral plots indicates the 95% confidence level. Periodicities in the wavelet plot significant at the 95% confidence level are outlined in bold. The cone of influence is displayed as a curved black line in the wavelet plots; anything that falls outside of this zone, shaded black, is subject to edge effects; (***b***) spectral and wavelet plot for Sr in the distance domain (from Smith *et al.* (2009), with the permission of Elsevier).

strong annual variation in Sr, which has also been used as a distinctive marker for using chemistry to determine interval chronologies by counting annual chemical layers (Smith *et al.*, 2009 and Fig. 16). P and Sr are typically antipathetic. This is likely to be due to the exclusion of Sr from defect sites occupied by P when P becomes available. Hence this variation is likely to reflect the annual hydrological cycle, rather than the temperature cycle directly.

A second type of annual textural lamination (other than IPL) is couplets of clear and cloudy calcite (*e.g.* Genty & Quinif, 1996). These are generally recognized to reflect seasonal variations in supersaturation, leading to different fabric development. In turn, these are driven by seasonal changes in cave ventilation, which is seen to be a very important factor in the physiology of many caves (Fairchild *et al.*, 2007). In the season with more extensive ventilation (normally, but not exclusively the summer) P_{CO_2} is normally (but not exclusively, Mattey *et al.*, 2008), lower resulting in a greater saturation index and quicker growth. In this faster growth season, there may be more extensive prior calcite precipitation; there may also be changes in fabric resulting in kinetic factors affecting trace-element incorporation, although this requires more documentation. One trace element known to respond directly to the *pH* changes induced by changing P_{CO_2} is sulphate. Busenberg and Plummer (1985) modelled sulphate as substituting for carbonate ion and derived a very small partition coefficient. Such a straightforward relationship would imply that sulphate would decrease by a factor of 10 in calcite per one *pH* unit increase following the *pH*-controlled decrease in the HCO_3/CO_3 ratio. Our unpublished experiments indicate that this model is qualitatively correct, but that in practice the change per *pH* unit is between two and three. Fairchild *et al.* (2010) illustrate examples of annual sulphate changes that reflect the effects of cave ventilation. In principle such variations could be used to establish the relative length of winter and summer seasons, but the Frisia *et al.* (2005) elemental mapping demonstrated that the S data are inherently noisy and so multiple replicate scans (*e.g.* Treble *et al.*, 2005) may be needed to test this hypothesis.

6. Conclusion

A variety of elemental and isotopic proxies have been used to establish varying atmospheric input to terrestrial systems over time and the list of useful proxies continues to grow. Both Sr isotopes and sulphate elemental and isotopic variations show particular promise for the future.

Existing publications based on theory, cave monitoring and speleothem analysis have established the basic ground rules related to Mg and Sr incorporation into speleothems and these have been helpful in establishing palaeohydrological information on timescales from decadal (Treble *et al.*, 2003) to centennial (McMillan *et al.*, 2005) to millennial (Hellstrom & McCulloch, 2000) to precessional (Cruz *et al.*, 2007) in duration. However, it continues to be vital to use a multi-proxy approach in speleothems because there are always rational exceptions to each 'rule' concerning proxy behaviour.

There is a growing body of evidence that dissolved and colloidal organic matter is a distinct agent with a pervasive influence over the patterns of trace-element incorporation within speleothem laminae. When present in sufficient quantities it can bind free ionic solutes, mobilize previously bound species and may alter the crystallographic properties of calcite forming in its presence. Further research is ongoing to establish the relationship between colloidal mobilization and capture within speleothems. Colloidally transported elements may provide information relating to growth rate and potentially to changes in precipitation/infiltration and vegetative cover above cave sites.

Impurity-pulse laminae may provide a marker of the start of the hydrologic year (Borsato *et al.*, 2007), and hence, direct evidence of growth rate (Smith *et al.*, 2009). IPL may potentially yield hydrological information, particularly in hydrologically sensitive settings or where climate displays a distinct seasonality (*e.g.* the two wet seasons observed in many sub tropical climates), sub-annual responses can be detected (Shopov *et al.*, 1994; Baker *et al.*, 1999; Fairchild *et al.*, 2010). The derivation of useful information from IPL remains a challenge that requires the integration of organic geochemical studies, high-resolution monitoring of natural organic tracers, and quantitative models of cave-karst aquifer hydrology and hydrogeochemistry, thereby, enabling discrimination between climatic and cave-specific factors (Fairchild *et al.*, 2007).

Sulphate appears to be an indicator of seasonal variations in ventilation forced by temperature changes. Future research involving elemental mapping on this and other trace species will be useful in understanding the variability in the signal to noise ratio of trace elements in natural deposits.

Acknowledgements

AH acknowledges a NERC studentship. Mohammed Baloushа is thanked for his help in obtaining the TEM images and chemical data.

References

Appelo, C.A.J. & Postma, D. (2005) *Geochemistry, Groundwater and Pollution (2nd edition)* A.A. Balkema, Leiden, The Netherlands, 649 pp.

Atteia, O. & Kozel, R. (1997) Particle size distributions in waters from a karstic aquifer: from particles to colloids. *Journal of Hydrology*, **201**, 102–119.

Atteia, O., Perret, D., Adatte, T., Kozel, R. & Rossi, P. (1998) Evolution of size distributions of natural particles during aggregation: modelling versus field results. *Colloids and Surfaces A: Physicochemical and Engineering Aspects*, **139**, 171–188.

Baker, A. & Genty, D. (1999) Fluorescence wavelength and intensity variations of cave waters. *Journal of Hydrology*, **217**, 19–34.

Baker, A., Smart, P., Edwards, R. & Richards, D. (1993) Annual growth banding in a cave stalagmite. *Nature*, **364**, 518–520.

Baker, A., Genty, D., Dreybrodt, W., Barnes, W.L., Mockler, N.J. & Grapes, J. (1998) Testing theoretically predicted stalagmite growth rate with Recent annually laminated samples: Implications for past stalagmite deposition. *Geochimica et Cosmochimica Acta*, **62**, 393–404.

Baker, A., Proctor, C.J. & Barnes, W.L. (1999) Variations in stalagmite luminescence laminae structure at Poole's Cavern, England, AD 1910–1996: calibration of a palaeoprecipitation proxy. *The Holocene*, **9**, 683–688.

Baker, A., Genty, D. & Fairchild, I.J. (2000) Hydrological characterisation of stalagmite drip waters at Grotte de Villars, Dordogne, by the analysis of inorganic species and luminescent organic matter. *Hydrology and Earth System Sciences*, **4**, 439–449.

Baker, A., Genty, D., Fairchild, I.J., Tan, M., Fuller, L., Jex, C. & Asrat, A. (2008) Annually laminated speleothems. *International Journal of Speleology*, **37**, 193–206.

Baker, A., Asrat, A., Fairchild, I.J., Leng, M.J., Thomas, L., Widmann, M., Jex, C.N., Dong, B., van Calsteren, P. & Bryant, C. (2010) Decadal-scale rainfall variability in Ethiopia recorded in an annually laminated, Holocene-age, stalagmite. (Submitted to *The Holocene*). DOI: HOL-09-0122.R1.

Baldini, J.U.L., McDermott, F., Baker, A., Baldini, L.M., Mattey, D.P. & Railsback, L.B. (2005) Biomass effects on stalagmite growth and isotope ratios: A 20th century analogue from Wiltshire, England. *Earth and Planetary Science Letters*, **240**, 486–494.

Banner, J.L., Musgrove, M., Asmerom, Y., Edwards, R.L. & Hoff, J.A. (1996) High-resolution temporal record of Holocene ground-water chemistry: Tracing links between climate and hydrology. *Geology*, **24**, 1049–1053.

Benedetti, M.F., VanRiemsdijk, W.H., Koopal, L.K., Kinniburgh, D.G., Gooddy, D.C. & Milne, C.J. (1996) Metal ion binding by natural organic matter: From the model to the field. *Geochimica et Cosmochimica Acta*, **60**, 2503–2513.

Borsato, A., Frisia, S., Fairchild, I.J., Somogyi, A. & Susini, J. (2007) Trace element distribution in annual stalagmite laminae mapped by micrometer-resolution X-ray fluorescence: implications for incorporation of environmentally significant species. *Geochimica et Cosmochimica Acta*, **71**, 1494–1512.

Bourg, I.C., Sposito, G. & Bourg, A.C.M. (2007) Modeling the acid-base surface chemistry of montmorillonite. *Journal of Colloid and Interface Science*, **312**, 297–310.

Buffle, J., Wilkinson, K.J., Stoll, S., Filella, M. & Zhang, J.W. (1998) A generalized description of aquatic colloidal interactions: The three-colloidal component approach. *Environmental Science & Technology*, **32**, 2887–2899.

Busenberg, E. & Plummer, L.N. (1985) Kinetic and thermodynamic factors controlling the distribution of SO_4^{2-} and Na^+ in calcites and selected aragonites. *Geochimica et Cosmochimica Acta*, **49**, 713–725.

Chen, C.C. & Hayes, K.F. (1999) X-ray absorption spectroscopy investigation of aqueous Co(II) and Sr(II) sorption at clay-water interfaces. *Geochimica et Cosmochimica Acta*, **63**, 3205–3215.

Collins, C.R., Sherman, D.M. & Ragnarsdottir, K.V. (1999) Surface complexation of Hg^{2+} on goethite: Mechanism from EXAFS spectroscopy and density functional calculations. *Journal of Colloid and Interface Science*, **219**, 345–350.

Cruz, F.R., Burns, S.J., Jercinovic, M., Karmann, I., Sharp, W.D. & Vuille, M. (2007) Evidence of rainfall variations in Southern Brazil from trace element ratios (Mg/Ca and Sr/Ca) in a Late Pleistocene stalagmite. *Geochimica et Cosmochimica Acta*, **71**, 2250–2263.

Davis, G., Ghabbour, E.A. & Cherkasskiy, A. (2001) Tight metal binding by solid phase peat and soil humic acids. In: *Humic Substances and Chemical Contaminants* (C.E. Clapp, P.M. Jardine, N. Senesi & M.H.B. Hayes, editors), 371–395. Soil Science Society of America, Madison, Wisconsin, USA.

Fairchild, I.J. & Treble, P.C. (2009) Trace elements in speleothems as recorders of environmental change. *Quaternary Science Reviews*, **28**, 449–468.

Fairchild, I.J., Borsato, A., Tooth, A.F., Frisia, S., Hawkesworth, C.J., Huang, Y., McDermott, F. & Spiro, B. (2000) Controls on trace element (Sr-Mg) compositions of carbonate cave waters: implications for speleothem climatic records. *Chemical Geology*, **166**, 255–269.

Fairchild, I.J., Baker, A., Borsato, A., Frisia, S., Hinton, R.W., McDermott, F. & Tooth, A.F. (2001) High-resolution, multiple-trace-element variation in speleothems. *Journal of the Geological Society, London*, **158**, 831–841.

Fairchild, I.J., Smith, C.L., Baker, A., Fuller, L., Spötl, C., Mattey, D., McDermott, F. & E.I.M.F. (2006a) Modification and preservation of environmental signals in speleothems. *Earth-Science Reviews*, **75**, 105–153.

Fairchild, I.J., Tuckwell, G.W., Baker, A. & Tooth, A.F. (2006b) Modelling of dripwater hydrology and hydrogeochemistry in a weakly karstified aquifer (Bath, UK): implications for climate change studies. *Journal of Hydrology*, **321**, 213–231.

Fairchild, I.J., Frisia, S., Borsato, A. & Tooth, A.F. (2007) Speleothems. In: *Geochemical Sediments and Landscapes* (D.J. Nash & S.J. McClaren, editors), 200–245. Blackwells, Oxford, UK.

Fairchild, I.J., Loader, N.J., Wynn, P.M., Frisia, S., Thomas, P.A., Lageard, J.G.A., de Momi, A., Hartland, A., Borsato, A., La Porta, N. & Susini, J. (2009) Sulfur fixation in wood mapped by synchrotron X-ray studies: implications for environmental archives. *Environmental Science and Technology*, **43**, 1310–1315.

Fairchild, I.J., Spötl, C., Frisia, S., Borsato, A., Susini, J., Wynn, P.M., Cauzid, J. & EIMF (2010) Petrology and geochemistry of annually laminated stalagmites from an Alpine cave (Obir, Austria): seasonal cave physiology. *Tufas and Speleothems* (H.M. Pedley & M. Rogerson, editors), Geological Society Special Publication (in press).

Frisia, S., Borsato, A., Fairchild, I.J. & McDermott, F. (2000) Calcite fabrics, growth mechanisms, and environment of formation in speleothems from the Italian Alps and southwestern Ireland. *Journal of Sedimentary Research*, **70**, 1183–1196.

Frisia, S., Borsato, A., Fairchild, I.J., McDermott, F. & Selmo, E.M. (2002) Aragonite-calcite relationships in speleothems (Grotte de Clamouse, France): environment, fabrics, and carbonate geochemistry. *Journal of Sedimentary Research*, **72**, 687–699.

Frisia, S., Borsato, A., Fairchild, I. J. & Susini, J. (2005) Variations in atmospheric sulphate recorded in stalagmites by synchrotron micro-XRF and XANES analyses. *Earth and Planetary Science Letters*, **235**, 729–740.

Frumkin, A. & Stein, M. (2004) The Sahara-East Mediterranean dust and climate connection revealed by strontium and uranium isotopes in a Jerusalem speleothem. *Earth and Planetary Science Letters*, **217**, 451–464.

Fuller, L., Baker, A., Fairchild, I.J., Spötl, C., Marca-Bell, A., Rowe, P. & Dennis, P.F. (2008) Isotope hydrology of dripwaters in a Scottish cave and implications for stalagmite palaeoclimate research. *Hydrology and Earth System Science*, **12**, 1065–1074.

Genty, D. & Quinif, Y. (1996) Annually laminated sequences in the internal structure of some Belgian stalagmites-importance for paleoclimatology. *Journal of Sedimentary Research*, **66**, 275–288.

Goede, A., McCulloch, M., McDermott, F. & Hawkesworth, C. (1998) Aeolian contribution to strontium and strontium isotope variations in a Tasmanian speleothem. *Chemical Geology*, **149**, 37–50.

Hartland, A., Fairchild, I.J., Lead, J.R., Dominguez-Villar, D., Gunn, J., Baalousha, M. & Ju-Nam, Y. (2010) The dripwaters and speleothems of Poole's Cavern: a review of recent and ongoing research. *Cave and Karst Science*, **36**, 37–46.

Hellstrom, J.C. & McCulloch, M.T. (2000) Multi-proxy constraints on the climatic significance of trace element records from a New Zealand speleothem. *Earth and Planetary Science Letters*, **179**, 287–297.

Henderson, G.M. (2002) New oceanic proxies for palaeoclimate. *Earth and Planetary Science Letters*, **203**, 1–13.

Howard, A.G. (1998) *Aquatic Environmental Chemistry*. Oxford University Press, Oxford, UK.

Huang, Y. & Fairchild, I.J. (2001) Partitioning of Sr^{2+} and Mg^{2+} into calcite under karst-analogue experimental conditions. *Geochimica et Cosmochimica Acta*, **65**, 47–62.

Huang, Y., Fairchild, I.J., Borsato, A., Frisia, S., Cassidy, N.J., McDermott, F. & Hawkesworth, C.J. (2001) Seasonal variations in Sr, Mg and P in modern speleothems (Grotta di Ernesto, Italy). *Chemical Geology*, **175**, 429–448.

Ihlenfeld, C., Norman, M.D., Gagan, M.K. & Drysdale, R.N. (2003) Climatic significance of seasonal trace element and stable isotope variations in a modern freshwater tufa. *Geochimica et Cosmochimica Acta*, **67**, 2341–2357.

Johnson, K.R., Hu, C., Belshaw, N.S. & Henderson, G.M. (2006) Seasonal trace-element and stable isotope variations in a Chinese speleothem: The potential for high-resolution paleomonsoon reconstruction. *Earth and Planetary Science Letters*, **244**, 394–407.

Karmann, I., Cruz, F.W., Viana, O. & Burns, S.J. (2007) Climate influence on trace element geochemistry of waters from Santana–Pérolas cave system, Brazil. *Chemical Geology*, **244**, 232–247.

Kinniburgh, D.G., van Riemsdijk, W.H., Koopal, L.K., Borkovec, M., Benedetti, M.F. & Avena, M.J. (1999) Ion binding to natural organic matter: competition, heterogeneity, stoichiometry and thermodynamic consistency. *Colloids and Surfaces A – Physicochemical and Engineering Aspects*, **151**, 147–166.

Lachniet, M.S. (2009) Climatic and environmental controls on speleothem oxygen-isotope values. *Quaternary Science Reviews*, **28**, 412–432.

Lead, J.R. & Wilkinson, K.J. (2006) Aquatic colloids and nanoparticles: Current knowledge and future trends. *Environmental Chemistry*, **3**, 159–171.

Marang, L., Eidner, S., Kumke, M.U., Benedetti, M.F. & Reiller, P.E. (2009) Spectroscopic characterization of the competitive binding of Eu(III), Ca(II), and Cu(II) to a sedimentary originated humic acid. *Chemical Geology*, **264**, 154–161.

Mason, H.E., Frisia, S., Tang, Y., Reeder, R.J. & Phillips, B.L. (2007) Phosphorus speciation in calcite speleothems determined from solid-state NMR spectroscopy. *Earth and Planetary Science Letters*, **254**, 313–322.

Mattey, D., Lowry, D., Duffet, J., Fisher, R., Hodge, E. & Frisia, S. (2008) A 53 year seasonally resolved oxygen and carbon isotope record from a modern Gibraltar speleothem: Reconstructed drip water and relationship to local precipitation. *Earth and Planetary Science Letters*, **269**, 80–95.

Mattey, D.P., Fairchild, I.J., Atkinson, T.C., Latin, J.-P., Ainsworth, M. & Durrell, R. (2010) Seasonal microclimate controls on calcite fabrics, stable isotopes and trace elements in modern speleothem from St. Michaels Cave, Gibraltar. *Tufas and Speleothems* (H.M. Pedley & M. Rogerson, editors). Special Publication, Geological Society, London (in press).

Mavrocordatos, D., Mondi-Couture, C., Atteia, O., Leppard, G.G. & Perret, D. (2000) Formation of a distinct class of Fe-Ca(-C-org)-rich particles in a complex peat-karst system. *Journal of Hydrology*, **237**, 234–247.

McCarthy, J.F. & McKay, L.D. (2004) Colloid transport in the subsurface: Past, present, and future challenges. *Vadose Zone Journal*, **3**, 326–337.

McCarthy, J.F. & Shevenell, L. (1998) Processes controlling colloid composition in a fractured and karstic aquifer in eastern Tennessee, USA. *Journal of Hydrology*, **206**, 191–218.

McDermott F., Schwarcz, H.P. & Rowe, P.J. (2005) 5. Isotopes in speleothems. In: *Isotopes in Palaeoenvironmental Research* (M.J. Leng, editor), 185–225. Springer, Dordrecht, The Netherlands.

McDonald, J., Drysdale, R. & Hill, D. (2004) The 2002–2003 El Niño recorded in Australian cave drip waters: Implications for reconstructing rainfall histories using stalagmites. *Geophysical Research Letters* **31**, L22202, doi:10.1029/2004GL020859.

McGillen, M. & Fairchild, I.J. (2005) An experimental study of the controls on incongruent dissolution of $CaCO_3$ under analogue glacial conditions. *Journal of Glaciology*, **51**, 383–390.

McKay, L.D., Gillham, R.W. & Cherry, J.A. (1993) Field Experiments in a Fractured Clay Till. 2. Solute and Colloid Transport. *Water Resources Research*, **29**, 3879–3890.

McKay, L.D., Sanford, W.E. & Strong, J.M. (2000) Field-scale migration of colloidal tracers in a fractured shale saprolite. *Ground Water*, **38**, 139–147.

McMillan, E.A., Fairchild, I.J., Frisia, S., Borsato, A. & McDermott, F. (2005) Annual trace element cycles in calcite-aragonite speleothems: evidence of drought in the western Mediterranean 1200–1100 yr BP. *Journal of Quaternary Science* **20**, 423–433.

Miorandi, R., Borsato, A., Frisia, S., Fairchild, I.J. & Richter, D.K. (2010) Epikarst hydrology and implications for stalagmite capture of climate changes at Grotta di Ernesto (N.E. Italy): results from long-term monitoring. *Hydrological Processes*, (in press).

Morse, J.W. & MacKenzie, F.T. (1990) *Geochemistry of Sedimentary Carbonates*. Elsevier, Amsterdam, 707 pp.

Peacock, C.L. (2009) Physiochemical controls on the crystal-chemistry of Ni in birnessite: Genetic implications for ferromanganese precipitates. *Geochimica et Cosmochimica Acta*, **73**, 3568–3578.

Pingitore, N.E. & Eastman, M.P. (1986) The coprecipitation of Sr^{2+} with calcite at 25°C and 1 atmosphere. *Geochimica et Cosmochimica Acta*, **50**, 2195–2203.

Plaschke, M., Rothe, J., Denecke, M.A. & Fanghanel, T. (2004) Soft X-ray spectromicroscopy of humic acid europium(III) complexation by comparison to model substances. *Journal of Electron Spectroscopy and Related Phenomena*, **135**, 53–62.

Proctor, C.J., Baker, A., Barnes, W.L. & Gilmour, M.A. (2000) A thousand year speleothem proxy record of North Atlantic climate from Scotland. *Climate Dynamics*, **16,** 815–820.

Richter, D.K., Gotte, T., Niggemann, S. & Wurth, G. (2004) REE^{3+} and Mn^{2+}-activated cathodoluminescence in lateglacial and Holocene stalagmites of central Europe: evidence for climatic processes? *The Holocene*, **14**, 759–767.

Roberts, M.S., Smart, P. & Baker, A., (1998) Annual trace element variations in a Holocene speleothem. *Earth and Planetary Science Letters*, **154**, 237–246.

Shopov, Y., Ford, D.C. & Schwarcz, H.P. (1994) Luminescent microbanding in speleothems. High resolution chronology and paleoclimate. *Geology*, **22**, 407–410.

Smith, C.L., Fairchild, I.J., Spötl, C., Frisia, S., Borsato, A., Moreton, S.G. & Wynn, P.M. (2009) Chronology-building using objective identification of annual signals in trace element profiles of stalagmites. *Quaternary Geochronology*, **4**, 11–21.

Stern, J.C., Sonke, J.E. & Salters, V.J.M. (2007) A capillary electrophoresis-ICP-MS study of rare earth element complexation by humic acids. *Chemical Geology*, **246**, 170–180.

Strawn, D.G. & Sparks, D.L. (1999) The Use of XAFS to distinguish between inner- and outer-sphere lead adsorption complexes on montmorillonite. *Journal of Colloid and Interface Science*, **216**, 257–269.

Strawn, D.G. & Sparks, D.L. (2000) Effects of soil organic matter on the kinetics and mechanisms of Pb(II) sorption and desorption in soil. *Soil Science Society of America Journal*, **64**, 144–156.

Stumm, W.M. & Morgan, J.J. (1996) *Aquatic Chemistry (3rd Edition).* Wiley-Interscience, New York, 1022 pp.

Tipping, E. (2001) *Cation Binding by Humic Substances.* Cambridge University Press, Cambridge, UK, 434 pp.

Tooth, A.F. & Fairchild, I.J. (2003) Soil and karst aquifer hydrological controls on the geochemical evolution of speleothem-forming drip waters, Crag Cave, southwest Ireland. *Journal of Hydrology*, **273**, 51–68.

Treble, P., Shelley, J.M.G. & Chappell, J. (2003) Comparison of high-resolution sub-annual records of trace elements in a modern (1911–1992) speleothem with instrumental climate data from southwest Australia. *Earth and Planetary Science Letters*, **216**, 141–153.

Treble, P.C., Chappell, J. & Shelley, J.M.G. (2005) Complex speleothem growth processes revealed by trace element mapping and scanning electron microscopy of annual layers. *Geochimica et Cosmochimica Acta*, **69**, 4855–4863.

Tyler, G. & Olsson, T. (2001) Concentrations of 60 elements in the soil solution as related to the soil acidity. *European Journal of Soil Science*, **52**, 151–165.

Verheyden, S., Keppens, E., Fairchild, I.J., McDermott, F. & Weis, D. (2000) Mg, Sr and Sr isotope geochemistry of a Belgian Holocene speleothem: implications for paleoclimate reconstructions. *Chemical Geology*, **169**, 131–144.

Villegas-Jiménez, A., Mucci, A. & Paquette, J. (2009) Proton/calcium ion exchange behavior of calcite. *Physical Chemistry Chemical Physics*, **11**, 8895–8912.

Wynn, P.M., Fairchild, I.J., Baker, A., Baldini, J.U.L. & McDermott, F. (2008) Isotopic archives of sulphate in speleothems. *Geochimica et Cosmochimica Acta*, **72**, 2465–2477.

Wynn, P.M., Fairchild, I.J., Frisia, S., Spötl, C., Baker, A., Borsato, A. & EIMF (2010) High-resolution sulphur isotope analysis of speleothem carbonate by secondary ionisation mass spectrometry. *Chemical Geology* (in press).

Xia, K., Bleam, W. & Helmke, P.A. (1997a) Studies of the nature of binding sites of first row transition elements bound to aquatic and soil humic substances using X-ray absorption spectroscopy. *Geochimica et Cosmochimica Acta*, **61**, 2223–2235.

Xia, K., Bleam, W. & Helmke, P.A. (1997b) Studies of the nature of Cu^{2+} and Pb^{2+} binding sites in soil humic substances using X-ray absorption spectroscopy. *Geochimica et Cosmochimica Acta*, **61**, 2211–2221.

Zhang, P.C., Brady, P.V., Arthur, S.E., Zhou, W.Q., Sawyer, D. & Hesterberg, D.A. (2001) Adsorption of barium(II) on montmorillonite: an EXAFS study. *Colloids and Surfaces A: Physicochemical and Engineering Aspects*, **190**, 239–249.

Zhou, H., Chi, B, Lawrence, M., Zhao, J., Yan, J., Greig, A. & Feng, Y. (2008) High-resolution and precisely date record of weathering and hydrological dynamics recorded by manganese and rare-earth elements in a stalagmite from Central China. *Quaternary Research*, **69**, 438–446.

EMU Notes in Mineralogy, Vol. 10 (2010), Chapter 8, 289–324

Dissolution, sorption/(re)precipitation, formation of solid solutions and crystal growth phenomena on mineral surfaces: implications for the removal of toxic metals from the environment

ATHANASIOS GODELITSAS[1,*] and JOSÉ M. ASTILLEROS[2]

[1] *Department of Mineralogy and Petrology, National and Kapodistrian University of Athens, Panepistimioupoli, Zographou 15784, Greece, e-mail: agodel@geol.uoa.gr*
[2] *Department of Crystallography and Mineralogy, Complutense University of Madrid, 28040 Madrid 28040, Spain*

Sorption of dissolved heavy metals, at low-temperature conditions, mediated by mineral surfaces, plays an important role in controlling both the dispersion and accumulation of these pollutants as well their geoavailability and bioavailabity in near-surface environments. In parallel to sorption, dissolution of mineral surfaces always occurs as an unavoidable coupled phenomenon at the mineral-water interface. The main sorption mechanisms, related to metal uptake from aqueous solutions, can be considered *ab*sorption (including ion-exchange), *ad*sorption and surface (co)/(re)precipitation leading to subsequent crystal-growth phenomena and formation of solid-solutions. The mineral reactivity depends on the chemical composition of the interacting solid, which in turn is related to particular physicochemical characteristics such as the porosity, the specific surface area (SSA), the cation exchange capacity (CEC), and the zeta-potential/pH_{PZ}. Of special interest, for industrial and environmental technology, is the interaction of carbonate (*e.g.* calcite), sulphate (*e.g.* gypsum) as well as aluminosilicate and Mn-oxide micro/nanoporous minerals (*e.g.* various zeolites and todorokite) with 'ultra' heavy toxic metals including actinides (*e.g.* Hg, Pb, Th, U).

1. General aspects of mineral-surface science, environmental mineralogy-geochemistry and nanogeoscience

Over the last two decades, technological developments have revitalized an area of research in mineralogy focused on the structure and reactivity of mineral surfaces (*e.g.* Vaughan & Pattrick, 1995; Brady, 1996). The combined utilization of advanced microscopic and spectroscopic techniques, such as atomic force microscopy (AFM), scanning tunneling microscopy (STM), transmission electron microscopy (TEM), secondary ion mass spectrometry (SIMS), laser-induced breakdown spectroscopy (LIBS), X-ray photoelectron spectroscopy (XPS), Rutherford backscattering spectroscopy (RBS) and extended X-ray absorption fine structure (EXAFS) has enabled the study of physicochemical processes occurring *in situ* and *ex situ* at mineral interfaces on a molecular scale (nanoscale). Mineral surface science is closely associated with the

DOI: 10.1180/EMU-notes.10.8

fields of molecular geochemistry and biogeochemistry in that it investigates geochemical processes at the molecular level (*e.g.* Brown *et al.*, 1999; Becker *et al.*, 2005; Casey *et al.*, 2005), and it has contributed to the recent establishment of nanogeoscience (including nanomineralogy and nanogeochemistry). Nanogeoscience is associated with the study of micro/nanopores in geological media and consequently to the study of micro/nanoporous minerals such as zeolites, Mn-oxides and clays including both natural and synthetic species (*e.g.* Wang *et al.*, 2003). Moreover, some of the research in nanogeoscience concerns the surface chemical behaviour of sparingly soluble salts and, in particular, carbonate minerals such as calcite (the trigonal $CaCO_3$ polymorph). This common mineral plays a major role in the global CO_2 cycle, participates in key biomineralization processes, and shows high reactivity in fluids, thereby controlling the geoavailability and bioavailability of certain contaminants. The processes that affect environmental availability of contaminants are strongly related to environmental mineralogy-geochemistry. Consequently it is of great importance to study aquatic and atmospheric nanoparticles, which act as the principal hosts and carriers of hazardous elements and radionuclides, as well as to characterize mineral nanoparticles and nanominerals (*e.g.* Fe-phases, Mn-phases, TiO_2 polymorphs) that exhibit peculiar structural and physicochemical properties (*e.g.* Hochella, 2006; Hochella *et al.*, 2008).

2. Mineral reactivity and chemical processes at the mineral-water interface (micro/nano-scaled critical zone)

Sorption is, in a broad sense, the change of mass of a chemical in the solid phase as a result of mass-transfer between the fluid and solid phases. The sorption of dissolved heavy metals at low temperatures, mediated by mineral surfaces, plays an important role in controlling both the dispersion and accumulation of these pollutants, as well their geoavailability and bioavailabity in near-surface environments (*e.g.* Stumm, 1992; Vaughan & Pattrick, 1995; Brady, 1996; Brown, 2001; Maurice, 2009). Dissolution of mineral surfaces always occurs in parallel with sorption at the mineral-water interface. The main sorption mechanisms, which differ in how metal uptake occurs from aqueous solutions, can be considered *ab*sorption, including ion exchange in the case of micro/nanoporous and layered minerals; *ad*sorption, represented by inner- and outer-sphere surface complexation; and surface (co)/(re)precipitation, leading to subsequent crystal-growth phenomena, such as the formation of solid solutions, overgrowth through heterogeneous nucleation and even topotactic reactions (Sposito, 1986). These mechanisms are depicted in Figure 1. As one might expect, the interfacial phenomena above depend crucially on the mineral-surface topography, correlated to macroscopic and micro/nanoscopic experimental observations (Fig. 2). In the case of heterogeneous nucleation, when the nanoscale surface structure of the substrate narrowly matches the structure of the precipitating phase, epitaxy (oriented overgrowth) may occur. Epitaxy is favored by good alignment between the adjacent lattices, and it is not feasible when there is extensive lattice misfit. Lattice misfit (d_m) measures the

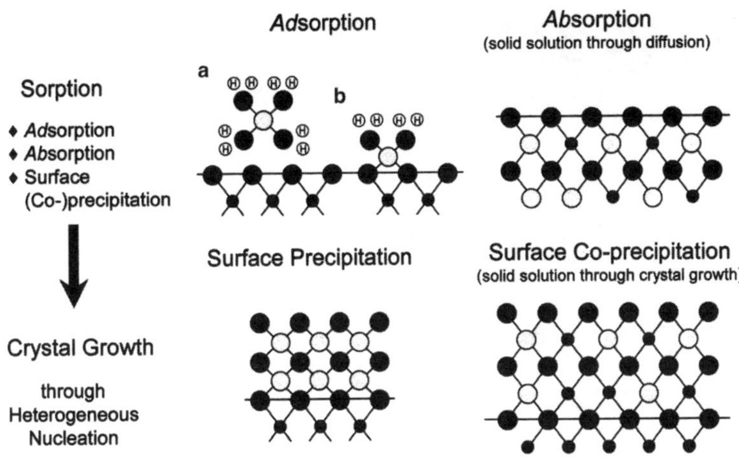

Fig. 1. Metal sorption mechanisms at the mineral-water interface (after Brown *et al.*, 1995).

mismatch between arrays of atoms at the interface and is calculated (Maurice, 2009) from

$$d_m = da/a \tag{1}$$

where a is the lattice parameter of the stress-free substrate crystal and da is the difference in lattice parameters between the substrate and the overgrowth. Misfits of $<10-15\%$ are generally thought to allow epitaxial overgrowth.

Among the uptake processes (sorption mechanisms) operating during the interaction of mineral surfaces with contaminant-bearing solutions, the most effective is the partitioning of the toxic elements into the solid phase by means of joint dissolution co-precipitation mechanisms. Thus, this process has been widely investigated as a potential method to remove toxic metals from polluted waters (*e.g.* Tesoriero & Pankow, 1996; Godelitsas *et al.*, 2003a; Prieto *et al.*, 2003; Köhler *et al.*, 2007). If a contaminant can incorporate within a solid phase rather than be *ad*sorbed at the surface, then it is less available for desorption and can be considered as immobilized, at least until the dissolution of the host phase due to a change in the aqueous chemistry occurs. Sparingly soluble salts have been demonstrated to be effective immobilizers for a wide spectrum of elements (*e.g.* Lorens, 1981; Davis *et al.*, 1987; Zachara *et al.*, 1991; Prieto *et al.*, 2003; Godelitsas *et al.*, 2003a,b; Heberling *et al.*, 2008). The joint dissolution co-precipitation process consists of several sequential steps, which can be summarized in the following way: (1) arrival of the dissolved contaminant to the vicinity of the mineral surface and, in parallel, (2) the dissolution of the mineral substrate (sorbent), which causes (3) a change in the chemistry of the mineral-aqueous interface, mainly an increase in the concentration of certain aqueous species, which in turn changes the solution *pH*. This can lead to the (4) increase in supersaturation of the aqueous solution with respect to a secondary phase that may result in (5) the precipitation (heterogeneous nucleation) of a contaminant-bearing precipitate. This sorbate is an adherent, crystalline

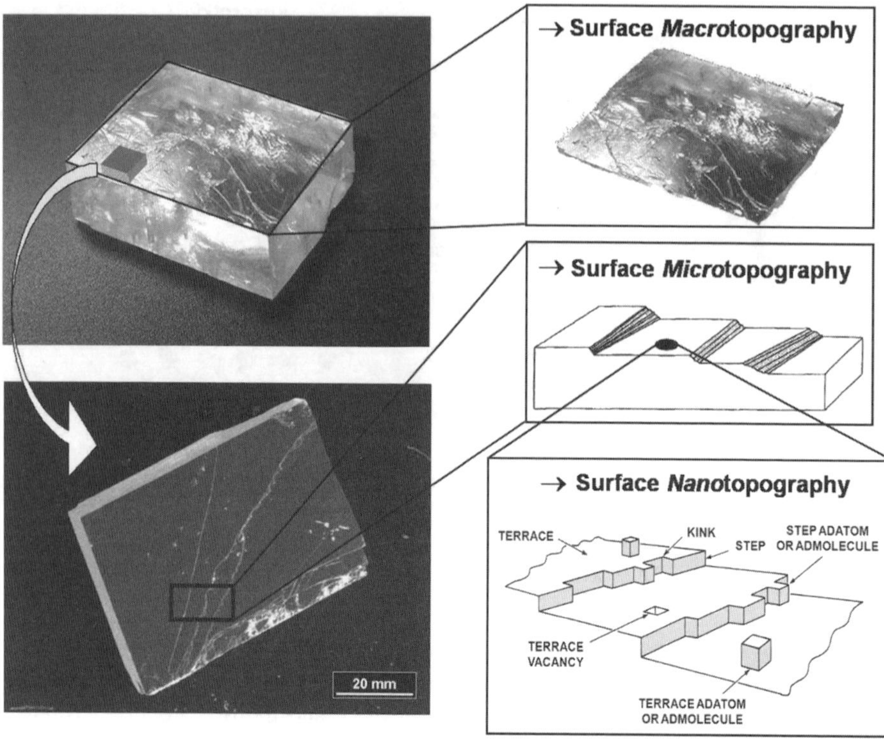

Fig. 2. Mineral surface topography on a decreasing scale, using the example of calcite.

phase that consists of chemical species from both the aqueous solution and the dissolution of the mineral.

This joint dissolution co-precipitation process generally entails the formation of a more or less extended solid solution. Chemically, a solid solution can occur when ions of one type substitute for ions of another type in a structure, forming a single crystalline phase. Such formation of a solid solution is described by the general reactions:

$$BA_{(s)} + H_2O_{(l)} \rightarrow A^{n-}_{(aq)} + B^{n+}_{(aq)} + H_2O_{(l)} \quad (2)$$

$$A^{n-}_{(aq)} + (1-x)B^{n+}_{(aq)} + xC^{n+}_{(aq)} \rightarrow B_{(1-x)}C_xA_{(s)} \quad (3)$$

To what extent the solid solution forms depends on thermodynamic conditions, *i.e.* temperature, pressure, and solubility product of the end members; the crystal chemical properties, such as the ionic size of the ions involved (Prieto, 2010 – chapter 1 in this volume); the growth rate; and even on crystallographic directions (Sunagawa, 2005; see Putnis, 2010 – chapter 2 in this volume). Equations 2 and 3 appropriately explain most cases of dissolution co-precipitation, though other situations are not infrequent, such as anionic substitutions (mainly oxyanions) (Prieto *et al.*, 1997; Fernández-González *et al.*, 1999; Prieto *et al.*, 2002; Andara *et al.*, 2005; Fernández-González *et al.*, 2006), coupled

2

Invoice Number:	4985
Customer Purchase Order Number:	

INVOICE

Quantiy	Description	Total
1	EMU Notes in Mineralogy 10: Ion partitioning in ambient-temperature aqueous systems - ORDERS BY INDIVIDUALS	£30.00
		Sub Total: £30.00
		VAT Cost: £0.00
		Delivery Cost: £20.95
		Total Cost: £50.95

Customer EC VAT Registration:

Payment received with thanks

Registered Charity: 233706
VAT Reg. No. GB 238 767 617

Russell Rajendra

From: russell@minersoc.org
Sent: 10 April 2018 17:05
To: Russell Rajendra
Cc: Admin
Subject: Mineralogical Society Shop Order Confirmation

MINERALOGICAL SOCIETY

To:

John Mahoney
892 S. Newcombe Way
Lakewood
CO
80226
USA

Email: jmahoney@mahoneygeochem.com

substitutions or the formation of a precipitate containing hydrous components (H_2O or OH) (Zachara *et al.*, 1991; Godelitsas *et al.*, 2003b). A more detailed and complete picture of the processes should include the *ad*sorption and/or even *ab*sorption of the contaminant on the mineral surface, which frequently occurs in parallel with the dissolution and co-precipitation processes and can directly affect their subsequent development.

The ability and effectiveness of mineral surfaces to remove certain toxic and hazardous elements from aqueous solutions by means of a dissolution-precipitation process is basically determined by the chemical reactivity of the minerals. Those minerals that are soluble or moderately soluble are good candidates for being able to remove contaminants effectively. They are able to cause a rapid change in the physicochemical conditions of the aqueous solutions and, consequently, trigger the precipitation of the insoluble solid phase containing the hazardous element. A second basic condition that should be fulfilled to ensure the effectiveness of the process is that the precipitate formed after the reaction should be insoluble under the prevailing physicochemical conditions. In addition, a suitable spatial arrangement and crystallographic relationship between substrate and overgrowth is required (Prieto *et al.*, 2003). Finally, the most advantageous situation occurs when the reacting mineral is naturally abundant and widespread. Among the geological materials available, moderately or sparingly soluble rock-forming salts emerge as potentially useful for taking up a broad spectrum of dissolved metals because these salts are ubiquitous.

In the next sections we will use examples to describe and review a common framework aimed at understanding the effectiveness of certain minerals as metal sequesterers. These materials will include quite insoluble geological materials such as silicates and aluminosilicates (quarz, feldspars and zeolites), oxide minerals and sparingly soluble salts such as carbonates (calcite and aragonite), phosphates (apatite), and sulphates (gypsum). The current trend in experimental studies shows that macroscopic (solution-based) measurements, combined with microscopic and/or surface spectroscopic experiments, offer a powerful approach to unravel mineral-solute interactions since they provide information on different scales about the products and the kinetics of these reactions (Brown *et al.*, 1999). Macroscopic experiments are usually carried out by measuring the extent of uptake as a function of solution *pH*, the concentration of sorbate in the aqueous solution, ionic strength, and time. Since dissolution-precipitation processes take place mainly at the solid-aqueous solution interface, an understanding of the surface micro/nanotopography and reactivity of the solid phase that acts as a substrate is also required. Microscopic *in situ* (AFM) and *ex situ* (AFM, SEM) surface techniques, and in some cases, spectroscopic surface techniques (XPS, RBS), allow us to distinguish and define the various processes occurring at the mineral-water interface. Moreover, a complete picture of the uptake process requires a basic knowledge of the thermodynamics of the system and the kinetics/mechanistics factors that determine the real effectiveness of the sorption process. In addition, complete removal of the metal from the aqueous media depends significantly on the physicochemical characteristics of the interacting solid, including the porosity, the specific surface area (SSA), the cation exchange capacity (CEC), and the zeta-potential (pH_{PZC}) (Figs 3–5).

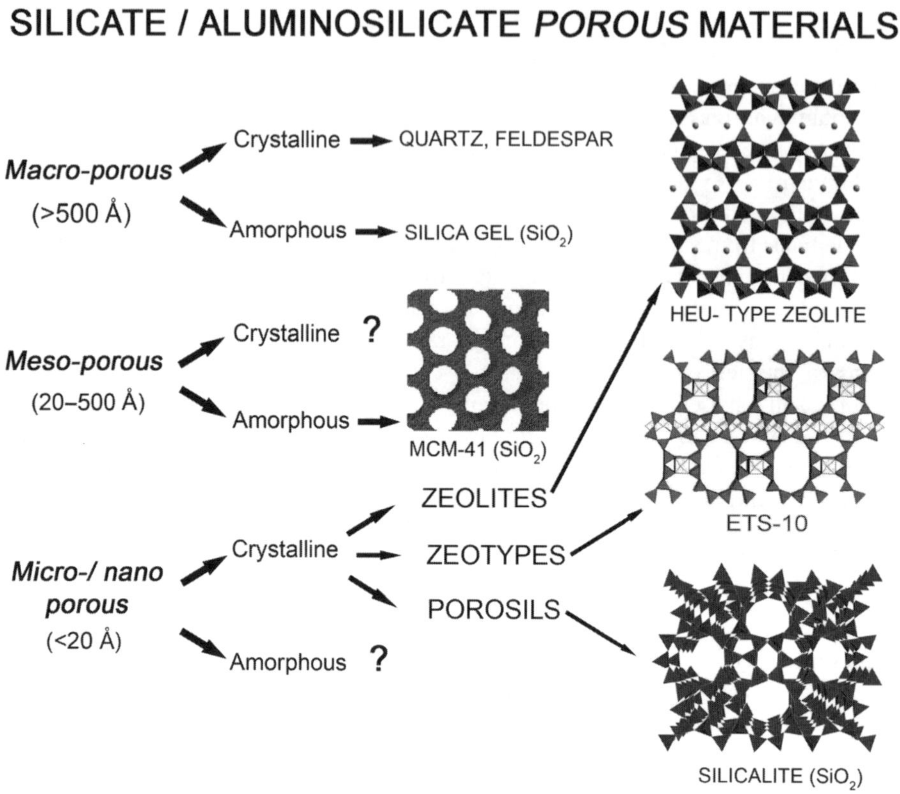

Fig. 3. Porosity of major silicate/aluminosilicate materials.

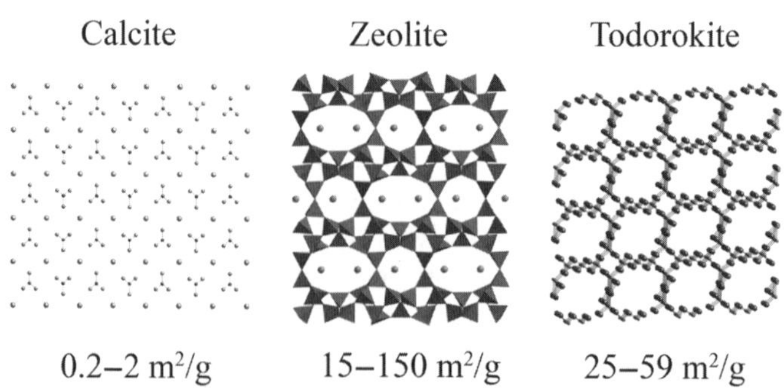

Fig. 4. Specific surface area (*SSA*) of characteristic macroporous and micro/nanoporous minerals with environmental importance.

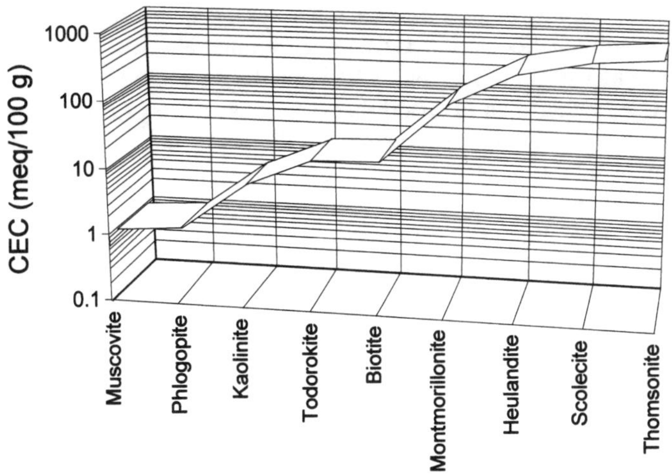

Fig. 5. Cation exchange capacity (CEC) of typical micro/nanoporous minerals, including layered minerals. These porous minerals can be used as metal sorbents.

3. Interaction of heavy metals with silicate/aluminosilicate and oxide-mineral surfaces

3.1. Silicates/aluminosilicates

The interaction at low temperature between dissolved heavy metals and primary rock-forming tectosilicate minerals, such as quartz and feldspars, is not macroscopically intense, and these materials can sorb small amounts of metal species on their surfaces. However, on the nanoscale, the phenomena are proportionally enhanced and peculiar sorption processes, along with classical adsorption/surface complexation and surface (co)/(re)precipitation mechanisms, may occur. The work of Manceau *et al.* (1999), based on extended X-ray absorption fine structure (EXAFS) spectroscopy, revealed that Co-rich trioctahedral clays can form upon sorption of Co^{2+} at the quartz-water interface. Chen *et al.* (2006), in their studies of metal sorption on quartz, have pointed out the significant contribution of Synchrotron Radiation (SR)-based spectroscopic studies for bridging the gap between macroscopic sorption experiments (based mainly on wet-chemistry measurements) and the real elucidation of nanoscale processes at the interface. In addition, Chardon *et al.* (2008) showed that the combination of macroscopic and surface microscopic and spectroscopic measurements, using AFM, XPS, and SIMS be the key to understanding the various *ad*sorption and surface-precipitation mechanisms governing the interaction of feldspar surface with Pb(II) and U(VI) in aqueous solutions. Evidently, the tectosilicate minerals, which have a rather low pH_{PZC}, show reduced ability to take up heavy metals from aqueous solution because of their low solubility and dissolution rate, as well as low micro-/nano-porosity, low SSA, and low CEC. On the other hand, several primary rock-forming phyllosilicates, including biotite, can sorb significant quantities of very toxic heavy metals, such as Hg(II) (Misaelides *et al.*, 1998a). Furthermore, geochemically stable secondary phyllosilicates with smaller crystal size, *i.e.* clay minerals, show enhanced sorptive properties attributed to high SSA and CEC, but they are used mostly to fix alkalis (for instance Cs^+) rather than to *ad*sorb heavy metals such as Cd (*e.g.* Sposito *et al.*, 1999; Vasconcelos *et al.*, 2008). In some cases, the interaction of clays with dissolved heavy metals may involve unusual 'phase transformation' phenomena due to dissolution followed by precipitation of a new metal-bearing phase. The prolonged exposure (several months) of kaolinite to dissolved Co was found to be responsible for the formation of Co-containing hydrotalcite, through the reaction of partially dissolved kaolinite with Co *ad*sorbed on the surface (Thompson *et al.*, 1999).

Zeolites, including HEU-type zeolites (heulandite-clinoptilolite series), which are the most abundant zeolites on Earth, are the only group of tectosilicate minerals that show substantial sorptive properties and real ion-exchange characteristics. In recent decades, zeolitic materials, which are cheap industrial minerals, were used in the development of environmental technology for the removal of heavy metals from contaminated aqueous media (*e.g.* Armbruster, 2001). The structure of zeolites is characterized by a three-dimensional aluminosilicate framework containing intracrystalline nanopores called channels, which normally host exchangeable cations (*e.g.* Ca^{2+}) and mobile H_2O molecules. The size of the channels (generally <20 Å), which regulates the sorptive properties of the materials, is related to the framework density (*FD*, in $T_{sites}/1000$ Å^3), while the CEC (in meq/100 g) depends on the Si/Al ratio and controls the number of negatively charged extraframework sites. The dissolved metallic cations interact with the zeolite crystals, and they are subsequently removed from the solution through different sorption mechanisms, primarily involving so-called 'ion exchange'. A typical 'ion-exchange' mechanism is thought to be a reversible chemical reaction in which an ion from solution is exchanged for a similarly charged ion that is initially bonded to a solid behaving as an 'ion exchanger'. In the case of zeolites, ion exchange is, practically speaking, a cation-exchange reaction. This reaction, in the case of a binary process involving A^{z_a} and B^{z_b} dissolved cations, can be simply expressed as $z_bA^{z_a} + z_a\overline{B}^{z_b} \leftrightarrow z_b\overline{A}^{z_a} + z_aB^{z_b}$, where $z_{a,b}$ represents the valence of the respective cations, and $\overline{B}$ & $\overline{A}$ re exchangeable cations in the solid (*e.g.* $2Na^+ + \overline{Ca}^{2+} \leftrightarrow 2\overline{Na}^+ + Ca^{2+}$). The binary exchange may be described by an ion-exchange isotherm that fundamentally defines the equilibrium at a specified temperature and solution normality, and further reveals on the basis of its shape the preference or selectivity of the zeolite for ion A^{z_a} relative to ion B^{z_b} in terms of a selectivity coefficient α.

The sorption of metallic cations from aqueous solutions by zeolite crystals via ion exchange seems to be affected by a combination of factors corresponding to the physicochemical characteristics of both the solution and the solid. First of all, ion exchange depends on temperature, solution *pH*, concentration of dissolved ions, and, in particular, on the relation between the hydrated ionic radii of the dissolved ions and the ratio of the zeolitic channels. It should be noted that the radius of a fully hydrated ion is proportional to the ionic charge (valence), but not proportional to the ionic radius without the H_2O shell. For example, Cs^+ (ionic radius 1.69 Å, hydrated radius 3.29 Å) is exchanged more easily than Li^+ (ionic radius 0.60 Å, hydrated radius 3.82 Å) or even Be^{2+}

(ionic radius 0.31 Å, hydrated radius 4.59 Å). Elevated temperature normally enhances the ion-exchange reactions, and *pH* is critical in the case of acidic solutions, because the small H^+ cations are much more mobile and reactive than any other metallic cations. Therefore, the optimum conditions for successful metal-loading involve the exchange of cations of low valence and large ionic radius (small hydrated radius) at high temperature and high *pH* in order to minimize the effect of H^+ cations. Other factors that should be taken into account for efficient ion exchange are the solid/liquid ratio, the size of the zeolite particles (high SSA favours ion exchange) and treatment time, which probably should be long and should involve renewal of the solution (step-by-step ion exchange).

Strictly speaking, the above statements are valuable mostly for the alkali and alkaline earth element cations, which are stable to further hydrolysis reactions in aqueous solutions. As a result, the corresponding partially or even fully exchanged (homoionic) forms of zeolites can be achieved by regulating the above factors. The situation is more complicated for many of the transition element cations, which are unstable and tend to undergo hydrolysis reactions. In most of these cases, hydrolysis is unavoidable because the *pH* increases rapidly due to simultaneous H^+ sorption, or because of the elevated temperature chosen for sufficient metal-loading. Therefore the uptake of transition elements such as Co, Ni, Hg, Th and U depends heavily on their aqueous chemistry and particularly on hydrolysis reactions that yield a variety of soluble and insoluble hydrolysis products (ionic and even neutral species); these hydrolysis products then interact with the microporous zeolitic solid. As a consequence, the sorption mechanisms do not involve ion exchange, which is a kind of bulk sorption, *i.e.* *ab*sorption into the channels of the tetrahedral framework. Rather, the mechanisms involve *ad*sorption, which can be specific and/or non-specific through inner-sphere and/or outer-sphere surface complexation, as well as surface precipitation/co-precipitation. The corresponding metal-loaded crystals (metal-forms) are usually non-homoionic because major extraframework cation sites are inaccessible to bulky aqueous complexes/hydrolysis products. In addition, these metal-forms can be non-stoichiometric with irregular metal distribution throughout the structure, or even real composite materials containing *ab*sorbed (ion-exchanged) metal cations in the micropores as well as metal ions and distinct metal phases (*e.g.* hydroxides) supported on the surface (*e.g.* Godelitsas & Armbruster, 2003, Fig. 6). The metal forms of the zeolites can further interact with organic molecules, a process that leads to complexes supported on the aluminosilicate zeolitic substrate (*e.g.* Godelitsas *et al.* 1999, 2001, 2003c, Fig. 7). The coordination of organometallic compounds can involve either their encapsulation inside the nanopores (intrazeolite complexes) or their arrangement on the surface of the crystals (surface complexes). The surface complexes are anchored by chemical bonds to active (Brønsted or Lewis) surface sites, whereas the intrazeolite ones are hosted in the channels, allowing the zeolite to act as a 'solid solvent'. In the latter case, the guest complexes principally retain all of their properties and play the role of 'inorganic molecular-scale microreactors' within the nanopores. Within these pores, they are often stable under conditions of dimerization, auto-oxidation, clustering and thermal degradation. Moreover, both types of complexes have novel properties that the organic molecules do not have on their own if oxygen atoms in the zeolite lattice participate

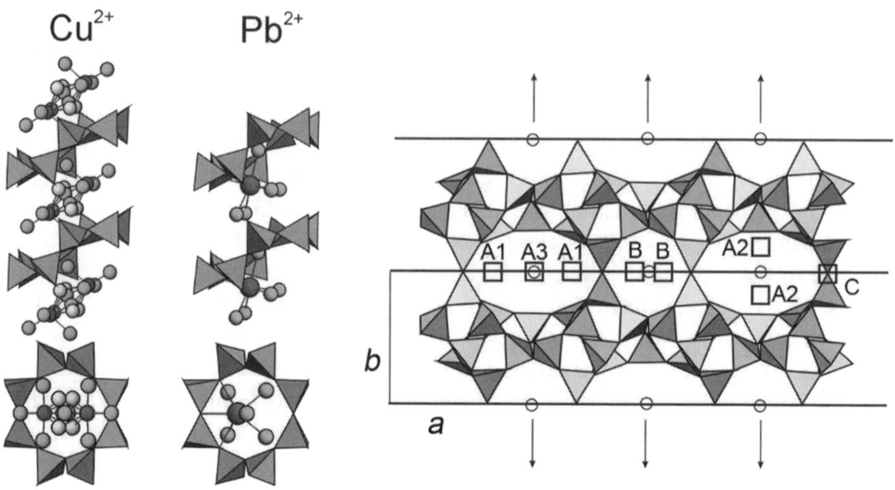

Fig. 6. The position of certain heavy-metal cations sorbed by ion exchange into the micro/nanopores of the natural HEU-type zeolite structure.

in the first coordination sphere. In fact, these solid systems constitute an entirely new class of composite materials of environmental and technological interest.

3.2. Oxides

The surface reactivity of other anhydrous oxide minerals is generally more complicated than that of quartz (SiO_2). Numerous studies have investigated the interaction of phases

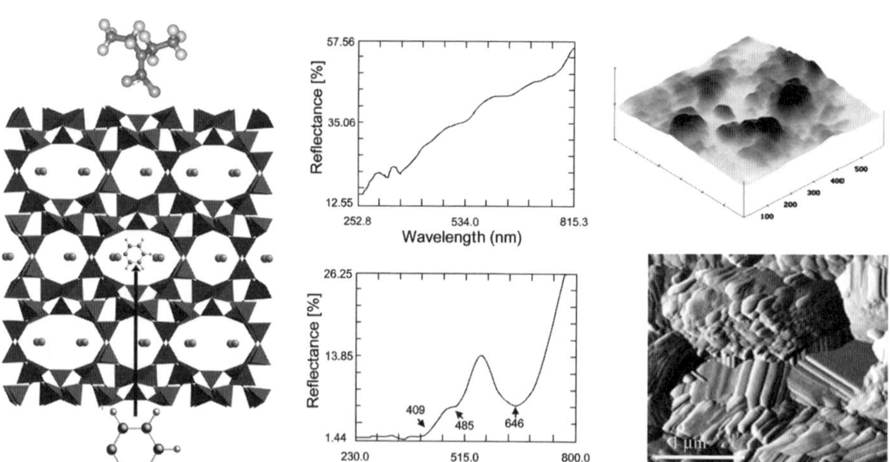

Fig. 7. Interaction of the metal form of the natural HEU-type zeolite with organic molecules forming surface and intrazeolite complexes (left). UV-Vis reflectance spectra, together with AFM images, indicating the nanoscale transformation of surface Co(III) hydroxide and/or oxyhydroxide precipitates into $Co[S_2CN(C_2H_5)_2]_3$ crystals (right).

such as Al_2O_3 and Fe_2O_3 with heavy metals in aqueous solutions. In most cases *ad*sorption of various metal hydroxy and hydrated species takes place in an electrical double layer (EDL) of the interface, leading to the formation of outer- and inner-sphere surface complexes (*e.g.* Stumm, 1992), which can be studied using advanced spectroscopic techniques (*e.g.* Brown *et al.*, 2008). Although these phenomena are important from the theoretical and geochemical points of view, they are not thought to be of significant interest to environmental science and technology. The best known exception is certain Mn oxides, such as todorokite, which are partially hydrated and possess a micro/nanoporous structure and a complex chemical composition. Todorokite, also referred to as a 'natural MnO_6 octahedral molecular sieve' (OMS-1, see *e.g.* Shen *et al.*, 1993), is a major constituent of marine Mn nodules (*e.g.* Turner *et al.*, 1982). It exhibits exceptional sorptive properties, like zeolites, giving it a great potential for future applications.

4. Interaction of heavy metals with carbonate, phosphate, sulphate and sulphide mineral surfaces

4.1. Carbonates

In recent decades, considerable experimental work has been carried out on the interaction of heavy metals, as well as other divalent metals, with carbonates. Carbonate materials, mainly calcite and aragonite (which is the second most important polymorph of $CaCO_3$), are frequently chosen because they are common constituent of soils, shallow aquifers, sediments, and rocks (*e.g.* Deer *et al.*, 1996). They are known to sorb a wide spectrum of trace metals from solution such as cadmium, lead, copper, strontium, zinc, manganese, rare earth elements (*REE*) and actinides (*e.g.* Davis *et al.*, 1987; Reeder & Grams, 1987; Zachara *et al.*, 1991; Stipp *et al.*, 1992; Tesoriero & Pankow, 1996; Prieto *et al.*, 1997; Godelitsas *et al.*, 2003a; Stipp *et al.*, 2003; Chada *et al.*, 2005; Marques Fernandes *et al.*, 2008). As a result, they can play a fundamental role in controlling the concentration, transport and fate of metals in many natural systems. Moreover, due to the low solubility of both calcite and aragonite ($K_{sp,calcite} = 10^{-8.48}$, $K_{sp,aragonite} = 10^{-8.336}$) most reactions in the carbonate system proceed at a rate compatible with experiments done in laboratory time. The studies involving carbonates cover some related topics, including thermodynamic modelling (*e.g.* Königsberger *et al.*, 1991; Rock *et al.*, 1994, McBeath *et al.*, 1998; Zhu, 2002; Prieto, 2009; Katsikopoulos *et al.*, 2009), measurements of distribution coefficients in co-precipitation reactions (*e.g.* Lorens, 1981; Reeder, 1996; Tesoriero & Pankow, 1996; Prieto *et al.*, 1997; Fernández-González *et al.*, 1999, Katsikopoulos *et al.*, 2008), and study of the sorption mechanisms onto the surface of carbonate materials (*e.g.* Davis *et al.*, 1987; Zachara *et al.*, 1991; Stipp *et al.*, 1992; Chiarello *et al.*, 1997; Godelitsas *et al.*, 2003a and b; Astilleros *et al.*, 2006; Pérez-Garrido *et al.*, 2007 and 2009). Here, we compare the uptake of some toxic divalent metals (cadmium, lead and mercury) by calcite and/or aragonite. The sorption reactions are promoted by a change in the physicochemical conditions of the mineral-aqueous

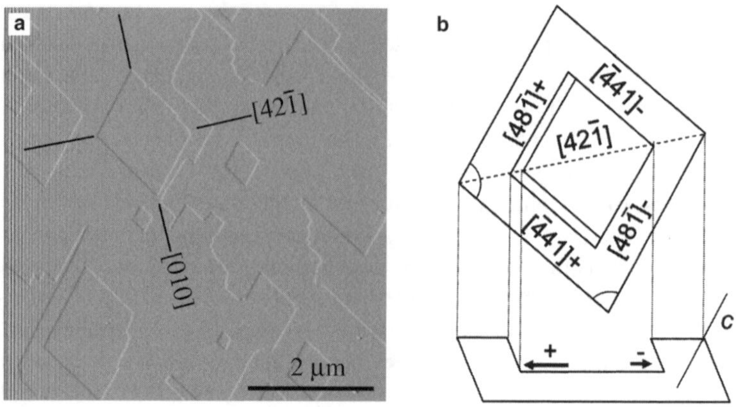

Fig. 8. AFM deflection image of a calcite (10$\bar{1}$4) surface dissolving in pure water. On the dissolution pit, four non-equivalent steps parallel to the <$\bar{4}$41> directions define a rhombus shape. The two adjacent steps ([48$\bar{1}$]$_-$ and [$\bar{4}$41]$_-$) extend over the underlying layer to define an acute angle, while the other two adjacent steps ([48$\bar{1}$]$_+$ and [$\bar{4}$41]$_+$) define an obtuse angle. The long and short diagonals of the rhombus run parallel to the [010] and [42$\bar{1}$] directions, respectively.

solution due to the surface dissolution of the carbonate mineral in nanoscale (Fig. 8). This leads to the release of carbonate ions into the solution, which is frequently accompanied by a drastic increase in the *pH*.

4.1.1. The uptake of cadmium by calcite and aragonite

Cadmium is not naturally abundant on the Earth's surface, but due to anthropogenic activities recent times, its concentration has increased to dangerous levels in air, water, and soil. In fact, the contamination of soils and agricultural environments by cadmium constitutes a threat to public health, as this metal easily incorporates into the food chain, leading to diseases of the kidneys, pancreas and the liver (Nriagu, 1981; Moore, 1991). Fortunately, cadmium reacts easily with CO_3^{2-} ions to form lead-insoluble otavite ($CdCO_3$, $K_{sp,otavite} = 10^{-12.1}$; Stipp *et al.*, 1993). For this reason, considerable experimental work on the interaction of cadmium with the common $CaCO_3$ polymorphs, calcite and aragonite, can be found in the literature (*e.g.* Davis *et al.*, 1987; Zachara *et al.*, 1991; Chiarello & Sturchio, 1994; Reeder, 1996; Tesoriero & Pankow, 1996; Martin-Garin *et al.*, 2003; Prieto *et al.*, 2003; Köhler *et al.*, 2007; Pérez-Garrido *et al.*, 2007). All the studies agree that the uptake of cadmium by calcite and aragonite is concomitant with surface dissolution and occurs *via* surface precipitation of a cadmium-rich (Cd,Ca)CO_3 solid solution with a calcite-type structure according to the reaction (Prieto *et al.*, 2003):

$$yCa^{2+} + zCd^{2+} + CaCO_3 \rightarrow (Cd_xCa_{(1-x)}CO_3)_{cal} + (y + x)Ca^{2+} + (z - x)Cd^{2+} \quad (4)$$

The fact that the solubility of otavite is approximately three orders of magnitude less than that of calcite results in an extremely preferential partitioning of cadmium into the solid phase. If a nearly ideal character is assumed for this solid solution (Davis *et al.*, 1987; Königsberger *et al.*, 1991; Tesoriero & Pankow, 1996), an assumption supported by the similar ionic radii of Cd^{2+} and Ca^{2+}, the equilibrium distribution coefficient for cadmium is approximated by the ratio of the solubility products of the end members, namely $D_{Cd} = K_{sp,calcite}/K_{sp,otavite} = 4168$. This value is calculated under equilibrium conditions, where the growth rate is zero. Although a number of kinetic/mechanistic effects during crystal growth can modify this value (Paquette & Reeder, 1995; Prieto *et al.*, 1997), a value for D_{Cd} in the range of 1000–4500 can be expected. This high value implies a strong tendency to form Cd-rich calcite precipitates ($Cd_{>0.99}Ca_{<0.01}CO_3$). At the beginning of the mineral–aqueous solution interaction, the fluid is undersaturated with respect to calcite, aragonite, and otavite. Both aragonite and calcite dissolve at measurable velocities in the presence of Cd-bearing aqueous solutions (aragonite dissolves faster because is more soluble than calcite) and this leads to a rapid increase in the supersaturation of the aqueous solutions with respect to otavite, due to the low solubility of this mineral. Therefore, both calcite and aragonite could be expected to take up large amounts of Cd. Macroscopic experiments carried out in stirred reactors corroborate this hypothesis in the case of aragonite.

Analytical measurements show that biogenic and abiogenic aragonite mineral grains (Prieto *et al.*, 2003) are extremely effective as sequesterers of Cd. During the early hours of the experiments, the Cd concentration for parent solutions of <1 mM falls dramatically even below the detection limit of atomic absorption spectroscopy (AAS). For prolonged reaction times, ICP-MS analyses yield Cd concentrations in the nanomolar range. In contrast to sorption onto aragonite, the uptake of cadmium by calcite is much more limited, slower and tends to become rapidly inefficient and negligible at higher concentrations of cadmium in the fluid.

The fact that calcite is slightly less soluble than aragonite cannot satisfactorily account for why calcite solids have lower sorption capacity. Independently of the starting mineral the final product should be practically the same. The explanation for this opposite sorption behaviour comes from the different relationships between the crystal structures of both the sorbant and the sorbate. When epitaxial relations exist between both, the surface of the sorbant can rapidly become passivated and as a result, further uptake of the pollutant becomes inefficient. The existence of epitaxial relationships depends on both the nature of the pollutant cation and the $CaCO_3$ polymorph in question. In the case of calcite, the substrate and precipitate ($CdCO_3$) are isostructural and the process occurs by oriented overgrowth (epitaxy) of layers, which protect the substrate from further dissolution. As a consequence, the process stops when only a small amount of cadmium has been removed from the fluid. However, on aragonite surfaces, the precipitating crystals grow in random orientations, allowing continuous access by the aqueous solution onto the substrate. Macroscopic measurements and microscopic *ex-situ* observations (SEM) successfully explain some key processes, such as the decrease in the rate of calcite dissolution in the presence of cadmium or the smaller Cd-uptake capacity of calcite compared to aragonite. However, to gain an insight into these processes, techniques are required that can provide information about the mechanisms operating during the mineral substrate–aqueous solution interaction. *In situ* AFM is unique in its ability to observe and directly record the dynamic evolution of reacting solid surface structures by monitoring changes in surface nanotopography. AFM experiments conducted on calcite $\{10\bar{1}4\}$ surfaces in contact with Cd-bearing aqueous solutions show the dissolution of the original surface and the simultaneous growth of layers of the $Cd_xCa_{(1-x)}CO_3$ solid solution, which passivate the calcite substrate (Pérez-Garrido *et al.*, 2007). With respect to calcite dissolution, the AFM images reveal a change in the shape of etch pits. When calcite is in contact with pure water, typical etch pits of rhombohedral shape form on microterraces (see Fig. 8). The retreat of the edges of these etch pits is highly anisotropic, with a pair of steps retreating rapidly and the other one retreating very slowly. The anisotropy of the retreat of the edges during dissolution is a consequence of the inequivalence of the two pairs (positive and negative) of opposing straight steps parallel to the $[\bar{4}41]$ and $[48\bar{1}]$ directions (Staudt *et al.*, 1994). Thus, while the 'negative' $[\bar{4}41]_-$ and $[48\bar{1}]_-$ steps overhang the underlying layer, thereby defining an acute angle (78°) that contains constrained kink sites, the 'positive' $[\bar{4}41]_+$ and $[48\bar{1}]_+$ steps define an obtuse angle (102°) that contains open kink sites. The obtuse (positive) steps are more reactive (DeLeeuw & Parker, 1999), and their retreat velocity is much faster than that of the acute (negative) steps. After the interaction of calcite substrate with Cd-bearing solutions they become progressively elongated along the $[42\bar{1}]$ direction and their edges develop scalloped contours (Fig. 9a). These effects are due to the preferential incorporation of Cd^{2+} at specific types of sites, causing the blocking of those sites. In particular, the stabilization of the $[42\bar{1}]$ step edge is consistent with a preferential adsorption of

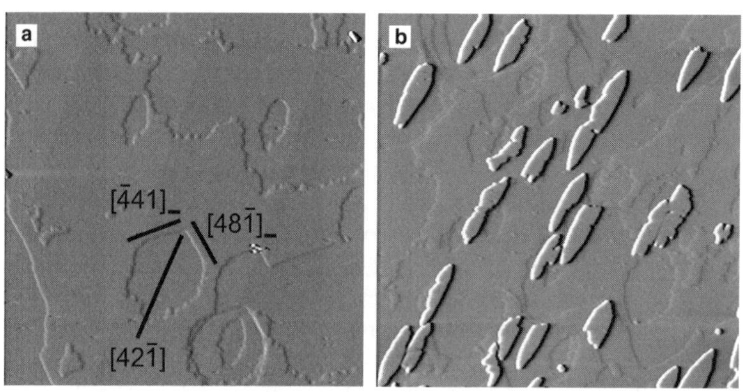

Fig. 9. AFM deflection image showing the calcite $(10\bar{1}4)$ surface in contact with a Cd-bearing (0.025 mM) aqueous solution. (***a***) The etch pits formed are elongated along $[42\bar{1}]$ and show straight steps parallel to $[\bar{4}41]_-$ and $[48\bar{1}]_-$ near the corner that they define. The rest of the edges appear scalloped and rough. (***b***) After several minutes of interaction, three-dimensional islands grow on a calcite surface in contact with a 0.025 mM $CdCl_2$ solution. The islands grow elongated along the $[42\bar{1}]$ direction, and the vertical distance is 2.76 nm (AFM pictures by C. Pérez-Garrido).

Cd^{2+} (bonded to CO_3^{2-} ions) at acute steps and an inhibition of dissolution at the (+/−) corners where obtuse and acute steps join. The adsorption of Cd^{2+} not only affects the shape of the etch pits, but also their expansion rate, which decreases when the concentration of Cd^{2+} in the aqueous solution increases. With respect to the sorbate, AFM provides information about aspects such as its geometry and spatial distribution. AFM images (Fig. 9b) show that the coating of the calcite surfaces occurs by the coalescence of $Cd_xCa_{(1-x)}CO_3$ islands growing epitaxially onto the calcite substrate. The islands have a constant thickness of 2.75 nm equivalent to 10 monolayers of otavite and are elongated along $[42\bar{1}]$, the direction of greatest lattice mismatch but also with the smallest stiffness. A similar behaviour has been observed during the growth of pseudokutnahorite (Lea *et al.*, 2003) and rhodochrosite islands (Pérez-Garrido *et al.*, 2009) on the calcite $\{10\bar{1}4\}$ surface. Finally, the progressive coalescence of the Cd-rich islands results in the formation of a nanometric epitaxial layer that armours the substrate from further dissolution and determines the end of the process at a "partial" pseudo-equilibrium endpoint.

4.1.2. The uptake of lead by calcite and aragonite

Several international organizations classify lead at the top of the priority list of the most hazardous substances; in fact, the U.S. Environmental Protection Agency places it immediately after As. It is still used extensively in many industrial applications such as batteries, pigments, ammunition, devices to shield X-rays, roofing and photographic material. As a result, the worldwide anthropogenic inputs of lead into freshwaters and the atmosphere are in the range of several thousands of metric tons/year, which gradually accumulate in the geo- and biosphere (Nriagu & Pacyma, 1988). Most of the work on the interaction of $CaCO_3$ with Pb is related to calcite, while the data for aragonite are very limited (*e.g.* Gamsjäger *et al.*, 1984; Petrović *et al.*, 1999; Sturchio *et al.*, 1997; Godelitsas *et al.*, 2003a). Macroscopic experiments on calcite and aragonite fragments (100–200 μm) show that the removal of lead from aqueous solutions is very fast and that the practical end of the sorption process (almost complete removal of the metal) is reached after ~1 min of interaction for low-concentration solutions (10 mg/l). Experiments with more concentrated solutions show that the uptake of lead is slower but still very rapid: 10 min for both calcite and aragonite for an initial concentration of 100 mg/l, and 1.5 h for calcite and 12 h for aragonite for an initial concentration of 1000 mg/l. The results of experiments with solutions of 1000 mg/l are presented in Fig. 10a. It is evident that the sorption of Pb and the dissolution of the calcium carbonate solids (expressed by the Ca release) occur in parallel until removal of the metal is almost complete. These data reveal the high selectivity of calcium carbonates for Pb, which is comparable only to that of calcium phosphates among common inorganic solids and minerals (see below).

Another important factor that should be taken into account is the crystal size, since the sorption capacity depends to a great extent on this parameter. Chemical data on the interaction of mm-sized calcite and aragonite crystals with solutions of 1000 mg/l show that amounts of Pb sorbed per mass of the solids are more than two orders of magnitude smaller than those recorded for the micrometer-sized materials over the

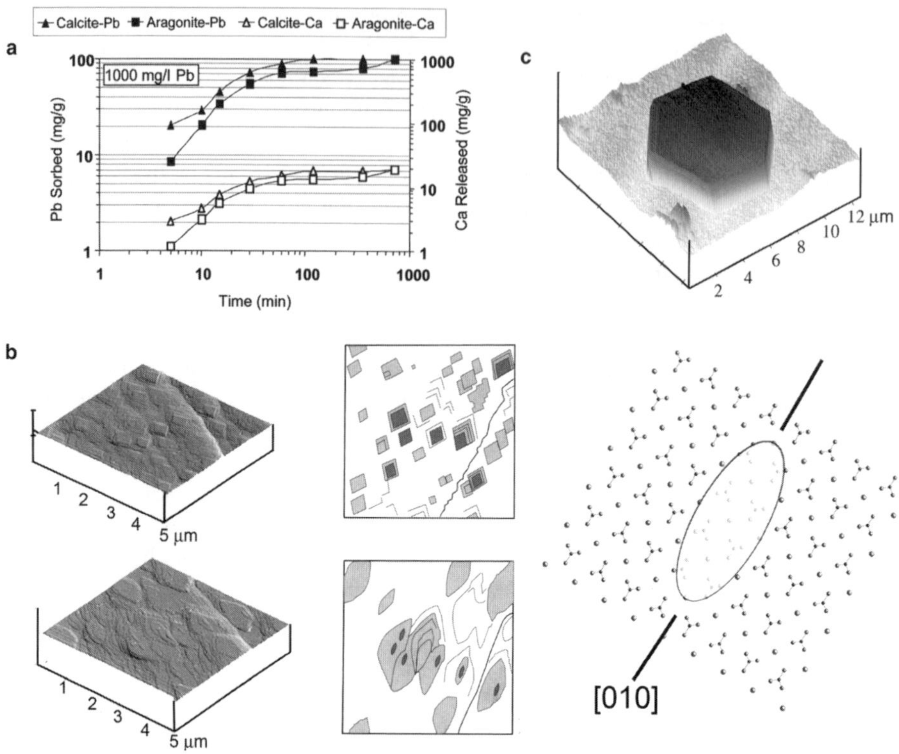

Fig. 10. (***a***) Macroscopic analytical data for the interaction of μm-sized calcite and aragonite crushed crystals (100–200 μm fragments) with aqueous solutions containing 1000 mg/l Pb. (***b***) *In situ* AFM results showing the dynamic evolution of the freshly cleaved (10$\bar{1}$4) calcite surface during the interaction with distilled water (upper) and 1000 mg/l Pb aqueous solution (lower). (***c***) *Ex situ* AFM image of heterogeneous nucleation of a Pb-carbonate on the surface of aragonite following interaction with $Pb^{2+}{}_{(aq)}$.

same period of time. This is reasonable, considering the greater reactive surface area of the crushed μm-sized crystals; the crushing correlates with the appearance of several kinds of mechanically induced surface macrodefects such as macrosteps, macrofractures, and broken edges. SEM-EDS investigation of all the solid experimental products obtained from the solutions of 100 and 1000 mg/l, demonstrate that massive surface precipitation leading to growth of lead carbonates takes place in the first minutes of interaction. Detailed powder X-ray diffraction (XRD) study showed that both cerussite ($PbCO_3$) and hydrocerussite ($Pb_3(CO_3)_2(OH)_2$) occur on the surface of calcite, but that only cerussite occurs on the surface of aragonite. The dissolution-precipitation process can be summarized in the following way: the slightly acidic conditions contribute to surface dissolution and release of excess CO_3^{2-} ions, which reacts rapidly with the dissolved lead to yield cerussite and/or hydrocerussite. The surface precipitation is enhanced by the increased initial concentration of lead and takes place according to the following reactions. For simplicity we have omitted the formation of solid solution.

$$CO_3^{2-} + Pb^{2+} \rightarrow PbCO_{3(s)} \quad (5)$$

$$2CO_3^{2-} + 3Pb^{2+} + Ca^{2+} + 2OH^- \rightarrow Pb_3(CO_3)_2(OH)_{2(s)} \quad (6)$$

The coexistence of both lead carbonate phases is due to the fact that the equilibrium in the reaction $3PbCO_{3(s)} + H_2O \rightarrow Pb(OH)_2 \cdot (PbCO_3)_{2(s)} + CO_{2(g)}$ lies very close to the partial pressure of CO_2 in the atmosphere ($P_{CO_2} = 10^{-3.5}$). Consequently, the shift to either cerussite or hydrocerussite is extremely sensitive to variations in this parameter. The dissolution of calcite in contact with a solution containing Pb can lead to a locally rapid increase in *pH* at the solid-water interface and a decrease in $[CO_2]_{aq}$. As a consequence, hydrocerussite can crystallize. A subsequent equilibration with atmospheric $CO_{2(g)}$ can lead to cerussite to be more stable than hydrocerussite.

In situ AFM investigation (Fig. 10b) demonstrates that the dissolution of the calcite substrate starts with the formation of rhombic etch pits (in water), which are rapidly broadened and further transformed to elongated oval etch pits along the [010] direction after subsequent addition of 1000 mg/l Pb solution. The stabilization of the [010] step edge is probably due to the preferential adsorption of Pb^{2+} at the opener (+/+) corners of this step, given that the ionic radius of Pb^{2+} is considerably larger than that of Ca^{2+}. The adsorption processes are also shown by the X-ray photoelectron spectroscopy (XPS) spectra, which show that during the first stages of the interaction (1–5 min), adsorption of Pb^{2+} ions occurs on the surface of calcium carbonates simultaneously with dissolution and before the initiation of any surface precipitation. After ~10 min of interaction in the AFM fluid cell, nucleation initially occurs inside the elongated oval etch pits and later onto the neighbouring microterraces and specifically at the edges of microsteps. These nuclei possess an ellipsoid shape, and their height varies between 10 and 50 Å. The growth of the nuclei rapidly leads to the development of crystals, bounded by well defined edges and showing a nearly perfect tabular hexagonal shape, which corresponds to either cerussite or hydrocerussite. After 1 month of interaction, the calcite surface is completely covered by a thick 'layer' composed of cerussite and hydrocerussite crystals of different sizes and morphologies.

The surface of aragonite, in contrast to that of calcite, exhibits rounded dissolution features with the absence of crystallographically oriented etch pits after interaction with solutions of 1000 mg/l Pb. XPS spectra confirmed that adsorption of Pb occurs simultaneously with dissolution over short interaction times and that it is more intense than on a calcite substrate. Under these conditions, lead carbonate nuclei also form on the (001) dissolved surface of aragonite, and furthermore nearly perfect tabular hexagonal crystals readily develop (Fig. 10c). Powder-XRD investigation of material removed from the surface of aragonite crystals interacted with Pb for ~20 min in the AFM fluid cell, confirming the presence of hydrocerussite exclusively. However, after 1 month of interaction the powder-XRD study indicates the existence of cerussite exclusively. The data above prove that over short interaction times, hydrocerussite forms on the surface of aragonite and is later replaced by cerussite or, more likely, the precursor hydrocerussite is transformed to cerussite. This phase transformation (hydrocerussite → cerussite), as well as the reverse (cerussite → hydrocerussite), are probably based on the thermodynamic considerations presented above. In addition to *ad*sorption of lead into the calcite and aragonite, as shown by *in situ* AFM and XPS, its surface precipitation, as shown by AFM and SEM and its *ab*sorption have also been detected. Rutherford-backscattering spectra (^{12}C-RBS) obtained for calcite samples allowed to interact for 1 min (prior to crystallization of lead carbonates), reveal the presence of Pb to a depth of ~430 Å and a greater depth (up to ~2 μm) than when the same experiment is carried out with aragonite. The deep penetration of Pb can be explained by the presence of water, which alters the entire surface structure through dissolution processes. Absorption of Pb occurs while the surface is dissolving, and it is enhanced by the appearance of dissolution-induced surface microdefects (vacancies, etch pits, microfractures). Moreover, the presence of additional surface macrodefects (macrofractures/cracks) may help to explain the rather deep apparent penetration of Pb. Figure 11 summarizes the dissolution/sorption processes taking place on the $\{10\bar{1}4\}$ surface in contact with Pb-bearing solutions. A similar picture may apply to aragonite–aqueous solution interaction. As a final conclusion we can state that the solid surface of calcium carbonate crystals can act as an efficient scavenger of Pb during its interaction with toxic concentrations of the metal in aqueous solution.

4.1.3. The uptake of mercury (II) by calcite

The primary sources of pollutant mercury in the environment are the emissions associated with fossil fuel burning and combustion processes such as municipal waste incineration (Fitzgerald & Lamborg, 2003). In solution its simplest form is the hydrated mercury (II) ion, Hg^{2+}, which can be readily methylated to give the highly toxic methylmercury (CH_3Hg^+). Methylmercury easily accumulates in biological organisms like freshwater and marine fish, posing a threat to human health (Howard, 1998). In spite of the element's toxicity, the literature concerning its interaction with carbonates is much less than that dedicated to other toxic elements such as cadmium and lead (Godelitsas *et al.*, 2003a). In this section we will show some micro/nanoscopic (AFM and SEM) and spectroscopic investigation (XPS) of the interaction of calcite $\{10\bar{1}4\}$ surfaces with $Hg(CH_3COO)_2$ in aqueous solutions. The results reveal phenomena with clear similarities to those described in the previous sections.

The interaction between calcite substrate and an aqueous solution of $Hg(CH_3COO)_2$ ($pH \sim 3.5$) containing 5 mM Hg(II) leads to the transformation of the initial H_2O-induced rhombic etch pits into deeper ones shaped like isosceles triangles with slightly curved sides (Fig. 12). The transformation of the etch pits is exceptionally rapid and intense (due to the small *pH*), and as a result, the entire surface microtopography of calcite is completely modified in the time it takes to scan the surface (~30 s). The AFM sequence shows that the dissolution is strongly anisotropic, with steps retreating faster along the $[42\bar{1}]$ direction than along the [010] direction (~12 nm s^{-1} and ~4 nm s^{-1}, respectively). Under these conditions, the selective adsorption of Hg^{2+} at

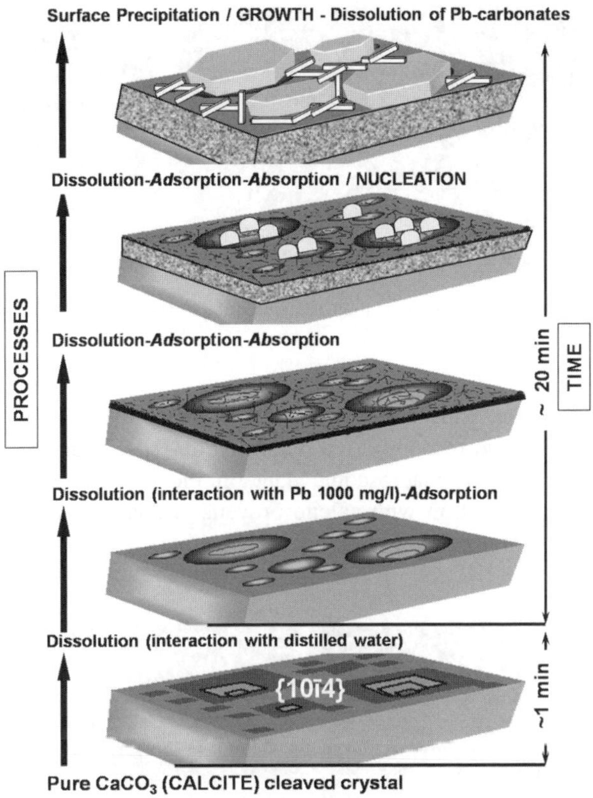

Fig. 11. Overview of the chemical processes taking place at the calcite-water interface during interaction with $Pb^{2+}{}_{(aq)}$, as revealed by a combination of surface *in situ* and *ex situ* microscopic and spectroscopic techniques (AFM, XPS, ^{12}C-RBS).

the opener (+/+) and (+/−) corners may account for the stabilization of the steps parallel to [010] and of the curved steps, respectively, as the ionic radius of Hg^{2+} ($rHg^{2+} = 1.02$ Å) is slightly larger than rCa^{2+} (1 Å). The *in situ* AFM observations are supported by the XPS spectra of the calcite surface obtained after 1 and 5 min of interaction with the Hg-bearing aqueous solution. They demonstrate that Hg(II) ions are being intensively *ad*sorbed on calcite while its surface is dissolving.

Further FESEM-EDS investigation of the calcite crystal removed from the AFM fluid cell show the existence of extended dissolution features on a micro-scale, too large to be observed by *in situ* AFM. Many of these large etch pits accommodate Hg-containing solid phases overgrown on the dissolved calcite surface. As shown in Figure 13a,b, obtained after 2 h of interaction, the aggregates of the Hg-containing flakes are accompanied by small spheres that are supported on the calcite surface. Longer periods of interaction (12 h) lead to the formation of aggregates that consist of scattered tufts with well-developed, elongated Hg-containing crystals (Fig. 13c). Powder-XRD

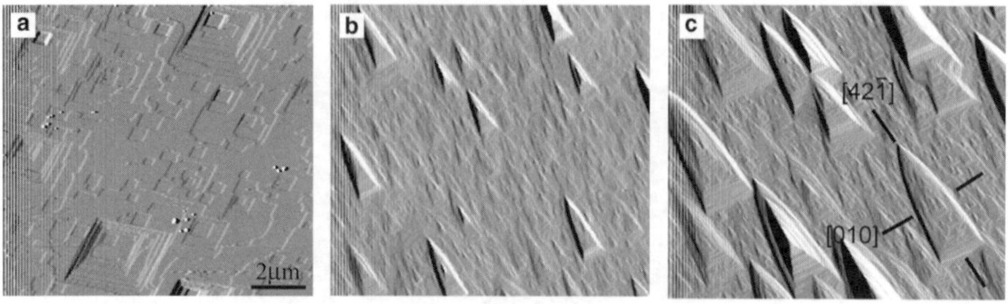

Fig. 12. *In situ* AFM results showing the evolution of calcite surface nanotopography during the first stages of its interaction with $Hg^{2+}{}_{(aq)}$.

patterns in combination with Fourier Transform infrared (FTIR) spectra confirm that these precipitates correspond to a montroydite-type hydrated Hg (II) oxide ($HgO \cdot nH_2O$). The lack of macroscopic data on the composition of the aqueous solutions after their interaction with calcite prevents a quantitative estimation of the Hg-sorption capacity of calcite. However, the possible hypothetical formation of a

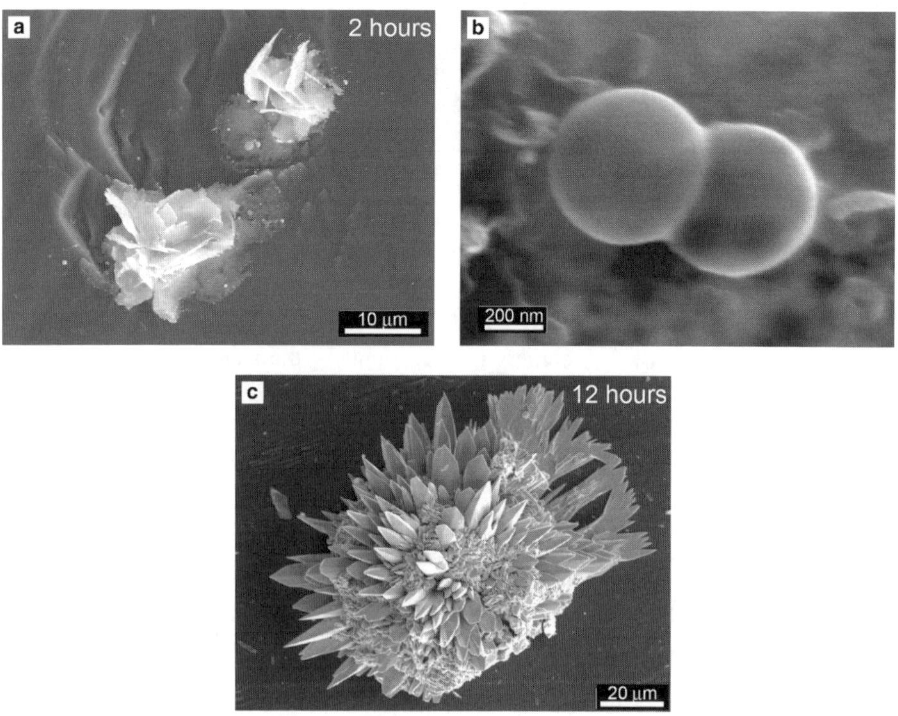

Fig. 13. SEM investigation of nucleation and crystal growth of Hg (II)-oxide (montroydite-type phases) onto a calcite surface after interaction with $Hg^{2+}{}_{(aq)}$.

Ca-Hg(II) carbonate solid solution such us (Hg,Ca)$CO_3 \cdot HgO$, which could sorb this metal, is unlikely. Indeed, $HgCO_3 \cdot HgO$ is unstable under Earth's atmospheric conditions. In addition, the stability field of montroydite, which is limited to circumneutral and high-*pH* conditions (Brookins, 1998), points to a limited effectiveness of calcite as Hg-sequesterer.

4.2. Phosphates

Despite their relatively low solubility and dissolution rates, certain phosphate minerals, and particularly Ca-phosphates represented by OH^-, F^-, and Cl^- apatites, show remarkable reactivity against heavy metals dissolved in aqueous solutions. Most of the work conducted concerns Pb, and some other metals such as Cd and Zn, and special emphasis has been placed on environmental technology for *in situ* immobilization of these contaminants in soils and sediments (*e.g.* Ma *et al.*, 1993; Chen *et al.*, 1997; Traina & Laperche, 1999; Peld *et al.*, 2004). The main metal-sorption mechanisms taking place at the mineral-water interface involve intense surface (co)/(re)precipitation of more insoluble metal carbonates (such as otavite, $CdCO_3$), hydroxycarbonates (such as hydrocerussite, $Pb_3(CO_3)_2(OH)_2$), phosphates (such as hopeite, $Zn_3(PO_4)_2 \cdot 4H_2O$), and most frequently hydroxyphosphates (such as hydroxypyromorphite, $Pb_{10}(PO_4)_6(OH)_2$). Which species are involved depends primarily on the solution *pH*. Direct observation of the interaction of Pb with apatite on the nanoscale using *in situ* AFM, revealed new insights into the simultaneous dissolution and surface precipitation processes leading to crystal growth of hydroxypyromorphite through both homogeneous and heterogeneous nucleation, depending on the initial $Pb^{2+}_{(aq)}$ concentration (Lower *et al.*, 1998a,b; Manecki *et al.*, 2000). In the case of heterogeneous nucleation, hydroxypyromorphite can overgrow epitaxially on the hydroxyapatite surface because both minerals are isostructural and the basal plane value of lattice misfit d_m is 6.06%. It is well known that surface microtopography is very important for heterogeneous nucleation, and preferential growth of hydroxypyromorphite is observed along microsteps and microcracks.

4.3. Sulphates

In comparison with carbonates and phosphates, the literature involving sulphates as metal-sequesterers is scarcer, even though some of them, such as gypsum ($CaSO_4{\cdot}2H_2O$) and anhydrite ($CaSO_4$), are considered rock-forming minerals (Deer *et al.*, 1996) and therefore good candidates to remove toxic metals. However, although the chemical behaviour of most sulphate phases is similar to that of calcium carbonates and calcium phosphates, in that most of them are very or relatively reactive in water, in most cases the subsequent formation of a sparingly soluble, toxic metal-bearing solid phase is not guaranteed. The reason lies in two main factors: first, the dissolution of most common sulphates is not accompanied by a sharp increase in *pH*, which could trigger a rapid formation of oxy/hydroxy insoluble compounds; and second, due to the high solubility of most potential metal sulphates, they are not able to precipitate after a reaction of sulphate mineral with the metal-bearing aqueous solution. Important exceptions to this general trend are anglesite ($PbSO_4$), which can precipitate after the interaction of dissolved Pb with sulphates (Astilleros *et al.*, 2010), due to its relatively low solubility product ($K_{sp} = 10^{-7.79}$; Parkhurst & Appelo, 1999); and the well documented cases of schwertmannite, $Fe^{3+}_{16}O_{16}(SO_4)_2(OH)_{12} \cdot nH_2O$ ($n = 10 - 12$), and jarosite, $KFe_3(SO_4)_2(OH)_6$, which are the main phases identified in acid mine drainage (AMD) and which can effectively remove toxic oxyanions such as arsenate and chromate (Fukushi *et al.*, 2004; Regenspurg & Peiffer, 2005; Asta *et al.*, 2009).

In this section, we will examine and compare the capacity of gypsum to sorb Pb and As(V) dissolved in water. Among sedimentary rock-forming minerals, gypsum emerges as potentially useful to take up a broad spectrum of dissolved metals because of (1) the high reactivity of its surfaces, which makes it much more soluble ($K_{sp} = 10^{-4.58}$; Parkhurst & Appelo, 1999) than calcium carbonate and apatite; (2) the relative non-toxicity of both dissolved calcium and sulphate in comparison to phosphate; and (3) its abundance in nature. In the case of both toxic elements, experiments suggest that the sorption reactions are promoted by the dissolution of gypsum and the subsequent interaction of the sulphate ions released with the toxic elements. We will also discuss the main factors that control the dissolution-precipitation reaction, including the stability of the resulting phases and the structural relationships between the gypsum substrate and the sorbate.

4.3.1. The uptake of lead by gypsum

Few studies have examined the interaction of sulphate ions with dissolved Pb, and most of them have focused on AMD and weathering of sulphidic mineral wastes (Bigham & Nordstrom, 2000; Blowes *et al.*, 2003; Lottermoser, 2007). In such environments, gypsum is one of the most abundant and widespread of the post-mining secondary minerals (Blowes *et al.*, 2003), and its dissolution can lead to an intense reaction with Pb to form more insoluble anglesite according to the following equations:

$$CaSO_4 \cdot 2H_2O_{(s)} + nH_2O_{(l)} \longrightarrow Ca^{2+}{}_{(aq)} + SO_4^{2-}{}_{(aq)} + (n+2)H_2O_{(l)} \tag{7}$$

$$SO_4^{2-}{}_{(aq)} + Pb^{2+}{}_{(aq)} \longrightarrow PbSO_{4(s)} \tag{8}$$

Moreover, the reaction of gypsum with lead to give anglesite has also been described in other environments, such as soils of shootings ranges where the flux of lead comes from shotgun pellets (Lin *et al.*, 1995). It is evident from these examples that gypsum may be a potential method to remove lead from polluted waters and soils.

Macroscopic experiments to test this possibility show two different sorption behaviours depending on the initial concentration of lead (Astilleros *et al.*, 2010). The interaction of mm-sized fragments of gypsum with solutions carrying high concentrations of Pb (100 and 1000 mg/l) caused a dramatic uptake of lead. As shown in Figure 14a, after just 5 min of interaction, more than 50% of lead was removed from the solution; and after 1 month, the final value was close to 3.0 mg/l. However, when solutions with lower concentrations (10 mg/l) were used, the Pb concentration decreased gradually.

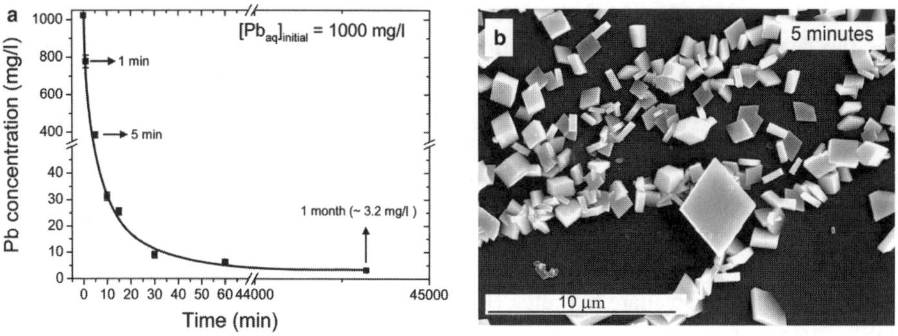

Fig. 14. (***a***) Evolution of Pb concentration as a function of time during the interaction of gypsum fragments with Pb-lead solutions ($[Pb_{aq}]_{initial}$ = 1000 mg/l). (***b***) SEM photomicrograph showing anglesite crystals (bright) growing onto the (010) surface of a gypsum grain after interaction with $[Pb_{aq}]_{initial}$ = 1000 mg/l solutions.

Indeed, no significant change was detected during a relatively long initial period (1 day). Afterwards, the Pb concentration reduction occurred very slowly to reach a final value of ~3.8 mg/l after 1 month of interaction. SEM reveals in the first case that after some minutes of interaction, the gypsum surface was covered by a discontinuous layer composed of rhombus-shaped anglesite crystals (Fig. 14b) oriented randomly with respect to the substrate. The fact that the precipitate grows randomly on the gypsum surface due to the lack of structural relationships between sorbate and sorbent favours the development of the process. In contrast, for lower concentrations of Pb (10 mg/l), no anglesite precipitation was observed, irrespective of the interaction time. RBS analysis shows that under these conditions, the slight decrease in lead concentration (from 10 to 3.8 mg/l) was due first to the adsorption of lead at the gypsum surface from the beginning of the interaction until ~3 days, and subsequently due to the formation of Pb-sulphate layers (surface precipitation).

Although the uptake capacity of gypsum is limited by its solubility product, gypsum appears to be a powerful sorbent for lead in environments where this metal reaches toxic concentrations. Comparison with other minerals shows that gypsum has a lower sorption capacity than calcium carbonates. However, under acidic conditions, where anglesite is relatively stable (Garrels & Christ, 1965), carbonates can easily dissolve and release the sorbed lead back into the solution. Moreover, natural attenuation of lead is likely to occur through anglesite precipitation in aqueous environments, especially in those related to AMD, where sulphate anions are continuously generated as a result of sulphide oxidation.

4.3.2. The uptake of As(V) by gypsum (at intermediate pH *values)*

Arsenic is highly toxic and can lead to a wide range of health problems in humans. It is carcinogenic, mutagenic, and teratogenic (Plant *et al.*, 2003) and is considered the most hazardous substance according to the U.S. Environmental Protection Agency.

Very small concentrations of this element in drinking water supplies are causing massive health problems to many millions of people in some poor and developing countries (Vaughan, 2006). As(V) is the most common form of arsenic in natural waters, with AsO_4^{3-}, $HAsO_4^{2-}$, $H_2AsO_4^-$, and $H_3AsO_4^0$ being the prevailing aqueous species at different *pH* ranges. The chemistry of arsenic is extremely complex and the interaction of these anions with calcium can lead to a diversity of arsenates. For instance, at basic *pH*, a vast number of these arsenates can be formed, including $Ca_5(HAsO_4)_2(AsO_4)_2 \cdot 9H_2O$ (guerinite or ferrarisite), $Ca_5(HAsO_4)_2(AsO_4)_2 \cdot 4H_2O$ (sainfeldite), $Ca_2NaH(AsO_4)_2 \cdot 6H_2O$, double salts such as $NaCaAsO_4 \cdot 7.5H_2O$ and amorphous calcium arsenate hydrates (Rodríguez *et al.*, 2008). The high reactivity of gypsum in water ensures a continuous supply of Ca and, therefore, under favourable conditions, it can react with the dissolved As(V) to produce a calcium arsenate salt.

In this section we will focus on the interaction of gypsum with As(V)-bearing solutions at intermediate *pH*s and 25°C. Under these conditions the interaction has been shown to result in surface precipitation of pharmacolite ($CaHAsO_4 \cdot 2H_2O$, $K_{sp,pharmacolite} = 10^{-4.68}$, Rodríguez-Blanco *et al.*, 2007). Such a process involves the release of solutes (Ca^{2+} and SO_4^{2-} ions) from the gypsum surface to the fluid and the reaction between the released Ca^{2+} and the aqueous $HAsO_4^{2-}$ ions to form pharmacolite nuclei and crystals according to the reactions:

$$CaSO_4 \cdot 2H_2O_{(s)} + nH_2O_{(l)} \longrightarrow Ca^{2+}{}_{(aq)} + SO_4^{2-}{}_{(aq)} + (n+2)H_2O_{(l)} \quad (7)$$

$$HAsO_4^{2-}{}_{(aq)} + Ca^{2+}{}_{(aq)} + nH_2O_{(l)} \longrightarrow CaHAsO_4 \cdot 2H_2O_{(s)} + (n-2)H_2O_{(l)} \quad (9)$$

In contrast to the results reported in the previous section, the interaction of gypsum with aqueous solutions carrying a high concentration of arsenic (100 mmol/l) leads to a slow and progressive decrease in the arsenic concentration, until it reaches an asymptotic value of 79 mmol/l. SEM reveals the rapid growth of large and thick pharmacolite crystals during the first stages of the interactions. The crystals are elongated and clearly show an epitaxial relationship with the gypsum substrate. The rapid growth of pharmacolite crystals leads to a rapid seal of the gypsum substrate and, as a consequence, it becomes completely isolated from the aqueous solution. Under these experimental conditions, the solution cannot gain access to the substrate anymore, and the process stops at a 'pseudo-equilibrium' end point. The conclusion to be drawn from this system is that gypsum is not a suitable remediation tool in this *pH* range for two main reasons. One is the high solubility of pharmacolite, which can only be precipitated when huge amounts of arsenic are present in the aqueous solution. Moreover, a change in the physicochemical conditions of the aqueous solution may lead to a rapid dissolution and release of the arsenic back into the solution. The second reason is the unsuitable spatial arrangement between substrate and overgrowth, which prevents the solution from gaining free access to the solution and prematurely stops the dissolution precipitation process.

4.4. Sulphides

Sulphides are also important minerals for the environment and their surfaces may sorb heavy metals from aqueous media. The typical sorption mechanism involves surface precipitation in nanoscale through redox chemical reactions. In the most characteristic case, Eggleston & Hochella (1991) used STM techniques in order to study the adsorption of aqueous Au species ($AuCl_4^-$) on (100) galena surfaces. They concluded that a direct coupling of Au reduction and sulphide surface oxidation regulates the initial formation of Au (Au^0) nano-islands along [110] directions. Pyrite is also considered to be a major scavenger of Hg(II), although sorption reversibility is possible by adding strong ligands such as J^-, $S_2O_3^{2-}$ and CN^- to the aqueous solution. However, the *ad*sorption of Hg(II) onto pyrite is not such a simple phenomenon, particularly at low *pH*, as revealed by XPS and XAS techniques, because surface oxidation causes the formation of Fe-(hydr)oxide coatings, perhaps allowing Hg(II) to create surface complexes with both oxides and pyritic S (Ehrhardt *et al.*, 2000; Behra *et al.*, 2001; Bower *et al.*, 2008).

5. Interaction of actinide elements with mineral surfaces

Most of the primary rock-forming tectosilicate minerals, such as quartz and feldspars, seem to be rather inactive towards actinides ('actinoids' according to recent IUPAC nomenclature) in aqueous solutions at low temperature. These minerals show limited ability to sorb small amounts of actinide species on their surfaces through *ad*sorption and surface precipitation (*e.g.* Misaelides *et al.*, 1998b). On the other hand, rock-forming phyllosilicates, such as biotite, exhibit greater sorption ability (*e.g.* Misaelides *et al.*, 2001), acting as suitable phases for actinide retention in massive unfractured crystalline rocks. In addition, it is generally established that secondary silicate and aluminosilicate micro/nanoporous minerals, such as clays and zeolites, show enhanced sorptive properties towards a variety of actinides, including transuranic elements. A relevant review was presented by Misaelides & Godelitsas (1999), who mentioned the importance of actinide hydrolysis processes (mainly depending on the *pH*) and the role of colloidal and pseudocolloidal systems. Payne *et al.* (1998) noticed that the sorption of U^{VI} by kaolinite is influenced by a large number of factors, including *pH*, ionic strength, P_{CO_2}, concentration of U and the presence of ligands such as phosphate and humic acid. Using EXAFS, Thompson *et al.* (1998) demonstrated the formation of multinuclear surface complexes or precipitates. At the same time, Kersting *et al.* (1999) reported that although Pu has been shown experimentally to strongly sorb to clays and zeolites, it can also exist as an intrinsic colloid, composed of Pu-oxide. Both types of colloids have the capacity to be transported by ground water. Recent investigations (Novikov *et al.*, 2006), based on nano-SIMS techniques, revealed that the above mentioned colloidal-facilitated aquatic transport of actinides, can be realized by means of Pu^{IV}-hydroxides and Pu^{IV}-/U^{VI}-carbonates sorbed on amorphous Fe-oxide nanoparticles. In that case, most Pu^{IV} may be the result of reduction of Pu^V to Pu^{IV} after adsorption onto Fe-oxide due to the high concentration of dissolved organic carbon at the interface. Similarly hydrolyzed Pu^{IV} species can be stabilized on oxidizing surfaces of Mn-oxide minerals, such as pyrolusite, which can be responsible for the preferential sorption of Pu in micro/nanoporous mineral-containing Yucca Mountain tuffs (Powell *et al.*, 2006).

Natural Mn-oxides, as well as Mn-hydroxides and Mn-oxyhydroxides, are highly reactive minerals that play important roles in elemental biogeochemical cycles. Mn oxides are among the most significant oxidants in the environment, and they are involved in a broad range of redox reactions with organic and inorganic chemical moieties. It is also important to note that many metal-reducing bacteria may use Mn oxides as the terminal electron acceptor for the oxidation of organic matter or H_2 in the absence of oxygen. In fact, micro-organisms can contribute to the oxidation of Mn^{II} through complicated surface-catalysed reactions, resulting in the formation of $Mn^{III,IV}$-oxide biominerals. As a consequence, biotic Mn^{II} oxidation is rapid and more favorable in nature than are abiotic (inorganic and/or organic) Mn^{II} oxidation processes. It is now known that the majority of low-temperature surface and superegene natural Mn oxides are biogenic, forming directly from biological Mn^{II} oxidation processes or through biological alteration of precursor Mn phases. The predominant oxidation states of Mn in nature are +II, +III and +IV. In aqueous media, Mn^{II} (as Mn^{2+} cation) is thermodynamically preferred in the absence of oxygen and at low *pH*, whereas Mn^{III} and Mn^{IV} are favored in the presence of oxygen and at higher *pH*. In the solid state, Mn^{II} occurs primarily as the main constituent in insoluble carbonates and phosphates and as minor constituents in various phases including $Mn^{III,IV}$ oxides. Mn^{III} is thermodynamically unstable in aqueous media, except in the case of distinct soluble organic complexes, and it disproportionates to Mn^{II} and Mn^{IV}. However, in the solid state, and particularly in the case of insoluble biogenic Mn oxides found in soils, sediments and supergene deposits, Mn^{III} coexists with Mn^{IV} and Mn^{II}. The Mn^{III}/Mn^{II} and Mn^{IV}/Mn^{II} couples have enhanced redox potentials, making Mn^{III} and Mn^{IV} phases among the stronger oxidants found in the environment. Besides, Mn^{III} is the crucial intermediate stage in the oxidation of Mn^{II} to Mn^{IV} catalysed by bacteria, and it is therefore considered to be the key criterion for characterizing biogenic Mn oxides. In particular, the main pathway for bacterial Mn^{II} oxidation is enzymatic, and although bacteria that oxidize Mn^{II} are phylogenetically diverse, they all use a multicopper oxidase-like enzyme to oxidize Mn^{II}. Thus, the oxidation of Mn^{II} to Mn^{IV} occurs *via* a soluble or enzyme-complexed Mn^{III} intermediate. The primary biogenic Mn^{IV}-oxide formed is an alkali- and alkaline earth-containing phyllomanganate of large surface area, close to δ-MnO_2 (which is the synthetic analogue of vernadite) or acid birnessite, which generally shows a basal plane *d*-spacing of 7 Å. During the growth of this rather poorly crystallized phase, additional heavy metal cations (*e.g.* Zn^{2+}, Ni^{2+}, Cu^{2+}, Co^{2+} and perhaps Pb^{2+}) as well as protons (H^+ and/or H_3O^+) can be scavenged as impurities by occupying vacant layer octahedral sites. In the presence of major Ca, a Mn oxide is grown, which exhibits greater crystallinity and a basal plane *d*-spacing of 10 Å. The initial biogenic Mn oxides are highly reactive catalysts for further Mn^{II} oxidation reactions and during the various biomineralization and post-biomineralization stages, several other secondary Mn oxides are formed, mostly micro/nanoporous and hydrated ones, such as todorokite- and hollandite-type phases (*e.g.* Tebo *et al.*, 2004).

According to Misaelides *et al.* (2002), todorokite can sorb considerable amounts of dissolved actinides (Th, U, Pa) through different sorption mechanisms, including surface precipitation of insoluble hydrolysis products, as revealed by XPS. Moreover, Webb *et al.* (2006) showed, using EXAFS and SR-XRD, that U^{VI} (UO_2^{2+}) is strongly incorporated into Mn-oxides during biogenesis. At low U concentrations a strong bidentate surface complex is formed, whereas at higher concentrations UO_2^{2+} incorporates into the micro/nanopores as a tridentate complex. However, in the case of Fe oxides, such as magnetite, it has been proved that U^{VI} is reduced to U^{IV} through direct electron transfer with structural Fe^{II}, leading to surface precipitation of hyperstoichiometric UO_2 (Scott *et al.*, 2005). Similar phenomena take place at the mineral-water interface during the interaction of actinides with sulphides. The main sorption mechanism, as has also been observed for transition-metal ions, concerns surface precipitation in nanoscale, including partial actinide-ion reduction and growth of distinct metal nanoparticles. Wersin *et al.* (1994) used near-surface sensitive analytical techniques (AES and XPS) and fourier transform infrared spectroscopy (FTIR) to resolve the formation of ~40 nm wide needles of U^{VI}/U^{IV} oxide (U_3O_8-type) on galena surfaces interacted with $U^{VI}{}_{(aq)}$ ($UO_2^{2+}{}_{(aq)}$) under anoxic conditions. More recently, Eglizaud *et al.* (2006) and Scott *et al.* (2007) also used XPS to conclude that U^{VI} interaction with pyrite involves *ad*sorption and partial reduction, leading to surface precipitation of $UO_{(2+x)\ (s)}$ $(0 < x < 2)$ and U^{VI}/U^{IV}-oxides. It should be noted that the stabilization of sorbed U-species with intermediate oxidation state (U^V) is rather uncertain with regard to Fe-, Mn-oxides and sulphides, though Ilton *et al.* (2005) showed that ferrous mica surfaces may reduce U^{VI} and preserve U^V over a broad range of solution compositions. However, Moyes *et al.* (2000) argued, on the basis of EXAFS, that muscovite behaves in a simpler way by removing $U^{VI}{}_{(aq)}$ through surface complexation/precipitation.

The interaction of dissolved actinides with carbonate mineral surfaces is another important issue inasmuch as the most common polymorph of $CaCO_3$ (calcite) is widespread in nature and the corresponding carbonate rocks (limestones, marls, travertines, marbles, *etc.*) occupy a major part of the Earth's landscape. Because of the high solubility and K_{sp} of calcite, the interaction of its surfaces with $UO_2^{2+}{}_{(aq)}$ is related to intense surface-dissolution processes and subsequent surface (co)/(re)precipitation and crystal growth phenomena (*e.g.* Carroll & Bruno, 1991; Carroll *et al.*, 1992). According to Schindler & Putnis (2004) and Schindler *et al.* (2004), distinct uranyl-hydroxy-hydrate and uranyl carbonate phases, related to schoepite, becquerelite and wyartite, can be overgrown on the $(10\bar{1}4)$ surface of calcite when it interacts with concentrated acidic and basic UO_2^{2+}-bearing solutions. EXAFS and GIXAFS (polarization-dependent EXAFS in GI geometry) have been used to examine *ad*sorption mechanisms at lower $UO_2^{2+}{}_{(aq)}$ concentrations. Under these conditions, uranyl triscarbonate-like *ad*sorption complexes can be formed (Elzinga *et al.*, 2004), whereas uranyl cations

($[O{=}U{=}O]^{2+}$) can be *ad*sorbed onto terrace step edges but not onto the surface plane (Denecke *et al.*, 2005). Of course, the *ad*sorption of actinide, and particularly of trans-uranic element traces, to calcite surfaces is strongly *pH*-dependent; it decreases at high *pH* due to carbonate complexation in solution (Zavarin *et al.*, 2005).

It is obvious that mineral surfaces play an important role in the transport of actinides, and particularly of U, in the environment, by carrying them through aquatic pathways (streams, rivers, underground waters) towards the final oceanic sinks. Uranium is considered to be conservative in oxygenated waters due to the formation of stable and soluble U^{VI}-carbonate complexes, $UO_2(CO_3)_3^{4-}{}_{(aq)}$, which can be reduced to insoluble U^{IV}-phases (*e.g.* UO_2) at the mineral/sediment-water interface under anoxic conditions that reflect the presence of organic compounds. The reduction of U^{VI} to U^{IV} occurs relatively late in the diagenetic sequence, after microbially induced dissolution of Fe oxides and Mn oxides (*e.g.* Klinkhammer & Palmer, 1991). In addition, the investigation of the interaction of the actinides with natural micro/nanoporous materials is of special significance for the nuclear industry because of the suitability of these materials for the treatment of liquid effluents and the final storage of radioactive waste (sorbents, backfill materials, constituents of disposal host rocks including the Yucca Mountain tuffs). Thus, low-cost industrial minerals, such as HEU-type zeolites (heulandite-clinoptilolite series) that constitute the most abundant zeolites on Earth, have been investigated for their ability to remove actinides from aqueous solutions (*e.g.* Pabalan *et al.*, 1998). However, the uptake of dissolved actinide-ions by natural zeolite crystals is not a simple phenomenon, since it depends on the physicochemical properties of both the aqueous solution and the interacting solid (Fig. 15). The sorption of U by the Na-form of HEU-type zeolite (NaHEU) powder was found to be rather limited given the CEC of the material (Godelitsas *et al.*, 1996a). If we suppose that only UO_2^{2+} ions can be sorbed through ion exchange, the material should theoretically be loaded with a maximum of ~408 mg/g UO_2^{2+} or ~360 mg/g U. The experimental results showed bulk U concentrations in the modified solids of between 2.84 and

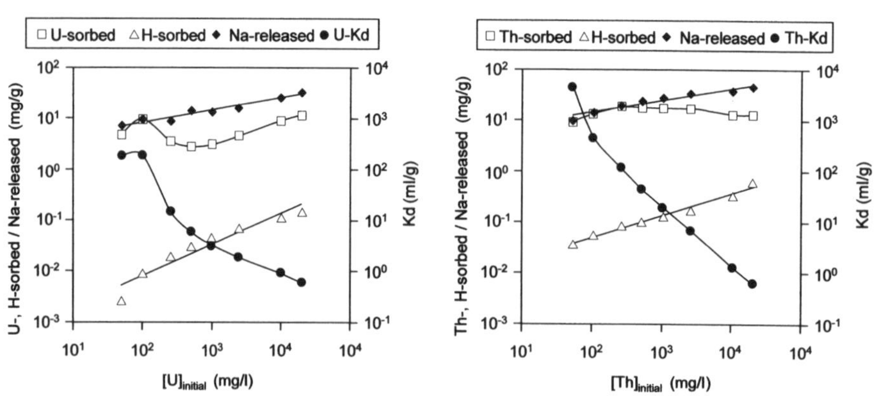

Fig. 15. $U(VI)_{(aq)}$ ($UO_2^{2+}{}_{(aq)}$) and $Th(IV)_{(aq)}$ ($Th^{4+}{}_{(aq)}$) sorption by pure NaHEU-type natural zeolite crystals.

11.68 mg/g. In addition, the K_d values were found to be up to 190.59 ml/g for a solution containing 100 mg/l U, indicating that the best metal removal achieved was equal to 48.6%. On the other hand, the amounts of Na released from the structure of the zeolite were found to be high in all cases. The sorption of H^+ ions, responsible for the increase in the initial *pH* values (from $pH_{initial} = 2.78–4.87$ to $pH_{final} = 3.04–6.39$) was also significant. It is evident that the sorption of H^+, along with the simultaneous release of Na^+, occur in parallel, while the sorption of U occurs independently (Godelitsas *et al.*, 1996a; Godelitsas & Armbruster, 2003). These parallel chemical processes are probably due to ion-exchange reactions, and they can be adequately described by the general function (H-sorbed) or (Na-released) = $a[U]^b_{initial}$ ($R^2 > 0.9$), where *a* and *b* are numerical factors. On the other hand, the sorption of U is more complicated, due to both the aqueous chemistry of the element and the properties of the microporous solid that is gradually modified during the interaction. In aqueous solutions of $pH < 4$, the most dominant species is the hydrated uranyl ion $[UO_2(H_2O)_6]^{2+}$, which exhibits a hexagonal bipyramidal structure with the U atom at the centre. At greater *pH* values, it is readily hydrolysed towards a variety of hydrolysis products of the general type $[(UO_2)_x(OH)_y]^{(2x-y)+}$. Hydrolysis is accelerated by the presence of the zeolite, which preferentially sorbs H^+ ions and subsequently increases the initial *pH*. For solutions with high U concentrations ($[U]_{initial} > 1000$ mg/l), the initial and final *pH* values are <4. In this case, hydrolysis is rather restricted, and the sorption of U may be attributed to ion exchange of UO_2^{2+} ions and other species such as $[UO_2(OH)^+]$. In the case of solutions with small U concentrations ($[U]_{initial} < 250$ mg/l), the initial and final *pH* values are >4, and the sorption of U can be explained primarily on the basis of *ad*sorption and surface precipitation mechanisms. Indeed, the examination of the corresponding modified zeolite samples by SEM-EDS confirmed that distinct U precipitates were deposited onto the surface of the crystals. Detailed powder XRD analysis gave evidence for small but well defined peaks attributed to hydrous U oxides of the type $UO_3 \times H_2O$. The argument for the presence of such a crystalline phase on the surface of HEU-type zeolites modified with U was strengthened by XPS, in the case of the mm-sized CaHEU crystals, taking into account the $U4f_{7/2}$ photopeak at 381.5 eV. In addition, the ^{12}C-RBS spectra strongly indicated the existence of a U-rich surface 'layer' approximately 250 nm thick due to *ad*sorption and surface precipitation (Misaelides *et al.*, 1996; Godelitsas & Armbruster, 2003). Relevant phenomena, involving sorption of H^+ along with the simultaneous release of Na^+, have also been observed in the case of HEU-type zeolite crystals interacting with Th^{4+} ions in aquous solutions (Godelitsas *et al.*, 1996b; Godelitsas & Armbruster, 2003).

Acknowledgments

This work has been partially financed by the Spanish government (under Contract CGL2007-65523-C02-01). Financial support has also been provided by the research group 910148 (UCM-Comunidad de Madrid, Spain).

References

Andara, A.J., Heasman, D.M., Fernández-González, A. & Prieto, M. (2005) Characterization and crystallization of $Ba(SO_4,SeO_4)$ solid solution. *Crystal Growth and Design*, **5**, 1371–1378.

Armbruster, Th. (2001) Clinoptilolite-heulandite: applications and basic research. In: *Zeolites and Mesoporous Materials at the Dawn of the 21st Century* (A. Galarnau, F. Di Renzo, F. Faujula & J. Vedrine, editors). Studies in Surface Science and Catalysis, **135**, 13–27. Elsevier, Amsterdam.

Asta, M.P., Cama, J., Martínez, M. & Giménez, J. (2009) Arsenic removal by goethite and jarosite in acidic conditions and its environmental implications. *Journal of Hazardous Materials*, **171**, 965–972.

Astilleros, J.M., Pina, C.M., Fernández-Díaz, L., Prieto, M. & Putnis, A. (2006) Nanoscale phenomena during the growth of solid solutions on calcite {10–14} surfaces. *Chemical Geology*, **225**, 322–335.

Astilleros, J.M., Godelitsas, A., Rodríguez-Blanco, J.D., Fernández-Díaz, L., Prieto, M., Lagoyannis, A. & Harissopulos, S. (2010) Interaction of gypsum with Pb-bearing aqueous solutions. *Applied Geochemistry*, **25**, 1008–1016.

Becker, U., Biswas, S., Kendall, T., Risthaus, P., Putnis, C. & Pina, C. (2005) Interactions between mineral surfaces nd dissolved species: from monovalent ions to complex organic molecules. *American Journal of Science*, **305**, 791–825.

Behra, P., Bonnissel-Gissinger, P., Alnot, M., Revel, R. & Ehrhardt, J.J. (2001) XPS and XAS study of the sorption of Hg(II) onto pyrite. *Langmuir*, **17**, 3970–3979.

Bigham, J.M. & Nordstrom, D.K. (2000) Iron and aluminum hydrosulfates from acid sulphate waters. In: *Sulfate Minerals: Crystallography, Geochemistry and Environmental Significance* (C.N. Alpers, J.L. Jambor & D.K. Nordstrom, editors). Reviews in Mineralogy and Geochemistry, **40**, 351–403. Mineralogical Society of America, Whasington, D.C. and the Geochemical Society, St. Louis, Missouri, USA.

Blowes, D.W., Ptacek, C.J., Jambor, J.L. & Weisener, C.G. (2003) The geochemistry of acid mine drainage. In: *Treatise on Geochemistry Vol. 9* (B. Sherwood Lollar, editor), 149–204. Elsevier-Pergamon, Oxford, UK.

Bower, J., Savage, K.S., Weinman, B., Barnett, M.O., Hamilton, W.P. & Harper, W.F. (2008) Immobilization of mercury by pyrite (FeS_2). *Environmental Pollution*, **156**, 504–514.

Brady, P.V. editor (1996) *The Physics and Chemistry of Mineral Surfaces.* CRC-Press, Boca Raton, Florida, USA, 368 pp.

Brookins, D.G. (1998) *Eh-pH Diagrams for Geochemistry*. Springer-Verlag, Berlin, 176 pp.

Brown, G.E. (2001) How minerals react with water. *Science*, **294**, 67–70.

Brown, G.E., Foster, A.L. & Ostergren, J.D. (1999) Mineral surfaces and bioavailability of heavy metals: a molecular-scale perspective. *Proceedings of the National Academy of Science of the United States of America*, **96**, 3388–3395.

Brown, G.E., Parks, G.A. & O'Day, P.A. (1995) Sorption at mineral-water interfaces: Macroscopic and microscopic perspectives. In: *Mineral Surfaces* (D.J. Vaughan & R.A.D. Pattrick, editors). Mineralogical Society Series, Chapman & Hall, London, **5**, 129–183.

Brown, G.E., Trainor, Th. & Chaka, A.M. (2008) Geochemistry of mineral surfaces and factors affecting their chemical reactivity. In: *Chemical Bonding at Surfaces and Interfaces* (A. Nilsson, L.G.M. Petterson & J.K. Nørskov, editors), 457–510. Elsevier, Amsterdam.

Carroll, S.A. & Bruno, J. (1991) Mineral-solution interactions in the U(VI)-CO_2-H_2O system. *Radiochimica Acta*, **52/53**, 187–193.

Carroll, S.A., Bruno, J., Petit, J.C. & Dran, J.C. (1992) Interactions of U(VI), Nd, and Th(IV) at the calcite-solution interface. *Radiochimica Acta*, **58/59**, 245–252.

Casey, W.H., Rustad, J.R., Banerjee, D. & Furrer, G. (2005) Large molecules as models for small particles in aqueous geochemistry research. *Journal of Nanoparticle Research*, **7**, 377–387.

Chada, V.G.R., Hausner, D.B., Strongin, D.R., Rouff, A.A. & Reeder, R. (2005) Divalent Cd and Pb uptake on calcite {10–14} cleavage faces: An XPS and AFM study. *Journal of Colloid and Interface Science*, **288**, 350–360.

Chardon, E.S., Bosbach, D., Bryan, N.D., Lyon, I.C., Marquardt, C., Roemer, J., Schild, D., Vaughan, D., Wincott, P.L., Wogelius, R. & Livens, F.R. (2008) Reactions of the feldspar surface with metal ions: sorption of Pb(II), U(VI) and Np(V), and surface analytical studies of the reaction with Pb(II) and U(VI). *Geochimica et Cosmochimica Acta*, **72**, 288–297.

Chen, C.C., Coleman, M.L. & Katz, L.E. (2006) Bridging the gap between macroscopic and spectroscopic studies of metal ion sorption at the oxide/water interface: Sr(II), Co(II), and Pb(II) sorption on quartz. *Environmental Science & Technology*, **40**, 142–148.

Chen, X., Wright, J.V., Conca, J.L. & Peurrung, L.M. (1997) Effects of pH on heavy metal sorption on mineral apatite. *Environmental Science & Technology*, **31**, 624–631.

Chiarello, R.P. & Sturchio, N.C. (1994) Epitaxial growth of otavite on calcite observed in situ by synchrotron X-ray scattering. *Geochimica et Cosmochimica Acta*, **58**, 5633–5638.

Chiarello, R.P., Sturchio, N.C., Grace, J.D., Geissbuhler, P., Sorensen, L.B., Cheng, L. & Xu, S. (1997) Otavite-calcite solid solution formation at the calcite-water interface studied in situ by synchrotron X-ray scattering. *Geochimica et Cosmochimica Acta*, **61**, 1467–1474.

Davis, J.A., Fuller, C.C. & Cook, A.D. (1987) A model for trace metal sorption processes at the calcite surface: adsorption of Cd^{2+} and subsequent solid solution formation. *Geochimica et Cosmochimica Acta*, **51**, 1477–1490.

Deer, W.A., Howie, R.A. & Zussman (1996) *An Iintroduction to the Rock-Forming Minerals (2nd edition).* Prentice Hall, New Jersey, 712 pp.

DeLeeuw, N.H. & Parker, S.C. (1999) Molecular dynamics simulation of crystal dissolution from calcite steps. *Physical Review B*, **60**, 13792–13799.

Denecke, M., Bosbach, D., Dardenne, K., Lindqvist-Reis, P., Rothe, J. & Yin, R.Z. (2005) Polarization dependent grazing incidence (GI) XAFS measurements of uranyl cation sorption onto mineral surfaces. *Physica Scripta*, **T115**, 877–881.

Eggleston, C.M. & Hochella, M.F. Jr. (1991) Scanning Tunneling Microcopy of galena (100) surface oxidation and sorption of aqueous gold. *Science*, **254**, 983–986.

Eglizaud, N., Miserque, F., Simoni, E., Schlegel, M. & Descostes, M. (2006) Uranium(VI) interaction with pyrite (FeS_2) Chemical and spectroscopic studies. *Radiochimica Acta*, **94**, 651–656.

Ehrhardt, J.J., Behra, P., Bonnissel-Gissinger, P. & Alnot, M. (2000) XPS study of the sorption of Hg(II) onto pyrite FeS_2. *Surface and Interface Analisys*, **30**, 269–272.

Elzinga, E.J., Tait, C.D., Reeder, R.J., Rector, K.D., Donohoe, R.J. & Morris, D.E. (2004) Spectroscopic investigation of the U(VI) sorption at the calcite-water interface. *Geochimica et Cosmochimica Acta*, **68**, 2437–2448.

Fernández-González, A., Martín-Díaz, R. & Prieto, M. (1999) Crystallisation of $Ba(SO_4,CrO_4)$ solid solutions from aqueous solutions. *Journal of Crystal Growth*, **200**, 227–235.

Fernández-González, A., Andara, A., Alía, J.M. & Prieto, M. (2006) Miscibility in the $CaSO_4 \cdot 2H_2O$ – $CaSeO_4 \cdot 2H_2O$ system: implications for the crystallization and dehydration behaviour. *Chemical Geology*, **225**, 256–265.

Fitzgerald, W.F. & Lamborg, C.H. (2003) Geochemistry of mercury in the environment. In: *Treatise on Geochemistry* (B. Sherwood Lollar, editor), **9**, 177–148. Elsevier-Pergamon, Oxford, UK.

Fukushi, K., Sato, T., Yanase, N., Minato, J. & Yamada, H. (2004) Arsenate sorption on schwertmannite. *American Mineralogist*, **89**, 1728–1734.

Gamsjäger, H., Fluch, A. & Swinehart, J.H. (1984) The effect of potential aqueous pollutants on the solubility of Pb^{2+} in cerussite–calcite phases. *Monatshefte für Chemie*, **115**, 251–259.

Garrels, R.M. & Christ, C.L. (1965) *Solutions, Minerals, and Equilibria.* Harper and Row, New York, 450 pp.

Godelitsas, A. & Armbruster, Th. (2003) HEU-type zeolites modified by transition elements and lead: A review. *Microporous and Mesoporous Materials*, **61**, 3–24.

Godelitsas, A., Misaelides, P., Filippidis, A., Charistos, D. & Anousis, I. (1996a) Uranium sorption from aqueous solutions by the sodium-form of HEU-type zeolite crystals. *Journal of Radioanalytical and Nuclear Chemistry*, **208**, 393–402.

Godelitsas, A., Misaelides, P., Charistos, D., Filippidis A. & Anousis, I. (1996b) Interaction of HEU-type zeolite crystals with thorium aqueous solutions, *Chemie der Erde*, **56**, 143–156.

Godelitsas, A., Charistos, Dwyer, J., Tsipis, C., Filippidis, A., Hatzidimitriou, A. & Pavlidou, E. (1999) Copper(II)-loaded HEU-type zeolite crystals: characterization and evidence of surface complexation with *N,N*-diethyldithiocarbamate anions. *Microporous and Mesoporous Materials*, **33**, 77–87.

Godelitsas, A., Charistos, D., Tsipis, A., Tsipis, C., Filippidis, A., Triantafyllidis, C., Manos, G. & Siapkas, D. (2001) Characterisation of zeolitic materials with a HEU-Type structure modified by transition metal elements: definition of acid sites in nickel-loaded crystals in the light of experimental and quantum-chemical results. *Chemistry – A European Journal*, **7**, 3705–3721.

Godelitsas, A., Astilleros, J.M., Hallam, K., Harissopoulos, S. & Putnis, A. (2003a) Interaction of calcium carbonates with lead in aqueous solutions. *Environmental Science & Technology*, **37**, 3351–3360.

Godelitsas, A., Astilleros, J.M., Hallam, K., Löns, J. & Putnis, A. (2003b) Microscopic and spectroscopic investigation of the calcite surface interacted with mercury (II) acetate aqueous solutions. *Mineralogical Magazine*, **67**, 1196–1204.

Godelitsas, A., Charistos, D., Tsipis, C., Misaelides, P., Filippidis, A. & Schindler, M. (2003c) Heterostructures patterned on aluminosilicate microporous substrates: crystallization of cobalt(III) *tris*(*N,N*-diethyldithiocarbamato) on the surface of a HEU-type zeolite. *Microporous and Mesoporous Materials*, **61**, 69–77.

Heberling, F., Denecke, M.A. & Bosbach, D. (2008) Neptunium(V) Coprecipitation with Calcite. *Environmental Science & Technology*, **42**, 471–476.

Hochella, M.F. (2006) The case of nanogeoscience. *Annals of the New York Academy of Sciences*, **1093**, 108–122.

Hochella, M.F., Lower, S.K., Maurice, P.A., Penn, L.R., Sahai, N., Sparks, D. & Twining, B.S. (2008) Nanominerals, mineral nanoparticles and Earth systems. *Science*, **319**, 1631–1635.

Howard, A.G. (1998) *Aquatic Environmental Chemistry*. Oxford University Press, Oxford, UK, 96 pp.

Ilton, E.S., Haiduc, A., Cahill, Ch. & Felmy, A. (2005) Mica surfaces stabilize pentavalent uranium. *Inorganic Chemistry*, **44**, 2986–2988.

Katsikopoulos, D., Fernández-González, A., Prieto, A.C. & Prieto, M. (2008) Co-crystallization of Co(II) with calcite: Implications for the mobility of cobalt in aqueous environments. *Chemical Geology*, **254**, 87–100.

Katsikopoulos, D., Fernández-González, A. & Prieto, M. (2009) Precipitation and mixing properties of the 'disordered' (Mn,Ca)CO_3 solid solution. *Geochimica et Cosmochimica Acta*, **73**, 6147–6161.

Kersting, A.B., Efurd, D.W., Finnegan, D.L., Rokop, D.J., Smith, D.K. & Thompson, J.L. (1999) Migration of plutonium in ground water at the Nevada Test Site. *Nature*, **397**, 56–59.

Klinkhammer, G.P. & Palmer, M.R. (1991) Uranium in oceans: Where it goes and why. *Geochimica et Cosmochimica Acta*, **55**, 1799–1806.

Köhler, S.J, Cubillas, Rodriguez-Blanco, J.D., Bauer, C. & Prieto, M. (2007) Removal of Cadmium from wastewaters by aragonite shells and the influence of other divalent cations. *Environmental Science & Technology*, **41**, 112–118.

Königsberger, E., Hausner, R. & Gamsjäger, H, (1991) Solid-solute phase equilibria in aqueous solution. V: The system $CdCO_3$-$CaCO_3$-CO_2-H_2O. *Geochimica et Cosmochimica Acta*, **55**, 3505–3514.

Lea, A.S., Hurt, T.T., El-Azab, A., Amonette, J.E. & Baer, D.R. (2003) Heteroepitaxial growth of a manganese carbonate secondary nano-phase on the $\{10\bar{1}4\}$ surface of calcite in solution. *Surface Science*, **524**, 63–77.

Lin, Z., Comet, B., Qvarfort, U. & Herbert, R. (1995) The chemical and mineralogical behaviour of Pb in shooting range soils from central Sweden. *Environmental Pollution*, **89**, 303–309.

Lorens, R.B. (1981) Sr, Cd, Mn and Co distribution coefficients in calcite as a function of calcite precipitation rate. *Geochimica et Cosmochimica Acta*, **45**, 553–561.

Lottermoser, B. (2007) *Mine Wastes: Characterization, Treatment and Environmental Impacts (2nd edition)*. Springer-Verlag, Berlin, 304 pp.

Lower, S.K., Maurice, P.A. & Traina, S.J. (1998a) Simultaneous dissolution of hydroxylapatite and precipitation of hydroxypyromorphite: direct evidence of homogeneous nucleation. *Geochimica et Cosmochimica Acta*, **62**, 1773–1780.

Lower, S.K., Maurice, P.A., Traina, S.J. & Carlson, E.H. (1998b) Aqueous Pb sorption by hydroxylapatite: applications of atomic force microscopy to dissolution, nucleation, and growth studies. *American Mineralogist*, **83**, 147–158.

Ma, Q.Y., Traina, S.J. & Logan, T.J. (1993) In situ lead immobilization by apatite. *Environmental Science & Technology*, **27**, 1803–1810.

Manceau, A., Schlegel, M., Nagy, K. & Charlet, L. (1999) Evidence for formation of trioactahedral clay upon sorption of Co^{2+} on quartz. *Journal of Colloid and Interface Science*, **220**, 181–197.

Manecki, M., Maurice, P.A. & Traina, S.J. (2000) Uptake of aqueous Pb by Cl-, F-, and OH-apatites: mineralogic evidence for nucleation mechanisms. *American Mineralogist*, **85**, 932–942.

Marques Fernandes, M., Stumpf, T., Rabung, T. Bosbach, D. & Fanghänel, Th. (2008) Incorporation of trivalent actinides into calcite: A time resolved laser fluorescence spectroscopy (TRLFS) study. *Geochimica et Cosmochimica Acta*, **72**, 464–474.

Martin-Garin, A., Van Cappellen, P. & Charlet, L. (2003) Aqueous cadmium uptake by calcite: A stirred flow-through reactor study. *Geochimica et Cosmochimica Acta*, **67**, 2763–2774.

Maurice, P.A. (2009) *Environmental Surfaces and Interfaces from the Nanoscale to the Global Scale.* J. Wiley & Sons, Hoboken, New Jersey, USA, 464 pp.

McBeath, M.K., Rock, P.A., Casey, W.H. & Mandell, G.K. (1998) Gibbs energies of formation of metal-carbonate solid solutions: Part 3. The $Ca_xMn_{1-x}CO_3$ system at 298 K and 1 bar. *Geochimica et Cosmochimica Acta*, **62**, 2799–2808.

Misaelides, P., Godelitsas, A., Kossionidis, S. & Manos, G. (1996) Investigation of chemical processes at mineral surfaces using accelerator-based and surface analytical techniques: heavy metal sorption on zeolite crystals. *Nuclear Instruments and Methods in Physics Research*, **B113**, 296–299.

Misaelides, P., Godelitsas, A., Stephan, A., Meijer, J., Rolfs, C., Harissopulos, S., Kokkoris, M. & Filippidis, A. (1998a) Application of proton microprobe and ^{12}C-Rutherford backscattering spectroscopy to the identification of Hg(II)-cations sorbed by granite minerals. *Radiochimica Acta*, **83**, 43–48.

Misaelides, P., Godelitsas, A., Noli, F., Kokkoris M. & Harissopoulos, S. (1998b) Application of radiochemical, accelerator-based and surface analytical techniques to the investigation of the U- and Th-interaction with tectosilicate minerals. In: *Actinides and the Environment* (P.A. Sterne, A. Gonis & A.A. Borovoi, editors), 303–306. Kluwer, Dordrecht, The Netherlands.

Misaelides, P. & Godelitsas, A. (1999) Interaction of actinides with natural microporous materials: a review. *Czechoslovak Journal of Physics*, **49**, 167–174.

Misaelides, P., Godelitsas, A., Harissopoulos, S. & Anousis, I. (2001) Interaction of granitic biotite with selected lanthanides and actinides. *Journal of Radioanalytical and Nuclear Chemistry*, **247**, 325–328.

Misaelides, P., Katranas, T., Godelitsas, A., Klewe-Nebenius, H. & Anousis, A. (2002) The chemical behavior of the natural microporous Mn-oxide todorokite in actinides (Th, U, Pa) aqueous solutions. *Separation Science and Technology*, **37**, 1109–121.

Moore, J.W. (1991) *Inorganic Contaminants of Surface Water Research and Monitoring Priorities.* Springer-Verlag, New York, 334 pp.

Moyes, L.N., Parkman, R.H., Charnock, J.M., Vaughan, D.J., Livens, F.R., Hughes, C.R. & Braithwaite, A. (2000) Uranium uptake from aqueous solution by interaction with goethite, lepidocrocite, muscovite, and mackinawite: An X-ray absorption spectroscopy study. *Environmental Science & Technology*, **34**, 1062–1068.

Nriagu, J.O. (1981) *Cadmium in the Environment. Part 2. Health Effects.* Wiley-Interscience, New York, 908 pp.

Nriagu, J.O. & Pacyna, J.M. (1998) Quantitative assessment of worldwide contamination of air, water, and soils by trace metals. *Nature*, **33**, 134–139.

Novikov, A.P., Kalmykov S.N., Utsunomiya, S., Ewing, R.C., Horreard, F., Merkulov, A., Clark, S., Tkachev, V. & Myasoedev, B. (2006) Colloidal transport of plutonium in the far-field of the Mayak production association, Russia. *Science*, **314**, 638–641.

Pabalan, R.T., Turner, D.R., Bertetti, F.P. & Prikryl, J.D. (1998) Uranium^{VI} sorption onto selected mineral surfaces. In: *Adsorption of Metals by Geomedia* (E.A. Jenne, editor), 99–130. Academic Press, San Diego, California, USA.

Paquette, J. & Reeder, R.J. (1995) Relationship between surface structure, growth mechanism, and trace element incorporation in calcite. *Geochimica et Cosmochimica Acta* **59**, 735–749.

Parkhurst, D.L. & Appelo, C.A. J. (1999) *User's guide to PHREEQC (Version 2). A computer program for speciation, batch-reaction, one-dimensional transport and inverse geochemical calculations.* US Geological Survey. Water-Resources Investigations Report **99-4259**.

Payne, T.E., Lumpkin, G.R. & Waite, T.D. (1998) Uranium^{VI} adsorption on model minerals. In: *Adsorption of Metals by Geomedia* (E.A. Jenne, editor), 75–97. Academic Press, San Diego, California, USA.

Peld, M., Tonsuaadu, K. & Bender, V. (2004) Sorption and desorption of Cd^{2+} and Zn^{2+} ions in apatite-aqueous systems. *Environmental Science & Technology*, **38**, 5626–5631.

Pérez-Garrido, C., Fernández-Díaz, L., Pina, C.M. & Prieto, M. (2007) In situ AFM observations of the interaction between calcite (10–14) and Cd-bearing aqueous solutions. *Surface Science*, **601**, 5499–5509.

Pérez-Garrido, C., Astilleros, J.M., Fernández-Díaz, L. & Prieto, M. (2009) In situ AFM study of the interaction between calcite {10–14} surface and supersaturated Mn^{2+}–CO_3^{2-} aqueous solutions. *Journal of Crystal Growth*, **311**, 4730–4739.

Petrović, M., Kaštelan-Macan, M. & Horvat, J.M. (1999) Interactive sorption of metal ions and humic acids onto mineral particles. *Water Air Soil Pollution*, **111**, 41–56.

Plant, J.A., Kinniburgh, D.G. Smedley, P.L., Fordyce, F.M. & Klinck, B.A. (2003) Arsenic and selenium. In: *Treatise on Geochemistry* (B. Sherwood Lollar, editor), **9**, 17–66. Elsevier-Pergamon, Oxford, UK.

Powell, B.A., Duff, M.C., Kaplan, D.J., Fjeld, R.A., Newville, M., Hunter, D.B., Bertsch, P.M., Coates, J.T., Eng, P., Rivers, M.L., Serkiz, S.M., Sutton, S.R., Triay, I.R. & Vaniman, D.T. (2006) Plutonium oxidation and subsequent reduction by Mn(IV) minerals in Yucca Mountain tuffs. *Environmental Science & Technol*ogy, **40**, 3508–3514.

Prieto, M. (2009) Thermodynamics of solid solution-aqueous solution systems. In: *Thermodynamics and Kinetics of Water-rock Interaction* (E.H. Oelkers & J. Schott, editors). Reviews in Mineralogy and Geochemistry, **70**, 47–85. Mineralogical Society of America, Washington, D.C. and the Geochemical Society, St. Louis, Missouri, USA.

Prieto, M. (2010) Thermodynamics of ion partitioning in solid solution-aqueous solution systems. In: *Ion Partitioning in Ambient-Temperature Aqueous Systems* (M. Prieto, editor), **10**, 1–42. EMU Notes in Mineralogy, the European Mineralogical Union and the Mineralogical Society of Great Britain & Ireland.

Prieto, M., Fernández-González, A., Putnis, A. & Fernández-Díaz, L. (1997) Nucleation, growth and zoning phenomena in crystallizing (Ba,Sr)CO_3, Ba(SO_4,CrO_4), (Ba,Sr)SO_4, and (Cd,Ca)CO_3 solid solutions from aqueous solutions. *Geochimica et Cosmochimica Acta*, **61**, 3383–3397.

Prieto, M., Fernández-González, A. & Martín-Díaz, R. (2002) Sorption of chromate ions diffusing through barite – hydrogel composites: Implications for the fate and transport of chromium in the environment. *Geochimica et Cosmochimica Acta*, **66**, 783–795.

Prieto, M., Cubillas, P. & Fernández-González, A. (2003) Uptake of dissolved Cd by biogenic and abiogenic aragonite: a comparison with sorption onto calcite. *Geochimica et Cosmochimica Acta*, **67**, 3859–3869.

Putnis, A. (2010) Effects of kinetics and mechanisms of crystal growth on ion-partitioning in solid solution–aqueous solution (SS-AS) systems. In: *Ion Partitioning in Ambient-Temperature Aqueous Systems* (M. Prieto, editor), **10**, 43–64. EMU Notes in Mineralogy, the European Mineralogical Union and the Mineralogical Society of Great Britain & Ireland.

Regenspurg, S. & Peiffer, S. (2005) Arsenate and chromate incorporation in schwertmannite. *Applied Geochemistry*, **20**, 1226–1239.

Reeder, R.J. (1996) Interaction of divalent cobalt, zinc, cadmium, and barium with the calcite surface during layer growth. *Geochimica et Cosmochimica Acta*, **60**, 1543–1552.

Reeder, R. & Grams, J.C. (1987) Sector zoning in calcite cement crystals: implications for trace element distributions in carbonates. *Geochimica et Cosmochimica Acta*, **51**, 187–194.

Rock, P.A., Casey, W.H., McBeath, M.K. & Walling, E.W. (1994) A new method for determining Gibbs energies of formation of metal-carbonate solid solutions: 1. The $Ca_xCd_{1-x}CO3$ (s) system at 298 K and 1 bar. *Geochimica et Cosmochimica Acta*, **58**, 4281–4291.

Rodríguez, J.D., Jiménez, A., Prieto, M., Torre, L. & García-Granda, S. (2008) Interaction of gypsum with As(V) – bearing aqueous solutions: Surface precipitation of guernerite, sainfeldite, and $Ca_2NaH(AsO_4)_2 \cdot 6H_2O$, a synthetic arsenate. *American Mineralogist*, **93**, 928–939.

Rodríguez-Blanco, J.D, Jiménez, A. & Prieto, M. (2007) Oriented overgrowth of pharmacolite ($CaHAsO_4 \cdot 2H_2O$) on gypsum ($CaSO_4 \cdot 2H_2O$). *Crystal Growth and Design*, **7**, 2756–2763.

Schindler, M. & Putnis, A. (2004) Crystal growth of schoepite on the (104) surface of calcite. *The Canadian Mineralogist*, **42**, 1667–1681.

Schindler, M., Hawthorne, F.C., Putnis, C. & Putnis, A. (2004) Growth of uranyl-hydroxy-hydrate and uranyl carbonate minerals on the (104) surface of calcite. *The Canadian Mineralogist*, **42**, 1683–1697.

Scott, T.B., Allen, G.C., Heard, P.J. & Randell, M.G. (2005) Reduction of U(VI) to U(IV) on the surface of magnetite. *Geochimica et Cosmochimica Acta*, **69**, 5639–5646.

Scott, T. B., Tort, O. R. & Allen, G. C. (2007) Aqueous uptake of uranium onto pyrite surfaces; reactivity of fresh versus weathered material. *Geochimica et Cosmochimica Acta*, **71/21**, 5044– 5053.

Shen, Y.F., Zerger, R.P., DeGuzman, R.N., Suib, S.L., McCurdy, L., Potter, D.I. & O'Young, C.L. (1993) Manganese oxide octahedral molecular sieves: preparation, characterization, and applications. *Science*, **260**, 511–515.

Sposito, G. (1986) Distinguishing adsorption from surface precipitation. In: *Geochemical Processes of Mineral Surfaces* (J. A. Davis & K. Hayes, editors). ACS Symposium Series, **323**, 217–228. American Chemical Society, Washington, D.C.

Sposito, G., Skipper, N.T., Sutton, R., Park, S.-H., Soper, A.K. & Greathouse, J.A. (1999) Surface geochemistry of the clay minerals. *Proceedings of the National Academy of Science of the United States of America*, **96**, 3358–3364.

Staudt, W.J., Reeder, R.J. & Schoonen, M.A.A. (1994) Surface structural controls on compositional zoning of SO_4^{2-} and SeO_4^{2-} in synthetic calcite crystals. *Geochimica et Cosmochimica Acta*, **58**, 2087–2098.

Stipp, S.L.S., Hochella, M.F., Parks, G.A. & Leckie, J.O. (1992) Cd^{2+} uptake by calcite, solid-state diffusion, and the formation of solid solutions: Interface processes observed with near-surface sensitive techniques (XPS, LEED, and AES). *Geochimica et Cosmochimica Acta*, **56**, 1941–1954.

Stipp, S.L.S, Parks, G.A., Nordstrom, D.K. & Leckie, J.O. (1993) Solubility-product constant and thermodynamic properties for synthetic otavite, $CdCO_3$(s), and aqueous association constants for the Cd (II)–CO_2–H_2O system. *Geochimica et Cosmochimica Acta*, **57**, 2699–2713.

Stipp, S.L.S., Lakshtanov, L.Z., Jensen, J.T. & Baker, J.A. (2003) Eu^{3+} uptake by calcite: Preliminary results from coprecipitation experiments and observations with surface sensitive techniques. *Journal of Contaminant Hydrology*, **61**, 33–43.

Stumm, W. (1992) *Chemistry of the Solid-Water Interface – Processes at the Mineral-Water and Particle-Water Interface in Natural Systems.* J. Wiley & Sons, New York, 448 pp.

Sturchio, N.C., Chiarello, R.P., Cheng, L., Lyman, P.F., Bedzyk, M.J., Qian, Y., You, H., Yee, D., Geissbuhler, P., Sorensen, L.B., Liang, Y. & Baer, D.R. (1997) Lead adsorption at the calcite-water interface: Synchroton X-ray standing wave and X-ray reflectivity studies. *Geochimica et Cosmochimica Acta*, **61**, 251–263.

Sunagawa, I. (2005) *Crystals: Growth, Morphology and Perfection.* Cambridge University Press, Cambridge, UK, 295 pp.

Tebo, B.M., Bargar, J.R., Clement, B.C., Dick, G.J., Murray, K.J., Parker, D., Verity, R. & Webb, S.M. (2004) Biogenic manganese oxides: Properties and mechanisms of formation. *Annual Review of Earth and Planetary Sciences*, **32**, 287–328.

Tesoriero, A.J. & Pankow, J.F. (1996) Solid solution partitioning of Sr^{2+}, Ba^{2+}, and Cd^{2+} to calcite. *Geochimica et Cosmochimica Acta*, **60**, 1053–1063.

Thompson, H.A., Parks, G.A. & Brown, G.E. (1998) Structure and composition of uranium^{VI} sorption complexes at the kaolinite-water interface. In: *Adsorption of Metals by Geomedia* (E.A. Jenne, editor), 349–370. Academic Press, San Diego, California, USA.

Thompson, H.A., Parks, G.A. & Brown, G.E. (1999) Ambient temperature synthesis, evolution, and characterization of cobalt-aluminium hydrotalcite-like solids. *Clays and Clay Minerals*, **47**, 425–438.

Traina, S.J. & Laperche, V. (1999) Contaminant bioavailability in soils, sediments, and aquatic environments. *Proceedings of the National Academy of Science of the United States of America*, **96**, 3365–3371.

Turner, S., Siegel, M.D. & Buseck, P.R. (1982) Structural features of todorokite intergrowths in manganese nodules. *Nature*, **296**, 841–842.

Vasconcelos, I.F., Haack, E.A., Maurice, P.A. & Bunker, B.A. (2008) EXAFS analysis of cadmium(II) adsorption to kaolinite. *Chemical Geology*, **249**, 236–250.

Vaughan, D.J. (2006) Arsenic. *Elements*, **2**, 71–75.

Vaughan, D. & Pattrick, R. editors (1995) *Mineral Surfaces*, The Mineralogical Society Series **5**, Chapman & Hall, London, 370 pp.

Wang, Y., Bryan, C., Xu, H. & Gao, H. (2003) Nanogeochemistry: geochemical reactions and mass transfers in naopores. *Geology*, **31(5)**, 387–390.

Webb, S.M., Fuller, C.C., Tebo, B.M. & Bargar, J.R. (2006) Determination of uranyl incorporation into biogenic Mn-oxides using X-ray absorption spectroscopy and scattering. *Environmental Science & Technology*, **40**, 771–777.

Wersin, P., Hochella, M.F., Persson, P., Redden, G., Leckie, J.O. & Harris, D.W. (1994) Interaction between aqueous uranium (VI) and sulfide minerals: spectroscopic evidence for sorption and reduction. *Geochimica et Cosmochimica Acta*, **58**, 2829–2843.

Zachara, J.M., Cowan, C.E. & Resch, C.T. (1991) Sorption of divalent metals on calcite. *Geochimica et Cosmochimica Acta*, **55**, 1549–1562.

Zavarin, M., Roberts, S.K., Hakem, N., Sawvel, A.M. & Kersting, A.B. (2005) Eu(III), Sm(III), Np(V), Pu(V), and Pu(IV) sorption to calcite. *Radiochimica Acta*, **93**, 93–102.

Zhu, C. (2002) Estimation of surface precipitation constants for sorption of divalent metals onto hydrous ferric oxide and calcite. *Chemical Geology*, **188**, 23–32.

EMU Notes in Mineralogy, Vol. 10 (2010), Chapter 9, 325–344

Water-CO_2-rock interaction during carbon sequestration

Eric H. Oelkers[1,*] and Sigurdur R. Gislason[2]

[1]*Geochemie et Biochemie Experimental, LMTG/CNRS UMR 5563, 14 ave Edouard Belin, 31400 Toulouse, France, e-mail: oelkers@lmtg.obs-mip.fr*
[2]*Institute of Earth Sciences, University of Iceland, Sturlugata 7, 101 Reykjavík, Iceland*
**Corresponding author*

Carbon capture and storage looks to be the most promising option to stem the rise in atmospheric CO_2 concentration and its consequences including global warming. Consideration of the various options available suggests that geological storage will probably dominate future carbon-storage efforts. The safest geological storage host is the formation of carbonate minerals. Carbonate mineral formation is favoured by the presence of reactive divalent-metal silicate minerals, in particular basalts and ultra-mafic rocks. In the absence of such rocks the most stable form of injected CO_2 is likely to be *via* solubility trapping, which involves its dissolution into formation waters. The quantity of CO_2 that can be stored *via* solubility depends on the composition of the formation waters and is greatly enhanced by increasing the *pH* and the formation of aqueous carbonate complexes.

1. Introduction

The past several decades have witnessed an unprecedented increase in the concentration of CO_2 in our atmosphere. The pre-industrial CO_2 concentration of the atmosphere is estimated to have been ~280 ppm by volume. This concentration had increased to roughly 355 ppm by 1960; and to 380 ppm at the beginning of this century. Global circulation models suggest that this increase in atmospheric CO_2 concentration is responsible for an increase in global temperature (Manabe & Stouffer, 1993; Johns *et al.*, 2003). Current and potential future global warming could have a large number of detrimental societal consequences including more intense precipitation events (Groisman *et al.*, 2005), rising sea levels in response to glacial melting (*e.g.* Manabe & Stouffer, 1994; Rignot, 1998), an increase in the number and extent of diseases that can infect humans (Patz *et al.*, 1996), and the extinction of terrestrial plants (Thomas *et al.*, 2004). A potential way to stem the increase of atmospheric CO_2 concentration due to human activity is the capture and storage of CO_2 through carbon sequestration (*cf.* Lackner, 2003; Oelkers & Schott, 2005; Broecker, 2008; Oelkers & Cole, 2008; Bénézeth *et al.*, 2009).

Another consequence of increasing atmospheric CO_2 concentration is ocean acidification. The *pH* of the ocean's surface water has already decreased by 0.1 units compared

DOI: 10.1180/EMU-notes.10.9

to pre-industrial values (Haugan & Drange 1996). Model calculations indicate that continued anthropogenic CO_2 emissions could lead to a *pH* decrease in the oceans by as much as 0.7 units (Calderia & Wickett, 2000; Orr *et al.*, 2005). This *pH* decrease would destabilize calcite and aragonite, the minerals that form the skeletons of many marine organisms (Andersson *et al.*, 2006). As a consequence, it has been suggested that calcifying organisms, including corals, coralline algae, molluscs and foraminifera, will have difficulties building their skeletons. This could have drastic implications for marine ecosystems, including disaster for some planktonic species, which would affect the entire food chain and vast expanses of coral reefs (Andersson *et al.*, 2006).

The recent increase in atmospheric CO_2 concentration can be attributed to human influence on the global carbon cycle. A simplified illustration of the global carbon cycle is shown in Figure 1 (see also Falkowski *et al.*, 2000). The atmosphere is a relatively small terrestrial carbon reservoir containing ~800 gigatons of carbon. A similar mass of carbon is estimated to be contained in vegetation, whereas that contained within soil is approximately twice that found in the atmosphere. The world's ocean is host to 39,000 gigatons of carbon, mostly as dissolved HCO_3^-. Carbonate rocks in the Earth's crust host ~60,000,000 gigatons of carbon and sedimentary rocks contain ~15,000,000 gigatons of organic carbon (Berner, 2004). The ocean and crustal sedimentary rocks therefore host 50 and 94,000 times more carbon than the Earth's atmosphere. Carbon

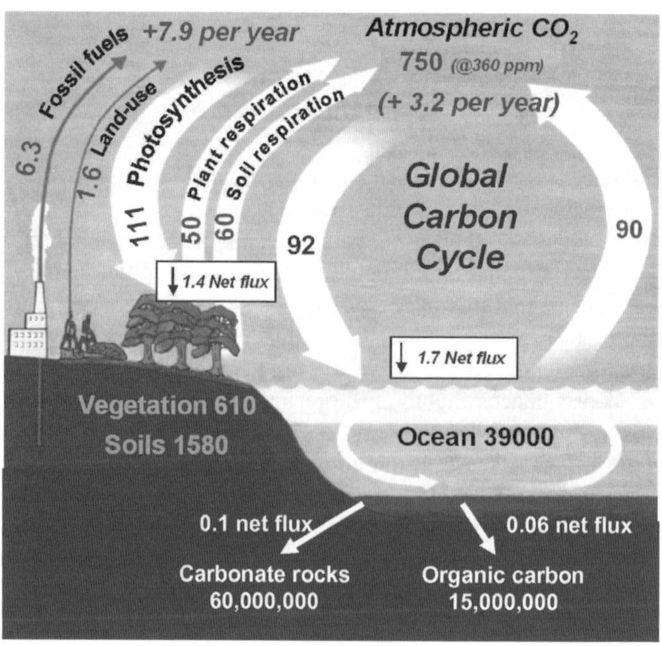

Fig. 1. Schematic diagram of part of the global carbon cycle. Fluxes (arrows) are reported in units of gigatons of carbon (Gt C) per year, whereas masses of the various reservoirs are presented in Gt C.

is passed from the atmosphere to vegetation, soils and the oceans at a remarkably fast rate. It is estimated that 92 gigatons of carbon or $>10\%$ of the atmosphere's CO_2 is dissolved into the oceans every year. This dissolution is close to balanced by the annual exsolution of 90 gigatons of carbon from ocean water back into the atmosphere, leaving a net flux of $\sim$1.7 gigatons of carbon from the atmosphere to the oceans each year. This net influx of carbon has the advantage of removing carbon from the atmosphere, but has the disadvantage of leading to ocean acidification. Similarly it is estimated that $\sim$111 gigatons of carbon are transferred from the atmosphere to plants and soils each year, but all but 1.4 gigatons are re-released to the atmosphere as plants and soils degrade. The residence time of carbon in each of these reservoirs is readily calculated from these fluxes. The average carbon-dioxide molecule resides in the atmosphere for 4 years, but carbon resides in vegetation/soils for 20 years, and the oceans for 430 years. Sedimentary rocks are by far a larger carbon reservoir than the atmosphere, ocean, soil or vegetation. The flux of carbon into carbonate rocks and organic matter in sedimentary rocks is, however, relatively slow, only 0.1 gigatons is transferred to carbonate rocks per year (Gaillardet *et al.*, 1999) and 0.06 gigatons into organic matter in sedimentary rocks (Berner, 2004). A small fraction of the organic matter in sedimentary rocks will eventually form fossil fuel. Some carbon does get returned to the atmosphere and oceans through volcanic activity and chemical weathering. The mass of carbon transferred to and from rocks is balanced on a geological time scale (Berner, 2004); the residence time of carbon in organic matter in sedimentary rock and of inorganic carbon in carbonate rocks is of the order of 250 and 600 million years, respectively.

The effect of human activity on the CO_2 concentration of the atmosphere stems from the burning of hydrocarbons and to a lesser degree from the manufacture of cement and land use (*e.g.* deforestation). It is estimated that human activity emitted 9.9 gigatons of carbon to the atmosphere in 2008 (Le Quéré *et al.*, 2009). In contrast to the carbon flux to the atmosphere from oceans, vegetation, and soils, human carbon emissions are not coupled directly to a sink. It is hoped that carbon capture and storage, sometimes referred to as CCS will provide such a compensating sink.

In the ideal case, carbon sequestration could again balance the carbon cycle. This would require the sequestration of 9.9 gigatons of carbon annually. 9.9 gigatons of carbon as carbon dioxide weighs 36.3 gigatons. As a gas at 1 bar and 0°C, this mass of carbon dioxide has the volume of 18,600 km^3. As a supercritical fluid at 150 bars of pressure and a temperature of 40°C this mass of carbon-dioxide has a volume of 50 km^3. The successful storage of such volumes requires geological-scale efforts.

2. Carbon sequestration options

Many of the options being explored for carbon storage come directly from the global carbon cycle. The major carbon reservoirs of the Earth being considered for carbon storage include vegetation, soils, oceans, and the crust (*e.g.* Lai, 2008, 2009). A short review of these storage options is listed below.

2.1. Biosequestration (storage through forestation)

Forests cover more than one third of the world's land area. Trees and other forest plants fix CO_2 through photosynthesis. Wood contains roughly 50 wt.% carbon (Lamlom & Savidge, 2003), and a typical tropical tree has a mass of 500–2000 kg (Hiratsuka *et al.*, 2005). It follows that carbon sequestration through forestation is as simple as planting trees. Forests, however, also can release CO_2 back to the atmosphere as dead trees decompose and when trees are burned (Dixon *et al.*, 1994). Forests need to be managed to ensure that the bulk of the carbon created by forestation remains fixed in the solid state. Critical to the long-term storage includes management (1) that ensures that the forests grow continuously, and (2) that ensures carbon harvested from forests is not returned to the atmosphere. The potential for forests to sequester CO_2, however, is limited. Nilsson & Schopfhauser (1995) estimated that only 345 million ha are available worldwide for new forestation; a forestation program of this scale would only fix 1.5 Gt C/year, which is 15% of the current anthropogenic carbon input to the atmosphere. Moreover, even attaining this limited land area for forestation is challenging due to pressure to use land for other purposes, in particular for agriculture and development.

2.2. Soil storage

The addition of carbon to soils as part of a carbon-storage program offers a big advantage as increasing carbon is soil could substantially increase crop yields and thus food security (Lai, 2004). Solid CO_2 could be added to soils in the form of biochar. Biochar is charcoal created by the partial burning of biomass. Once created, biochar could be added directly to soils. At least some biochar can have long lifetimes in soil; evidence from soils from the Amazon show that significant concentrations of biochar can persist over thousand-year timeframes (Lehmann *et al.*, 2003). Though no doubt useful, it is unlikely that the addition of biochar to soils can itself provide a solution to global carbon storage needs due to the required scale. The creation of sufficient biochar to store completely the 9.9 gigatons of anthropogenic carbon would require harvesting roughly 10% of the world's forests per year.

2.3. Ocean storage

Ocean storage involves the injection of captured CO_2 into the ocean, usually at depths >1000 m, where it would be isolated from the atmosphere (Marchetti, 1977; Herzog *et al.*, 1991; Adams & Caldeira, 2008). At depths greater than ~ 3000 m, CO_2 is denser than seawater; CO_2 injected at greater depths would sink. Injected CO_2 would subsequently dissolve into the ocean and become part of the global carbon cycle. There are, however, two major disadvantages of ocean storage. First, as mentioned above, the dissolution of CO_2 into seawater leads to ocean acidification. Estimates suggest that parts of the ocean could experience a drop in pH of at least 0.7 units during ocean sequestration (Calderia & Wickett, 2000). The second disadvantage of ocean storage is the relatively short residence time of carbon in the ocean. The residence time of carbon in the ocean is ~ 430 years. The vast majority of CO_2 injected into the oceans would therefore be expected to be released back into the atmosphere within 1000 years.

2.4. Geological storage

Due to the limits in size and short residence times of the other storage options it seems likely that a large percentage, if not the vast majority, of future carbon sequestration efforts will be aimed at geological storage. Geological storage relies on the injection of CO_2 into porous rock formations (Holloway, 2001; Friedmann, 2007; Benson & Cole, 2008). CO_2 storage formations include sedimentary basins, depleted oil reservoirs and non-economic coal beds (Bruant *et al.*, 2002; White *et al.*, 2003; Kharaka *et al.*, 2006). CO_2 has been injected into declining oil fields for >40 years, to increase oil recovery (Blunt *et al.*, 1993; Steeneveldt *et al.*, 2006). Typically, 10–15% additional recovery of the original oil in place is possible. Several industrial-scale geological CO_2 storage programs are already underway, including the Norwegian Sleipner project in the North Sea (Korbøl & Kaddour, 1995) and the Weyburn project in Canada (Emberley *et al.*, 2004).

The major challenge of geological carbon storage is keeping the injected carbon in the subsurface. The simplest way to inject CO_2 is as a separate phase. As a gas at room temperature and pressure CO_2 is very buoyant in H_2O. The critical point of CO_2 is at 73.8 bars and 31.1°C. Above the critical pressure and temperature, CO_2 will be a super-critical fluid. Although more dense as a supercritical fluid, CO_2 is nevertheless less dense than H_2O at most pressure and temperature conditions found in the crust. The relative density of CO_2 *vs.* water as a function of pressure and temperature is shown in Figure 2. Supercritical CO_2 is ~20% less dense than H_2O along a typical geothermal gradient of 25°C/km. Supercritical CO_2 will, therefore tend to migrate upwards, potentially escaping to the atmosphere. Cap-rocks are essential to ensure that this CO_2

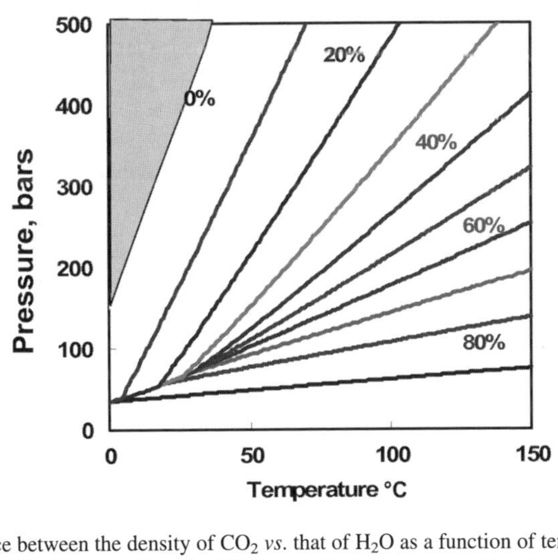

Fig. 2. The difference between the density of CO_2 *vs.* that of H_2O as a function of temperature and pressure. The density of CO_2 and H_2O are equal at the 0% line. At high pressures and low temperatures, CO_2 is more dense than H_2O, and will tend to sink. In contrast, along a typical 25°C/km geothermal gradient at hydrostatic pressure, supercritical CO_2 is ~20–25% less dense than H_2O and it will tend to rise to the surface.

remains in the subsurface. CO_2 in this form is thus viewed as being relatively insecure. The long-term stability and security of geological storage depends on the long-term integrity of this cap-rock and the degree to which reactions between the injected CO_2 and its host rocks and fluid are advantageous or detrimental to carbon storage.

3. The fate of CO_2 injected into geological formations: H_2O-CO_2-rock interaction

After injection, CO_2 can go through a series of transformation, as illustrated in Figure 3. With time, some of the injected CO_2 will become trapped in capillaries and some or all of the injected CO_2 will dissolve into the pore waters. The dissolution rate of carbon-dioxide into water is rapid, as proven by the rapid drop in *pH* of pure water from 7 to ~5.6 when left in contact with the atmosphere. As the dissolution rate of CO_2 into H_2O is rapid, the degree by which it dissolves is limited by the availability of water in the system, how much water comes in contact with the injected CO_2, and the solubility and diffusivity of CO_2 in the aqueous phase. The solubility of CO_2 at any given pressure and temperature is related to its aqueous speciation. The dissolution reaction of CO_2 by water can be described by:

$$CO_2(g) + H_2O = H_2CO_3 \tag{1}$$

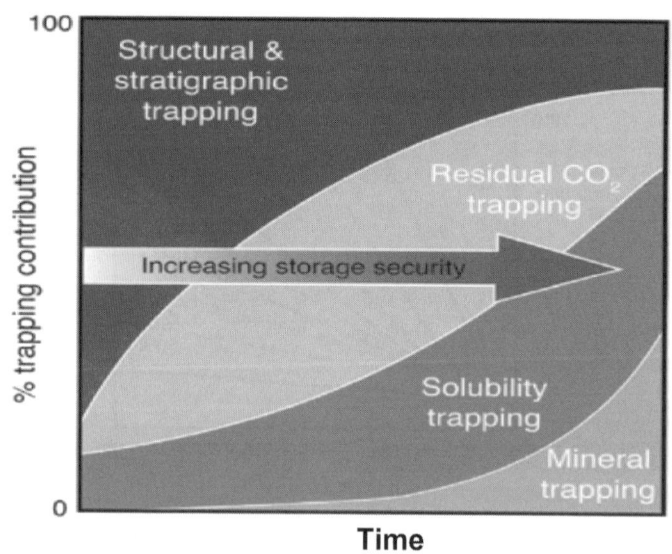

Fig. 3. Schematic illustration of transformations occurring in CO_2 following its injection into a geological formation as part of carbon-storage efforts (after Metz *et al.*, 2005).

where H_2CO_3 is aqueous carbonic acid. This acid can dissociate according to

$$H_2CO_3 = HCO_3^- + H^+ \quad (2)$$

and

$$HCO_3^- = CO_3^{2-} + H^+ \quad (3)$$

The distribution of carbonate species in the CO_2-H_2O system as a function of *pH* is shown in Figure 4. The neutral H_2CO_3 species dominates at $pH < 6$, the HCO_3^- species dominates at $6 < pH < 10$, and the CO_3^{2-} species dominates at $pH > 10$. Due to the formation of the HCO_3^- and CO_3^{2-} species, the solubility of CO_2 increases dramatically with increasing *pH*. The solubility of CO_2 in water as a function of *pH* is shown in Figure 5. As can be seen, the solubility of CO_2 increases by a factor of 15 by increasing *pH* from 4 to 8 and by a factor of more than 200 by increasing *pH* from 4 to 9. The dissociation of carbonic acid in the pore fluids, however, releases protons and acidifies the pore solution. The *pH* of H_2O and sea water in equilibrium with CO_2 at 25°C is shown as a function of CO_2 partial pressure in Figure 6. The *pH* of saturated H_2O and sea water at a CO_2 partial pressure of 50 bars, is 3.07 and 3.36, respectively. As seen in Figure 5, this acid *pH* tends to limit the solubility of CO_2 compared to what it would be in neutral to basic conditions.

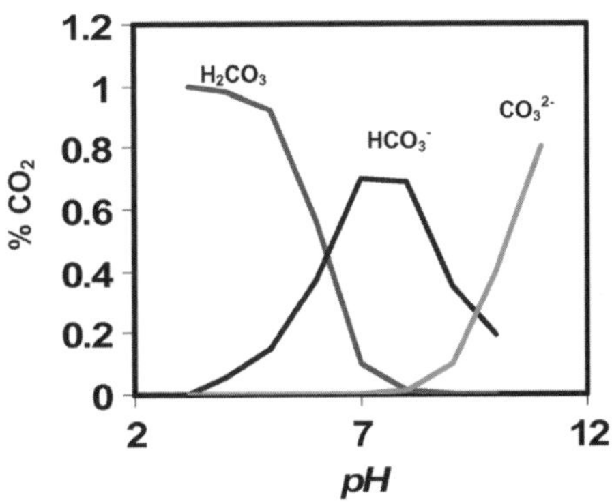

Fig. 4. Distribution of aqueous CO_2 as a function of *pH* at 25°C and 1 bar as calculated using PHREEQC (Parkhurst & Appelo, 1998).

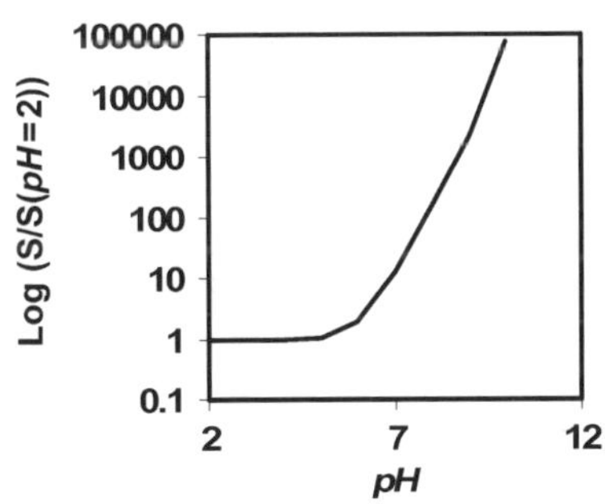

Fig. 5. Relative solubility of CO_2 in water at 25°C as a function of *pH vs.* its value at $pH = 2$, as calculated using PHREEQC (Parkhurst & Appelo, 1998).

A further consequence of the low *pH* of CO_2-rich waters is that acidic pore fluids tend to be reactive; both the solubilities and dissolution rates of a large number of minerals increase dramatically with increasing *pH* at acid conditions (*cf.* Oelkers *et al.*, 2009; Schott *et al.*, 2009). The degree to which reactions among CO_2, the pore fluid and the host solids is advantageous or detrimental to carbon storage efforts depends on the mineralogy, geometry, and hydrology of the system. Gaus (2010) outlined a number of locations where significant CO_2-pore-fluid-rock interactions can occur:

(1) Reactions with well materials at the injection point

Wellbore cements are commonly Ca-rich. Portlandite cement, for example, contains 45–75% tri-calcium silicate, $(CaO)_3SiO_2$. This material can readily react with CO_2 to form calcite ($CaCO_3$) in accordance with:

$$(CaO)_3SiO_2 + 3CO_2 = 3CaCO_3 + SiO_2 \qquad (4)$$

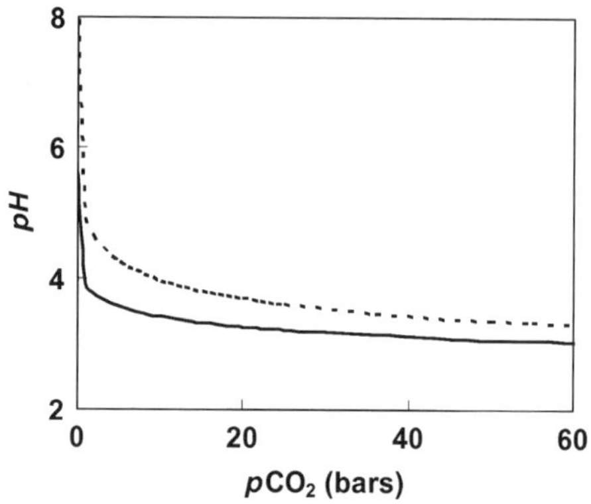

Fig. 6. pH of carbon-dioxide-saturated H_2O and seawater as a function of the partial pressure of CO_2 as calculated using PHREEQC (Parkhurst & Appelo, 1998). The solid and dotted curves correspond to results for H_2O and seawater, respectively.

This reaction can effect greatly the permeability and porosity near the injection point as this reaction is strongly volume positive (Wigand *et al.*, 2009). The precipitation of calcite near the injection point could block flowpaths leading to pressure build-up. Regnault *et al.* (2009) showed that at temperatures from 80 to 120°C, all portlandite cement was rapidly converted into calcite in the presence of liquid water, but this reaction could be arrested by the precipitation of a passivating layer of calcite crystals on the portlandite surface.

(2) Reactions with caprocks

Reactions among minerals in caprocks and carbon-dioxide charged aqueous fluids could facilitate the escape of CO_2 upwards through mineral dissolution or lower its permeability through mineral precipitation (Gaus *et al.*, 2005). The precipitation of carbonate minerals has been observed to block potential CO_2 leakage pathways in natural analogues by May (2005). The degree to which reactions could occur will depend strongly on the quantity of fluid that seeps into the cap-rock, which itself depends on its porosity and the existence of flow-paths into the cap-rock including the presence of pre existing faults or those formed by the injection process. The reactions that could occur thus need to be assessed case by case.

(3) Reactions with porous reservoir rocks

A large number of chemical reactions are expected to occur among CO_2-H_2O and host formation rocks. These reactions will be explored in more detail below.

If CO_2 is injected into carbonate-rich rocks such as limestone or calcite-cemented sedimentary rocks, carbonate minerals such as calcite, dolomite, and siderite could rapidly dissolve in accord with:

$$\underset{\text{Carbonate Minerals}}{\mathbf{(Fe, Ca, Mg)CO_3}} + H^+ \rightarrow (Fe, Ca, Mg)^{2+} + HCO_3^- \tag{5}$$

This reaction releases additional carbonate into solution. Carbonate dissolution reactions tend to be very rapid in acidic solutions; many might recall that a common field test for the presence of calcite in a rock is to drop dilute HCl solution onto the rock and look for the liberation of CO_2 bubbles. Because carbonate minerals dissolve orders of magnitude more quickly than silicate minerals (compare rates reported by Chou *et al.*, 1989; Alkattan *et al.*, 1998; Oelkers & Gislason, 2001; Schott *et al.*, 2009), the dissolution of carbonate minerals may dominate the neutralization of acidic CO_2-rich fluids even in rocks that have contain relatively little carbonate. For example, the neutralization of CO_2-saturated water during a field injection test in the Palisades sill was dominated by carbonate dissolution despite the fact that this rock formation, a dolerite, is dominated by silicates (Assayag *et al.*, 2009). Substantial carbonate dissolution was observed in the time frame of several weeks. The degree to which carbonate released into solution by reaction in equation 5 by carbonic acid is ever re-precipitated into solid phases will depend on the specific mineralogy of the system.

Protons created by the dissolution of CO_2 will also promote the dissolution of silicate minerals. For example, the dissolution of olivine, pyroxene, and Ca-anorthite, by these protons can be described by

$$\underset{\text{Olivine}}{\mathbf{(Mg, Fe)_2SiO_4}} + 4H^+ \rightarrow 2(Mg^{2+}, Fe^{2+}) + 2H_2O + SiO_{2(aq)} \tag{6}$$

$$\underset{\text{Pyroxene}}{\mathbf{(Mg, Fe, Ca)SiO_3}} + 2H^+ \rightarrow (Mg^{2+}, Fe^{2+}, Ca^{2+}) + H_2O + SiO_{2(aq)} \tag{7}$$

and

$$\underset{\text{Ca-plagioclase}}{\mathbf{CaAl_2Si_2O_8}} + 8H^+ \rightarrow Ca^{2+} + 2Al^{3+} + 2SiO_{2(aq)} + 4H_2O \tag{8}$$

These silicate dissolution reactions have two important consequences. First, they consume protons and thus tend to neutralize the acid formed by the original dissolution of carbon dioxide. If this silicate dissolution neutralizes *pH*, the HCO_3^- and CO_3^{2-} species will be stabilized relative to the neutral H_2CO_3 species (see Figure 4) and the solubility of CO_2 in water will increase. It follows that the neutralization of carbon-dioxide charged pore fluids by the dissolution of silicate rocks will increase the efficiency of solubility trapping as well as stabilize the dissolved CO_2 present in the aqueous phase.

A second consequence of the dissolution of silicate minerals is that they can release divalent metals into solution. These metals could react with dissolved carbonate to precipitate carbonate minerals by the reverse of the reaction in equation 5

$$(Fe, Ca, Mg)^{2+} + HCO_3^- \rightarrow \underset{\text{Carbonate Minerals}}{\mathbf{(Fe, Ca, Mg)CO_3}} + H^+ \qquad (9)$$

Carbonate minerals that could form from divalent cations include the common carbonates calcite, dolomite, magnesite and siderite. The formation of carbonate minerals by this process is commonly termed carbonation. Carbonate minerals are believed to provide a long-term safe storage host (Lackner *et al.*, 1995). The coupling of silicate-mineral dissolution to carbonate-mineral precipitation in accord with reactions 6–9 is the process by which atmospheric carbon is converted into rocks during weathering processes (*e.g.* Gaillardet *et al.*, 1999; Gislason *et al.*, 2009). The degree to which a particular carbonate mineral does actually form depends on the physical conditions of the system. By combining equations 6–9, overall mineral carbonation reactions can be generated. The Gibbs free energy of a number of representative mineral-carbonation reactions is shown as a function of temperature in Figure 7; a larger number of such figures can be found in Marini (2007). When the Gibbs free energy of reaction is negative, the reaction is favoured; the reaction is not favoured when the Gibbs free energy is positive. Mineral carbonation is generally favoured thermodynamically at temperatures below 200°C, for a CO_2 partial pressure of 1 bar. An exception is for the formation of the mineral dawsonite, $NaAlCO_3(OH)_2$, which is less stable than the divalent metal carbonates; the formation of this mineral is not favoured at temperatures above ~40°C at a CO_2 partial pressure of 1 bar. Increasing CO_2 partial pressure increases the temperature range over which mineral carbonation reactions are favoured. For example, the carbonation of anorthite is favoured at temperatures up to nearly 350°C at a CO_2 partial pressure of 200 bars. The fact that a reaction is thermodynamically favoured, however, does not guarantee that the reaction will proceed. Products other than carbonate formation could be favoured (*e.g.* clay-mineral precipitation could consume dissolved Mg or Fe). In addition, the precipitation of carbonate-product minerals could be kinetically inhibited. Whereas calcite can readily precipitate at low temperatures, magnesite-precipitation kinetics are sluggish at temperatures of <80°C and may not precipitate from supersaturated conditions at low temperatures (Arvidson & Mackenzie, 1999; Pokrovsky & Schott, 1999; Saldi *et al.*, 2009). As such, aqueous carbonate released by the dissolution of dolomite by CO_2-charged waters may not be completely fixed by a subsequent precipitation of a magnesium carbonate mineral. The degree to which injected CO_2 will be fixed in the solid phase *via* carbonation reactions is related to the availability of silicate minerals containing those divalent metals that form stable carbonates. Unfortunately a large number of porous rocks, notably most sedimentary rocks, contain few divalent metal-cation-bearing silicate minerals. The reason for this is that natural weathering is extremely efficient at converting silicates into carbonates on a geological time scale. Most divalent cations that can readily be converted into carbonate minerals are converted during weathering and the subsequent formation of sedimentary rocks (see Berner & Berner, 1996). For example, little to no Ca is hosted by silicate rocks in sedimentary rocks (Garrels & Mackenzie, 1971). As such, the major metals released by silicate dissolution in sedimentary basins (other than Si)

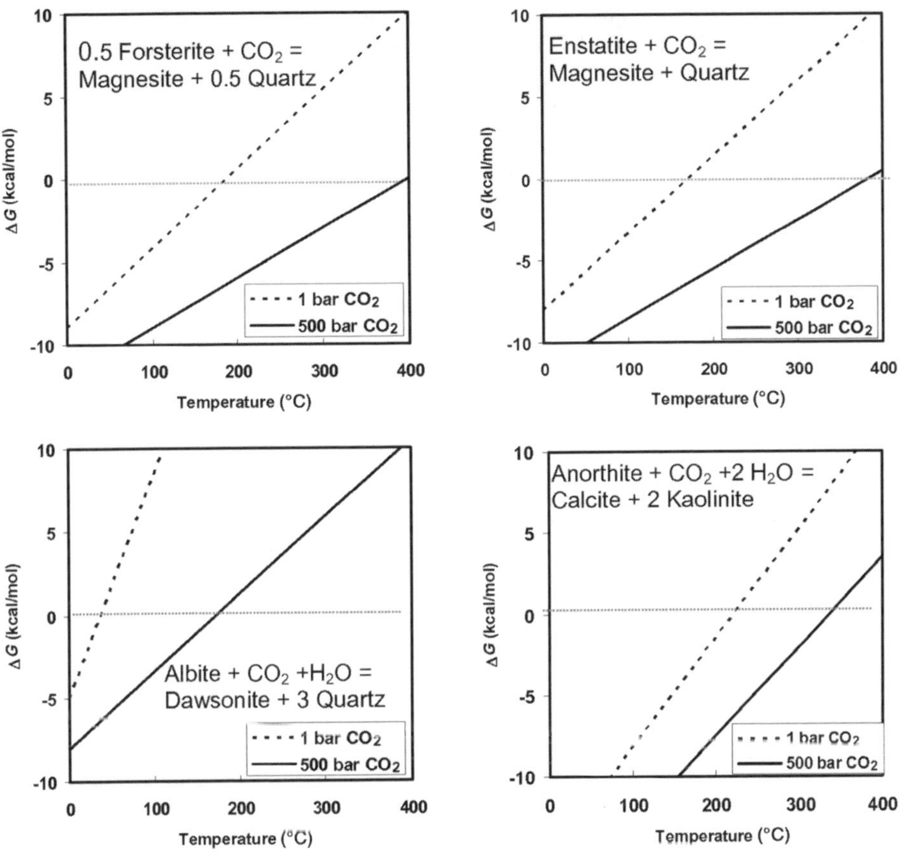

Fig. 7. Molar Gibbs free energy change of the indicated carbonation reactions (after Marini, 2007). Mineral carbonation is only favoured if the Gibbs free energy is negative.

include Al and monovalent cations such as Na and K. A number of studies have argued that in such cases dissolved carbon could be fixed through the precipitation of a rare Na–Al bearing carbonate mineral called dawsonite (*e.g.* Xu *et al.*, 2005). Dawsonite has the formula $NaAl(CO_3)(OH)_2$. Laboratory experiments and modelling, however, indicate that dawsonite is probably a short-lived, ephemeral phase in such systems (Hellevang *et al.*, 2005; Bénézeth *et al.*, 2007). As such, although sedimentary basins could provide pore space and a cap-rock favourable for the storage of CO_2 in the form of a separate phase or dissolved in water, it seems unlikely that substantial dissolved carbonates will be converted into minerals if injected into most sedimentary basins. This conclusion is in agreement with that of Gilfillan *et al.* (2009) who concluded, based on noble gas and carbon isotope tracers, that dissolution of CO_2 into formation water was the sole major carbon sink in seven gas fields with siliciclastic- or carbonate-dominated lithologies.

An alternative injection host, which could provide metals favouring the formation of stable carbonate minerals, is divalent-metal-rich igneous or metamorphic rocks. The degree to which these rocks favour the rapid formation of carbonate minerals depends on their composition. The Ca + Mg release rates from crystalline and glassy silicate rocks are shown as function of the Si:O ratio of the rock in Figure 8. Si-poor rocks (*e.g.* basalts and ultramafic rocks) are found to release Ca + Mg 2–3 orders of magnitude more quickly than Si-rich rocks (granites and rhyloites). As such, a number of studies have argued that porous basalts and ultramafic rocks may be the best target reservoir rock for long-term CO_2 storage (McGrail *et al.*, 2006; Goldberg *et al.*, 2008; Kelemen & Matter, 2008; Oelkers *et al.*, 2008).

A significant amount of other evidence has demonstrated the potential of CO_2 sequestration in mafic and ultramafic rocks. First, despite the fact that $<10\%$ of the Earth's continental surface is covered by basalt, it takes up $\sim 33\%$ of all the CO_2 consumed in during natural weathering of Earth's surface silicates (Dessert *et al.*, 2003). Second, several well documented examples of carbonation of basaltic rocks are known, such as through CO_2 metasomatism in a basalt-hosted petroleum reservoir (Roger *et al.*, 2006), hydrothermal alteration (Arnórsson, 1989; Gudmundsson & Arnórsson, 2002; Neuhoff *et al.*, 1999), and surface weathering of basalts (Gislason & Eugster, 1987; Gislason *et al.*, 1996, 2006; Stefánsson & Gislason, 2001).

The rate and extent to which injected carbon dioxide can be fixed within carbonates can be enhanced by increasing the dissolution rates of metal-bearing silicate minerals. These dissolution rates can be enhanced in several ways including increasing the mineral-fluid interfacial surface area, avoiding the formation of pervasive mineral coatings of secondary minerals on the surfaces of dissolving silicates (Cubillas *et al.*, 2005; Stockman *et al.*, 2008), increasing the temperature (Gislason &

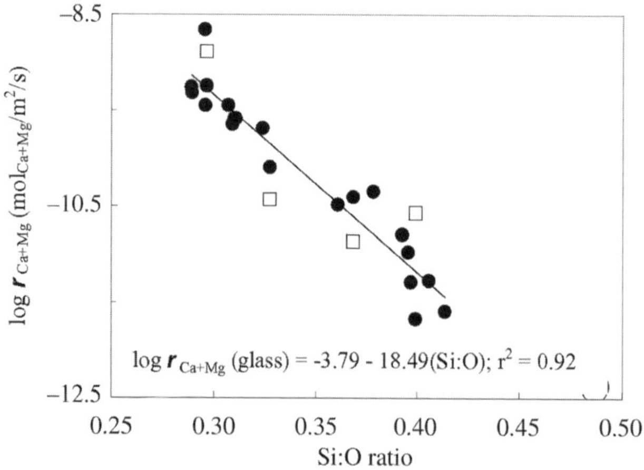

Fig. 8. Far-from-equilibrium elemental release rates of Ca + Mg from natural glasses and rocks at 25°C and *pH* 4 as a function of their Si:O ratio. The trendlines represent least-squares fits of the elemental release rates of natural glasses and rocks (after Wolff-Boenisch *et al.*, 2006).

Table 1. Percent volume change of some representative mineral carbonation reactions. Mineral volumes required for this calculation were taken from Helgeson *et al.* (1978).

Mineral carbonation reaction	**Percent volume change**
0.5 Forsterite + CO_2 = Magnesite + 0.5 Quartz	79.7
Enstatite + CO_2 = Magnesite + Quartz	62.1
Wollastonite + CO_2 = Calcite + Quartz	49.3
Anorthite + CO_2 +2 H_2O = Calcite + 2 Kaolinite	35.4
Albite + CO_2 +H_2O = Dawsonite + 3 Quartz	26.2
1/3 Chryostile + CO_2 = Magnesite + 2/3 Quartz + 2/3 H_2O	2.1
Laumonite + CO_2 = Calcite + Kaolinite + 2 Quartz + 2 H_2O	−12.3

Oelkers, 2003), and by alerting the pore-fluid composition (*cf.* Wolff-Boenisch *et al.*, 2004).

An important limit to the extent of carbonation in any subsurface system, including basaltic and ultramafic rocks, is that the volumes of mineral carbonation reactions tend to be large and positive. The percent change in volume of the reactants and products of selected minerals during carbonation is shown in Table 1. The carbonation of forsterite and enstatite to form magnesite and quartz results in an increase in the volume of solids by >60%. The carbonation of anorthite and albite results in a volume increase of >25%. The carbonation of the serpentine mineral chryostile results in a volume increase of only 2%. In only a few special cases (*e.g.* the carbonation of zeolites) does the volume of mineral carbonation reactions decrease. The consequences of the increase in volume during mineral carbonation include the clogging of pores and fractures, which could hinder further CO_2 injection, or other unwanted processes such as fracturing or changes in ground elevation (Marini, 2007).

Insight into the rates and chemical reactions that occur during geological carbon storage can be obtained through geochemical modelling (*e.g.* Xu *et al.*, 2004, 2005; Knauss *et al.*, 2005; Zerai *et al.*, 2006; Gysi & Stefánsson, 2008). An overview of geochemical modelling applied to characterizing CO_2-H_2O-rock interaction is provided by Marini (2007). In this overview, Marini (2007) reviews the kinetic-rate data required for making quantitative geochemical estimates and also presents the results of a large number of calculations illustrating the fate and consequences of CO_2-H_2O-rock interaction in a variety of host rocks. The results of one such reaction are shown in Figure 9. The results in this figure correspond to the predicted fate of CO_2 injection into a serpentinite at a temperature of 60°C and a partial CO_2 pressure of 100 bars. The rock is assumed to contain the minerals chrysotile, magnetite and kaolinite. This example illustrates many of the major processes expected to occur as a result of CO_2-H_2O-rock interaction. The dissolution of the host-rock formation starts immediately upon injection, and continues throughout the 1000 year simulation. After some time the formation fluid becomes saturated with secondary minerals which are predicted to form as they become saturated; precipitating minerals are assumed to be in local equilibrium with their co-existing pore fluids. Chalcedony, an oxide of silicon, is predicted to begin to precipitate in <1 day following the injection, and the carbonate minerals magnesite and dolomite are predicted to start precipitating within a few days. The iron

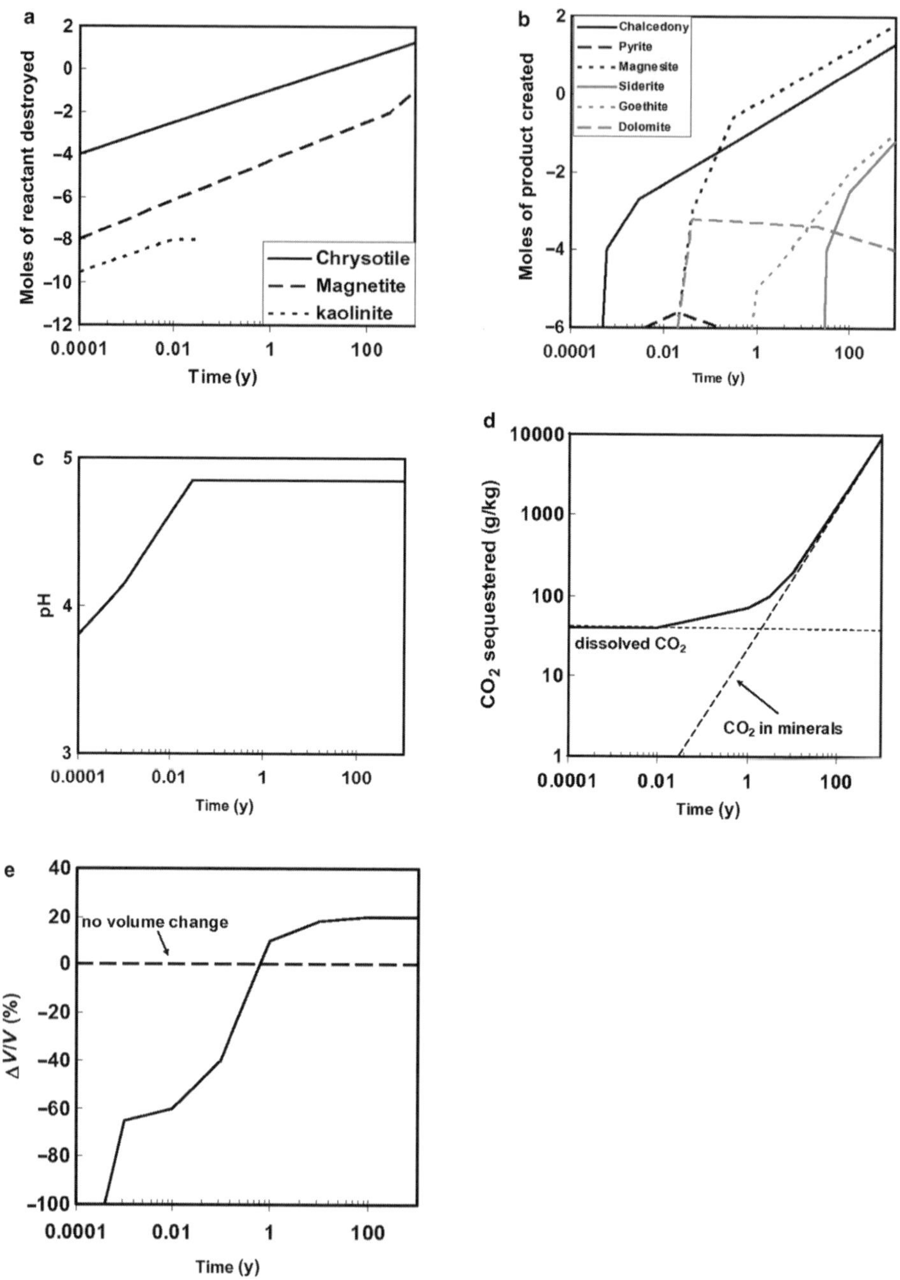

Fig. 9. Predicted chemical and mineralogical evolution of a water saturated sepentinite host rock during the first 1000 years after its continuous exposure to CO_2 at a pressure of 100 bars at a temperature of 60 °C (after Marini, 2007). Results have been normalized to one kg H_2O and assume excess of mineral reactants: (***a***) moles of minerals dissolved, (***b***) moles of secondary minerals precipitated, (***c***) aqueous solution *pH*, (***d***) grams of CO_2 sequestered per kg of water assuming it is constantly saturated with 100 bars CO_2 and in contact with fresh rock, and (***e***) net volume change normalized to the volume of minerals dissolved.

hydroxide goethite is estimated to begin to precipitate after 1 year. This secondary mineral consumes Fe without forming stable carbonates, limiting somewhat the efficiency of the carbonation process. The combination of dissolution and precipitation reactions leads to an increase in aqueous solution *pH* from 3.8 to 4.8 with time. Carbon storage is initially dominated by CO_2 dissolution into the fluid phase, but becomes dominated by carbonate mineral precipitation with time. Eventually, as secondary carbonate minerals form, the net result of the dissolution and precipitation reactions leads to an increase in solid phase volume, which could lead to clogging of pores and decreasing permeability.

The application of geochemical models such as those described above have significant uncertainties for a number of reasons including (1) a poor understanding of the reactive surface area of minerals, which makes calculation of mineral dissolution rates highly uncertain, and (2) a poor understanding of the kinetics of mineral precipitation reactions. This first uncertainty affects greatly the rate at which the calculated reaction path progresses with time. The second uncertainty forces geochemical models to make assumptions concerning mineral precipitation rates. Common assumptions include (1) local equilibrium of secondary phases, or (2) that mineral-precipitation rates can be calculated using dissolution-rate equations. Experimental observations show that both of these assumptions are generally incorrect, and many critical minerals will not precipitate at low temperatures including magnesite, dawsonite, and quartz. Such geochemical models should be considered as works in progress, particularly in terms of predicting the conditions and extent of minerals precipitation.

Another current limitation of geochemical models stems from the current state of their geochemical databases. Although these databases can calculate the solubility of major minerals with reasonable accuracy, thermodynamic data for most minor elements and adsorption/desorption reactions are lacking (Oelkers *et al.*, 2009). This makes it difficult to estimate the degree to which toxic metals might be released due to the interaction between acidic CO_2-charged water with formation rocks (*cf.* Flaathen *et al.*, 2009). Similarly, these model calculations typically do not include provision for the potential toxic organic compounds which could be mobilized by the interaction between CO_2-charged formation waters and residual oil if injecting into former petroleum reservoirs (*cf.* Kharaka *et al.*, 2009).

4. Conclusions and perspectives

Consideration of the various options available suggests that geological storage will probably dominate future carbon-storage efforts. The injection of CO_2 into porous formations in the subsurface and its subsequent dissolution into pore fluids will create reactive acidic fluids. These fluids will react with the formation rock and perhaps its adjacent cap-rock. The first step of this interaction will involve dissolution, initially of reactive phases such as carbonate minerals and then of silicates. These dissolution reactions will increase the stability of aqueous CO_2, enhancing its solubility. These dissolution reactions could also reduce cap-rock integrity. The degree to which secondary minerals could eventually precipitate in this system and the extent by which such minerals could serve to store carbon through carbonation reactions depends heavily on the composition of the host-rock formation. The lack of divalent-metal-bearing silicate minerals makes it unlikely that most sedimentary rocks will promote mineral carbonation. The rock formations most favourable for *in situ* mineral carbonation are porous basalts and ultramafic rocks. Although such rocks may provide the most secure long-term carbon storage option, it seems likely that much CO_2 will be injected into the closest porous and permeable rock formation to the point source (*e.g.* power plant) because (1) installing pipelines for CO_2 transport is costly, and (2) it may not be legal in many countries to inject CO_2 into the subsurface where it might migrate below land not owned by the owner of the CO_2. As such, many future injection sites may not be ideal. A critical factor going forward will be the monitoring of CO_2 after its injection into the subsurface. Although seismic methods can track CO_2 when it is present as a separate phase, there is a great need for geochemical methods to monitor carbon after it dissolves into the fluid phase. Such methods will be required to verify the safe and long-term storage of CO_2 especially in environments where mineral carbonation will be limited.

Acknowledgements

The authors thank all the graduate students that have worked with them on carbon storage issues including Giuseppe Saldi, Allison Stephenson, Quentin Gautier, Helgi Alfredsson, Edda Sif Aradóttir, Therese Flaathen, Teresa Roncal-Herrero, Snorri Gudbrandsson, Alexander Gysi, Mahnaz Rezvani Khalilabad, and Gabrielle J. Stockmann, and Iwona M. Galeczka. They also thank the colleagues and friends that have helped them to refine their understanding of carbon storage, in particular Domenik Wolff-Boenisch, Wallace S. Broecker, Luigi Marini, Hólmfridur Sigurdardóttir, Juerg M. Matter, Andri Stefansson, Bergur Sigfússon, Martin Stute, Susan Stipp, Kevin Burton, Jacques Schott, Pascale Bénézeth, Oleg Pokrovsky and Eydís S. Eiriksdottir.

References

Adams, E.E. & Caldeira, K. (2008) Ocean storage of CO_2. *Elements*, **4**, 317–322.

Alkattan, M., Oelkers, E.H., Dandurand, J.L. & Schott, J. (1998) An experimental study of calcite and limestone dissolution rates as a function of pH from -1 to 3 and temperature from 25 to 80°C. *Chemical Geology*, **151**, 199–214.

Andersson, A.J., MacKenzie, F.T. & Lerman, A. (2006) Costal ocean CO_2-carbonic acid-carbonate sediment system of the Anthropocene. *Global Biogeochemical Cycles*, **20**, GB1S92.

Arnórsson, S. (1989) Deposition of calcium carbonate minerals from geothermal waters – theoretical considerations. *Geothermics*, **18,** 33–40.

Arvidson, R.S. & Mackenzie, F.T. (1999) The dolomite problem: Control of precipitation kinetics by temperature and saturation state. *American Journal of Science*, **299,** 257–288.

Assayag, N., Matter, J., Ader, M., Goldberg, D. & Agrinier, P. (2009) Water-rock interaction during a CO_2 injection field-test: Implication on host rock dissolution and alteration effects. *Chemical Geology*, **265**, 227–235.

Benson, S.M. & Cole, D.R. (2008) Reducing greenhouse gas emissions with carbon dioxide capture and sequestration in deep geological formations. *Elements*, **4**, 323–328.

Bénézeth, P., Palmer, D.A., Anovitz, L.M. & Horita, J. (2007) Dawsonite synthesis and reevaluation of its thermodynamic properties from solubility measurements: Implications for trapping of CO_2. *Geochimica et Cosmochimica Acta*, **71**, 4438–4455.

Bénézeth, P., Menez, B. & Norel, C. (2009) CO_2 geological storage: Integrating geochemical, hydrodynamical, mechanical and biological processes from the pore to the reservoir scale. *Chemical Geology*, **265**, 1–2.

Berner, E.K. & Berner, R.A. (1996) *Global Environment: Water, Air, and Geochemical Cycles*. Prentice Hall, New Jersey, USA, 376 pp.

Berner, R.A. (2004) *The Phanerozoic Carbon Cycle.* Oxford University Press, Oxford, UK, 150 pp.

Blunt, M., Fayers, F.J. & Orr, F.M. (1993) Carbon-dioxide in enhanced oil recovery. *Energy Conversion and Management*, **34**, 9–11.

Broecker, W.S. (2008) CO_2 capture and storage: Possibilities and perspectives. *Elements*, **4,** 295–297.

Bruant, R.G., Guswa, A.J., Celia, M.A. & Peters, C.A. (2002) Safe storage of CO_2 in deep saline aquifers. *Environmental Science & Technology*, **36**, 240A–245A.

Calderia, K. & Wickett, M.E. (2003) Anthropogenic carbon and ocean pH. *Nature*, **425**, 365.

Chou, L., Garrels, R.M. & Wollast, R. (1989) Comparative study of the kinetics and mechanisms of dissolution of carbonate minerals. *Chemical Geology*, **78**, 269–282.

Cubillas, P., Köhler, S., Prieto, M. & Oelkers, E.H. (2005) Experimental determination of the dissolution rates of calcite, aragonite and bivalves. *Chemical Geology*, **216,** 59–77.

Dessert, C., Dupré, B., Gaillardet, J., Francois, L.M. & Allégre, C.J. (2003) Basalt weathering laws and the impact of basalt weathering on the global carbon cycle. *Chemical Geology*, **202,** 257–273.

Dixon, R.K., Brown, S., Houghton, R.A., Solomon, A.M., Trexler, M.C. & Wisniewski, J. (1994) Carbon pools and flux of global forest ecosystems. *Science*, **263**, 185–190.

Emberley, S., Hutcheon, I., Shevalier, M., Durocher, K., Gunter, W.D. & Perkins, E.H. (2004) Geochemical monitoring of fluid-rock interaction and CO_2 storage at the Weyburn CO_2-injection enhanced oil recovery site, Saskatchewan, Canada. *Energy*, **29**, 1393–1401.

Falkowski, P. & 16 co-authors (2000) The global carbon cycle: A test of our knowledge of Earth as a system. *Science*, **290**, 291–296.

Flaathen, T.K., Gislason, S.R., Oelkers, E.H. & Sveinbjornsdottir, A.E. (2009) Chemical evolution of the Mt. Hekla, Iceland, groundwaters: A natural analogue for CO_2 sequestration in basaltic rocks. *Applied Geochemistry*, **24**, 463–474.

Friedmann, S.J. (2007) Geological carbon dioxide sequestration. *Elements*, **3**, 179–184.

Gaillardet, J., Dupré, B., Louvat, P. & Allégre, C.J. (1999) Global silicate weathering and CO_2 consumption rates deduced from the chemistry of the large rivers. *Chemical Geology*, **159**, 3–30.

Garrels, R.M. & Mackenzie, F.T. (1971) *Evolution of Sedimentary Rocks.* W.W. Norton and Company Inc., New York, 397 pp.

Gaus, I. (2010) Role and impact of CO_2-rock interactions during CO_2 storage in sedimentary rocks. *International Journal of Greenhouse Gas Control*, **4**, 73–89.

Gaus, I., Azaroual, M. & Czernichowski-Lauriol, I. (2005) Reactive transport modelling of the impact of CO_2 injection on the clayey cap rock at Sleipner (North Sea). *Chemical Geology*, **217**, 319–337.

Gilfillan, S.M.V. & 10 co-authors (2009) Solubility trapping in formation water as dominant CO_2 sink in natural gas fields. *Nature*, **458**, 614–618.

Gislason, S.R. & Eugster, H.P. (1987) Meteoric water-basalt interactions. II: A field study in N.E. Iceland. *Geochimica et Cosmochimica Acta*, **51**, 2841–2855.

Gislason, S.R. & Oelkers, E.H. (2003) The mechanism, rates and consequences of basaltic glass dissolution: II. An experimental study of the dissolution rates of basaltic glass as a function of pH and temperature. *Geochimica et Cosmochimica Acta*, **67**, 3817–3832.

Gislason, S.R., Arnórsson, S. & Ármannsson, H. (1996) Chemical weathering of basalt in SW Iceland: Effects of runoff, age of rocks and vegetative/glacial cover. *American Journal of Science*, **296,** 837–907.

Gislason, S.R., Oelkers, E.H. & Snorrason, Á. (2006) The role of river-suspended material in the global carbon cycle. *Geology*, **34**, 49–52.

Gislason, S.R., Oelkers, E.H., Eiriksdottir, E.S., Kardjilov, M.I., Gisladottir, G., Sigfusson, B., Snorrason, A., Elefsen, S.O., Hardardottir, J., Torssander, P. & Oskarsson, N. (2009) Direct evidence of the feedback between climate and weathering. *Earth and Planetary Science Letters*, **277,** 213–222.

Goldberg, D.S., Takahashi, T. & Slagle, A.L. (2008) Carbon dioxide sequestration in deep-sea basalt. *Proceedings of the National Academy of Science*, **105,** 9920–9925.

Groisman, P.Y., Knight, R.W., Easterling, D.R., Karl, T.R., Hegerl, G.C. & Razuvaev, V.N. (2005) Trends in intense precipitation in the climate record. *Journal of Climate*, **18**, 1326–1350.

Gudmundsson, B. & Arnórsson, S. (2002) Geochemical monitoring of the Krafla and Námafjall geothermal areas, N-Iceland. *Geothermics*, **31**, 195–243.

Gysi, A.P. & Stefánsson, A. (2008) Numerical modeling of CO_2-water-basalt interaction. *Mineralogical Magazine*, **72**, 55–59.

Haugan, P.M. & Drange, H. (1996) Effects of CO_2 on the ocean environment. *Energy Conversion and Management*, **37**, 1019–1022.

Helgeson, H.C., Delaney, J.M., Nesbitt, H.W. & Bird, D.K. (1978) Summary and critique of the thermodynamic properties of rock forming minerals. *American Journal of Science*, **278**, 1–229.

Hellevang, H., Aagaard, P., Oelkers, E.H. & Kvamme, B. (2005) Can dawsonite permanent trap CO_2? *Environmental Science and Technology*, **39**, 8281–8287.

Herzog, H., Golomb, D. & Zemba, S. (1991) Feasibility, modelling, and economics of sequestering power plant CO_2 emissions in the deep ocean. *Environmental Progress*, **10**, 64–74.

Hiratsuka, M., Toma, T., Mindawati, N., Heriansyak, I., & Morikawa, Y. (2005) Biomass of a man-made forest of timber tree species in the humid tropics of West Java, Indonesia. *Journal of Forest Research*, **10**, 487–491.

Holloway, S. (2001) Storage of fossil fuel-derived carbon dioxide beneath the surface of the Earth. *Annual Review of Energy and the Environment*, **26**, 145–166.

Johns, T.C. & 11 co-authors (2003) Anthropogenic climate change for 1860 to 2100 simulated with the HadCM3 model under updated emissions scenarios. *Climate Dynamics*, **20**, 583–612.

Kelemen, P. & Matter, J.M. (2008) In situ carbonation of peridotite for CO_2 storage. *Proceedings of the National Academy of Science*, **105**, 17295–17300.

Kharaka, Y.K., Cole, D.R., Hovorka, S.D., Gunter, W.D., Knauss, K.G. & Freifeld, B.M. (2006) Gas-water-rock interactions in Frio Formation following CO_2 injection: Implications for the storage of greenhouse gases in sedimentary basins. *Geology*, **34**, 577–580.

Kharaka, Y.K., Thordsen, J.J., Hovorka, S.D., Nance, H.S., Cole, D.R., Phelps, T.J., & Knauss, K.G. (2009) Potential environmental issues of CO_2 storage in deep saline aquifers: Geochemical results from the Frio-1 brine pilot test, Texas, USA. *Applied Geochemistry*, **24**, 1106–1112.

Knauss, K.G., Johnson, J.W. & Steefel, C.I. (2005) Evaluation of the impact of CO_2, co-contaminant gas, aqueous fluid and reservoir–rock interactions on the geologic sequestration of CO_2. *Chemical Geology*, **217,** 339–350.

Korbøl, R. & Kaddour, A. (1995) Sleipner Vest CO_2 disposal – Injection of removed CO_2 into the Utsira Formation. *Energy Conversion and Management*, **36**, 509–512.

Lackner, K.S. (2003) A guide to CO_2 sequestration. *Science*, **300**, 1677–1678.

Lackner, K.S., Wendt, C.H., Butt, D.P., Joyce, E.L. & Sharp, D.H. (1995) Carbon-dioxide disposal in carbonate materials. *Energy*, **20**, 1153–1170.

Lai, R. (2004) Soil carbon sequestration impacts on global climate change and food security. *Science*, **304**, 1623–1627.

Lai, R. (2008) Sequestration of atmospheric CO_2 in global carbon pools. *Energy and Environmental Science*, **1**, 86–100.

Lai, R. (2009) Sequestering atmospheric carbon dioxide. *Critical Reviews in Plant Sciences*, **28**, 90–96.

Lamlom, S.H. & Savidge, R.A. (2003) A reassessment of carbon content in wood: Variation within and between 41 North American Species. *Biomass and Bioenergy*, **25**, 381–388.

Lehmann, J., da Silva, J.P., Steiner, C., Nehls, T., Zech, W. & Glaser, B. (2003) Nutrient availability and leaching in an archaeological Anthrosol and a Ferralsol of the Central Amazon basin: fertilizer, manure and charcoal amendments. *Plant and Soil*, **249**, 343–357.

Le Quéré, C. & 30 co-authors (2009) Trends in the sources and sinks of carbon dioxide. *Nature Geoscience*, **2**, 831–836.

Manabe, S. & Stouffer, R.J. (1993) Century-scale effects of increased atmospheric CO_2 on the ocean-atmosphere system. *Nature*, **364**, 215–218.

Manabe, S. & Stouffer, R.J. (1994) Multiple-Century response of a coupled ocean-atmosphere model to an increase of atmospheric carbon dioxide. *Journal of Climate*, **7**, 5–23.

Marchetti, C. (1977) Geo-engineering and CO_2 problem. *Climatic Change*, **1**, 59–68.

Marini, L. (2007) *Geological Sequestration of Carbon Dioxide: Thermodynamics, Kinetics, and Reaction Path Modeling*. Elsevier, Amsterdam, 453 pp.

May, F. (2005) Alteration of wall rocks by CO_2-rich water ascending in fault zones: natural analogues for reactions induced by CO_2 migrating along faults in siliciclastic reservoir and cap rocks. *Oil and Gas Science and Technology*, **60**, 19–32.

McGrail, B.P., Schaef, H.T., Ho, A.M., Chien, Y.J., Dooley, J.J. & Davidson, C.L. (2006) Potential for carbon dioxide sequestration in flood basalts. *Journal of Geophysical Research*, **111**, B12201.

Metz, B., Davidson, O., de Coninck, H., Loos, M. & Meyer, L. (editors) (2005) *IPCC Special Report on Carbon Dioxide Capture and Storage*. Cambridge University Press, Cambridge, UK, 431 pp.

Neuhoff, P.S., Fridriksson, T., Arnorsson, S. & Bird, D.K. (1999) Porosity evolution and mineral paragenesis during low-grade metamorphism of basaltic lavas at Teigarhorn, Eastern Iceland. *American Journal of Science*, **299**, 467–501.

Nilsson, S. & Schopfhauser, W. (1995) The carbon-sequestration potential of a global afforestation program. *Climatic Change*, **30**, 267–293.

Oelkers, E.H. & Cole, D.R. (2008) Carbon dioxide sequestration. A solution to a global problem. *Elements*, **4**, 305–310.

Oelkers, E.H. & Gislason, S.R. (2001) The mechanism, rates and consequences of basaltic glass dissolution: I. An experimental study of the dissolution rates of basaltic glass as a function of aqueous Al, Si and oxalic acid concentration at 25°C and pH = 3 and 11. *Geochimica Cosmochimica Acta*, **65**, 3671–3681.

Oelkers, E.H. & Schott, J. (2005) Geochemical aspects of CO_2 sequestration. *Chemical Geology*, **217**, 183–186.

Oelkers, E.H., Gislason, S.R. & Matter, J. (2008) Mineral Carbonation of CO_2. *Elements*, **4**, 331–335.

Oelkers, E.H., Bénézeth, P. & Pokrovski, G.S. (2009) Thermodynamic databases for water-rock interaction. In: *Thermodynamics and Kinetics of Water-rock Interaction* (E.H. Oelkers & J. Schott, editors). Reviews in Mineralogy and Geochemistry, **70**, 1–46. Mineralogical Society of America, Washington, D.C. and the Geochemical Society, St. Louis, Missouri, USA.

Orr, J.C. & 26 co-authors (2005) Anthropogenic ocean acidification over the twenty-first century and its impact on calcifying organisms. *Nature*, **437**, 681–686.

Parkhurst, D.L. & Appelo, C.A.J. (1999) *User's guide to PHREEQC (Version 2) – A computer program for speciation, batch-reaction, one-dimensional transport and inverse geochemical calculations*. U.S. Geological Survey Water Resources Investigations Report **99-4259**, U.S. Geological Survey, Denver, USA, 312 pp.

Patz, J.A., Epstein, P.R., Burke, T.A. & Balbus, J.M. (1996) Global climate change and emerging infectious diseases. *Journal of the American Medical Association*, **275**, 217–223.

Pokrovsky, O.S. & Schott, J. (1999) Processes at the magnesium-bearing carbonates solution interface. II. Kinetics and mechanism of magnesite dissolution. *Geochimica et Cosmochimica Acta*, **63**, 881–897.

Regnault, O., Lagneau, V. & Schneider, H. (2009) Experimental measurement of portlandite carbonation kinetics with supercritical CO_2. *Chemical Geology*, **265**, 113–121.

Rignot, E.J. (1998) Fast recession of a West Antarctic glacier. *Science*, **281**, 549–551.

Roger, K.L., Neuhoff, P.S., Pedersen, A.K. & Bird, D.K. (2006) CO_2 metasomatism in a basalt-hosted petroleum reservoir, Nuussuaq, West Greenland. *Lithos*, **92**, 55–82.

Saldi, G.D., Jordon, G., Schott, J. & Oelkers, E.H. (2009) Magnesite growth rates as a function of temperature and saturation state. *Geochimica et Cosmochimica Acta*, **73**, 5646–5657.

Schott, J., Oelkers, E.H. & Pokrovsky, O.S. (2009) The link between mineral dissolution/precipitation kinetics and solution chemistry. In: *Thermodynamics and kinetics of water-rock interaction* (E.H. Oelkers &

J. Schott, editors) Reviews in Mineralogy and Geochemistry, **70**, 207–238. Mineralogical Society of America, Washington, D.C. and the Geochemical Society, St. Louis, Missouri, USA.

Steeneveldt, R., Berger, B. & Torp, T.A. (2006) CO_2 capture and storage – Closing the knowing-doing gap. *Chemical Engineering Research and Design*, **84**, 739–763.

Stefánsson, A. & Gíslason, S.R. (2001) Chemical weathering of basalts, SW Iceland: effect of rock crystallinity and secondary minerals on chemical fluxes to the ocean. *American Journal of Science*, **301**, 513–556.

Stockmann, G., Wolff-Boenisch, D., Gislason, S.R. & Oelkers, E.H. (2008) Dissolution of diopside and basaltic glass: the effect of carbonate coating. *Mineralogical Magazine*, **72**, 135–139.

Thomas C.D. & 18 co-authors (2004) Extinction risk from climate change. *Nature*, **427**, 145–148.

White, C.M., Strazisar, B.R., Granite, E.J., Hoffman, J.S. & Pennline, H.W. (2003) Separation and capture of CO_2 from large stationary sources and sequestration. I. Geological formations – Coalbeds and deep saline aquifers. *Journal of the Air & Waste Management Association*, **53**, 645–715.

Wigand, M., Kaszuba, J.P., Cary, J.W. & Hollis, W.K. (2009) Geochemical effects of CO_2 sequestration on fractured wellbore cement at the cement/caprock interface. *Chemical Geology*, **265**, 122–133.

Wolff-Boenisch, D., Gislason, S.R. & Oelkers, E.H. (2004) The effect of fluoride on the dissolution rates of natural glasses at pH 4 and 25°C. *Geochimica et Cosmochimica Acta*, **68**, 4571–4582.

Wolff-Boenisch, D., Gislason, S.R. & Oelkers, E.H. (2006) The effect of crystallinity on dissolution rates and CO_2 consumption capacity of silicates. *Geochimica et Cosmochimica Acta*, **70**, 858–870.

Xu, T., Apps, J. & Pruess, K. (2004) Numerical simulation of CO_2 disposal by mineral trapping in deep aquifers. *Applied Geochemistry*, **19**, 917–936.

Xu, T., Apps, J.A. & Pruess, K. (2005) Mineral sequestration of carbon dioxide in a sandstone-shale system. *Chemical Geology*, **217**, 295–318.

Zerai, B., Saylor, B.Z. & Matisoff, G. (2006) Computer simulation of CO_2 trapped through mineral precipitation in the Rose Run Sandstone, Ohio. *Applied Geochemistry*, **21**, 223–240.

EMU Notes in Mineralogy, Vol. 10 (2010), Chapter 10, 345–376

Solid-solution formation and the long-term safety of nuclear-waste disposal

DIRK BOSBACH

Institut für Energieforschung – Sicherheitsforschung und Reaktortechnik (IEF-6), Forschungszentrum Jülich, 52425 Jülich, Germany
e-mail: d.bosbach@fz-juelich.de

The disposal of nuclear waste in geological formations poses major scientific and social challenges to be met in the coming decades. One of the key issues is the long-term safety of a waste-repository system over extended periods of time (up to 10^6 years). Demonstrating the safety over geological time scales requires a sound understanding of the migration and retention of radionuclides in the geosphere. Predicting the geochemical evolution of a waste-repository system over such time scales cannot be based on 'simple' extrapolations of empirical phenomena, but requires molecular-level process understanding. In recent years it has been demonstrated that understanding fundamental reactions can improve the reliability of geochemically based long-term predictions. A sound understanding of radionuclide binding to minerals by sorption reactions is an important issue when discussing the long-term safety of nuclear disposal.

The need of the radiogeochemistry community for molecular-level system understanding of structural incorporation of key radionuclides and subsequent solid-solution formation will be discussed on the basis of two prominent examples: (1) the $(Ra,Ba)SO_4$ solid solution series; and (2) the uptake of trivalent actinides and lanthanides by calcite. The current status of our knowledge of these systems will be discussed critically and knowledge gaps for the long-term safety assessment will be identified.

1. Introduction

Solid solutions are frequently encountered within the nuclear fuel cycle such as in the form of: (1) $(U, Pu)O_2$ MOX nuclear fuel; (2) spent nuclear fuel (after reactor operation), which contains the fission products and the actinides; and (3) waste forms, which contain the radionuclides after Pu extraction and reprocessing such as borosilicate glasses and crystalline waste forms. Here, the focus will be on the back end of the nuclear fuel cycle. Nuclear waste is produced by various industrial, medical and military activities. Nuclear waste is classified as low and intermediate level waste (LILW), which may be subdivided into short-lived and long-lived waste, and high-level waste (HLW) (IAEA, 1994).

While in almost all countries with an advanced nuclear programme the disposal of LILW has been solved, the disposal of HLW has not yet been implemented in any country producing this type of waste. HLW is a type of nuclear waste which is created by the operation of nuclear reactors: spent nuclear fuel (SNF) and waste materials from reprocessing (such as borosilicate glasses with up to 20% waste

DOI: 10.1180/EMU-notes.10.10

loading). In some countries, HLW is produced in nuclear weapons production. These military aspects of HLW disposal are not specifically considered here. HLW contains the fission products (mostly β-emitters) and actinide elements (mostly α-emitters) with an activity of $>10^{14}$ Bq/m^3. Some of the radionuclides have very long half-lives, $t_{1/2}$, (^{99}Tc: 0.211E6 a, ^{79}Se: 0.295E6 a, ^{93}Zr: 1.53E6 a, ^{135}Cs: 2.3E6 a, ^{129}I: 15.7E6 a, ^{237}Np: 2.14E6 a, ^{239}Pu: 24.0E3 a, ^{240}Pu: 6.563E3 a, ^{242}Pu: 0.375E6 a, ^{243}Am: 7.37E3 a, ^{245}Cm: 8.5E3 a).

HLW contains 99% of the radioactivity of all radioactive waste but contributes less than 10% of the total volume of all radioactive waste. A typical nuclear reactor (*e.g.* 1500 MWe) produces 25–30 tons of spent nuclear fuel per year. After reprocessing and subsequent vitrification, the volume of the HLW borosilicate waste glass would be $\sim$3 m^3.

Significant heat production is linked to the radioactive decay in particular of two medium-lived fission products ^{137}Cs (30.23 a) and ^{90}Sr (28.9 a). This adds another challenge to the disposal of HLW.

Three disposal options are studied in detail: (1) final (permanent) disposal in a deep geological formation is thought suitable for ensuring isolation over time-frames of several 100,000 years without any maintenance; (2) partitioning and transmutation (P&T) of the radionuclides by nuclear reactions in accelerator-driven systems (ADS) or (fast) nuclear reactors offers the possibility to convert highly radiotoxic and long-lived radionuclides into short-lived or stable isotopes; (3) the long-term temporary storage of HLW, *e.g.* several centuries, offers the option to retrieve the waste in case new disposal methods are discovered in the future. However, this strategy requires maintenance over extended time-frames (*e.g.* several hundred years).

There seems to be a consensus in almost all countries producing HLW that the final disposal of HLW in a deep geological formation is considered to be the best option. Even with transmutation implemented, there would be a need for a geological repository for HLW. It was recognized early in the 'nuclear age' that a strategy was required that would keep the waste isolated from the environment over very long time frames. It was also judged to be ethically correct that the generation that had produced the waste should provide for its safe and permanent disposal (OECD NEA, 1999).

Many countries have implemented a nuclear-waste management programme, which is aimed at a nuclear waste repository for HLW in a deep geological formation on a national level. Recently, the European Commission (EC) has launched a technology platform for the implementation of geological disposal (IGD-TP) for HLW (www.igdt-p.eu). Their vision is to start operation of the first European HLW repositories in 2025. The principle types of geological host formations are sedimentary (*e.g.* clayrock), crystalline (*e.g.* granite) and salt domes. Considering the relevant properties of these geological formations with respect to nuclear disposal, they have favourable and less favourable properties: thermal conductivity, heat resistance, permeability, dissolution behaviour, sorption behaviour, strength, deformation behaviour. It is difficult to judge on the basis of a one-to-one comparison of the properties of such a list what the best geological formation for a nuclear waste repository is. In fact, less favourable properties may be compensated by introducing a technical (wasteform, container) and/or geotechnical barrier (backfill material) to support the geological barrier. Ultimately, in any case, a multi-barrier system will ensure the isolation of the radiotoxic inventory.

The intention of this contribution is to outline the needs of the radiogeochemistry community concerned with the long-term safety of nuclear waste disposal with respect to solid-solution formation in aqueous environments as well as to provide an overview of work that has been done. In particular, two examples will be discussed in more detail: (1) The $Ra_xBa_{(1-x)}SO_4$ solid solution system; and (2) the structural incorporation of trivalent actinides and lanthanides into calcite. It is not intended to give an excessive overview of all studied solid solution–aqueous solution (SSAS) systems with relevance to nuclear-waste disposal.

2. Long-term safety of nuclear-waste disposal

The disposal of HLW in deep geological formations poses major scientific and social challenges for the next decades, independently of political decisions to stop or continue the use of nuclear energy. Perhaps one of the most critical issues is related to the long-term safety of a nuclear-waste repository system over extended periods of time (up to 10^6 years). The national organization responsible for the disposal of HLW produces a document called the safety case as part of the license application for a waste repository, which addresses, in particular, the long-term safety aspect. Recently, safety cases for disposal concepts in clayrock formations were compiled in various European countries (Nagra, 2002; ANDRA, 2005; Ondraf/Nirond, 2001).

The main objectives of the safety case are to demonstrate (1) that a geological environment is suitable for a HLW repository, (2) that a repository can be practically constructed and operated in such an environment and (3) that the long-term safety is assured. The main challenge for the safety case is to demonstrate that for each radionuclide it is possible to predict with a sufficiently high degree of certainty how the initial mass of each radionuclide present in a waste package redistributes itself in space, in particular in the host formation and over- and underlying geological formations over the safety assessment timeframe (typically extending up to 10^6 years).

Here, the focus will be on the long-term safety issues associated with geochemical aspects that control the potential release of radionuclides from the HLW into the environment. In principle, one has to distinguish the evolution of a repository system in the presence of water and the absence of water. Without water, the release mechanisms for radionuclides from the HLW into the environment are extremely limited. The long-term isolation of the waste is with a high degree of certainty assured. On the contrary, in the presence of water, complex dissolution/corrosion processes of the waste matrix and subsequent release of radionuclides may occur. Simultaneously, new phases will form and the released radionuclides may be bound to these secondary alteration phases. The so called source term for the near field of a nuclear-waste repository system defines the concentration of radionuclides in aqueous solution, which is either controlled by the solubility of the radionuclides and associated solid phases or by the dissolution and subsequent radionuclide release kinetics. The mobility of radionuclides in the far field of a waste repository – the geosphere – is another key issue. In principle, three geochemistry-related aspects determine the transport of radionuclides in the geosphere: (1) aqueous chemistry of radionuclides; (2) sorption reactions onto/into minerals of geological barrier and other materials of the technical and geotechnical barrier; and (3) diffusion- and advection-driven transport in porous media (hydraulic constraints).

In particular, sorption reactions can result in an effective retention of dissolved radionuclides and their release in the biosphere/environment is prevented. Sorption reactions are phenomenologically described by a distribution coefficient, K_d [l/kg], which relates the concentration of dissolved radionuclides linearly to the radionuclides sorbed onto/ into a solid material:

$$m_{\mathrm{sorbed}}[\mathrm{mol/kg}] = K_d[\mathrm{l/kg}] \times m_{\mathrm{dissolved}}[\mathrm{mol/l}], \tag{1}$$

where m_{sorbed} and $m_{\mathrm{dissolved}}$ represent the amount of radionuclide sorbed and dissolved. Such a partition coefficient does not provide any information about the nature of the sorption reaction and treats it as a constant. The applicability of such an approach is naturally limited to specific conditions and cannot be extrapolated to other geochemical conditions.

There seems to be a consensus in the scientific community that the assessment of the long-term safety including the geochemical evolution of a waste repository system should be guided by equilibrium thermodynamics and geochemical insights. Furthermore, a molecular-level process understanding can help to improve the confidence in available data on radionuclide behaviour in the geosphere beyond a simple phenomenological description.

In principle, four molecular-level sorption mechanisms have been identified in recent years: (1) outer-sphere surface complexation (physisorption); (2) inner-sphere surface complexation (chemisorption); (3) ion exchange; and (4) structural incorporation and subsequent solid-solution formation. Adsorption and ion exchange phenomena and the associated molecular-level reaction mechanisms have been studied extensively in the past. Currently, the retention of actinides and fission products is thought to be primarily controlled (in laboratory time intervals) by ion exchange and adsorption reactions at external mineral surfaces. However, incorporation of radionuclides into the bulk structure of host minerals also represents a potentially efficient retention mechanism, the importance of which increases with the reaction progress time and is therefore highly relevant for the long-term safety of nuclear-waste disposal.

Without a doubt, solid solutions are ubiquitous in nature (Geiger, 2001) whereas the formation of pure compounds is rather an exception. Solid solutions have been studied for decades in Earth sciences, and a sound understanding of chemical components' mixing behaviour has been developed in the framework of classical thermodynamics. However, these studies were typically focused on high-temperature systems ($T \gg 100°C$) for obvious reasons. Only a limited number of solid solution systems has been studied under ambient conditions in aqueous environments ($0 < T < 100°C$). Further, the SSAS systems studied involve, in most cases, only simple substitutions (no charge difference between substituting ions), such as the well known $Ra_xBa_{(1-x)}SO_4$ solid-solution system. Many systems have been studied, in which dissolved trace elements partition preferentially into a solid phase. If a thermodynamically favourable solid solution forms, then the solubility of the incorporated trace element will be significantly reduced. The reduced solubility, compared to that of pure compounds (end-members), has a high relevance for the migration of trace elements/radionuclides through the geosphere. In contrast to simple substitutions, structural incorporation of actinides and specific fission products in host minerals involves, in many cases, heterovalent substitutions and various ionic radionuclide species. This has been shown in a number of publications that have appeared in recent years. A sound understanding of such SSAS systems is lacking, and its application to the long-term safety assessment of nuclear-waste disposal is very limited so far. In fact, solid-solution formation has not been included in any long-term safety assessment so far. However, recent research activities indicate a demand for such retention mechanisms for certain radionuclides and an improved confidence in the solid solution–aqueous solution concept. A recent publication by the Nuclear Energy Agency has compiled a state of the art report of chemical thermodynamics of solid solutions of interest in nuclear-waste management (Bruno *et al.*, 2007).

3. Actinides and mobile fission and activation products

Considering all radionuclides being generated by nuclear fission in a reactor, only a limited number of radionuclides are of concern when discussing long-term safety issues of nuclear-waste disposal in a deep geological formation. As an example, during the operation of a nuclear reactor, 1 ton of nuclear fuel ($= {}^{235}U + {}^{238}U$, 3.5% ${}^{235}U$) is converted within 3 years with a burn-up of 33 GWd/tU into 9 kg Pu, 0.5 kg Np, 0.12 kg Am and 0.04 kg Cm (+35 kg fission products, 4.6 kg ${}^{236}U$) as well as 8 kg ${}^{235}U$ and 943 kg ${}^{238}U$ remaining. A large number of fission and activation products as well as actinides have relatively short half-lifes and do not have to be considered regarding the long-term safety of a nuclear-waste repository. This is perhaps a positive aspect of nuclear waste – its radiotoxicity will reduce with time. However, some radionuclides have very long half-lives and need to be considered in long-term safety assessments.

One group of radionuclides are U, Pu and the minor actinides Np, Am and Cm. Many sorption experiments with tri- and tetravalent actinides, which is the expected redox-state in most repository concepts (in all European repository concepts), with minerals typical of clayrocks or granites (not necessarily for salt domes), indicate high partition coefficients of $\geq 10^3$ [l/kg]. With such high K_d values, the actinides are confined in any multi-barrier system of a nuclear-waste repository. Some aqueous species form very stable complexes in solution, which compete with sorption reactions and increase the solubility of the actinides. Furthermore, tri- and tetravalent actinides bind efficiently to typical groundwater colloids and they also tend to form eigen-colloids. These reactions enhance actinide migration in the geosphere. Due to the extremely high radiotoxicity of the actinides, a sound understanding of their geochemical behaviour under repository-relevant conditions is of prime importance in order to minimize uncertainties in long-term safety assessments and to improve the confidence in the safety case.

A second group of radionculides represents some long-lived fission and activation products: ^{14}C, ^{36}Cl, ^{79}Se, ^{129}I, and ^{135}Cs. Although these radionuclides are significantly less radiotoxic than the actinides, they dominate the potential long-term exposure risk from a nuclear-waste repository (Grambow, 2008). This is due to the fact that some of these radionuclides form under waste repository-relevant conditions of anionic species, Se(-II)/Se(IV), I(-I), Cl(-I), which do not sorb to many minerals and the corresponding K_d value is close to zero. Furthermore, due to a lack of or uncertainty about the required thermodynamic data for some of the relevant species, K_d values of zero are conservatively assumed. As a consequence, these radionuclides dominate the dose in performance assessment calculations for certain scenarios. Furthermore, for specific scenarios, other fission products may cause challenges in safety assessments. One prominent example is ^{226}Ra ($t_{1/2} = 1600$ a), which will be discussed in more detail below. In case water gets access to the waste matrix, the concentration of the radionuclides released will be at trace concentration levels due to the relatively slow corrosion/dissolution rates of spent fuel as well as HLW borosilicate glass under repository-relevant conditions. Furthermore, the solubility of most actinide compounds is very low (under reducing conditions!). Consequently, with respect to solid-solution formation, in most cases dilute systems may be expected. Furthermore, due to structural incompatibilities of some of the above-mentioned radionuclides, only limited mixing may be expected. An extended mixing gap also arises from the fact that the expected temperatures in the vicinity of a waste repository will be relatively low, $<100°C$ (in the case of salt domes $<200°C$). These geochemical boundary conditions raise a number of experimental and analytical challenges which will be discussed in the next section.

4. Experimental and analytical aspects

The theoretical background of the thermodynamics of SSAS systems is well established (Gamsjaeger *et al.*, 2000; Glynn, 2000; Glynn & Reardon, 1990; Lippmann, 1980) and advanced numerical methods have been developed to describe such systems quantitatively (Bruno *et al.*, 2007). However, applying these concepts successfully to 'real world' problems such as the long-term safety assessment of a nuclear-waste repository requires actual thermodynamic data for the relevant systems as a baseline for geochemical modelling. Such thermodynamic data can be obtained either from laboratory experiments with synthetic compounds or from analysis of naturally occurring solid-solution phases. Here, the focus will be on experimental aspects of how to derive thermodynamic data, in particular from co-precipitation experiments, using synthetic systems at relatively low (waste-repository relevant) temperatures of $<100°C$.

4.1. Co-precipitation experiments in aqueous solution

Synthesizing solid-solution compounds is, for obvious reason, a prerequisite for any further study. Co-precipitation experiments in aqueous solution are ultimately linked to fundamental aspects of crystal growth and the actual experimental procedures. Although with respect to geochemical modelling one is ultimately interested in thermodynamic data (which are independent of the reaction path), co-precipitation experiments are usually done in a supersaturated non-equilibrium system. Therefore, it is important to demonstrate that experimentally obtained data from such an experiment on element partitioning from aqueous solution into a solid solution phase represent equilibrium conditions.

In the following, various complications will be outlined, which may obscure experimental data on SSAS systems. Most of these complications are related to kinetic aspects of co-precipitation experiments at repository-relevant temperatures. They arise from the fundamental problem that one wants to derive thermodynamic data on SSAS equilibria from supersaturated systems which are not in equilibrium (otherwise the solid solution phase would not form).

One key aspect of co-precipitation is related to a solution-boundary-layer process during mineral precipitation in aqueous solution (Rimstidt *et al.*, 1998): Dissolved species have to be transported from the bulk solution to the mineral surface, where they attach/adsorb to certain surface sites and eventually become part of the crystal structure. In the case of a rather static solution (without mechanical mixing/stirring) the compounds will reach the growing mineral surface *via* diffusion. If the mineral precipitation rate is faster than the diffusion rate, the solution adjacent to the mineral surface will become depleted in the species that become incorporated into the growing crystal. As a consequence, a concentration gradient will evolve and the chemical composition of the bulk solution, which is usually analysed, differs from the chemical composition of the solution in contact with the growing mineral surface.

Applying this concept to co-precipitation, it is obvious that the ratio of the species involved in the solution in contact with the growing mineral surface might change, depending on their diffusion coefficients and the actual distribution coefficient. In case the co-precipitation rate is greater than the diffusion rate of the involved species as well as their diffusion coefficients, the bulk solution analysed does not represent the solution in contact with the mineral surface, in particular with respect to the activity ratio of competing ions in a solid solution. This effect is more pronounced the larger the precipitation rate and the larger or smaller the distribution coefficient is.

Another process also affects co-precipitation even when one can establish an experimental setup where radionuclide uptake is surface-reaction controlled (thus excluding diffusion control). Mineral precipitation in aqueous solution is driven by the chemical potential and this is of course, in principle, also true for co-precipitation experiments. However, at high precipitation rates, dissolved species might become incorporated into a growing mineral due to physical trapping rather than thermodynamics (Wang & Xu, 2001). Adsorbed radionuclides can become overgrown before they desorb, depending on their adsorption/desorption behaviour and the precipitation dynamics on a growing mineral surface (*e.g.* kink-site formation). Near-equilibrium distribution coefficients can be significantly more (*e.g.* Mn^{2+}/calcite) or less than 1 (*e.g.* Sr^{2+}/calcite) (Curti, 1999; Rimstidt *et al.*, 1998). In both cases, at high precipitation rates the distribution coefficients tend to increase/decrease towards 1. Various studies have shown that the distribution coefficient can change by 1 or 2 orders of magnitude due to this kinetic trapping effect. Since the incorporation of radionuclides at high precipitation rates is entirely controlled by kinetic trapping, the composition of the precipitated compound may even have an intermediate composition that lies within the mixing gap. In such a case, the measured composition may lead to erratic thermodynamic data.

Closely linked to these two kinetic effects on co-precipitation is the chemical homogeneity of synthesized solid solution phases. Compositional zoning is a common phenomenon in natural as well as synthetic compounds precipitated in the presence of a radionuclide (or a trace metal).

Within a single crystal, internal chemical heterogeneities can define sector zones. 'Sector zoning' is linked to differential trace-component uptake as a consequence of preferential incorporation at specific surface sites on a growing mineral surface underneath (Fouke & Reeder, 1992; Paquette & Reeder, 1990, 1995). Ultimately, the mineral-surface structure controls the incorporation of the trace component. Sector zoning is macroscopic evidence for non-equilibrium conditions. It occurs under a small or moderate degree of supersaturation below the roughening transition (without heterogeneous nucleation). Treating sector-zoned minerals as homogeneous phases yields only an average value which cannot easily be used to derive thermodynamic data.

Many natural and synthetic minerals exhibit 'oscillatory zoning', with their chemical composition varying along a core-to-rim profile (L'Heureux & Jamtveit, 2002; Prieto *et al.*, 1993, 1997; Putnis *et al.*, 1992). Such spatio-temporal patterns arise by self-organization. Non-linear couplings of the parameters affecting diffusion-controlled mineral precipitation in aqueous solution can result in self-organized patterns. Again, such minerals are not chemically homogeneous and cannot easily be used to derive thermodynamic data. However, the surface of such zoned minerals can be in equilibrium with the adjacent aqueous solution.

Co-precipitation experiments are usually performed in more or less supersaturated aqueous solutions. In addition to the solid-solution phase of interest, other phases may precipitate simultaneously. Depending on the crystallinity, particle size, *etc.* of such an additional phase and the analytical effort as well as the detection limit of the methods used, one might overlook the presence of certain phases. However, the radionuclide of interest might have a high affinity to that phase. Quite often co-precipitation experiments are performed without studying the precipitated solid in detail. In fact, the solid composition is derived from the compositional changes of the aqueous solution assuming that an homogeneous solid has formed. In the case when an additional, yet unidentified phase, precipitates, experimental data will be obscured with respect to SSAS thermodynamics.

Precipitation experiments in supersaturated solutions may involve the formation of meta-stable precursor phases. The solid phase of interest will form after a solid state phase transformation of dissolution/re-precipitation processes. Of course, this also true for co-precipitation experiments. In fact, the presence of radionuclides may even favour the formation of the precursor phases. A very prominent example is the precipitation of calcite ($CaCO_3$) and its potential precursor phase, vaterite (Wolf *et al.*, 2000). The transformation of vaterite into calcite occurs at room temperature within hours. As a consequence one would not necessarily detect that vaterite had occurred in a co-precipitation experiment. The distribution coefficient as determined from the composition of the aqueous phase and the composition of the calcite would be controlled by a number of parameters other than the thermodynamics of calcite: the interaction of the radionuclide of interest with vaterite during co-precipitation and the behaviour of the radionuclide during the vaterite/calcite phase transformation. Another prominent example represents goethite/hematite and its precursor phase ferrihydrite. In contrast to the $CaCO_3$ system, ferrihydrite can be kept metastable at room temperature.

In addition to these experimental aspects under non-equilibrium conditions, there seems to be a consensus that once a solid solution has been co-precipitated with a certain atomistic configuration, there is no possible reconfiguration *via* solid-state diffusion, in order to achieve a configuration that represents equilibrium. Kinetic effects such as phase homogeneity, exsolution, compositional zoning, *etc.* as outlined above, will be kept frozen at temperature conditions of interest here (<100°C).

However, over time such non-equilibrium SSAS systems will evolve towards equilibrium, at least with respect to the surface *via* dissolution/re-precipitation processes. Depending on the reactivity of the solid phase, this might take quite long (considering typical timescales for laboratory studies). Without a doubt, thermodynamic equilibrium with an homogeneous solid phase is for many systems difficult to achieve at room-temperature conditions. However, this approach requires that the entire mineral surface is homogeneous, at least within the topmost monolayer in contact with the aqueous solution.

Various (co)precipitation techniques have been established to synthesize solid solution phases in aqueous solution. With the above potential complications outlined, the following synthesis procedures will be discussed critically:

(1) Probably the most common way to synthesize solid-solution phases is by simple 'batch type' experiments. At least two different solutions containing the required species will be mixed, thus producing a supersaturated solution with respect to the solid solution of interest. For example, in co-precipitation experiments with calcite as one end member, one would start with one solution containing $CaCl_2$ and possibly the radionuclide of interest. A second solution will be prepared with *e.g.* $NaHCO_3$. The solubility of $CaCl_2$ and $NaHCO_3$ is significantly greater than that of $CaCO_3$ and therefore, upon mixing these two solutions, one produces a supersaturated solution with respect to calcite. With such an experimental setup, one starts with a high degree of supersaturation. Finally, such a SSAS system will evolve towards equilibrium and the surface of the synthesized material may be in equilibrium with the aqueous solution. However, the co-precipitated solid phase will be chemically zoned in case the distribution coefficient deviates from 1.

Similarly, solid compounds (*e.g.* $CaCl_2$, $NaHCO_3$ and a radionuclide of interest in case one wants to synthesize calcite-based phases) are quite often added to an aqueous solution, where they dissolve and a supersaturated solution with respect to calcite evolves. Such a setup has often been used in hydrothermal reaction vessels, since it can be used in a closed system. However, the degree of supersaturation is controlled by the balance of dissolution of the initial compounds and the (co)precipitation of the solid solution phase. Again, with such a setup, the solid-solution phase might be chemically zoned with the surface potentially in equilibrium with the aqueous phase.

These two types of experiments include homogeneous/heterogeneous nucleation of the solid-solution phase (at a high degree of supersaturation). In order to avoid nucleation, one can perform such experiments with a suspension of seed crystals, thus inducing crystal growth/co-precipitation. Introducing such seed crystal requires a careful characterization of their surfaces. The presence of impurities should be excluded. Also, one should ensure that one does not use freshly ground particles as seed crystals, because such particles tend to interact strongly with almost any species in the aqueous solution due to their high-energy surface sites. These high-energy sites are generated at crystal surfaces due to the mechanical stress during grinding (surface defects). Preconditioning of such crystals in an aqueous solution with similar chemical composition being used in the actual co-precipitation experiment is a good strategy to remove these surface high-energy sites.

A typical feature of such simple batch-type experiments is the fact that the chemical composition of the aqueous solution will change during the synthesis procedure as a consequence of the co-precipitation process. The solid-solution phases may be chemically zoned.

(2) This drawback can be overcome by 'batch type constant composition methods' where one monitors compositional changes in the aqueous solution and adds a sufficient amount of required reactants in order to keep the chemical composition of the aqueous phase constant. Compositional changes in the aqueous phase are typically measured in real-time *via* a *pH* electrode (in case co-precipitation involves a change in *pH*), ion-selective electrodes or even simple conductivity measurements.

A potential drawback of these types of experiments is that one has to add the required compounds drop-wise in the form of small amounts of very highly concentrated solutions in order to keep the volume in the reaction vessel almost constant. However, adding a highly concentrated solution into the reaction vessel might induce nucleation processes, for example, and the experiment would temporarily be not well controlled.

(3) Another elegant way of synthesizing solid-solution phases is by 'counter diffusion experiments; (Prieto *et al.*, 1993, 1997). A U-tube is filled with a porous medium such as a silica gel and the two ends of the U-tube are filled with two solutions. The species in the two reservoirs will diffuse through the porous medium towards each other and precipitation occurs when a supersaturated solution has developed. With such a setup, co-precipitation is controlled by diffusion. It has been demonstrated that the solution composition in the pore space can be determined. However, it is not clear to what extent the silica gel affects the presence of dissolved species (either by complexation in solution or surface adsorption).

(4) Many of the drawbacks involved in the above-mentioned techniques can be overcome by performing co-precipitation experiment in a 'mixed flow reactor'. A reaction

vessel is fed by three or more independent solution reservoirs, containing the required species for co-precipitation. The reactor contains a suspension which is stirred with a suspended stirring bar. One can prove experimentally that the stirring rate is sufficient for surface-reaction-controlled co-precipitation (thus excluding diffusion). Since the suspension in the reactor is constantly fed by the reservoirs at a constant rate, the chemical composition of the aqueous solution in the reactor remains constant. Under these conditions the surfaces of the seed crystals will be conditioned and steady-state conditions will evolve. This setup can also be adapted to hydrothermal conditions ($T > 100°C$).

This experimental procedure ensures that homogeneous solid-solution phases can be synthesized in aqueous solution at repository-relevant temperatures. Also, distribution coefficients can be obtained under well defined experimental conditions and in particular, kinetic effects can be studied systematically.

With respect to experimental studies on SSAS systems it seems crucial that one understands the reaction mechanisms involved that occur throughout an experiment on a molecular level. In particular, when working under low- (room) temperature conditions in aqueous systems, kinetic effects, the formation of pre-cursor phases and metastability effects have to be taken into account critically and ruled out before treating experimental results as thermodynamic data. Ultimately, it depends on what kind of further experiments will be done with the co-precipitated solid-solution phases. Such co-precipitation experiments combined with or complemented by solubility measurements (thermodynamic equilibrium solubility, Lippmann, 1980, primary saturation solubility, Garrels & Wollast, 1978; Glynn & Reardon, 1990, stoichiometric saturation solubility, Thorstenson & Plummer, 1977) allow us to derive SSAS thermodynamics for the uptake of radionuclides even at (ultra)trace concentration levels. Besides the availability of state-of-the-art spectroscopic and microscopic techniques for sample characterization, understanding and controlling co-precipitation and solubility experiments is a key issue when discussing SSAS systems. In the following two sections, two examples will be discussed in more detail, in particular with respect to the applicability for long-term safety assessments for nuclear waste disposal.

5. Radium and barite

A number of 'classical' papers, including Marie Curie's work on the discovery of radium as well as H.A. Doerner and W.M. Hoskins' work on radium partitioning into barite have clearly demonstrated that radium can be effectively removed from aqueous solution by co-precipitation with barite (Curie *et al.*, 1898; Doerner & Hoskins, 1925; Hahn, 1936; Jucker & Treadwell, 1954; Gordon & Rowley, 1957; Langmuir & Riese, 1985).

The formation of 'radiobarite' scale during oil production is a well known issue (Kraemer & Reid, 1984; Hamlat *et al.*, 2003). Radiobarite is a sparingly soluble precipitate with a high radium content and associated radiation level (up to 1 kBq/g), which is thought to be a $Ra_xBa_{(1-x)}SO_4$ solid solution. Radiobarite is also found as a hydrothermal precipitate in nature (Randa *et al.*, 2009). Contact autoradiograms of such natural radiobarites show the spatial distribution of radium within the barite crystals. Although the radium concentration is very inhomogeneous within the crystals, the radium is not only limited to the surface volume. Radium-bearing barite is also frequently found in uranium mining tailings, where it may have a strong environmental impact due to its radiotoxicity. However, the concentration of radium in natural groundwaters as well as surface waters is usually extremely small ($<10^{-11}$ mol/l), which is, to a large extent a consequence of the small radium content in the rocks of the Earth's upper crust. Furthermore, the solubility of $RaSO_4$, which may be the solubility-limiting radium phase, is quite low. It is interesting to note that radium concentrations in natural systems stay below the solubility limit of $RaSO_4(s)$. $Ra_xBa_{(1-x)}SO_4$ solid-solution formation has been proposed to explain these phenomenological observations (Sturchio *et al.*, 1993; Grundl & Cape, 2006).

The aim of this section is to outline the interest of the radiogeochemistry community in solubility control with respect to long-term safety issues of nuclear-waste disposal, to briefly summarize the current knowledge on Ra uptake by barite and to indicate knowledge gaps.

Why would one want to consider the formation of $Ra_xBa_{(1-x)}SO_4$ solid solutions with respect to long-term safety aspects of nuclear-waste disposal? Significant amounts of barium are formed during nuclear fission of ^{235}U during reactor operation as fission products (mainly ^{134}Ba, ^{135}Ba, ^{138}Ba). ^{226}Ra forms within the decay chain of ^{238}U: ^{238}U (α: 4.47E9 a) ^{234}Th (β^-: 24.1 d) ^{234}Pa (β^-: 6.7 h) ^{234}U (α: 2.44E5 a) ^{230}Th (α: 7.7E4 a) ^{226}Ra (α: 1600 a). Natural groundwaters contain varying amounts of sulphate. Site investigations in various European countries indicate a range of 10^{-5}–10^{-1} mol/l of sulphate (Opalinus clay, Switzerland: 2.8E−2 molal) (Lauber *et al.*, 2000); Callovian-Oxfordian, France: 3.39E−2 molal (Gaucher *et al.*, 2006); Äspö, Sweden: 5.8E−3 molal (Metz *et al.*, 2003); Grimsel, Switzerland: 50.8E-6 molal (Schäfer *et al.*, 2004); Brine I, Germany: 0.18 molal water; Brine III, Germany: 0.06 molal water (Grambow *et al.*, 2000; Kienzler & Loida, 2001).

The formation of $Ra_xBa_{(1-x)}SO_4$ solid solutions with relevance to the long-term safety of nuclear-waste disposal may involve: (1) the co-precipitation of Ba and Ra which was simultaneously released from the waste matrix (spent fuel or HLW borosilicate glass); and (2) the dissolution of barite, which may be found as a constituent mineral of the geological barrier or which had formed as a secondary phase earlier, within the near field, and subsequent re-precipitation of a $Ra_xBa_{(1-x)}SO_4$ solid solution. Barite minerals may be found in sedimentary clayrock systems as well as a fracture-filling mineral in crystalline/granitic formations.

Aqueous speciation of radium depends on the actual geochemistry of the groundwater. Depending on the groundwater composition, the mono-sulphate, $RaSO_4(aq)$, mono-carbonate, $RaCO_3(aq)$, mono-hydroxo, $RaOH^+$, and mono-chloride, $RaCl^+$ complexes need to be considered (Langmuir & Riese, 1985; Hummel *et al.*, 2002, Table 1). The solubility product of $RaSO_4(s)$ was determined by Langmuir & Riese (1985) and Paige *et al.* (1998) to be in the range of $\log K_{sp}(RaSO_4) = -10.21$ to -10.38 (20–25°C). It turns out that the only experimental data available go back to the work of Lind *et al.* (1918) and Nikitin & Tomatscheff (1933). In the radiogeochemistry community, a value of -10.26 has been widely accepted (Hummel *et al.*, 2002). Also, the available data on the stability of $RaSO_4^0$ and $RaCl^+$ complexes are used frequently by the radiogeochemistry community (Hummel *et al.*, 2002). The solubility product of barite $\log K_{sp}$(barite) $= -9.9704$ and the aqueous speciation of Ba at room temperature are known with a sufficient level of certainty (Blount, 1977; Hummel *et al.*, 2002).

Complementary to the aqueous system, the solid (solution) phase(s) have been studied extensively. Barite has an orthorhombic structure (*Pnma*, $a = 8.87$ Å, $b = 5.45$ Å, $c = 7.14$ Å; Colville, 1967). Each unit cell consists of two $BaSO_4$ layers parallel to {001}, which are symmetrically related to each other *via* a 2_1 diad axis perpendicular to {001}. Barium is coordinated by 12 oxygen atoms at different distances belonging to different sulphate groups (Fig. 1). This structural configuration indicates significant structural flexibility. The only published study on the structure of $RaSO_4(s)$ is based on a powder X-ray diffraction pattern, that was indexed using the *Pnma* space group ($a = 9.16$ Å, $b = 5.5545$ Å, $c = 7.30$ Å; Weigel & Trinkl, 1968). A detailed structure refinement has not been performed. Intermediate compositions of the $Ra_xBa_{(1-x)}SO_4$ solid-solution series have not been characterized structurally and the applicability of Vegard's rule cannot be confirmed experimentally. Excess volumes of mixing are also unknown.

Barium and radium have similar ionic radii (Ba: 1.61 Å, Ra: 1.70 Å for 12-fold O coordination; Shannon, 1976). Therefore, the formation of a solid solution strongly suggested. However, due to a lack of direct structural data of the $Ra_xBa_{(1-x)}SO_4$ solid-solution series, it would be worthwhile to have a closer look at isostructural systems.

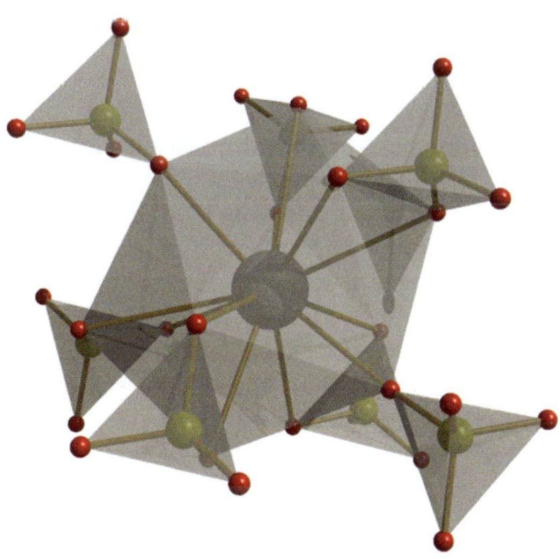

Fig. 1. Structure of barite, $BaSO_4$. Ba (and Ra in isostructural $RaSO_4$) is coordinated by 12 oxygen atoms.

Barite, $BaSO_4$, celestine, $SrSO_4$, anglesite, $PbSO_4$ and hashemite, $BaCrO_4$ have the same structure type, which is thought to be isostructural with $RaSO_4(s)$. The formation of binary solid-solution series among these four minerals is well known (Glynn, 2000). In particular, the $Sr_xBa_{(1-x)}SO_4$ solution series has been studied extensively (Felmy *et al.*, 1993; Prieto *et al.*, 1997; Becker *et al.*, 2000; Glynn, 2000; Prieto *et al.*, 2000; Pina *et al.*, 2004). A structure refinement of the celestine end-member structure confirmed the *Pmna* space group ($a = 8.34$ Å, $b = 5.36$ Å, $c = 7.30$ Å,

Garske & Peacor, 1965). X-ray diffraction data on $Sr_xBa_{(1-x)}SO_4$ powders with various Sr/Ba ratios indicate that the unit cell parameters *a* and *c* have linear behaviour for the whole solid-solution range, suggesting close to ideal behaviour according to Vegard's rule (Becker *et al.*, 2000). In contrast, *b* tends to deviate slightly, positively, from linearity.

The $Sr_xBa_{(1-x)}SO_4$ binary solid-solution series was, for a long time, considered to form a complete mixing series at room temperature in aqueous systems (Hanor, 2000 and references therein). However, the chemical composition of natural samples shows a strong bimodal distribution into Ba-rich and Sr-rich end members. Partition coefficients in the range $10^{-1.5}-10^{-4}$ were obtained from co-precipitation experiments (Goldschmidt, 1938; Starke, 1964; Prieto *et al.*, 1997), which implies strong partitioning of Ba into the solid-solution phase. However, this observation is related to the large difference in the solubility product of the end members (log K_{sp} (celestine) = -6.6319; Hummel *et al.*, 2002) rather than a non-ideal mixing behaviour.

On the contrary, a number of experimental studies have indicated a non-ideal mixing behaviour (Starke, 1964; Galinier *et al.*, 1989; Felmy *et al.*, 1993), which could be described by a regular solid-solution model with a Margules interaction parameter *W* ranging from 2000–6045 kJ/mol. Recently, first-principles calculations of the mixing properties of the barite–celestite series indicate that this solid solution behaves as non-ideal and non-regular (Becker *et al.*, 2000; Prieto *et al.*, 2000). At 25°C, the enthalpy of mixing shows a minimum around a mol fraction of $SrSO_4$, X_{SrSO4} of 0.5 due to an ordering tendency. The free energy of mixing shows two minima that delimit a wide and symmetric miscibility gap. The excess free energy of mixing requires a Guggenheim expansion series of five terms to be described, where the terms with odd exponents are zero as a consequence of the symmetric distribution of the mixing properties with composition. The unit-cell parameters *a* and *c* were calculated to have almost linear behaviour whereas *b* tends to deviate positively from linearity, in agreement with experimental values. In conclusion, despite extensive experimental as well as computational work done on the $Sr_xBa_{(1-x)}SO_4$ binary solid-solution series, some uncertainties still exist, including its ideal or non-ideal mixing behaviour.

Considering the differences (for 12-fold O coordination, Shannon, 1976) in ionic radii of Sr (1.44 Å)/Ba (1.61 Å) on the one hand and Ba (1.61 Å)/Ra (1.70 Å) on the other hand, it seems reasonable to argue that one can expect the $Ra_xBa_{(1-x)}SO_4$ solid-solution series to be closer to ideal mixing compared to the $Sr_xBa_{(1-x)}SO_4$ solid-solution series. The information available on the structure of $RaSO_4(s)$ is compatible with and supports the idea of a simple substitution scheme in the $Ra_xBa_{(1-x)}SO_4$ solid-solution series with no significant ordering phenomena. The concept of a regular, almost ideal, solid solution is sufficient to describe this system. Zhu (2004) reviewed the available data on the binary solid-solution systems isostructural with barite, including the $Ra_xBa_{(1-x)}SO_4$ solid-solution series. He derived one Margules interaction parameter of 887 kJ/mol based on the experimental data of Doerner & Hoskins (1925). It is intriguing to note, that Doerner & Hoskins (1925) concluded in their classical paper on co-precipitation of Ra with barite, that their experimental observations on element partitioning were not compatible with exclusive Ra adsorption reactions and that the observed Ra uptake by barite is only compatible when explained within the framework of crystal growth and structural incorporation.

Two geochemical model calculations will be used as examples to illustrate the influence of $Ra_xBa_{(1-x)}SO_4$ solid solution formation on Ra solubility: (1) in principle and (2) under repository-relevant conditions. The PHREEQC version 2.16.0 code (Parkhurst & Appelo, 1999) and the PSI/Nagra thermodynamic database (Hummel *et al.*, 2002) were used. The results are illustrated in a Lippmann diagram (Fig. 2) with the total solubility product, $\Sigma\Pi$, defined as:

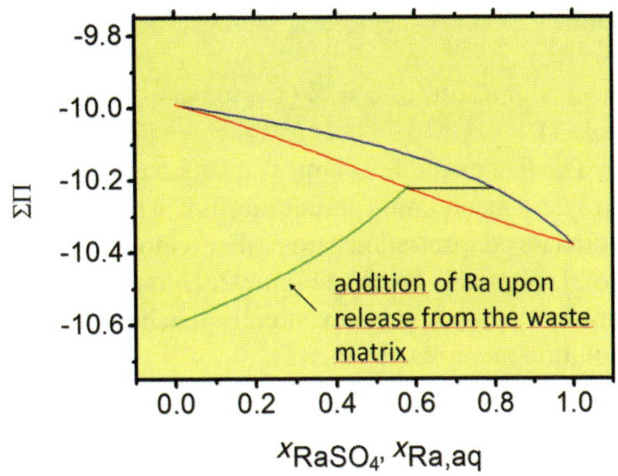

Fig. 2. Lippmann diagram for the $Ra_xBa_{(1-x)}SO_4$ solid solution series, $K_{sp}(BaSO_4) = 10^{-9.97}$, $K_{sp}(RaSO_4) = 10^{-10.38}$, assuming a regular solid solution model ($W_G = 887$ kJ/mol). Starting conditions: a(Ba): 5.0 E-6 molal, a(SO_4) = 5.0 E−6 molal. The total solubility product increases upon addition of Ra until the solutus curve is reached. The final equilibrium a(Ra) is 6.7E−6 molal.

$$\Sigma\Pi = \{SO_4^{2-}\}(\{Ra^{2+}\} + \{Ba^{2+}\}) \tag{2}$$

$\{Ra^{2+}\}$, $\{Ba^{2+}\}$ and $\{SO_4^{2-}\}$ represent the activities of the respective ions in aqueous solution. The solidus curve defined as:

$$\Sigma\Pi_{eq} = K_{RaSO_4(s)} x_{RaSO_4(s)} f_{RaSO_4(s)} + K_{barite} x_{barite} f_{barite} \tag{3}$$

with the solid-phase activity coefficients $f_{RaSO_4(s)}$ and f_{barite}, the solubility constant $K_{RaSO_4(s)}$ and K_{barite} and the respective mol fractions $x_{RaSO_4(s)}$ and x_{barite}. The solutus curve is defined as:

$$\Sigma\Pi_{eq} = 1/(x_{Ra,aq}/K_{RaSO_4(s)} f_{RaSO_4(s)} + x_{Ba,aq}/K_{barite} f_{barite}) \tag{4}$$

Here, the aqueous activity product is defined as:

$$x_{Ra,aq} = \{Ra^{2+}\}/(\{Ra^{2+} + \{Ba^{2+}\}) \tag{5}$$

The potential non-ideality is described with a regular solid-solution model and an interaction parameter of 887 kJ/mol as published by Zhu (2004). It seems reasonable to assume that the $Ra_xBa_{(1-x)}SO_4$ solid solution is very close to ideality with an interaction parameter ranging from 0 kJ/mole to 887 kJ/mol. Ra solubilities will be given for these extreme cases. The activity coefficient f of the solid phase is defined as:

$$RT \ln f_{RaSO_4(s)} = W(1 - x_{RaSO_4(s)}) \tag{6a}$$

and

$$RT \ln f_{barite} = W(1 - x_{barite}) \tag{6b}$$

The first example assumes an aqueous solution with 0.1 molal NaCl background electrolyte concentration in equilibrium with barite at room temperature. The barium and sulphate concentrations are 5.0E−6 molal. A Lippmann diagram for the $Ra_xBa_{(1-x)}SO_4$ solid solution is given in Figure 2. Upon addition of Ra to this system the total solubility product increases and eventually reaches the solutus curve and a $Ra_xBa_{(1-x)}SO_4$ solid solution precipitates (Fig. 2).

A calculation for more 'realistic' (nuclear-waste repository-relevant conditions) uses a sulphate concentration of 1.0E−2 molal and a NaCl background electrolyte concentration of 0.1 molal at room temperature. If such an aqueous solution is in equilibrium with barite, the Ba concentration will be 1.05E−8 molal. If Ra is introduced into such a model system, a $Ra_{0.012}Ba_{0.988}SO_4$ solid solution forms, assuming ideal mixing. The Ra concentration in the aqueous phase in equilibrium with this solid solution is 8.5E−11 molal. If a regular solid solution model with a Margules interaction parameter of 887 kJ/mol is assumed, the Ra solubility is slightly higher (1.2E−10 molal). Without any solid-solution formation and $RaSO_4(s)$ as the Ra solubility-limiting phase, the Ra concentration would be 4.1E−9. The Ra solubility is reduced by two orders of magnitude due to solid-solution formation under geochemical conditions relevant for the long-term safety of nuclear-waste disposal.

Two extreme cases are intended to illustrate the effect of the interaction parameter. If the interaction parameter is assumed to be unrealistically high, W = 3000 kJ/mol, the calculations give a Ra equilibrium concentration of 2.7E−10 molal. An unrealistically low value of W = −3000 kJ/mol gives a Ra concentration of 2.6E−11. The Ra equilibrium concentrations for these two extreme cases vary within one order of magnitude but for the highest calculated concentration it is still one order of magnitude lower than the solubility of pure $RaSO_4(s)$.

To conclude, the available experimental data on thermodynamic properties of radium in aqueous systems are very limited and many published thermochemical data for solid radium compounds are estimated. However, the thermodynamic data on $RaSO_4(s)$ as well as the stability constants for the $RaSO_4^0$ and $RaCl^+$ complexes are considered by the radiogeochemistry community to be reliable (*e.g.* Hummel *et al.*, 2002). The experimental data available strongly suggest that in the aqueous system radium-barium-sulfate, a binary solid solution $Ra_xBa_{(1-x)}SO_4$ forms, which may be described as a regular (almost ideal) solid solution. As already mentioned, Ra concentrations in natural waters never reach the solubility limit of $RaSO_4(s)$. In fact, in many cases, the observed Ra concentration can be modelled by assuming a $Ra_xBa_{(1-x)}SO_4$ solid solution as the Ra solubility-limiting phase. The observed Ra concentrations never exceed the Ra solubility as defined by a $Ra_xBa_{(1-x)}SO_4$ solid solution.

Taking these arguments all together it seems justifiable to consider $Ra_xBa_{(1-x)}SO_4$ solid-solution formation under waste-repository conditions and as the Ra-solubility-limiting phases in long-term safety assessment calculations for certain scenarios.

However, in almost all performance assessment calculations it is conservatively assumed that the solubility of Ra is controlled by $RaSO_4(s)$, which is the most insoluble radium phase under geochemical conditions typical for the near field of a waste repository. Some open questions and issues may explain why nuclear-waste management agencies tend to hesitate to include solid solution–aqueous solution equilibria in their long-term safety assessments. (1) What is the influence of other aqueous species on Ra uptake by barite? Long-term safety assessment calculations need to consider $Ra_xBa_{(1-x)}SO_4$ solid-solution formation in complex natural aqueous environments, with many competing cations present. Experimental data indicate that the presence of Sr promotes the structural incorporation of Ra into barite (Goldschmidt, 1938). However, this synergistic effect is so far not understood on a molecular level. (2) Reliable thermodynamic data on other aqueous species, in particular carbonate and hydroxo complexes, with radium. (3) Thermodynamic data for temperatures greater than room temperature. Depending on the concept for a nuclear waste repository, temperatures up to 200°C may be expected. (4) Applicability of solid-solution thermodynamics to describe $Ra_xBa_{(1-x)}SO_4$ solid solutions which form during co-precipitation (at least close to equilibrium) has been clearly demonstrated. However, in the case where a Ra-containing solution interacts with barite, it has not been demonstrated that the aqueous solution equilibrates completely with barite *via* dissolution (of barite) and re-precipitation (of a $Ra_xBa_{(1-x)}SO_4$ solid solution) or if solid solution formation is limited only to a thin near-surface volume.

6. Trivalent actinides and calcite

Calcite is a common mineral in most geological formations currently studied for the final disposal of HLW. It is a major component of clay rocks such as the Opalinus clay in Switzerland with 11% calcite (Bradbury & Baeyens, 1998) and the Bure argillite in France (ANDRA, 2005). It is also a key secondary alteration phase of cement-based structural components within a waste-repository system (Adler *et al.*, 1999). Therefore, co-precipitation/solid-solution formation and adsorption with actinides is an important aspect of a geochemical-based long-term safety assessment.

Calcite crystals formed in nature do contain significant amounts of rare earth elements (*REE*). The distribution of *REE* in marine limestone exhibit similar distribution patterns as normal seawater, which has led to the idea of direct *REE* co-precipitation (Parekh *et al.*, 1977; Tanaka *et al.*, 2004). In hydrothermally formed calcite crystals from Iceland, up to 100 ppm Eu was detected. A concave-normalized *REE* pattern was found with positive anomalies for Eu and Tb (Habermann *et al.*, 2001), which are supposed to reflect the *REE* composition of the hydrothermal fluids from which they were formed, thus also indicating direct co-precipitation. However, since the geochemical boundary conditions during calcite formation in natural systems are in many cases not well constrained, SSAS equilibria cannot be derived easily.

Co-precipitation of various trace metals with carbonate minerals has been studied experimentally for many years (Curti, 1999; Rimstidt *et al.*, 1998). In most co-precipitation studies, trace element uptake/incorporation by calcite was described in terms of an empirical partition coefficient. Only a few studies were actually intended to work out the SSAS thermodynamics. Nevertheless, a substantial amount of thermodynamic data for divalent metal-ion substitution into calcite as well as aragonite is available (Glynn, 2000). Furthermore, sorption studies with actinides (U(VI), Th(IV), Np(V), Pu(V)/Pu(VI) and Cm(III)) have been published (Carroll *et al.*, 1992; Shanbhag & Morse, 1982; Stumpf & Fanghänel, 2002; Zavarin *et al.*, 2005, Heberling *et al.*, 2005, 2008).

Here, the main focus will be on experimental studies on structurally incorporated trivalent f-elements, primarily the actinides Pu(III), Am(III) and Cm(III) and in particular chemical homologues, but non-radioactive, *REE*. Uptake of trivalent actinides, An(III), into a calcite host structure will be quantified in terms of a Henderson-Kracek partition coefficient, *D*:

$$D = \frac{x_{\mathrm{An(III)}}/x_{calcite}}{[\mathrm{An}^{3+}]/[\mathrm{Ca}^{2+}]} \qquad (7)$$

$x_{\mathrm{An(III)}}$ and $x_{calcite}$ represent the molar fractions of the actinide and Ca^{2+} in the precipitating solid. $[An^{3+}]$ and $[Ca^{2+}]$ are the molar concentrations in the aqueous solution. Where free-drift batch-type sorption/uptake experiments are performed, the chemical composition of the aqueous solution will change and only a heterogeneous Doerner-Hoskins partition coefficient can be derived.

Due to the similar ionic radii, structural incorporation of trivalent actinides into calcite presumably occurs on a calcite lattice site *via* replacement of Ca^{2+}. However, this apparently implies a complex heterovalent substitution mechanism, which needs to be characterized on a molecular level before SSAS thermodynamics can be derived. Ultimately, experimental data on SSAS equilibria and spectroscopic data should provide a coherent picture.

In the following, a critical overview of published experimental studies on An(III)/*REE* co-precipitation will be discussed. The first systematic experimental study on *REE* co-precipitation with calcite was published by Terakado & Masuda (1988). They used a free-drift batch-type experimental setup to study *REE* co-precipitation at room temperature (20–25°C). A Doerner-Hoskins partition coefficient of ~10 for all *REE* was derived. However, inherent to free-drift experiments is the fact that the solution composition changes throughout the experiment as a consequence of precipitation of a solid. Consequently, it is difficult to relate the partition coefficient obtained to a particular solution composition. Also, it was noted by Zhong & Mucci (1995) that Terakado & Masuda (1988) established a very high initial degree of supersaturation at the beginning of their experiments. Under these conditions, nucleation of calcite occurred before subsequent calcite crystal growth. It is not clear whether *REE* partitioning during nucleation follows the same rules as during crystal growth close to equilibrium, which primarily involves the attachment of *REE*s to kink sites on a calcite surface. Tsuno *et al.* (2002) observed that the presence of *REE* in a highly supersaturated calcium carbonate solution affects precipitation. At 25°C, instead of pure calcite a mixture of calcite and vaterite, a metastable polymorph of $CaCO_3$, precipitated. At slightly higher temperatures (50°C) almost 90% of the precipitated $CaCO_3$ was vaterite.

Vaterite represents a precursor phase which typically transforms into calcite and is quite often overlooked in experimental studies (due to the time span between the actual precipitation experiment and the application of analytical techniques. It was shown that the presence of *REE* in the aqueous solution and high degrees of supersaturation promotes the temporary formation of vaterite. This introduces a significant complication to *REE* co-precipitation experiments intended to derive data on *REE* partitioning and solid solution thermodynamics on the *REE*-calcite system. In such a case the macroscopically observed *REE* uptake by the precipitating carbonate phases is controlled by REE/calcite as well as *REE*/vaterite interactions. The behaviour of *REE* during the vaterite–calcite phase transition is not clear. There are two other metastable $CaCO_3$ phases, ikaite and calciumcarbonate-monohydrate, which may also precipitate from a supersaturated calcium carbonate solution, but have not been studied extensively.

Zhong & Mucci (1995) obtained *REE* partitioning data into calcite from (artificial) sea water using a constant addition experiments at 25°C. In contrast to the free drift experiments by Terakado & Masuda (1988), Zhong & Mucci (1995) established steady-state conditions in their experiments. Also, it seems reasonable to assume that the co-precipitation kinetics was surface-reaction controlled (and not transport/diffusion controlled), although this was not explicitly demonstrated on the basis of the apparent activation energy, for example, for co-precipitation or a non-linear rate *vs.* supersaturation dependence. Since the solution composition was kept constant throughout the experiment, an homogenous *REE*-containing calcite phase precipitated onto calcite seed crystals, where the *REE* incorporation was actually controlled by a reaction at the calcite-solution interface. The conditions were carefully chosen in order to ensure that the possible precipitation of $(REE)_2(CO_3)_3$ and $(REE)OHCO_3$ phases can be excluded.

A mean partition coefficient of 2340 for La and 65 for Yb was observed. The decreasing partitioning with increasing Z from La to Yb was explained by the size match between *REE* and Ca^{2+} on a calcite crystal structure site (Fig. 3). Within the calcite structure, Ca^{2+} occupies a 6-fold oxygen coordinate structure site, with an ionic radius of 1.00 Å, which is very similar to the ionic radius for 6-fold oxygen coordinated La with 1.032 Å. The lighter *REE* have a smaller ionic radius down to 0.868 Å for Yb (Eu: 0.947 Å).

In general, *REE* partitioning data in Zhong & Mucci's study are also dependent on the *REE* concentration in the aqueous solution. Log(*D*) increased linearly with $\log([REE]_{aq})$. For Eu, the partition coefficient varied between 200 at 10.0E−9 mol Eu and 1400 at 170.0E−9 mol Eu. This variation was, however, not linked with the variation in co-precipitation rate. In contrast to co-precipitation experiments with various divalent metal ions, *REE* co-precipitation seems not to show any dependence on the precipitation kinetics, at least in the investigated range of precipitation rates (3.3E−10 to 4.5E−8 mol g^{-1} s^{-1}). It should be emphasized though, that the trivalent *REE* inhibit calcite precipitation significantly, due to their strong adsorption behaviour. Many divalent metal ions do not show this strong inhibitory effect. The precipitation rate effect on the partition coefficient is related to the link between metal ion adsorption/incorporation kinetics and calcite growth (Ca^{2+}/CO_3^{2-} attachment to crystal surface).

Adsorption and coprecipitation of *REE*/calcite show similar trends, though *REE* partitioning (D) decreases from La to Lu significantly more compared to the distribution coefficient for adsorption (K_d). This indicates that co-precipitation is not simply related to adsorption but involves more complex interfacial reactions (*e.g.* dehydration, etc.) during crystal growth. From these observations it is not clear whether the measured *REE* partition coefficients represent equilibrium conditions, which could be a prerequisite for deriving thermodynamic data.

With respect to the Eu(III) incorporation or substitution mechanism, several macroscopic observations give at least some hints: (1) $\log(D_{\mathrm{REE}})$ *vs.* $\log(D_{\mathrm{Na}})$, (2) D_{REE} *vs.* $[\mathrm{REE}]_{aq}$, (3) $\log(D_{\mathrm{REE}})$ *vs.* $-\log(K_{sp,\mathrm{REE\text{-}carbonate}})$. Zhong & Mucci (1995) observed a correlation between the *REE* concentration in the precipitated carbonate and the partitioning of Na^+ from their artificial seawater solution. They suggest that this at least 'indicates that Na^+ might substitute for a Ca^{2+} parallel to the incorporation of Eu^{3+} in order to compensate for the charge. The observation was also taken as a hint that an actual calcite-like *REE*-containing solid solution had formed by a simple coupled substitution scheme:

$$2\mathrm{Ca}^{2+}_{calcite} \Longleftrightarrow \mathrm{Na}^{+}_{calcite} + \mathrm{Eu}^{3+}_{calcite} \tag{8}$$

For the light *REE* they also observed a correlation between the partitioning coefficient and the *REE* concentration in the aqueous solution. Apparently, the heavy *REE* did not show this clear correlation. They explained this observation by a two-step incorporation reaction during co-precipitation. Upon attachment of a REE^{3+} onto a calcite surface site, there will be charge-balance restrictions with respect to the possible precipitating ions in the vicinity. Only a monovalent or another REE^{3+} combined with a vacancy are allowed to attach next to a REE^{3+} surface site. Surface selectivity controls the correlation between partitioning coefficient and aqueous *REE* concentration. Since the heavy *REE* do not have such a high affinity to calcite surfaces (as indicated by their adsorption behaviour), they do not show this correlation. As long as the supersaturation/precipitation rate is not too high, this concept explains short-range ordering phenomena for light *REE* during co-precipitation from aqueous solution. Furthermore, a clear negative linear correlation of *REE* partitioning coefficient with solubility of *REE* carbonates was found. For thermodynamic equilibrium one expects, that the distribution coefficient is related to the ratio of the solubility products of the *REE*-carbonate and calcite $D_{\mathrm{REE}} = K^3_{sp,calcite}/K_{sp,(\mathrm{REE})_2(\mathrm{CO}_3)_3}$. However, Zhong & Mucci (1995) clearly state that they do not think that their observed *REE* partitioning data necessarily represent thermodynamic distribution coefficients. However, the apparent correlation is quite striking.

Lakshtanov & Stipp (2004) studied the co-precipitation of Eu(III) with calcite using a constant addition method also. They used a similar experimental setup as Zhong & Mucci (1995), although the investigated range of precipitation rate was larger (3.3E−10 to 4.5E−8 mol g^{-1} s^{-1}). The partition coefficient for a Eu concentration range between 0.1E−9 and 6.9E−9 mol was found to be 770 $\pm$ 290. Zhong & Mucci obtained a partition coefficient of 200 for 10.0E−9 mol Eu and of 1400 for 170.0E−9 mol Eu. Like Zhong & Mucci (1995), Lakshtanov & Stipp (2004) did not observe the dependence of Eu partitioning into calcite to be a function of the precipitation rate.

In addition to various precipitation rates, Lakshtanov & Stipp (2004) used two different samples with a difference in particle size (10–20 μm and 2–100 μm). It is well known from the crystal-growth community (Chernov, 1984) that precipitation rates are affected by particle size, especially if the precipitation rate is transport-controlled (rather than reaction controlled). It is argued that since there is no partition coefficient dependence on the precipitation rate, the measured partition coefficients do represent equilibrium conditions. Based on this assumption, various substitution mechanisms are tested with respect to the measured partition coefficient:

$$3Ca^{2+}_{calcite} \Longleftrightarrow 2Eu^{3+}_{calcite} + vac_{calcite} \tag{9}$$

Here, $vac_{calcite}$ represents a vacancy at the Ca site within the calcite structure.

$$Ca^{2+}_{calcite} \Longleftrightarrow EuOH^{2+}_{calcite} \tag{10}$$

$$2Ca^{2+}_{calcite} \Longleftrightarrow Na^{+}_{calcite} + Eu^{3+}_{calcite} \tag{11}$$

Following equation 9, Lakshtanov & Stipp (2004) showed that the experimental data followed the expected correlation between the aqueous Eu^{3+}/CO_3^{2-} concentration and the solubility product of the $Eu_2(CO_3)_3$ end member and the Eu mole fraction in the precipitated solid according to: $\{Eu^{3+}\}^2\{CO_3^{2-}\}^3 = K_{sp,Eu_2x(CO_3)_3}X^2_{Eu}$. Their experimental data follow this correlation, thus suggesting this substitution mechanism. Furthermore, since the experimental data also were compatible with the relation according to:

$$\{Eu^{3+}\}\{OH^-\}\{CO_3^{2-}\} = K_{sp,EuOHCO_3}/K_{sp,EuOH}X_{EuOHCO_3}, \tag{12}$$

a substitution mechanism according to equation 10 was also assumed. However, in contrast to Zhong & Mucci (1995), Lakshtanov & Stipp (2004) did not observe any correlation between Na^+ and Eu^{3+} partitioning. Consequently they ruled out a substitution mechanism according to equation 11. In summary, from these co-precipitation experiments, a complex substitution scheme is proposed, based on a ternary system with the end-members $CaCO_3$-$EuOHCO_3$-$Eu_2(CO_3)_3$.

Tanaka *et al.* (2004) published an experimental study on *REE* partitioning into calcite from constant addition experiments, also similar to Zhong & Mucci (1995). However, although they do not present partitioning data, but rather partitioning data normalized to Gd, it becomes clear that the results do not confirm earlier published data. Zhong & Mucci (1995) found a decreasing partitioning coefficient with increasing Z from La (2340) to Yb (65), whereas Tanaka *et al.* found a normalized *REE* pattern which shows the highest relative partition coefficient for Sm and significantly smaller partition coefficients for lighter or heavier REE. They claim that this is due to the differences in the chemical composition of the aqueous solution (compared to Zhong & Mucci, 1995). It is stated that in their experiments $REECO_3^+(aq)$ is the most abundant species, whereas under the experimental conditions used by Zhong & Mucci (1995)

$(REE)(CO_3)_2^-$ (aq) is the dominant species, thus explaining the difference in *REE* patterns. Unfortunately, the exact chemical composition of the aqueous solution is not presented in detail and this statement cannot be verified. However, the $(REE)CO_3^+(aq)/$ $(REE)(CO_3)_2^-(aq)$ does not change very much as a function of the CO_3^{2-} concentration (within the investigated *pH* range). Furthermore, Tanaka *et al.* used a higher Eu concentration, up to 100 ppb (= 680.0E−9 mol compared to 10.0E−9 and 170.0E−9mol in Zhong & Mucci's study). They concluded that *REE* partitioning into calcite is actually controlled by *REE* carbonate complexation in solution. They did not provide a concept for the molecular-level substitution mechanism. Also, since the chemical composition of the aqueous solution and of the precipitated solid-solution phase are not documented in detail, it is not possible to derive thermodynamic data. *REE* fractionation during calcite co-preciptation exhibits different trends between the Zhong & Mucci (1995) study and the Tanaka *et al.* (2004) study.

More recently, Curti *et al.* (2005) derived a thermodynamic model of the Eu-calcite SSAS system based on batch-type Eu sorption experiments at *pH* 13 combined with the experimental data from Zhong & Mucci (1995) as well as from Lakshtanov & Stipp (2004). It is assumed that in their sorption experiments, equilibrium was established between the aqueous solution and the re-precipitated/recrystallized Eu-containing calcite surface coating. Furthermore, the published *REE* coprecipitation experiments are also assumed to represent equilibrium, since there is apparently no correlation between precipitation rate and partition coefficient (see above). The Eu partitioning data could not be described by simple binary SSAS systems. Furthermore, it was not possible to fit the experimental data in terms of a SSAS system with the end members $CaCO_3$, $Eu_2(CO_3)_3$ and $EuNa(CO_3)_2$, as did Lakshtanov & Stipp (2004). Instead, they used an advanced thermodynamic 'inverse' modelling approach (DualTh GEM, Curti, 1999; Kulik, 2006) and predicted ΔG values for hypothetical end members $EuO(OH)$, $EuH(CO_3)_2$ and $EuO(CO_3)_{0.5}$. Only an ideal ternary solid solution with these end members allowed modelling of the experimental data consistently. Their thermodynamic approach leads to the assumption of incorporation of two distinctively different Eu species: (1) involving a coupled substitution $2Ca^{2+} \Leftrightarrow H^+ + Eu^{3+}$ and (2) the incorporation of an Eu-oxy-hydroxo species. The predicted Eu-oxy-hydroxo species involves significant structural relaxation within the calcite host structure, with a 9-fold O coordination of Eu. However, this structural model has not been confirmed by direct measurements, although time-resolved laser fluorescence spectroscopy data (see below) seem to be compatible with some of these assumptions. The EXFAS data by Elzinga *et al.* (2002) and Withers *et al.* (2003) were performed with *REE*-doped crystals that were grown at much lower-*pH* conditions.

To summarize, these studies have provided a substantial amount of data on empirical *REE* partitioning into calcite during calcite precipitation from aqueous solution at room temperature. Also, first attempts have been made to derive the molecular-level substitution mechanism and a thermodynamic solid-solution model. However, without direct structural information about the molecular-level incorporation and substitution mechanism, the interpretation of macrosopic observations remains ambiguous, especially as there seems to be no coherent substitution model in the studies presented above.

In the following, spectroscopic studies of structurally incorporated trivalent f-elements in calcite will be evaluated with respect to the substitution mechanism(s) which are responsible for the experimentally observed element partitioning (as outlined above). The local coordination of some *REE* and An(III) in a calcite host structure was studied using extended X-ray absorption fine structure (EXAFS) spectroscopy, time-resolved laser fluorescence spectroscopy (TRLFS) (for Eu(III) and Cm(III)) and IR spectroscopy (Piriou *et al.*, 1997; Elzinga *et al.*, 2002; Stumpf & Fanghänel, 2002; Withers *et al.*, 2003).

Elzinga *et al.* (2002) studied Nd, Sm, Dy, Yb incorporated in calcite crystals co-precipitated from aqueous solution at room temperature using EXAFS. The *REE*-doped calcite crystals were prepared by a batch-type synthesis procedure, where precipitation is essentially transport/diffusion controlled. Although, the *pH* remained constant in their experiments, the Ca and *REE* concentration of the aqueous solution changed during the synthesis. It was assumed that the structural-incorporation mechanism does not change depending on the element concentrations through the experiments. Some of the crystals studied were chemically homogeneous, whereas some crystals did exhibit compositional zoning, suggesting some kinetic effects during co-precipitation.

EXFAS spectra indicate that the *REE*s studied occupy a Ca lattice site within the calcite structure. For Dy and Yb they found the expected 6-fold oxygen coordination, whereas Nd and Sm seem to be 7-fold oxygen coordinated. Also, Nd-O and Sm-O distances are longer than consistent with the sum of the ionic radii. 7-fold *REE*-oxygen coordination implies a significant distortion of the local calcite structure, which could be a consequence of a bidentate ligation of a carbonate group. They also point out that due to the similarities in ionic radius, a similar behaviour can be expected for Am^{3+} and Cm^{3+}. There was no indication of a *REE* in the 2nd coordination shell.

Withers *et al.* (2003) characterized Nd-doped calcite crystals, which were prepared by the same procedure as Elzinga *et al.* (2002, see above) with temperature-dependent near-IR and EXFAS spectroscopy. Their EXAFS data suggest that Nd occupies a Ca lattice site with a coordination number of 6.6 oxygen with a substantial local lattice dilation. Furthermore, a large Debye-Waller factor was obtained which indicates 'some' difference from the ideal calcite structure. From IR measurements, they found a rather broad $^4F_{3/2}$ line without discrete structures (which is typical for other studied Nd-doped crystals). This observation would be expected for a broad and continuous distribution of ligand-field environments, which are typical for glassy rather than crystalline host materials. Furthermore, it is pointed out that broad absorption bands have been observed frequently in *REE* co-precipitated crystals as a consequence of a locally strained structure due to incorporated molecular water. Their IR spectra indicate H_2O-related absorption bands in their Nd-doped calcite crystals. However, these water absorption bands could also be linked to fluid inclusions. They conclude that the incorporation of a $NdOH^{2+}$ species would probably explain the observed IR spectra. It should be mentioned that small amounts of metastable vaterite were detected by powder XRD. It is hypothesized that structurally incorporated Eu^{3+} in calcite should have a 6-fold oxygen coordination. Therefore, such a Eu-containing calcite would be more stable than a Nd-containing calcite with its significant local structural deformation. However, this hypothesis is certainly very speculative.

Marques-Fernandes *et al.* (2008a) studied Cm(III)-doped calcite crystals by TRLFS. Their samples were prepared by co-precipitation near equilibrium in a mixed-flow reactor under steady-state surface-reaction-controlled conditions. The Cm(III) fluorescence signal indicated the presence of three Cm(III) species. Considering the fluorescence emission peak position and the and the corresponding fluorescence emission lifetimes, the Cm(III) species could be identified as: (1) a surface-sorbed Cm(III) species; (2) a structurally incorporated Cm(III) species; and (3) probably a structurally incorporated $CmOH^{2+}$ species, which is only present in case of high-*pH* conditions during co-precipitation. Site-selective TRLFS on Eu(III)-doped calcite revealed even more structural detail (Marques-Fernandes *et al.*, 2008b; Schmidt *et al.*, 2008). By means of selective excitation of the $^5D_0 \rightarrow {}^7F_0$ transition at low temperatures ($T < 20$ K). The non-degeneracy of the Eu^{3+} ions' fluorescing states gives rise to a single emission line for each species and facilitates determination of the number of non-equivalent sites. Three different Eu^{3+} species could be identified: (1) Eu^{3+} incorporated into the calcite structure on the (nearly) octahedral Ca^{2+} site (point group C_{3i}); (2) Eu^{3+} incorporated into the calcite structure, but the crystal field shows a much lower symmetry; and (3) Eu^{3+} adsorbed onto the calcite surface having 1–2 H_2O ligands left in its first coordination sphere and showing very little symmetry.

The structurally incorporated Cm(III) and Eu(III) species could only be identified if Na (NaCl as a background electrolyte) was present during co-precipitation. In case KCl was used, no structural incorporation was found. This observation clearly proves that a simple coupled substitution scheme $2Ca^{2+}_{calcite} \Leftrightarrow Na^{+}_{calcite} + An^{3+}_{calcite}$ is the dominant substitution scheme during co-precipitation of trivalent actinides, An(III) and lanthanides with calcite, under waste-repository-relevant conditions. This was only possible by combining state of the art spectroscopic techniques with well controlled co-precipitation experiments (Figs 4, 5, 6).

The identification of the molecular-level substitution mechanism needs to be complemented by thermodynamic data, in particular with the solubility product of the $NaEu(CO_3)_2$ end-member and the non-ideality of the solid solution including the maximum concentration for trivalent actinides and lanthanides (mixing gap). This information will be used as baseline information for future thermodynamic modelling (Vinograd *et al.*, 2009).

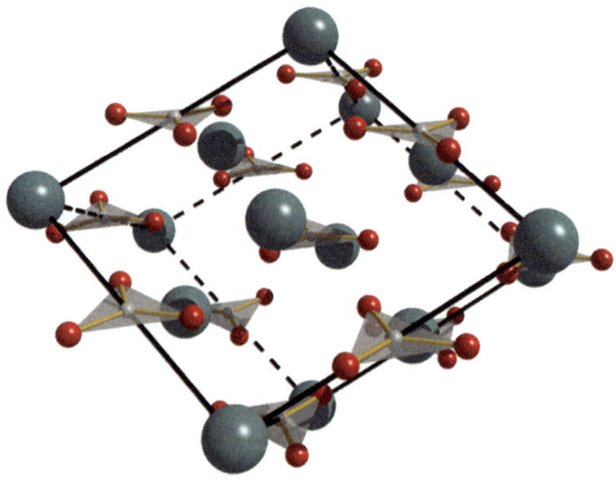

Fig. 3. Structure of calcite. Ca is coordinated by six oxygen atoms, which belong to six different carbonate groups.

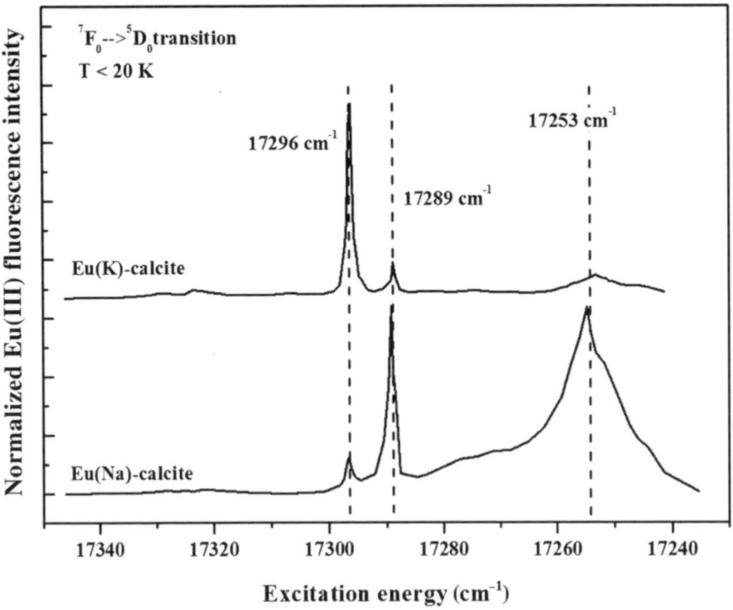

Fig. 4. Excitation spectra of the $^7F_0 \rightarrow {}^5D_0$ transition in Eu-doped calcite in the presence of K; and Eu-doped calcite in the presence of Na. Both spectra were recorded at temperatures below 20 K. F_{em} (EuIII): normalized EuIII fluorescence intensity; E_{exc}: excitation energy (after Stumpf *et al.*, 2008).

7. Summary and conclusions

During the last two decades, there has been tremendous progress in our understanding of radionuclide sorption onto/into minerals including structural incorporation and subsequent solid-solution formation with respect to geochemical conditions relevant for the long-term safety assessment of nuclear-waste disposal. In addition to the systems discussed here, a number of relevant solid solution–aqueous solution systems are currently being studied by the radiogeochemistry community, including radionculide uptake by powellite, $CaMoO_4$ (Bosbach *et al.*, 2004; Vinograd *et al.*, 2008), clay minerals (Finck *et al.*, 2008, 2009), calcium-silicate-hydrate (CSH) and ettringite cement phases (Kulik & Kersten, 2001), pyrite, FeS_2 (Liu *et al.*, 2008; Scheinost *et al.*, 2008) and layered-double-hydroxide (LDH) phases (Curtius & Kattilparampil, 2005; Curtius *et al.*, 2008, 2009). Since 2008, the German Helmholtz Association (HGF) has funded a virtual institute on 'Advanced Solid–Aqueous RadioGeochemistry' which is focusing on structural incorporation of radionuclides into relevant host minerals in the geosphere and the near-field of a nuclear-waste repository system as baseline information for improving long-term safety assessments with six international partners (Karlsruhe Institute of Technology, Germany [formerly Karlsruhe University and Research Centre Karlsruhe], University of Frankfurt, Germany, University of Oviedo,

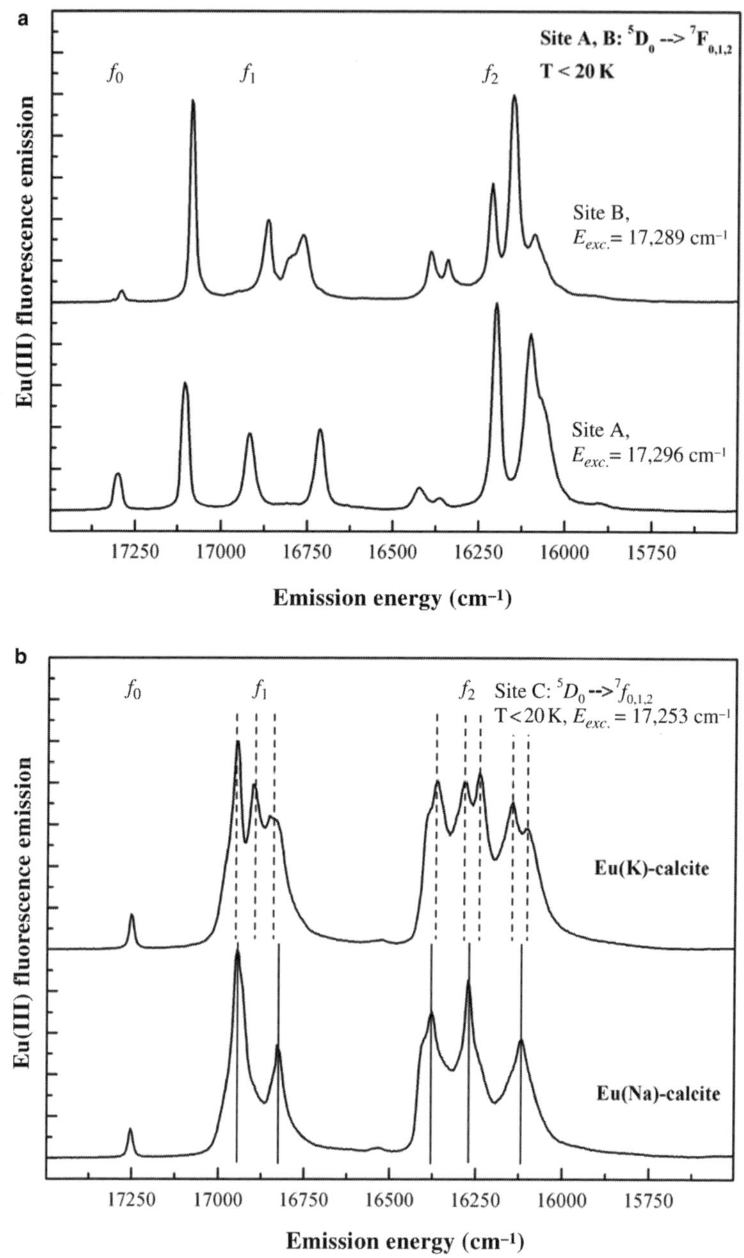

Fig. 5. (***a***) Fluorescence emission spectra of two distinct Eu sites in Eu-doped calcite in the presence of Na. Both spectra are strongly structured, showing maximum splitting in all bands. (***b***) Fluorescence emission spectra of a 3rd site in Eu-doped calcite in the presence of Na and K. The increased multiplicity of the signals for Eu(K) calcite is due to reduced symmetry caused by the lack of Na^+ (after Stumpf *et al.*, 2008).

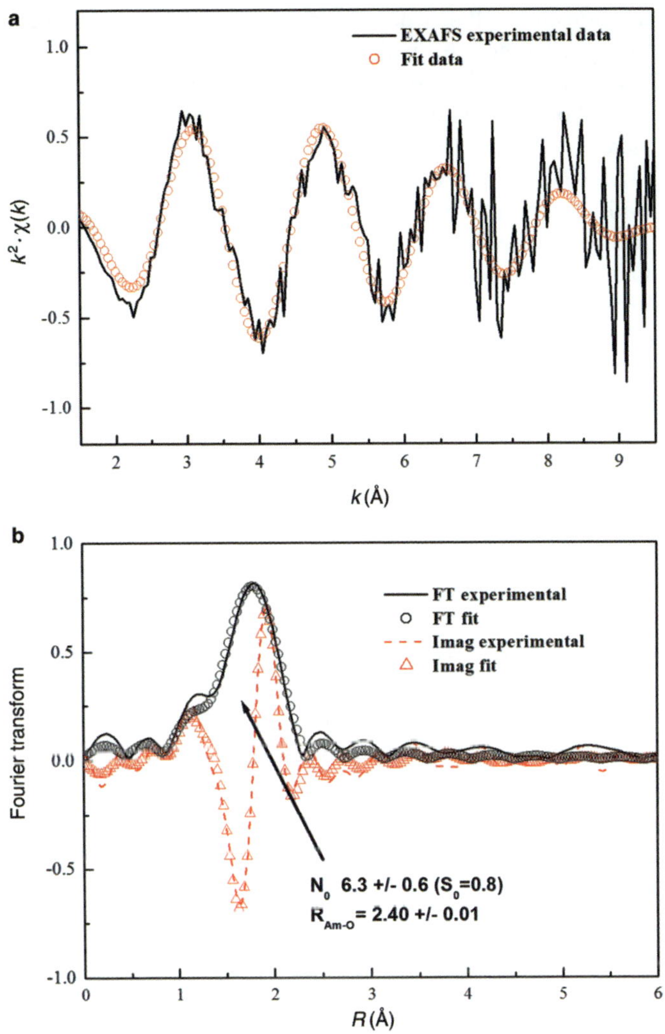

Fig. 6. L_{III}-edge k2-weighted EXAFS spectrum (***a***) and corresponding FT (***b***) of Am(III) in doped calcite (after Stumpf *et al.*, 2006).

Spain, Paul Scherrer Institute, Switzerland, and Forschungszentrum Jülich, Germany). The Nuclear Energy Agency (NEA) has published a state of the art report on solid-solution thermodynamics (Bruno *et al.*, 2007), thus demonstrating the interest of the radiogeochemistry community in data on relevant SSAS systems.

The long-term safety of HLW disposal in a deep geological formation is probably the most critical issue when discussing the pros and cons of nuclear energy. Current knowledge indicates that the radioactive inventory of HLW disposal in a waste repository in a deep geological formation will be isolated for at least 10,000 years based on technical/engineering concepts. The current challenge is to derive a chain of arguments regarding the safety beyond 10,000 years. Fortunately, only relatively small volumes of HLW need to be disposed of (compared to other sources for electricity production *e.g.* coal firing power plants) and its composition is very well known. With respect to the long-term safety assessment ($\gg$10,000 years), it is clear that it is not possible to predict exactly what will happen in the next 1 million years within a nuclear-waste repository system, in the sense of a weather forecast. However, for specific scenarios it is possible to assess the long-term safety over such geological time-scales.

The development of relevant scenarios – What *may* happen in the next 1 million years? – is a very important component for the long-term safety assessment. Due to the phenomenological complexity of the evolution of a nuclear-waste repository system and the large space and time scales that must be modelled, performance-assessment models do not take into account the full mathematical complexity of our understanding of every relevant process (*e.g.* complexation, adsorption, solid-solution formation, etc.). All performance-assessment models rely on simplified mathematical representations of each process which are chosen so as to reproduce the essential characteristics of the phenomena. The safety case for a repository system contains a whole set of long-term predictions for all relevant scenarios, thus demonstrating its long-term safety.

For specific scenarios, it is essential to include structural incorporation of certain radionuclides into minerals in order to model radionuclide migration within a nuclear waste repository system adequately in order to demonstrate its long-term safety. Basic research on relevant solid solution–aqueous solution systems is needed in order to fill knowledge and data gaps.

References

Adler, M., Mäder, U. & Waber, N. (1999) High-pH alteration of argilaveous rocks: An experimental study. *Schweizerische Mineralogische und Petrographische Mitteilungen*, **79**, 445–454.

ANDRA (2005) *Dossier Argile 2005: Safety Evaluation of a Geological Repository.*

Becker, U., Fernández-González, A., Prieto, M., Harrison, R.E. & Puntis, A. (2000) Direct calculation of thermodynamic properties of the barite/celestite solid solution from molecular principles. *Physics and Chemistry of Minerals*, **27**, 291–300.

Blount, C.W. (1977) Barite solubilities and thermodynamic quantities up to 300°C and 1400 bars. *American Mineralogist*, **62**, 942–957.

Bosbach, D., Rabung, T., Dardenne, K. & Fanghänel, T. (2004) Trivalent actinide coprecipitation with powellite ($CaMoO_4$): Secondary solid solution formation during HLW borosilicate-glass dissolution. *Radiochimica Acta*, **92**, 639–643.

Bradbury, M.H. & Baeyens, B. (1998) A physicochemical characterisation and geochemical modelling approach for determining porewater chemistries in argillaceous rocks. *Geochimica et Cosmochimica Acta*, **62**, 783–795.

Bruno J., Bosbach D., Kulik D. & Navrotsky A. (2007) *Chemical thermodynamics of solid solutions of interest in radioactive waste management: A state-of-the art report.* Chemical Thermodynamics Series, **10**, 292. OECD Nuclear Energy Agency, Paris.

Carroll, S.A., Bruno, J., Petit, J.C. & Dran, J.C. (1992) Interactions of U(VI) Nd and Th(IV) at the calcite-solution interface. *Radiochimica Acta*, **58**, 245–252.

Chernov, A.A. (1984) *Modern Crystallography III. Crystal Growth.* Springer-Verlag, Berlin, 246 pp.

Colville, A.A. (1967) A refinement of the structure of barite. *American Mineralogist*, **52**, 1877–1880.

Curie, M.P., Curie, Mme. P. & Bémont, M.G. (1898) On a new, strongly radioactive substance, contained in pitchblende. *Comptes Rendus*, **127**, 1215–1217.

Curti, E., Kulik, D.A. & Tits, J. (2005) Solid solutions of trace Eu(III) in calcite: thermodynamic evaluation of experimental data over a wide range of pH and pCO_2. *Geochimica et Cosmochimica Acta*, **69**, 1721–1737.

Curti, E. (1999) Coprecipitation of radionuclides with calcite: estimation of partition coefficients based on a review of laboratory investigations and geochemical data. *Applied Geochemistry*, **14**, 433–445.

Curti, E., Kulik, D.A. & Tits, J. (2005) Solid solutions of trace Eu(III) in calcite: Thermodynamic evaluation of experimental data over a wide range of pH and pCO2. *Geochimica et Cosmochimica Acta*, **69**, 1721–1737.

Curtius, H., Ufer, K. & Dardenne, K. (2009) Preparation and characterisation of Zr-IV-containing Mg-Al-Cl layered double hydroxide. *Radiochimica Acta*, **97**, 423–428.

Curtius, H., Paparigas, Z. & Kaiser, G. (2008) Sorption of selenium on Mg-Al and Mg-Al-Eu layered double hydroxides. *Radiochimica Acta*, **96**, 651–655.

Curtius, H. & Kattilparampil, Z. (2005) Sorption of iodine on Mg-Al-layered double hydroxide. *Clay Minerals*, **40**, 455–461.

Doerner, H.A. & Hoskins, W.M. (1925) Co-precipitation of radium and barium sulfates. *Journal of American Chemical Society*, **47**, 662–675.

Elzinga, E.J., Reeder, R.J., Withers, S.H., Peale, R.E., Mason, R.A., Beck, K.M. & Hess, W.P. (2002) EXAFS study of rare-earth element coordination in calcite. *Geochimica et Cosmochimica Acta*, **66**, 2875–2885.

Felmy, A.R., Rai, D. & Morre, D.A. (1993) The solubility of (Ba,Sr)SO4 precipitates: Thermodynamic equilibrium and reaction path analysis. *Geochimic et Cosmochimica Acta*, **57**, 4345–4363.

Finck, N., Stumpf, T., Walther, C. & Bosbach, D. (2008) TRLFS characterization of Eu(III)-doped synthetic organo-hectorite. *Journal of Contaminant Hydrology*, **102**, 253–262.

Finck, N., Schlegel, M.L. & Bosbach, D. (2009) Sites of Lu(III) sorbed to and coprecipitated with hectorite. *Environmental Science & Technology* **43**, 8807–8812.

Fouke, B.W. & Reeder, R.J. (1992) Surface Structural Controls on Dolomite Composition – Evidence from Sectoral Zoning. *Geochimica et Cosmochimica Acta*, **56**, 4015–4024.

Galinier, C., Dandurand, J.L., Souissi, F. & Schott, J. (1989) Sur le caracters non-ideal des solution solides (Ba,Sr)SO4: mise en evidence et determinationdes parameters thermodynmique par des essays de dissolution a 25°C. *Comptes Rendus de l'Academi des Sciences (Series II)*, **308**, 1363–1368.

Gamsjaeger, H., Koenigsberger, E. & Preis, W. (2000) Lippmann diagrams: theory and application to carbonate systems. *Aquatic Geochemistry*, **6**, 119–132.

Garrels, R. & Wollast, R. (1978) Discussion of 'Equilibrium criteria for two-component solid reacting with fixed composition in an aqueous phase. Example: The magnesian calcites'. *American Journal of Science*, **278**, 1469–1474.

Garske, D. & Peacor, D.R. (1965) Refinement of structure of celestite $SrSO_4$. *Zeitschrift für Kristallographie*, **121**, 204–206.

Gaucher, E., Blanc, P., Bardot, F., Braibant, G., Buschaert, S., Crouzet, C., Gautier, A. Girard, J.P., Jacquot, E., Lassin, A., Negrel, G., Tournassat, Vinsot, A. & Altmann, S. (2006) Modelling the porewater chemistry of the Callovian-Oxfordian formation at a regional scale. *Comptes Rendus Geoscience*, **338**, 917–930.

Geiger, C. (editor) (2001) *Solid Solutions in Silicate and Oxide Systems*. EMU Notes in Mineralogy, **3**, Eötvös University Press, Budapest.

Glynn, P.D. (2000) Solid solution solubilities and thermodynamics: Sulfates, carbonates and halides. In: *Sulfate Minerals: Crystallography, Geochemistry and Environmental Significance* (J.L. Jambor & D.K. Nordstrom, editors). Reviews in Mineralogy and Geochemistry, **40**, 481–511. Mineralogical Society of America, Washington D.C. and the Geochemical Society, St. Louis, Missouri, USA.

Glynn, P.D. & Reardon, E.J. (1990) Solid-solution aqueous-solution equilibria: thermodynamic theory and representation. *American Journal of Science*, **290**, 164–201.

Goldschmidt, B. (1938) Sur la précipitation mixte des sulfates de baryum et strontium. *Comptes Rendus Hebdomdaires des séances de l'Academie des Sciences*, **206**, 1110.

Gordon, L. & Rowley, K. (1957) Coprecipitation of radium with barium sulfate. *Analytical Chemistry*, **29**, 34–37.

Grambow, B., Loida, A., Martinez-Esparza, A., Díaz-Arocas, P., de Pablo, J., Paul, J.L., Marx, G., Glatz, J.P., Lemmens, K., Ollila, K. & Christensen, H. (2000) *Long-term safety of radioactive waste disposal. Source Term for Performance Assessment of Spent Fuel as a Waste Form.* Wissenschaftliche Berichte FZKA 6420, Forschungszentrum Karlsruhe, Karlsruhe, Germany, 355 pp.

Grambow, B. (2008) Mobile fission and activation products in nuclear waste disposal. *Journal of Contaminant Hydrology*, **102**, 180–186.

Grundl, T. & Cape, M. (2006) Geochemical factors controlling radium activity in a sandstone aquifer. *Ground Water*, **44**, 518–527.

Habermann, D., Niklas, J.R., Meijer, J., Stephan, A. & Gotte, T. (2001) Structural point defects in 'Iceland spar' calcite. *Nuclear Instruments and Methods in Physics Research B*, **181**, 563–569.

Hahn, O. (1936) *Applied Radiochemistry*. Cornell University Press, Ithaca, New York, 257 pp.

Hamlat, M.S., Kardi, H. & Fellag, H. (2003) Precipitate containing norm in the oil industry: modeling and laboratory experiments. *Applied Radiation and Isotopes*, **59**, 95–99.

Hanor, J.S. (2000) Barite–celestite geochemistry and environments of formation. In: *Sulfate Minerals: Crystallography, Geochemistry and Environmental Significance* (J.L. Jambor & D.K. Nordstrom, editors). Reviews in Mineralogy and Geochemistry, **40**, 193–275. Mineralogical Society of America, Washington D.C. and the Geochemical Society, St. Loius, Missouri, USA.

Heberling, F., Denecke, M.D. & Bosbach, D. (2005) Neptunium(V) co-precipitation with calcite. *Environmental Science & Technology*, **42**, 471–476.

Heberling, F., Brendebach, B. & Bosbach, D. (2008) Neptunium(V) adsorption to calcite. *Journal of Contaminant Hydrology*, **73**, 246–252.

Hummel, W., Berner, U.R., Curti, E., Pearson, F.J. & Thoenen, T. (2002) *Nagra/PSI chemical thermodynamic data base 01/01*. Universal Publishers, Parkland, Florida, USA.

IAEA (1994) *Classification of Radioactive Waste: a Safety Guide.* Safety Series 111-G-1.1. RADWASS Programme, Vienna.

Jucker, H. & Treadwell, W.D. (1954) Über die Mitfällung von Radium mit Bariumsulfat. *Helvetica Chimica Acta*, **37**, 2002–2010.

Kienzler, B. & Loida, A. (2001) *Endlagerrelevante Eigenschaften von hochradioaktiven Abfallprodukten. Charakterisierung und Bewertung – Empfehlungen des Arbeitskreises HAW-Produkte*. Forschungszentrum Karlsruhe Wissenschaftliche Berichte FZKA 6651.

Kulik, D.A. (2006) Dual-thermodynamic estimation of stoichiometry and stability of solid solution end members in aqueous-solid solution systems. *Chemical Geology*, **225**, 189–212.

Kulik, D.A. & Kersten, M. (2001) Aqueous solubility diagrams for cementitious waste stabilization systems: 2. End-member stoichiometries of ideal calcium silicate hydrate solid solutions. *Journal of the American Ceramic Society*, **84**, 3017–3026.

Kraemar, T.F. & Reid, D.F. (1984) The occurrence and behaviour of radium in saline formation water of the Gulf coast region. *Chemical Geology (Isotope Geoscience)*, **2**, 153–174.

Lakshtanov, L.Z. & Stipp, S.L.S. (2004) Experimental study of europium(III) coprecipitation with calcite. *Geochimica et Cosmochimica Acta*, **68**, 819–827.

L'Heureux, I. & Jamtveit, B. (2002) A model of oscillatory zoning in solid solutions grown from aqueous solutions: Applications to the $(Ba,Sr)SO_4$ system. *Geochimica et Cosmochimica Acta*, **66**, 417–429.

Langmuir, D. & Riese, A.C. (1985) The thermodynamic properties of radium. *Geochimica et Cosmochimica Acta*, **49**, 1593–1601.

Lauber, M., Baeyens, B. & Bradbury, M.H. (2000) *Physico-chemical Characterisation and Sorption Measurements of Cs, Sr, Ni, Eu, Th, Sn and Se on Opalinus Clay from Mont Terri.* Paul Scherrer Institut Bericht PSI 00-10,Villigen, Switzerland.

Lind, S.C., Underwood, J.E. & Whittemore, C.F. (1918) The solubility of pure radium sulfate. *Journal of the American Chemical Society*, **40**, 465.

Lippmann, F. (1980) Phase diagrams depicting aqueous solubility of binary mineral systems. *Neues Jahrbuch für Mineralogie mit Abhandlungen*, **139**, 1–25.

Liu, X., Fattahi, M., Montavan, G. & Grambow, B. (2008) Selenide retention onto pyrite under reducing conditions. *Radiochimica Acta*, **96**, 473–479.

Marques-Fernandes, M., Schmidt., M., Stumpf, T., Walther, C., Bosbach, D., Klenze, R. & Fanghändel, T. (2008a) Site-selective time-resolved laser fluorescence spectroscopy of Eu^{3+} in calcite. *Journal of Colloid and Interface Science*, **321**, 323–331.

Marques-Fernandes, M., Stumpf, T., Rabung, R., Bosbach, D. & Fanghänel, T. (2008b) Incorporation of trivalent actinides into calcite: A time resolved laser fluorescence spectroscopy (TRLFS) study. *Geochimica et Cosmochimica Acta*, **72**, 464–474.

Metz, V., Kienzler, B. & Schüßler, W. (2003) Geochemical evaluation of different groundwater-host rock systems for radioactive waste disposal. *Journal of Contaminant Hydrology*, **61**, 265–279.

Nagra (2002) *Project Opalinus Clay, Demonstration of disposal feasibility for spent fuel, vitrified high-level waste with long-lived intermediate-level waste.* Safety Report, Technical report 02–05.

Nikitin, B. & Tolmatscheff, P. (1933) Ein Beitrag zur Gültigkeit des Massenwirkungsgesetzes. II. Quantitative Bestimmung der Löslichkeit des Radiumsulfats in Natriumsulfatlösungen und in Wasser. *Zeitschrift für physikalische Chemie A*, **167**, 260–272.

OECD Nuclear Energy Agency (1999) *Progress towards geological disposal of radioactive waste: Where do we stand?* OECD Nuclear Energy Agency, Paris.

Ondraf/Nirond (2001) Safety assessment and feasibility. Interim Report 2 (Safir 2), Nirond 2001–05, France.

Paige, C.R., Kornicker, W.A., Hileman, O.E. & Snodgrass, W.J. (1998) Solution equilibria for uranium ore processing: The $BaSO_4$-H_2SO_4-H_2O system and the $RaSO_4$-H_2SO_4-H_2O system. *Geochimica et Cosmochimica Acta*, **62**, 15–23.

Parekh, P.P., Möller, P., Dukski, P. & Bausch, W.M. (1977) Distribution of trace elements between carbonate and non-carbonate phases in limestone. *Earth and Planetary Science Letters*, **34**, 39–50.

Paquette, J. & Reeder, R.J. (1990) New Type of compositional zoning in calcite – Insights into crystal-growth mechanisms. *Geology*, **18**, 1244–1247.

Paquette, J. & Reeder, R.J. (1995) Relationship between surface-structure, growth-mechanism, and trace-element incorporation in calcite. *Geochimica et Cosmochimica Acta*, **59**, 735–749.

Parkhurst, D.L. & Appelo, C.A.J. (1999) *User's guide to PHREEQC (Version 2) – A computer program for speciation, batch-reaction, one-dimensional transport, and inverse geochemical calculations.* U.S Geological Survey Water-Resources Investigations Report **99-4259**, 310.

Pina, C.M., Putnis, A. & Astilleros, J.M. (2004) The growth mechanisms of solid solutions crystallising from aqueous solution. *Chemical Geology*, **204**, 145–161.

Piriou, B., Fedoroff, M., Jeanjean, J. & Bercis, L. (1997) Characterization of the sorption of Europium(III) on calcite by site-selective and time-resolved luminescence spectroscopy. *Journal of Colloid and Interface Science*, **194**, 440–447.

Prieto, M., Putnis, A. & Fernandez-Diaz, L. (1993) Crystallization of solid-solutions from aqueous-solutions in a porous-medium – Zoning in $(Ba,Sr)SO_4$. *Geological Magazine*, **130**, 289–299.

Prieto, M., Fernández-Gonzalez, A., Putnis, A. & Fernández-Diaz, L. (1997) Nucleation, growth and zoning phenomena in crystallising $(Ba,Sr)CO_3$, $Ba(SO_4,CrO_4)$, $(Ba,Sr)SO_4$, and $(Cd,Ca)CO_3$ solid solutions from aqueous solutions. *Geochmica et Cosmochimica Acta*, **61**, 3383–3397.

Prieto, M., Fernández-González, A., Becker, U. & Putnis, A. (2000) Computing Lippmann diagrams from direct calculation of mixing properties of solid solutions: Application to the barite-celestite system. *Aquatic Geochemistry*, **6**, 133–146.

Putnis, A., Fernández-Díaz, L. & Prieto, M. (1992) Experimentally produced oscillatory zoning in the $(Ba,Sr)SO_4$ solid-solution. *Nature*, **358**, 743–745.

Randa, Z., Ulrych, J., Turek, K., Mihaljevic, M., Adamovic, J. & Mizera, J. (2009) Radiobarites from the Cenozoic volcanic region of the Bohemian Massif: radiochemical study, history, and lead isotopic composition. *Journal of Radioanalytical and Nuclear Chemistry*, 10.1007/s10967-009-0095-6.

Rimstidt, J.D., Balog, A. & Webb, J. (1998) Distribution of trace elements between carbonate minerals and aqueous solutions. *Geochimica et Cosmochimica Acta*, **62**, 1851–1863.

Schäfer, T., Geckeis, H., Bouby, M., & Fanghänel, T. (2004) U, Th, Eu and colloid mobility in a granite fracture under near-natural flow conditions. *Radiochimica Acta*, **92**, 731–737.

Scheinost, A., Kirsch, R., Banerjee, D., Fernandez-Marques, A., Zaenker, H., Funke, H. & Charlet, L. (2008) X-ray absorption and photoelectron spectroscopy investigation of selenite reduction by Fe(II)-bearing minerals. *Journal of Contaminant Hydrology*, **102**, 228–245.

Schmidt, M., Stumpf, T., Marques-Fernandes, M., Walther, C. & Fanghänel, T. (2008) Charge compensation in solid solutions. *Angewandte Chemie International Edition*, **47**, 5846–5850.

Shanbhag, P.M. & Morse, J.W. (1982) Americium interaction with calcite and arragonite surfaces in seawater. *Geochimica et Cosmochimica Acta*, **46**, 241–249.

Shannon, R.D. (1976) Revised effective ionic radii and systematic studies of interatomic distances in halides and chalcogenides. *Acta Crystallographica A*, **32**, 751–767.

Starke, R. (1964) Die Strontiumgehalte der Baryte. Freiberger Forschungshefte C150.

Stumpf, T. & Fanghänel, T. (2002) A time resolved laser fluorescence spectroscopy (TRLFS) study of the interaction of trivalent acinides (Cm(III)) with calcite. *Journal of Colloid and Interface Science*, **249**, 119–125.

Sturchio, N., Bohlke, J. & Markun, F. (1993) Radium geochemistry of geothermal waters, Yellowstone National Park, Wyoming, USA. *Geochmica et Cosmochimica Acta*, **57**, 1203–1214.

Tanaka, K., Ohta, A. & Kawabe, I. (2004) Experimental REE partitioning between calcite and aqueous solution at 25°C and 1 atm: Constraints on the incorporation of seawater REE into seamount-type limestone. *Geochemical Journal*, **38**, 19–32.

Terakado, Y. & Masada, A. (1988) The coprecipitation of rare-earth elements with calcite and aragonite. *Chemical Geology*, **69**, 103–110.

Thorstenson, D.C. & Plummer, L.N. (1977) Equilibrium criteria for two-component solids reacting with a fixed composition in aqueous phase. *American Journal of Science*, **277**, 1203–1223.

Tsuno, H., Kagi, H. & Akagi, T. (2002) High yield of vaterite precipitation induced by trace lanthanum ion from supersaturated solution of calcium carbonate at 50°C. *Chemistry Letters*, **31**, 960–961.

Vinograd, V., Bosbach, D., Winkler, B. & Gale, J.D. (2008) Subsolidus relations in $Ca_2Mo_2O_8$-powellite solid solution predicted from static lattice energy calculations and Monte Carlo simulations. *Physical Chemistry Chemical Physics*, **10**, 3509–3518.

Vinograd, V., Luchitskaia, M., Joswig, W., Stumpf, T., Bosbach, D. & Winkler, B. (2009) Solid solution in the system calcite-$NaEu(CO_3)_2$. *Geochimica et Cosmochimica Acta*, **73**, A1386.

Wang, Y. & Xu, H. (2001) Prediction of trace metal partitioning between minerals and aqueous solutions: A linear free energy correlation approach. *Geochimica et Cosmochimica Acta*, **65**, 1529–1543.

Weigel, F. & Trinkl, A. (1968) Zur Kristallchemie des Radiums, II. Radiumsalze vom Typ $RaXO_4$. *Radiochmicia Acta*, **9**, 140–144.

Withers, S.H., Peale, R.E. & Schulte, A.F. (2003) Broad distribution of the crystal field environments for Nd^{3+} in calcite. *Physics and Chemistry of Minerals*, **30**, 440–448.

Wolf, G., Konigsberger, E., Schmidt, H.G., Konigsberger, L.C. & Gamsjager, H. (2000) Thermodynamic aspects of the vaterite-calcite phase transition. *Journal of Thermal Analysis and Calorimetry*, **60**, 463–472.

Zavarin, M., Roberts, S.K., Hakem, N. & Kersting, A.B. (2005) Eu(III), Sm(III), Np(V), Pu(V), and Pu(VI) sorption to calcite. *Radiochimica Acta*, **93**, 93–102.

Zhong, S. & Mucci, A. (1995) Partitioning of rare earth elements (REEs) between calcite and seawater solutions at 25°C and 1 atm, and at high dissolved REE concentrations. *Geochimica et Cosmochimica Acta*, **59**, 443–453.

Zhu, C. (2004) Coprecipitation in the barite isostructural family: 1. Binary mixing properties. *Geochimica et Cosmochimica Acta*, **68**, 3327–3337.

EMU Notes in Mineralogy, Vol. 10 (2010), Chapter 11, 377–397

Ion partitioning and the geochemistry of coral skeletons: solving the mystery of the vital effect

ANNE L. COHEN* and GLENN A. GAETANI

Department of Geology and Geophysics, Woods Hole Oceanographic Institution, Woods Hole, MA 02543 USA, e-mail: acohen@whoi.edu
**Corresponding author*

Much of our understanding of Earth's climate history is based on interpretation of geochemical variability within the $CaCO_3$ tests and skeletons of marine organisms. Geochemical climate proxies are typically cast in terms of equilibrium thermodynamics, but there are important differences between the compositions of carbonates accreted by living organisms and predictions for carbonate minerals in equilibrium with seawater. These differences are commonly attributed to 'vital effects' thought to be caused by biological modification of the calcifying environment and of crystal-growth kinetics. If this were true, then biologically modified crystal chemistry may be unpredictable or challenging to model mathematically, making it difficult to extract accurate climate information from biogenic carbonates with any degree of confidence. Our goal with this paper is to demonstrate a systematic approach to the identification, characterization and understanding of 'vital effects' using coral skeletons as an example. We show, through insights gained from abiogenic aragonites precipitated experimentally from seawater under controlled conditions, that many of the so-called 'vital effects' in coral skeletal geochemistry are actually characteristic of abiogenic aragonites and can be described mathematically in terms of predictable physicochemical processes. By comparing elemental ratios (Mg/Ca, Sr/Ca, Ba/Ca) of abiogenic aragonite with that of coral aragonite, we show that Rayleigh fractionation and crystal-growth rate exert the dominant controls on the elemental chemistry of corals, and that the contribution of temperature is relatively small. Building on these insights, we have developed a new approach to extracting temperature information from coral skeletons that is fundamentally different from conventional palaeo-thermometry, by-passing 'vital effects' through the simultaneous use of multiple element ratios to reliably extract that small component of skeletal variability that is driven solely by temperature.

1. Introduction

The compositions of $CaCO_3$ tests and skeletons of many marine organisms vary in response to ocean conditions, and much of our understanding of past climate variability is based on the quantitative interpretation of these changes in terms of physical and chemical parameters of the ocean. For example, the magnesium-to-calcium (Mg/Ca) ratios of calcite tests of foraminifera and the strontium-to-calcium (Sr/Ca) ratios of aragonite skeletons of tropical corals are commonly taken to be geochemical proxies for temperature, and have been used to reconstruct ocean conditions during important periods in Earth's climate history, including the Last Glacial Maximum (*e.g.* Guilderson *et al.*, 1994; Elderfield & Ganssen, 2000). Such proxy-based reconstructions form the

DOI: 10.1180/EMU-notes.10.11

basis for much of our understanding of the behaviour of the Earth's climate system on decadal through orbital timescales, and provide crucial insights into the sensitivity of Earth's climate system to changes in boundary conditions, such as concentrations of atmospheric CO_2. Despite their importance and wide application, however, the majority of geochemical proxies are empirically based and we know surprisingly little about the mechanisms that control the geochemical composition of biogenic (biologically precipitated) carbonates.

Geochemical proxies are typically cast in terms of equilibrium thermodynamics, but there are important differences between the compositions of biogenic carbonates and predictions for carbonate minerals in equilibrium with seawater. For example, at 25°C the average Sr/Ca ratio of *Porites* coral skeleton, calculated on the basis of 37 Sr/Ca–temperature relationships compiled by Corrège (2006), is 9.04 ± 0.17 mmol/mol, whereas thermodynamics predicts that the Sr/Ca ratio of abiogenic aragonite in metastable equilibrium with seawater is 0.82 mmol/mol (Plummer & Busenberg, 1987). Thus, the Sr/Ca ratio of coral skeleton is more than 10 times greater than it would be if coral aragonite was precipitated in equilibrium with seawater. Abiogenic aragonite precipitated experimentally from highly supersaturated seawater provides a good morphological and compositional match to biogenic aragonite, confirming the important role played by kinetics, but significant discrepancies remain. For example, the Sr/Ca ratio of 9.43 ± 0.17 mmol/mol determined experimentally by Gaetani and Cohen (2006) for abiogenic aragonite is within the range predicted for *Porites* skeleton, but the experimentally determined relationship between Sr/Ca ratio and temperature (-0.038 ± 0.004 mmol/mol °C^{-1}) is much weaker than is the case for *Porites* skeleton (-0.061 ± 0.009 mmol/mol °C^{-1}).

The reliability of palaeo-temperature reconstructions derived from biogenic skeletons is further confounded by the large variability in E/Ca-temperature relationships derived from living specimens, as well as compositional variability occurring on timescales too short to be attributable to temperature. For example, Saenger *et al.* (2008) found that corals of a single genus (*Montastrea*) grown under identical temperature conditions can have very different relationships between Sr/Ca ratio and water temperature (Fig. 1). The Sr/Ca ratios, derived from four different *Montastrea* skeletons, range from 8.73 to 9.37 mmol/mol at 25°C, with Sr/Ca-temperature relationships ranging from -0.075 ± 0.010 to -0.099 ± 0.015 mmol/mol per °C. Depending on which of these relationships is used, Little Ice Age palaeo-temperatures derived from Sr/Ca ratios of fossil corals of the same species, collected at the same site, can range from 0°C to 5°C cooler than today. The application of microbeam techniques to carry out analyses of biogenic skeletons on finer and finer length-scales have revealed more and larger-amplitude discrepancies between measured coral skeletal composition and that expected on the basis of known environmental conditions at the time of crystal/skeletal growth (*e.g.* Cohen *et al.*, 2001; Cohen *et al.*, 2002; Allison & Finch, 2004; Meibom *et al.*, 2004; Meibom *et al.*, 2006). For example, Meibom *et al.* (2006) reported Mg/Ca fluctuations of 0.5–1 mmol/mol over micron length-scales (~daily timescales) within coral skeleton. Using Mitsuguchi *et al's.* (1996) Mg/Ca-SST calibration equation, this is equivalent to a temperature range of ~8°C, more than four times

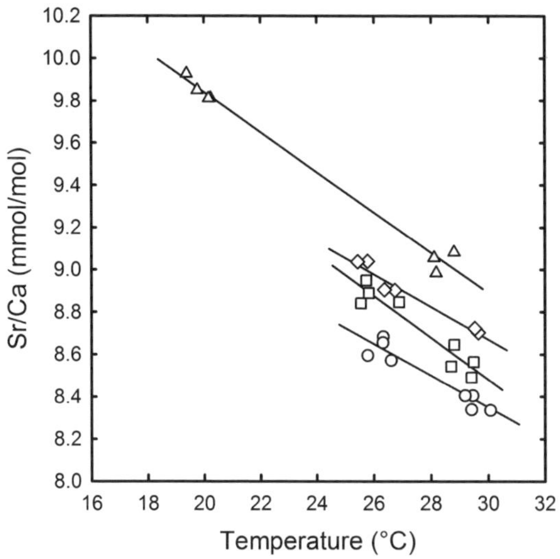

Fig. 1. Relationships between Sr/Ca ratio and temperature, derived from skeletons of the Atlantic coral genus *Montastrea* collected from St. Croix (circles, squares, diamonds) and Bermuda (triangles). These relationships vary systematically as a function of mean annual skeletal extension rate (adapted from Saenger *et al.*, 2008).

larger than the observed diurnal temperature fluctuations experienced by the coral. Such discrepancies are not limited to coral skeletons. For example, Curry and Marchitto (2008) used a secondary ion mass spectrometer (SIMS) ion microprobe to make intra-test measurements of Mg/Ca ratios in the calcitic benthic foraminfera *Cibicidoides pachyderma*. Within a single test, Mg/Ca ratios ranged from ~2.66 mmol/mol to 5.81 mmol/mol. Using the Mg/Ca temperature relationship established from whole-test Mg/Ca averages, this represents a temperature range of >30°C.

These inconsistencies between measured and expected compositions of biogenic skeletons have been recognized for many decades and are collectively referred to as 'vital effects', a term for which there are two working definitions: (1) biological processes linked to skeleton building that distort the climate signal; and (2) "a name which hides the fact that we don't understand them" (Broecker, 1982). The challenge posed by 'vital effects' for practitioners of proxy-based climate reconstructions is to identify and quantify the mechanisms that create them, and to use this understanding to develop reliable approaches to extracting climate records from biogenic carbonates. If the origin of 'vital effects' is wholly physiological, skeletal composition would probably be either unpredictable or challenging to model mathematically, making it unlikely that accurate climate information could be extracted with any degree of confidence. For example, Meibom *et al.* (2007) suggested that large variations in the Mg/Ca ratio of coral skeletons observed on micron length-scales and, importantly, unaccompanied by similarly large variations in Sr/Ca ratios must reflect "differential transport and incorporation of major and trace elements into the skeleton". In other words, biologically induced changes in the composition of the calcifying fluid or in the availability of ions for calcification (on hourly, weekly, seasonal and inter-annual timescales) are the primary drivers of compositional variability in coral skeletons. If this were indeed the case, it might be very difficult to use coral skeletal composition as a reliable indicator of ocean conditions.

Our goal with this paper is to demonstrate a systematic approach to the identification, characterization and understanding of 'vital effects', with a focus on coral skeletons. We show, through comparisons of coral skeletons with data obtained from abiogenic aragonite precipitated experimentally from highly supersaturated seawater under controlled conditions, that many of the so-called 'vital effects' on coral composition, such as those described by Meibom *et al.* (2007) and others, are also characteristic of abiogenic aragonites and can be described mathematically in terms of predictable physicochemical processes. To date we have identified three principal factors controlling the Mg/Ca, Sr/Ca, and Ba/Ca ratios of aragonite crystals grown from seawater: (1) water temperature, (2) crystal growth rate, and (3) Rayleigh fractionation (*e.g.* Cohen *et al.*, 2006; Gaetani & Cohen, 2006; Gabitov *et al.*, 2008; Holcomb *et al.*, 2009). In abiogenic aragonites precipitated from an infinite reservoir, the impact of Rayleigh fractionation on compositional variability is relatively small. In corals however, it appears that the influence of Rayleigh fractionation is dominant, at least on seasonal timescales. By comparison, we find that the contribution of water temperature to compositional variability in both coral skeletons and abiogenic aragonites is relatively small. Despite this, we have been able to use our understanding of coral skeletogenesis to develop a new, more reliable approach to extracting ocean temperature from the composition of coral skeleton. This approach to palaeo-thermometry combines the Rayleigh equation with experimentally determined aragonite-seawater partition coefficients to solve for temperature on the basis of information derived from combining multiple element ratios (Gaetani *et al.*, 2010). Unlike conventional coral thermometers, our Rayleigh-based multi-element approach does not rely on an initial calibration of coral skeletal composition to the instrumental temperature record. Rather, by viewing coral skeletogenesis as a biologically mediated but fundamentally physicochemical process it becomes possible to reliably extract that small component of skeletal variability that is driven solely by temperature.

2. Comparison of coral skeleton with abiogenic aragonite

Coral calcification takes place at the interface between the living polyp and the existing skeleton. The schematic cross-section through a coral colony shown in Figure 2 illustrates the geometric relationships among the coral animals (polyps), the skeleton they accrete and the site of calcification located between the base of the polyp tissue and the top of the existing skeleton. In living corals, the entire surface of the skeleton is covered by several layers of tissue so that, for much of the time, the surface of the growing skeleton is not in direct contact with the external seawater. This is important because it allows the coral to isolate the calcifying medium from the external seawater and to manipulate conditions at the calcification site. Current physicochemical models of coral calcification posit that corals transport seawater to the calcification site, even

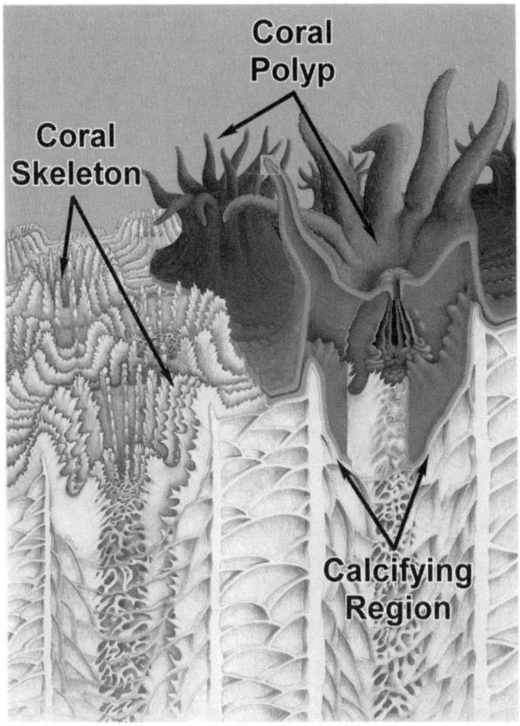

Fig. 2. Schematic cross section through a scleractinian coral showing the geometric relationship between the coral polyp, the skeleton and the calcifying region (adapted from Veron, 1986).

though the exact route or mechanism has not yet been identified (*e.g.* Cohen & McConnaughey, 2003). The most compelling evidence for the presence of seawater at the calcification site comes from the observation that when tracer dyes, such as calcein and alizarin Red-S, are introduced into the seawater surrounding the coral they are rapidly and uniformly deposited in the growing skeleton (*e.g.* Braun & Erez, 2004; Allemand *et al.*, 2010). Coral cell membranes are impermeable to these molecules, so that there must be either (1) a direct passage or channel that periodically links the calcification site with the external seawater, or (2) active transport of seawater-filled vacuoles across cell membranes to the site of calcification. Once seawater enters the site of calcification, its composition must be manipulated in order for aragonite crystals to nucleate and grow. Several lines of evidence indicate that corals achieve this by elevating the saturation state of the calcifying fluid.

Clues to the nature of the coral calcification process are found in the morphologies and compositions of the two distinct crystal types that make up the skeleton (Fig. 3). First, clusters of sub-micron sized, granular crystals occupy the centres of calcification and fine bands that are found throughout the skeleton. Second, bundles of fibrous crystals, several microns in length, make up the bulk of the skeletal mass (*e.g.* Constantz,

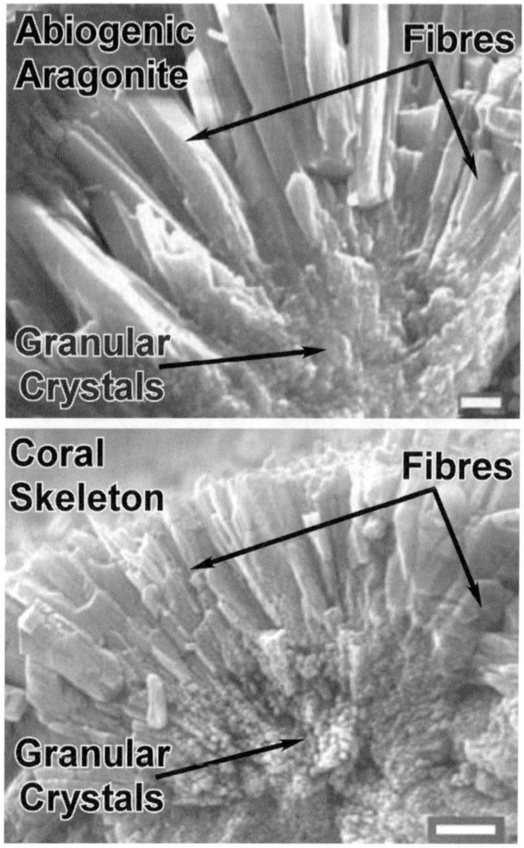

Fig. 3. Comparison of aragonite crystals produced experimentally from a highly supersaturated seawater solution (upper) and by corals (lower). In both cases, bundles of aragonite fibres radiate out from a central region occupied by aggregates of sub-micron-sized granular crystals. Granular crystals form at the highest saturation states and are thought to represent an interval of relatively rapid crystal growth (Holcomb *et al.*, 2009). Fibrous aragonite grows when the saturation state of the seawater is relatively low, indicating slower crystal-growth rates. The scale bars are 1 μm in both (adapted from Cohen & Holcomb, 2009).

1986, 1989; Cohen *et al.*, 2001; Holcomb *et al.*, 2009). Several independent studies, using *in situ* microbeam analytical techniques to selectively analyse the fine granular crystals and fibres have shown that these two crystal types are compositionally distinct (*e.g.* Cohen *et al.*, 2001; Adkins *et al.*, 2003; Allison *et al.*, 2005; Meibom *et al.*, 2006; Blamart *et al.*, 2007; Holcomb *et al.*, 2009; Juillet-Leclerc *et al.*, 2009; Rollion-Bard *et al.*, 2010). The Sr/Ca ratios tend to be slightly larger in regions of the coral skeleton occupied by fine granular crystals relative to adjacent skeleton occupied by fibrous crystals, whereas Mg/Ca and Ba/Ca ratios are significantly larger in these regions. Similarly, isotopic compositions of the two crystal types are distinct: the boron (δ^{11}B), oxygen (δ^{18}O) and carbon isotope ratios (δ^{13}C) tend to be lighter (depleted) in regions of the coral skeleton made up of fine granular crystals relative to adjacent skeleton comprised of fibre bundles. Understanding the growth conditions that produce these two distinct crystal types is a critical first step towards understanding the origin of their morphological and compositional differences in particular, and coral 'vital effects' in general.

Important insights have come from the study of abiogenic aragonite precipitated experimentally from highly super-saturated seawater. The results from these experiments demonstrate that abiogenic aragonites precipitated at low temperatures (15–80°C) from a highly super-saturated seawater solution consist of a combination of granular and fibrous crystals that are nearly identical to those found in coral skeleton (Fig. 3). In precipitation experiments carried out at 25°C, Holcomb *et al.* (2009) introduced isotope spikes (^{84}Sr, ^{86}Sr, ^{137}Ba) into the seawater at various times during the nucleation and growth of the aragonite crystals. By locating the spikes in these crystals, using the SIMS ion microprobe they were able to show that the fine granular crystals form when the saturation state of the seawater, with respect to aragonite, is almost seven times greater than ambient ($\Omega_{Arag} > 25$). Conversely, fibrous aragonite grows when the saturation state of the seawater is relatively low (over an Ω_{Arag} range of ~6–20). Further, Holcomb *et al.* (2009) showed that granular and fibrous crystals are compositionally distinct in abiogenic aragonite just as they are in coral skeleton. Granular crystals have significantly larger Mg/Ca and Ba/Ca ratios than the surrounding fibres, whereas Sr/Ca ratios are only slightly elevated.

The strong correspondence between the abiogenic aragonite precipitated experimentally from seawater and aragonite accreted by corals, in terms of both their morphology and composition, suggests that corals probably produce the crystals that comprise their skeleton by a similar mechanism: seawater is brought into the region between the polyp and skeleton, isolated from the external environment, and its saturation state is elevated until aragonite crystals nucleate and grow (*e.g.* Cohen & McConnaughey, 2003). Indeed, Holcomb et *al.* (2009) proposed that diurnal cycles in the saturation state of the calcifying fluid produces the alternating layers of granular and fibrous crystals found in coral skeletons. Importantly, the similarity between experimental and coral aragonites makes it plausible to use data from abiogenic aragonite precipitation experiments to gain insights into the factors that control the composition of biogenic aragonite accreted by corals.

3. Influence of temperature on aragonite composition

Conventional element-ratio thermometry assumes that coral skeleton aragonite precipitates from an infinite reservoir of seawater, and that compositional variability is attributable to temperature-dependent element partitioning (*e.g.* Smith *et al.*, 1979; Mitsuguchi *et al.*, 1996). This is at odds with strict biological control models for coral calcification, which hold that the site of calcification is completely sealed off from the external seawater environment, and that ions are actively transported to the site of calcification *via* ion channels and enzyme pumps in the cell membranes of the coral polyp (*e.g.* Meibom *et al.*, 2007; Reynaud *et al.*, 2007). The likelihood of either process being dominant in coral biomineralization can be evaluated through comparisons of experimentally determined element partitioning for abiogenic aragonite and the composition of coral skeleton.

Gaetani & Cohen (2006) investigated the extent to which coral skeleton composition is consistent with precipitation of aragonite from seawater using experiments carried out at temperature of 15 to 75°C. They determined that the aragonite-seawater exchange coefficient for Mg/Ca ($K_D^{Mg/Ca} = (C_{Mg}/C_{Ca})^{\text{Aragonite}}/(C_{Mg}/C_{Ca})^{\text{Seawater}}$ where C_i is the concentration of element i in either aragonite or seawater) is very small, whereas $K_D^{Sr/Ca}$ and $K_D^{Ba/Ca}$ are both >1. Their results also demonstrated that $K_D^{Mg/Ca}$, $K_D^{Sr/Ca}$ and $K_D^{Ba/Ca}$ all decrease with increasing temperature. The composition of abiogenic aragonite precipitated from an infinite reservoir of seawater calculated using these exchange coefficients has important similarities with coral skeleton (Fig. 4). The Mg/Ca ratio of coral skeleton is smaller than that of the seawater by approximately a factor of 1000, which is consistent with $K_D^{Mg/Ca} \approx 0.001$, while Sr/Ca and Ba/Ca ratios are similar to or slightly higher seawater ratios, consistent with $K_D^{Sr/Ca} \approx 1.2$ and $K_D^{Ba/Ca} \approx 2.3$. However, in detail, there are also important differences. The Mg/Ca, Sr/Ca and Ba/Ca ratios calculated for abiogenic aragonite are all larger than

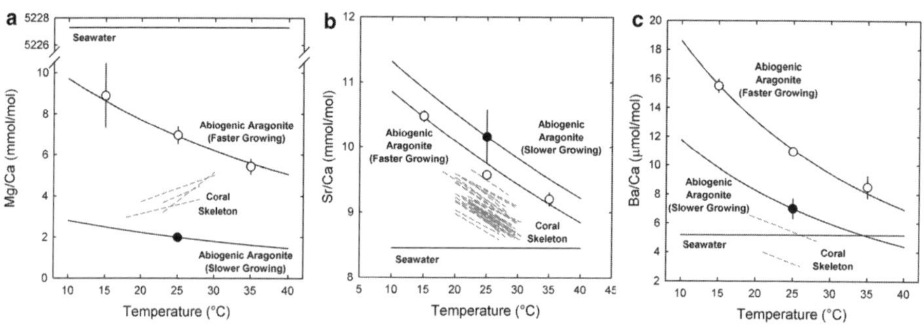

Fig. 4. Comparisons of (***a***) Mg/Ca ratios, (***b***) Sr/Ca ratios, and (***c***) Ba/Ca ratios in coral skeleton (dashed grey curves) with that of aragonite precipitated from an infinite reservoir of seawater (solid black curves) at different crystal-growth rates. The composition of abiogenic aragonite was calculated for seawater having a Mg/Ca ratio of 5.2 mol/mol, Sr/Ca ratio of 8.5 mmol/mol, and Ba/Ca ratio of 5.2 µmol/mol. Experimentally determined $K_D^{Mg/Ca}$, $K_D^{Sr/Ca}$, and $K_D^{Ba/Ca}$ values for faster crystal growth are from Gaetani & Cohen (2006), while values for slower growth were calculated from data for fibrous aragonite at 25°C from Holcomb *et al.* (2009) with temperature dependence estimated from the data of Gaetani & Cohen (2006). Relationships between the Mg/Ca ratio and temperature for coral skeleton are from Mitsuguchi *et al.* (1996), Sinclair *et al.* (1998), Fallon *et al.* (1999) and Watanabe *et al.* (2001). The relationships between Sr/Ca ratio and temperature for coral skeleton are from Smith *et al.* (1979), Beck *et al.* (1992), de Villiers *et al.* (1994), Shen *et al.* (1996), Mitsuguchi *et al.* (1996), Alibert & McCulloch (1997), Bessat (1997), Heiss *et al.* (1997), Gagan *et al.* (1998), Sinclair *et al.* (1998), Fallon *et al.* (1999), McCulloch *et al.* (1999), Corrège *et al.* (2000), Linsley *et al.* (2000), Wei *et al.* (2000), Marshall & McCulloch (2001), Quinn & Sampson (2002), Marshall & McCulloch (2002), Fallon *et al.* (2003), Allison & Finch (2004), Linsley *et al.* (2004), Felis *et al.* (2004), Zinke *et al.* (2004) and Corrège (2006). The relationships between the Ba/Ca ratio and temperature for coral skeleton are from Lea *et al.* (1989) and Gaetani & Cohen (2006).

those in coral skeleton and, more importantly, the predicted temperature dependence for the Mg/Ca ratio is negative whereas the Mg/Ca ratio of the biogenic aragonite correlates positively with temperature. In order to fully reconcile compositional variations in coral skeleton with abiogenic aragonite, factors other than temperature need to be considered. These are (1) the influence of crystal growth rate on element partitioning and (2) Rayleigh fractionation.

4. Influence of crystal-growth rate on aragonite composition

Kinetics represents the principal control on element distribution during carbonate precipitation at low temperature, so that partition coefficients vary with crystal growth rate (*e.g.* Lorens, 1981; Watson, 2004; Gaetani & Cohen, 2006; Gabitov *et al.*, 2008). The precipitation experiments conducted by Gaetani & Cohen (2006) produced spherulitic crystals, morphologically similar to those that comprise coral skeleton *i.e.* granular centres surrounded by bundles of acicular fibres. It is expected that as the crystal growth rate decreases, the Mg/Ca and Ba/Ca ratios of aragonite will decrease significantly and the Sr/Ca ratio will remain relatively constant (Gaetani & Cohen, 2006; Gabitov *et al.*, 2008; Holcomb *et al.*, 2009). This effect can be quantified using exchange coefficients calculated from the composition of fibrous aragonite reported for Experiment 1 of Holcomb *et al.* (2009) which, because of the slower rate at which Na_2CO_3 solution was added (4.8 ml/h *vs.* 60 ml/h), represent smaller crystal-growth rates relative to those studied by Gaetani & Cohen (2006). As crystal growth rate decreases, there are substantial decreases in both $K_D^{Mg/Ca}$ (from ~0.001 to. ~0.0004) and $K_D^{Ba/Ca}$ (from ~2.3- to ~1.4), whereas $K_D^{Sr/Ca}$ does not change significantly (~1.2).

Compositional differences between the granular and fibrous crystals that comprise coral skeleton are explicable in terms of differences in growth rate. Granular crystals that occupy centres of calcification and fine daily growth bands, which plausibly form at relatively large growth rates, typically have slightly larger Sr/Ca and significantly larger Mg/Ca and Ba/Ca ratios than the adjacent fibrous crystals (Holcomb *et al.*, 2009). Furthermore, the existence of large cycles in Mg/Ca ratios on micron length-scales unaccompanied by equally large variations in Sr/Ca ratios (*e.g.* Meibom *et al.*, 2008) are consistent with changes in crystal-growth rate associated with alternating layers of granular and fibrous crystals. Thus, even though in previous studies it has been concluded that these compositional differences within coral skeletons are attributable to physiologically controlled 'vital effects' (*e.g.* Blamart *et al.*, 2007; Meibom *et al.*, 2008; Juillet-Leclerc *et al.*, 2009; Rollion-Bard *et al.*, 2010), they are entirely consistent with abiogenic aragonite precipitated experimentally from seawater in the absence of any biological controls.

However, compositional variability within the bundles of aragonite fibres that make up the bulk of the coral skeleton cannot be explained solely by variations in crystal growth rate. The composition of abiogenic aragonite, calculated using the exchange coefficients from Experiment 1 of Holcomb *et al.* (2009), is compared with that of aragonite fibre bundles in coral skeleton in Figure 4. At this relatively small crystal growth rate, the Mg/Ca ratio of abiogenic aragonite is smaller than that of coral skeleton, while the Ba/Ca ratio is a slightly greater and the Sr/Ca ratio doesn't changed significantly. While these exchange coefficients do produce a closer match to coral skeleton, it does not appear possible that an exact match can be obtained simply by adjusting the crystal growth rate due to the differing responses of $K_D^{Mg/Ca}$, $K_D^{Sr/Ca}$ and $K_D^{Ba/Ca}$. Also, variations in crystal-growth rate cannot explain the positive correlation between Mg/Ca ratio and water temperature, combined with decreasing Sr/Ca and Ba/Ca ratios, found in coral skeleton. Therefore, while the influences of temperature and crystal growth rate on element partitioning appear to play important roles in determining the composition of coral skeleton, they cannot fully reproduce the observed compositional variability. However, the compositions of coral skeleton and abiogenic aragonite can be fully reconciled once the influence of Rayleigh fractionation is taken into account.

5. Rayleigh fractionation and compositional variability in coral skeleton

Physicochemical models for coral biomineralization implicitly require that Rayleigh fractionation play some role in determining the compositional variability in coral skeleton. In order to overcome the nucleation barrier and allow abiogenic aragonite to precipitate, the polyp must isolate seawater from the external environment and manipulate conditions at the site of calcification. As noted above, several lines of evidence indicate that corals achieve this by elevating the saturation state of the calcifying fluid, although some role for organic molecules in lowering the nucleation barrier cannot be entirely ruled out. The important point is that, in this scenario, aragonite does not precipitate from an infinite reservoir of seawater. Rather, crystal growth occurs in an isolated space and the composition of the calcifying fluid changes continuously in response to precipitation of aragonite. This process is governed by the Rayleigh distillation equation (Rayleigh, 1896).

It is the behaviour of the Mg/Ca ratio of coral skeleton that provides the key insight into the importance of Rayleigh fractionation. As discussed earlier, the relationship between Mg/Ca ratio and temperature in coral skeletons is the opposite of that predicted from abiogenic aragonite. During precipitation from an infinite reservoir, the Mg/Ca ratio of abiogenic aragonite is controlled by element partitioning and the composition of seawater. Since $K_D^{Mg/Ca}$ decreases with increasing temperature, the Mg/Ca ratio of the aragonite does also. However, it is well established that in coral skeleton the Mg/Ca ratio increases with increasing temperature (Mitsuguchi *et al.*, 1996; Sinclair *et al.*, 1998; Fallon *et al.*, 1999; Wei *et al.*, 2000; Watanabe *et al.*, 2001; Reynaud *et al.*, 2007). It is this discrepancy, more than anything else, which separates compositional variability in coral skeleton from that predicted from experimentally precipitated abiogenic aragonite. Neither temperature nor crystal-growth rate (alone or in combination) is capable of explaining the combined variability of Mg/Ca, Sr/Ca and Ba/Ca ratios in coral skeletons.

The increase in coral skeleton Mg/Ca ratio with increasing temperature, combined with decreases in both Sr/Ca and Ba/Ca ratios, can be explained by a Rayleigh process in which the mass of aragonite precipitated from an isolated 'batch' of calcifying fluid increases with increasing temperature. Such a temperature-dependent increase in the mass of aragonite precipitated is consistent with results from the *Acropora* culturing experiments of Reynaud *et al.* (2007), in which calcification rate increases by nearly a factor 6 as temperature increases from 21 to 29°C. This change in calcification rate is accompanied by a 30% increase in the Mg/Ca ratio. Therefore, the apparently anomalous temperature dependence of the Mg/Ca ratio of coral skeleton is actually related to its compatibility. Because $K_D^{Mg/Ca}$ is $\ll 1$, whereas $K_D^{Sr/Ca}$ and $K_D^{Ba/Ca}$ are both greater >1, the Mg/Ca ratio of the calcifying fluid increases continuously as aragonite precipitates, whereas the Sr/Ca and Ba/Ca ratios both decrease. The larger the mass of aragonite that precipitates, the stronger this effect becomes. Therefore, when the mass of aragonite precipitated from each 'batch' of calcifying fluid increases with increasing temperature, it produces a negative correlation between Mg/Ca and temperature and strengthens the apparent temperature dependence of Sr/Ca and Ba/Ca ratios. Results from precipitation calculations demonstrate that seasonal changes in the amount of aragonite precipitated by a coral from a fixed volume of calcifying fluid, combined with the effect of seasonal temperature cycles on element partitioning, can reproduce the Mg/Ca, Sr/Ca and Ba/Ca variations observed in the coral skeleton (Fig. 5). Therefore, the most important control on the composition of coral skeleton is Rayleigh fractionation, with temperature contributing only a small proportion of the total compositional variability.

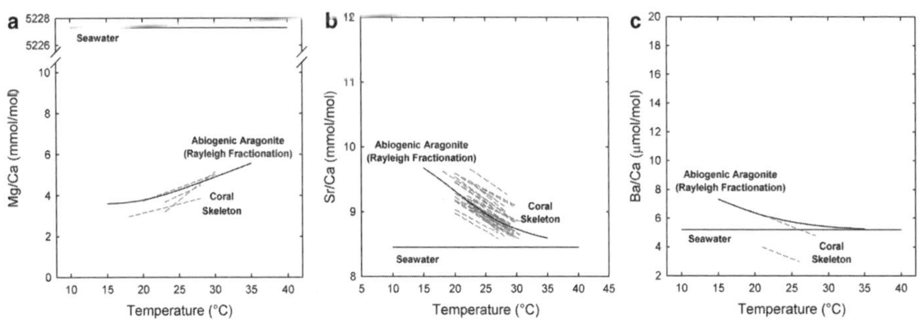

Fig. 5. Comparisons of (*a*) Mg/Ca ratios, (*b*) Sr/Ca ratios, and (*c*) Ba/Ca ratios in coral skeleton (dashed grey curves) with that of aragonite precipitated from an isolated reservoir of seawater (solid black curves). The positive correlation between Mg/Ca and temperature results from an increase in the mass of aragonite precipitated from each 'batch' of calcifying fluid. The composition of abiogenic aragonite was calculated for seawater having a Mg/Ca ratio of 5.2 mol/mol, Sr/Ca ratio of 8.5 mmol/mol, and Ba/Ca ratio of 5.2 μmol/mol. Experimentally determined $K_D^{Mg/Ca}$, $K_D^{Sr/Ca}$, and $K_D^{Ba/Ca}$ values were calculated from data for fibrous aragonite at 25°C from Holcomb *et al.* (2009) with temperature dependence estimated from the data of Gaetani & Cohen (2006). Relationships between water temperature and Mg/Ca, Sr/Ca and Ba/Ca ratios of coral skeleton are the same as in Figure 4.

6. A Rayleigh-based, multi-element approach to coral palaeo-thermometry

Recognition of the role of Rayleigh fractionation in coral biomineralization provides the basis for a new approach to coral thermometry that is fundamentally different from existing empirical calibrations. By combining analyses of multiple elemental ratios (*e.g.* Mg/Ca, Sr/Ca and Ba/Ca) from a given coral skeleton with experimentally determined partition coefficients for abiogenic aragonite, it becomes possible to construct a mathematically over-constrained system of Rayleigh equations that describe element fractionations during skeletogenesis (Gaetani *et al.*, 2010). Unlike conventional coral thermometers, this approach does not rely on calibrations involving living corals. Rather, considering coral skeletogenesis as a biologically mediated, physicochemical process provides a means to extract reliable temperature information from the skeleton using a set of experimentally determined partition coefficients and the Rayleigh equation. Moreover, because this approach is based on a model that quantitatively describes the 'vital effect', it may be applicable both across species and to corals growing in vastly different environments.

The conceptual basis for Rayleigh-based, multi-element coral thermometry is illustrated in Figure 6. The concentration of any minor element, relative to Ca, in aragonite precipitated from an isolated 'batch' of calcifying fluid is described by the following solution to the Rayleigh distillation equation:

$$\frac{\bar{C}_i^{\text{Aragonite}}}{\bar{C}_{Ca}^{\text{Aragonite}}} = \frac{C_i^0}{C_{Ca}^0}\frac{1 - FL^{D_i^{\text{Aragonite-Seawater}}}}{1 - FL^{D_{Ca}^{\text{Aragonite-Seawater}}}} \tag{1}$$

where $\bar{C}_i^{\text{Aragonite}}$ is the average concentration of element i in the aragonite, C_i^0 is the concentration of element i in the calcifying fluid as precipitation begins, FL is the mass fraction of the initial fluid remaining when aragonite precipitation ends, and $D_i^{\text{Aragonite-Seawater}}$ is the Nernst aragonite-seawater partition coefficient ($D_i^{\text{Aragonite-Seawater}} = C_i^{\text{Aragonite}}/C_i^{\text{Seawater}}$, where C_i^j is the weight concentration of element i in phase j). According to equation 1, elemental ratios in aragonite precipitated during a Rayleigh process are determined by three factors: (1) fluid composition (in this case seawater), (2) partition coefficients, which depend on both temperature and crystal-growth rate, and (3) the mass of aragonite that precipitates from a 'batch' of calcifying fluid ($1 - FL$). If it is assumed that crystal-growth rate does not vary significantly during precipitation of fibrous aragonite, partition coefficients are dependent only on temperature. In this case a given elemental ratio, *e.g.* a Sr/Ca ratio of 8.97 mmol/mol, does not represent a unique temperature but, rather, a curve on a plot of temperature *vs.* the mass of aragonite precipitated (Fig. 6). This is the reason that thermometers based on a single-element ratio are unreliable. Individual corals living at identical conditions may precipitate different amounts of aragonite at a given temperature, leading to variable relationships

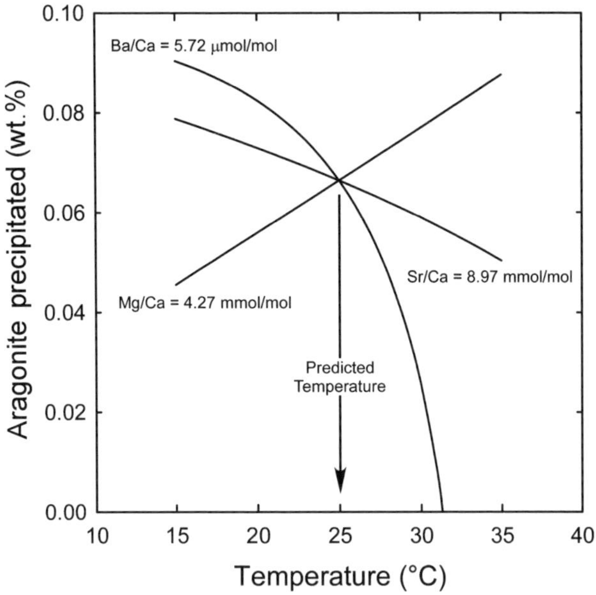

Fig. 6. Curves of constant Mg/Ca, Sr/Ca and Ba/Ca ratios calculated for aragonite precipitated from seawater using equation 1 and aragonite-seawater exchange coefficients calculated from the experimental data of Holcomb *et al.* (2009). These curves intersect at only one combination of temperature and mass of aragonite precipitated, providing the basis for Rayleigh-based, multi-element thermometry.

between Sr/Ca ratio and temperature. However, if the number of element ratios that are measured is increased, so that it is also known that the Mg/Ca ratio of the aragonite is 4.27 mmol/mol and the Ba/Ca ratio is 5.72 μmol/mol, a series of temperature *vs.* *FL* curves can be constructed that intersect at a unique point. It is this intersection that allows temperatures to be reliably extracted from coral skeleton using Rayleigh-based, multi-element thermometry. By taking the approach of solving for both temperature and *FL*, rather than assuming that Sr/Ca depends only on temperature, the uncertainty associated with single-element ratio thermometry is eliminated.

Note that equation 1 differs from the solution to the Rayleigh distillation equation used by Elderfield *et al.* (1996) and by Gagnon *et al.* (2007) to model biomineralization in foraminfera and deep-sea corals, respectively. Those studies relied on an equation that describes the instantaneous solid composition throughout precipitation of a 'single batch' of calcifying fluid:

$$\frac{C_i^{\text{Aragonite}}}{C_{Ca}^{\text{Aragonite}}} = K_D^{i/Ca}\frac{C_i^0}{C_{Ca}^0}\left(\frac{C_{Ca}^{\text{Seawater}}}{C_{Ca}^0}FL\right)^{K_D^{i/Ca}-1} \tag{2}$$

where C_i^j is the instantaneous concentration of element i in phase j (*e.g.* Eq. 1.5.9 of Albarède, 1995). Given that $FL \approx 1$, equation 2 was simplified to:

$$\frac{C_i^{\text{Aragonite}}}{C_{Ca}^{\text{Aragonite}}} = K_D^{i/Ca} \frac{C_i^0}{C_{Ca}^0} \left(\frac{C_{Ca}^{\text{Seawater}}}{C_{Ca}^0} \right)^{K_D^{i/Ca} - 1} \quad (3)$$

Equation 1 describes the average composition of aragonite precipitated from a 'batch' of calcifying fluid. It is, therefore, more appropriate for modelling seasonal variations in the composition of coral skeleton than equation 3 because seasonal variability occurs over length scales much greater than does precipitation of a single batch of calcifying fluid. Furthermore, calculating the average composition of aragonite precipitated from individual batches of fluid takes into account the averaging that occurs naturally at the length scales on which coral skeleton is sampled.

Carrying out Rayleigh-based, multi-element thermometry using equation 1 requires a quantitative understanding of the temperature dependence of element partitioning between aragonite and seawater. Therefore, expressions describing the temperature dependence of $D_{Mg}^{\text{Aragonite-Seawater}}$, $D_{Ca}^{\text{Aragonite-Seawater}}$, $D_{Sr}^{\text{Aragonite-Seawater}}$, and $D_{Ba}^{\text{Aragonite-Seawater}}$ were derived by fitting experimental data, and substituted into equation 1 to derive a temperature-dependent Rayleigh equation for each elemental ratio. The partitioning expressions are as follows:

$$\ln D_{Mg}^{\text{Aragonite-Seawater}} = -4.86370 + \frac{1706.2460}{(T + 273.15)} \qquad r^2 = 0.9891 \quad (4)$$

$$\ln D_{Ca}^{\text{Aragonite-Seawater}} = 8.2384 - \frac{233.4766}{(T + 273.15)} \qquad r^2 = 0.8199 \quad (5)$$

$$D_{Sr}^{\text{Aragonite-Seawater}} = 2199.6710 - 7.1050 \times T \qquad r^2 = 0.8995 \quad (6)$$

$$\ln D_{Ba}^{\text{Aragonite-Seawater}} = -0.6959 + \frac{2658.4031}{(T + 273.15)} \qquad r^2 = 0.9823 \quad (7)$$

where T is temperature in Celsius.

Equation 1 cannot be solved explicitly for temperature, so that a global minimization procedure is used. First, the Mg/Ca, Sr/Ca and Ba/Ca ratios are calculated for each analysed spot on the coral skeleton using a set of initial guesses for T, FL and calcifying fluid composition. Constant values were used as starting points for T and FL, combined with the Mg/Ca, Sr/Ca and Ba/Ca ratios for Vineyard Sound seawater (Gaetani & Cohen, 2006). Element ratios predicted by the Rayleigh equations on the basis of these starting guesses are then compared to measured values by calculating the sum of the squared percentage difference between the two:

$$\Sigma(100 \times (i/Ca_{\text{Rayleigh}} - i/Ca_{\text{Measured}})/i/Ca_{\text{Measured}})^2 \quad (8)$$

where i/Ca_{Rayleigh} and i/Ca_{Measured} are the element-to-calcium ratio predicted by the Rayleigh equation and measured in the coral skeleton, respectively. The initial guesses for T, FL, C^0_{Mg}/C^0_{Ca}, C^0_{Sr}/C^0_{Ca}, and C^0_{Ba}/C^0_{Ca} are then incrementally modified using the Nelder-Mead simplex algorithm of Lagarias *et al.* (1998) until equation 8 reaches a global minimum (*i.e.* the smallest possible value). This algorithm does a better job of avoiding local minima (*i.e.* places where changing any variable increases the value of equation 8, but the smallest possible value has not been achieved) when FL is assumed to be a polynomial function of T:

$$FL = c_1 + c_2(T + 273.15)^{-1} + c_3(T + 273.15) + c_4(T + 273.15)^2 \quad (9)$$

where T is temperature in Celsius and c_i are regression coefficients.

The precision and accuracy of our Rayleigh-based, multi-element approach to coral thermometry can be demonstrated using the compositions of *Acropora* sp. specimens cultured at 21–29°C by Reynaud *et al.* (2007). Analyses of Mg/Ca and Sr/Ca ratios were reported for 10–11 individual nubbins (small, live coral samples) in each of five experiments. The number of known values is 2 × 52 (ratios) = 104 and the number of unknown variables in the Rayeigh equations is 52 (temperatures) + 7 (regression coefficients) = 59, so that the system of equations is mathematically over-constrained. The Sr/Ca ratio of newly grown *Acropora* skeleton decreases monotonically from 9.59 ± 0.02 mmol/mol at 21°C to 9.16 ± 0.01 mmol/mol at 29°C, while Mg/Ca increases from 3.76 ± 0.03 to 4.90 ± 0.04 mmol/mol. The measured calcification rate increases from 0.061 ± 0.007% day^{-1} at 21°C to 0.35 ± 0.03% day^{-1} at 29°C.

If the Mg/Ca and Sr/Ca ratios of the *Acropora* calcifying fluid are insensitive to temperature (*i.e.* they are the same in each experiment), each of the Mg/Ca-Sr/Ca ratio pairs measured by Reynaud *et al.* (2007) corresponds to a unique combination of mass of aragonite precipitated and temperature. The latter should correspond to the temperature at which the corals were cultured. This is illustrated in Figure 7 using global minimization results for the culturing experiment carried out at 27°C. If the mass of aragonite precipitated from each 'batch' of calcifying fluid is held constant, both Mg/Ca and Sr/Ca ratios vary as a function of temperature alone (solid curves). If temperature is held constant, both ratios vary with the mass of aragonite precipitated (dashed curves). There is only one combination of temperature and aragonite precipitated (intersection of solid and dashed curves) that reproduces both the Mg/Ca (4.62 ± 0.08 mmol/mol) and Sr/Ca (9.20 ± 0.06 mmol/mol) ratios of coral skeleton from the culturing experiments. This corresponds to 0.155 ± 0.003 wt.% aragonite precipitated, or precipitation of ~93% of the Ca^{2+} initially in the seawater, and a temperature of 26.75 ± 0.25°C, which is within the uncertainty of the experimental temperature. Global minimization results for 21–29°C are shown in Figure 8. The Mg/Ca and Sr/Ca ratios of the *Acropora* skeletons are accurately reproduced by Rayleigh fractionation (Fig. 8a,b), and there is excellent agreement between temperatures predicted by Rayleigh-based, multi-element thermometry and those at which the *Acropora* were cultured, with a mean uncertainty of ± 0.35°C (Fig. 8c).

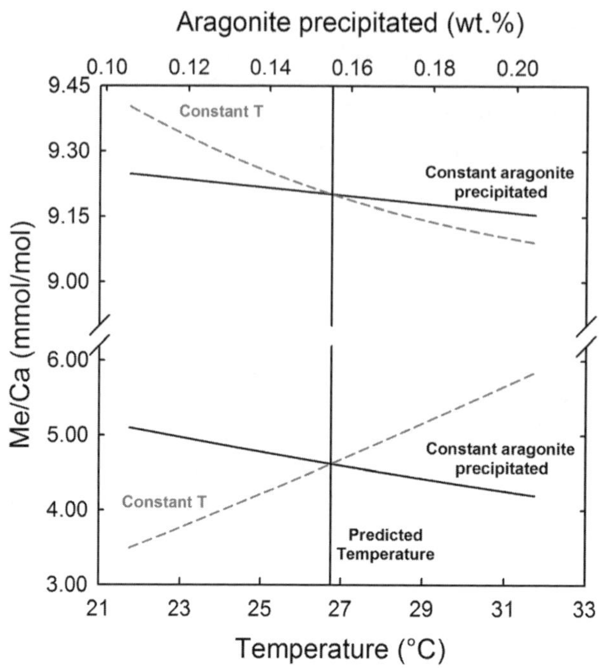

Fig. 7. Results from Rayleigh-based, multi-element thermometry on *Acropora* cultured at 27°C by Reynaud *et al.* (2007). The upper part of figure shows results for Sr/Ca and the lower part results for Mg/Ca. There is only one combination of temperature and aragonite precipitated that reproduces both Mg/Ca and Sr/Ca ratios of coral skeleton from the culturing experiments. This corresponds to 0.155 wt.% of aragonite precipitated at 26.75°C. The solid curves show how Mg/Ca and Sr/Ca ratios vary with temperature when 0.155 wt.% aragonite is precipitated. Dashed curves show how Mg/Ca and Sr/Ca ratios vary with the mass of aragonite precipitated at a temperature of 26.75°C.

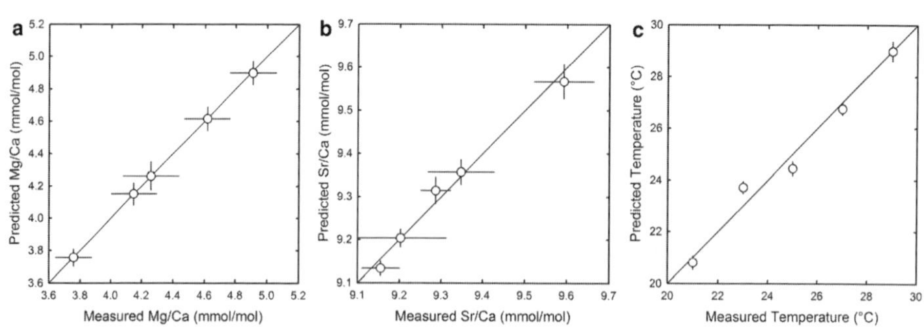

Fig. 8. Comparisons of predicted and measured (***a***) Mg/Ca ratios, (***b***) Sr/Ca ratios and (***c***) temperature from Rayleigh-based, multi-element coral thermometry carried out on *Acropora* cultured in the laboratory at 21–29°C by Reynaud *et al.* (2007). Solid lines are 1:1 relationships. All uncertainties are 2 standard errors.

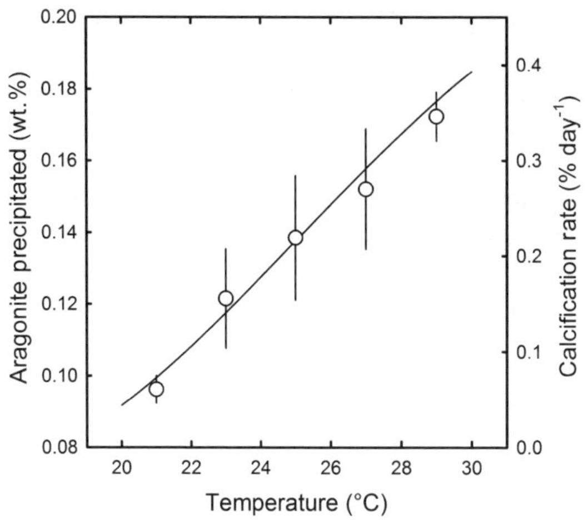

Fig. 9. Relationship between temperature and mass of aragonite precipitated from each 'batch' of calcifying fluid derived from Rayleigh-based, multi-element coral thermometry for *Acropora* cultivated in the laboratory at 21–29°C by Reynaud *et al.* (2007) (solid curve). Shown for comparison is the change in calcification rate reported for these corals by Reynaud *et al.* (2007) (open circles).

The aragonite precipitated ranges from 0.098 $\pm$ 0.002 wt.% ($\sim$81% of initial Ca^{2+} from the seawater) at 21°C to 0.176 $\pm$ 0.003 wt.% ($\sim$95% of initial Ca^{2+} from the seawater) at 29°C (Fig. 9).

7. Conclusions

Comparisons of morphological and compositional variability in abiogenic aragonites produced experimentally from highly supersaturated seawater with that of aragonites produced by corals have provided important insights to the biomineralization mechanisms and the factors governing compositional variability in coral skeletons. Through experimental manipulation of temperature and crystal-growth rates, we have identified three principal factors controlling the Mg/Ca, Sr/Ca, and Ba/Ca ratios of abiogenic aragonite grown directly from seawater. These are water temperature, crystal growth rate, and Rayleigh fractionation (through variations in the mass of aragonite precipitated from a volume of fluid). In the abiogenic aragonites, Mg/Ca, Sr/Ca and Ba/Ca ratios covary systematically and predictably, all increasing with increasing crystal growth rate and all decreasing with increasing seawater temperature. In coral skeletons, many of the compositional anomalies previously identified as 'vital effects' can be explained by the same processes *i.e.* variations in crystal growth rate and Rayleigh fractionation, without invoking differential transport of Mg^{2+} and other ions, or organic matrix control of crystal growth. Indeed, compositional differences between granular (faster-growing) and fibrous (slower-growing) crystals in coral skeletons are consistent with growth-dependent variability observed in abiogenic aragonites. On the other hand, trends in Mg/Ca, Sr/Ca and Ba/Ca ratios of coral skeletons on seasonal timescales cannot be explained solely in terms of temperature or crystal-growth rate, or the two in combination. Rather, the increase in Mg/Ca and simultaneous decrease in Sr/Ca and Ba/Ca that occur with increasing temperature are consistent with Rayleigh fractionation as the dominant control on these timescales. Identifying the processes involved in driving compositional variability in coral skeleton and the origin of the 'vital effect' has enabled us to develop a fundamentally different approach to coral palaeo-thermometry that is based on extracting the small component of the geochemical signal that is driven solely by temperature. If successful, the same approach can be applied to a broad range of biogenic $CaCO_3$ tests and skeletons: using abiogenic aragonite and calcite precipitation experiments as a framework within which to identify, understand and conquer the ubiquitous 'vital effect'.

Acknowledgements

The authors extend their gratitude to Nobuchimi Shimizu and the Northeast Ion Microprobe Facility for technical support on SIMS analyses of abiogenic aragonites and coral skeletons, to Jurek Blusztajn and the WHOI ICP-MS Facility for analyses of seawater samples, and to Michael Holcomb for fruitful discussions about abiogenic and coral aragonite geochemistry. This work was funded by NSF OCE-0823527 to ALC and GAG.

References

Albarède, F. (1995) *Introduction to Geochemical Modeling.* Cambridge University Press, Cambridge, UK, 554 pp.

Alibert, C. & McCulloch, M.T. (1997) Strontium/calcium ratios in modern *Porites* corals from the Great Barrier Reef as a proxy for sea surface temperature: calibration of the thermometer and monitoring of ENSO. *Paleoceanography*, **12**, 345–363.

Allemand, D., Tambutté, É., Zoccola, D. & Tambutté, S. (2010) Coral calcification, cells to reefs. In: *Coral and Coral Reefs* (Z. Dubinsky, editor). Springer, Berlin (in press).

Allison, N. & Finch, A.A. (2004) High-resolution Sr/Ca records in modern Porites lobata corals: Effects of skeletal extension rate and architecture. *Geochemistry, Geophysics, Geosystems*, **5**, Q05001.

Allison, N., Finch, A.A., Newville, M. & Sutton, S.R. (2005) Strontium in coral aragonite: 3. Sr coordination and geochemistry in relation to skeletal architecture. *Geochimica et Cosmochimica Acta*, **69**, 3801–3811.

Beck, J.W., Edwards, R.L., Ito, E., Taylor, F.W., Recy, J., Rougerie, F., Joannot, P. & Henin, C. (1992) Sea-surface temperature from coral skeletal strontium/calcium ratios. *Science*, **257**, 644–647.

Bessat, F. (1997) Variabilité hydro-climatique et croissance corallienne en Polynésie française: exemples de l'île de Mooréa et de l'atoll de Mururoa. Thèse de doctorat, Université Paris I – Panthéon – Sorbonne.

Blamart, D., Rollion-Bard, C., Meibom, A., Cuif, J.P., Juillet-Leclerc, A. & Dauphin, Y. (2007) Correlation of boron isotopic composition with ultrastructure in the deep-sea coral Lophelia pertusa: Implications for biomineralization and paleo-pH. *Geochemistry, Geophysics, Geosystems*, **8**, Q12001.

Braun, A. & Erez, J. (2004) Preliminary observations on sea water utilization during calcification in scleractinian corals. *Eos Transactions AGU*, **85**, Abstract B14B–04.

Broecker, W. S. (1982) Ocean chemistry during glacial time. *Geochimica et Cosmochimica Acta*, **46**, 1689–1705.

Cohen, A.L. & Holcomb, M. (2009) Why corals care about ocean acidification: uncovering the mechanism. *Oceanography*, **22**, 118–127.

Cohen, A.L. & McConnaughey, T.A. (2003) Geochemical perspectives on coral mineralization. In: *Biomineralization* (P.M. Dove, J.J. De Yoreo & S. Weiner, editors). Reviews in Mineralogy and Geochemistry, **54**, 151–187. Mineralogical Society of America, Washington D.C. and the Geochemical Society, St. Louis, Missouri, USA.

Cohen, A.L., Layne, G.D., Hart, S.R. & Lobel, P.S. (2001) Kinetic control of skeletal Sr/Ca in a symbiotic coral: implications for the paleotemperature proxy. *Paleoceanography*, **16**, 20–26.

Cohen, A.L., Owens, K.E., Layne, G.D. & Shimizu, N. (2002) The effect of algal symbionts on the accuracy of Sr/Ca paleotemperatures from coral. *Science*, **296**, 331–333.

Cohen, A.L., Gaetani, G.A., Lundälv, T., Corliss, B.H. & George, R.Y. (2006) Compositional variability in the cold-water scleractinian, *Lophelia pertusa*: new insights into "vital effects". *Geochemistry, Geophysics, Geosystems*, **7**, Q12004.

Constantz, B.R. (1986) Coral skeleton construction: A physiochemically dominated process. *Palaios*, **1**, 152–157.

Constantz, B.R. (1989) Skeletal organisation in Caribbean Acropora spp. (Lamarck). In: *Origin, Evolution and Modern Aspects of Biomineralization in Plants and Animals* (R.E. Crick, editor). Plenum Press, New York, pp. 175–200.

Corrège, T. (2006) Sea surface temperature and salinity reconstructions from coral geochemical tracers. *Palaeogeography Palaeoclimatology Palaeoecology*, **232**, 408–428.

Corrège, T., Delcroix, T., Récy, J., Beck, J.W., Cabioch, G. & Le Cornec, F. (2000) Evidence for stronger El Niño Southern Oscillation (ENSO) events in a mid-Holocene massive coral. *Paleoceanography*, **15**, 465–470.

Curry, W. B. & Marchitto, T.M. (2008) A secondary ionization mass spectrometry calibration of Cibicidoides pachyderma Mg/Ca with temperature. *Geochemistry, Geophysics, Geosystems*, **9**, Q04009.

de Villiers, S., Shen, G.T. & Nelson, B.K. (1994) The Sr/Ca-temperature relationship in coralline aragonite: Influence of variability in $(Sr/Ca)_{seawater}$ and skeletal growth parameters. *Geochimica et Cosmochimica Acta*, **58**, 197–208.

Elderfield, H. & Ganssen, G. (2000) Past temperature and $\delta^{18}O$ of surface ocean waters inferred from foraminiferal Mg/Ca ratios. *Nature*, **405**, 442–445.

Elderfield, H., Bertram, C.J. & Erez, J. (1996) A biomineralization model for the incorporation of trace elements into foraminiferal calcium carbonate. *Earth and Planetary Science Letters*, **142**, 409–423.

Fallon, S.J., McCulloch, M.T., van Woesik, R. & Sinclair, D.J. (1999) Corals at their latitudinal limits: laser ablation trace element systematics in Porites from Shirigai Bay, Japan. *Earth and Planetary Science Letters*, **172**, 221–238.

Fallon, S.J., McCulloch, M.T. & Alibert, C. (2003) Examining water temperature proxies in Porites corals from the Great Barrier Reef: a cross-shelf comparison. *Coral Reefs*, **22**, 389–404.

Felis, T., Lohmann, G., Kuhnert, H., Lorenz, S.J., Scholz, D., Paetzold, J., Al-Rousan, S.A. & Al-Moghrabi, S. M. (2004) Increased seasonality in Middle East temperatures during the last interglacial period. *Nature*, **429**, 164–168.

Gabitov, R.I., Gaetani, G.A., Watson, E.B., Cohen, A.L. & Ehrlich, H.L. (2008) Experimental determination of growth rate effect on U^{6+} and Mg^{2+} partitioning between aragonite and fluid at elevated U^{6+} concentration. *Geochimica et Cosmochimica Acta*, **72**, 4058–4068.

Gaetani, G.A. & Cohen, A.L. (2006) Element partitioning during precipitation of aragonite from seawater: A framework for understanding paleoproxies. *Geochimica et Cosmochimica Acta*, **70**, 4617–4634.

Gaetani, G.A., Cohen, A.L., Wang, Z. & Crusius, J. (2010) A Rayleigh-based, multi-element approach to coral paleothermometry. *Geochimica et Cosmochimica Acta*, (submitted).

Gagan, M.K., Ayliffe, L.K., Hopley, D., Mortimer, G.E., Chappell, J., McCulloch, M.T. & Head, M.J. (1998) Temperature and surface-ocean water balance of the mid-Holocene tropical western Pacific. *Science*, **279**, 1014–1018.

Gagnon, A.C., Adkins, J.F., Fernández, D.P. & Robinson, L.F. (2007) Sr/Ca and Mg/Ca vital effects correlated with skeletal architecture in a scleractinian deep-sea coral and the role of Rayleigh fractionation. *Earth and Planetary Science Letters*, **261**, 280–295.

Guilderson, T.P., Fairbanks, R.G. & Rubenstone, J.L. (1994) Tropical temperature variations since 20,000 years ago: modulating interhemispheric climate change. *Science*, **263**, 663–665.

Heiss, G.A., Camoin, G.F., Eisenhauer, A., Wischow, D., Dullo, W.C. & Hansen, B. (1997) Stable isotope and Sr/Ca signals in corals from the Indian Ocean. In: *8th International Coral Reef Symposium* (H.A. Lessios & I.G. MacIntyre, editors). Smithsonian Tropical Research Institute, Panama, Panama City, pp. 1713–1718.

Holcomb, M., Cohen, A.L., Gabitov, R.I. & Hutter, J.L. (2009) Compositional and morphological features of aragonite precipitated experimentally from seawater and biogenically by corals. *Geochimica et Cosmochimica Acta*, **73**, 4166–4179.

Juillet-Leclerc, A., Reynaud, S., Rollion-Bard, C., Cuif, J.P., Dauphin, Y., Blamart, D., Ferrier-Pagès, C. & Allemand, D. (2009) Oxygen isotopic signature of the skeletal microstructures in cultured corals: Identification of vital effects. *Geochimica et Cosmochimica Acta*, **73**, 5320–5332.

Lagarias, J.C., Reeds, J.A., Wright, M.H. & Wright, P.E. (1998) Convergence properties of the Nelder-Mead simplex method in low dimensions. *SIAM Journal of Optimization*, **9**, 112–147.

Lea, D.W., Shen, G.T. & Boyle, E.A. (1989) Coralline barium records temporal variability in Equatorial Pacific upwelling. *Nature*, **340**, 373–376.

Linsley, B.K., Wellington, G.M. & Schrag, D.P. (2000) Decadal sea surface temperature variability in the subtropical South Pacific from 1726 to 1997 A.D. *Science*, **290**, 1145–1148.

Linsley, B.K., Wellington, G.M., Schrag, D.P., Ren, L., Salinger, M.J. & Tudhope, A.W. (2004) Geochemical evidence from corals for changes in the amplitude and spatial pattern of South Pacific interdecadal climate variability over the last 300 years. *Climate Dynamics*, **22**, 1–11.

Lorens, R.B. (1981) Sr, Cd, Mn and Co distribution coefficients in calcite as a function of calcite precipitation rate. *Geochimica et Cosmochimica Acta*, **45**, 553–561.

Marshall, J.F. & McCulloch, M.T. (2001) Evidence of El Niño and the Indian Ocean Dipole from Sr/Ca derived SSTs for modern corals at Christmas Island, eastern Indian Ocean. *Geophysical Research Letters*, **28**, 3453–3456.

Marshall, J.F. & McCulloch, M.T. (2002) An assessment of the Sr/Ca ratio in shallow water hermatypic corals as a proxy for sea surface temperature. *Geochimica et Cosmochimica Acta*, **66**, 3263–3280.

McCulloch, M.T., Tudhope, A.W., Esat, T.M., Mortimer, G.E., Chappell, J., Pillans, B., Chivas, A.R. & Omura, A. (1999) Coral record of equatorial sea-surface temperatures during the penultimate deglaciation at Huon Peninsula. *Science*, **283**, 202–204.

Meibom, A., Cuif, J.P., Hillion, F., Constantz, B.R., Juillet-Leclerc, A., Dauphin, Y., Watanabe, T. & Dunbar, R.B. (2004) Distribution of magnesium in coral skeleton. *Geophysical Research Letters*, **31**, L23306.

Meibom, A., Mostefaoui, S., Cuif, J.P., Dauphin, Y., Houlbreque, F., Dunbar, R.B. & Constantz, B.R. (2007) Biological forcing controls the chemistry of reef-building coral skeleton. *Geophysical Research Letters*, **34**, L02601.

Meibom, A., Yurimoto, H., Cuif, J.P., Domart-Coulon, I., Houlbreque, F., Constantz, B. R., Dauphin, Y., Tambutté, É., Tambutté, S., Allemand, D., Wooden, J. & Dunbar, R.B. (2006) Vital effects in coral skeletal composition display strict three dimensional control. *Geophysical Research Letters*, **33**, L11608.

Meibom, A., Cuif, J.P., Houlbreque, F., Mostefaoui, S., Dauphin, Y., Meibom, K.L. & Dunbar, R.B. (2008) Compositional variations at ultra-structure length scales in coral skeleton. *Geochimica et Cosmochimica Acta*, **72**, 1555–1569.

Mitsuguchi, T., Matsumoto, E., Abe, O., Uchida, T. & Isdale, P.J. (1996) Mg/Ca thermometry in coral skeletons. *Science*, **274**, 961–963.

Plummer, L.N. & Busenberg, E. (1987) Thermodynamics of aragonite-strontianite solid-solutions: Results from stoichiometric solubility at 25 and 76°C. *Geochimica et Cosmochimica Acta*, **51**, 1393–1411.

Quinn, T.M. & Sampson, D.E. (2002) A multiproxy approach to reconstructing sea surface conditions using coral skeleton geochemistry. *Paleoceanography*, **17**, 1062.

Rayleigh, J.W.S. (1896) Theoretical considerations respecting the separation of gases by diffusion and similar processes. *Philosophical Magazine*, **42**, 77–107.

Reynaud, S., Ferrier-Pagès, C., Meibom, A., Mostefaoui, S., Mortlock, R., Fairbanks, R.G. & Allemand, D. (2007) Light and temperature effects on Sr/Ca and Mg/Ca ratios in the scleractinian coral *Acropora* sp. *Geochimica et Cosmochimica Acta*, **71**, 354–362.

Rollion-Bard, C., Blamart, D., Cuif, J.P. & Dauphin, Y. (2010) *In situ* measurements of oxygen isotopic composition in deep-sea coral, *Lophelia pertusa*: Re-examination of the current geochemical models of biomineralization. *Geochimica et Cosmochimica Acta*, **74**, 1338–1349.

Saenger, C., Cohen, A.L., Oppo, D.W. & Hubbard, D. (2008) Interpreting sea surface temperature from strontium/calcium ratios in Montastrea corals: Link with growth rate and implications for proxy reconstructions. *Paleoceanography*, **23**, PA3102.

Shen, C.C., Lee, T., Chen, C.Y., Wang, C.H., Dai, C.F. & Li, L.A. (1996) The calibration of D[Sr/Ca] versus sea surface temperature relationship for *Porites* coral. *Geochimica et Cosmochimica Acta*, **60**, 3849–3858.

Sinclair, D.J., Kinsley, L.P.J., and McCulloch, M.T. (1998) High resolution analysis of trace elements in corals by laser ablation ICP-MS. *Geochimica et Cosmochimica Acta*, **62**, 1889–1901.

Smith, S.V., Buddemeier, R.W., Redalje, R.C. & Houck, J.E. (1979) Strontium-calcium thermometry in coral skeletons. *Science*, **204**, 404–407.

Veron, J.E.N. (1986) *Corals of Australia and the Indo-Pacific*. University of Hawaii Press, Honolulu, 644 pp.

Watanabe, T., Winter, A. & Oba, T. (2001) Seasonal changes in sea surface temperature and salinity during the Little Ice Age in the Caribbean Sea deduced from Mg/Ca and $^{18}O/^{16}O$ ratios in corals. *Marine Geology*, **173**, 21–35.

Watson, E.B. (2004) A conceptual model for near-surface kinetic controls on the trace-element and stable isotope composition of abiogenic calcite crystals. *Geochimica et Cosmochimica Acta*, **68**, 1473–1488.

Wei, G., Sun, M., Li, X., & Nie, B. (2000) Mg/Ca, Sr/Ca and U/Ca ratios of a porites coral from Sanya Bay, Hainan Island, South China Sea and their relationships to sea surface temperature. *Palaeogeography Palaeoclimatology Palaeoecology*, **162**, 59–74.

Zinke, J., Dullo, W.C., Heiss, G.A. & Eisenhauer, A. (2004) ENSO and Indian Ocean subtropical dipole variability is recorded in a coral record off southwest Madagascar for the period 1659 to 1995. *Earth and Planetary Science Letters*, **228**, 177–194.

EMU Notes in Mineralogy, Vol. 10 (2010), Chapter 12, 399–415

Ion partitioning and trace-element proxies in foraminifera and coccoliths

HEATHER M. STOLL

Department of Geology, University of Oviedo, C/Arias de Velasco, s/n, 33005 Oviedo, Spain
e-mail: hstoll@geol.uniovi.es

Foraminifera and coccolithophorids are protists and marine algae, respectively, and produce most of the carbonate in the open ocean. The trace-element composition in these shells represents the longest-term archive of past ocean conditions. The trace-element composition of foraminifera may record spatial or temporal changes in ocean chemistry in cases where ion partitioning is relatively constant compared to the spatial or temporal chemical heterogeneity in the trace-element/Ca ratio in the ocean. As an example, variation in Cd/Ca ratios in deepwaters, sensitive to deepwater circulation pathways, has been reconstructed from benthic foraminiferal Cd/Ca ratios. Cyclical changes in river input of Ba, due to changes in precipitation patterns, has been inferred from Ba/Ca ratios in surface dwelling foraminifera. In contrast, thermodynamic controls appear to strongly influence Mg partitioning in calcite, with increased Mg incorporation at higher temperatures forming the basis of the Mg/Ca paleothermometer. Seawater *pH* strongly influences the speciation of trace elements B and U and their resulting incorporation in foraminiferal calcite. Foraminiferal growth rate appears to influence the partitioning of Sr into the calcite, potentially due to surface enrichment effects. The segregated intracellular calcification of coccolithophores in special vesicles gives rise to unique biological effects in effective partitioning in Sr and B, which are hypothesized to result from variations in cellular transport of these elements.

1. Introduction

The production of calcium carbonate in the ocean is predominantly modulated by organisms. In the open oceans today, foraminifera (unicellular animals) and coccolithophorids (microalgae) produce the great majority of the carbonate. Both groups of organisms produce calcite predominantly, although a few species of benthic foraminifera produce aragonitic shells. In the modern ocean, carbonate production by both groups is roughly comparable, whereas prior to the mid-Cenozoic, coccolithophorids produced >90% of the carbonate sediment found in most open ocean settings, including the great Mesozoic chalk sequences covering many parts of Europe. This chapter focuses exclusively on calcite biominerals from foraminifera and coccoliths because of the overwhelming dominance of this mineralogy. The reader is referred to the chapter by Cohen (2010) for treatment of ion partitioning in coralline biogenic aragonites.

DOI: 10.1180/EMU-notes.10.12

The successive stratigraphic deposition of foraminiferal and coccolith carbonate in ocean basins over very long geological successions, together with the advent of drilling of deep ocean sediments in the 1960s, has led to continued and intensified interest in using the isotopic and elemental chemistry of these carbonates to reconstruct changes in Earth-surface processes over time. The ocean is a great melting pot of surface geochemical cycling, the chemistry of which reflects the temporal variations in processes such as weathering of continental materials, hydrothermal alteration of mid-ocean ridges, climate, ocean circulation, and biological activity in the ocean. Can the ion partitioning in foraminifera and coccoliths serve as faithful recorders of these processes?

Foraminifera and coccolithophorid algae biomineralize with distinct mechanisms. Foraminifera precipitate carbonate using endocytosis of seawater for Ca and dissolved inorganic carbon (DIC) supply. Seawater is transported in a vacuole to the area where a new chamber is to be added and *pH* is elevated prior to calcite precipitation (de Nooijer *et al.*, 2009a, 2009b). Planktonic foraminifera add a chamber on average once every 24–48 h. A primary organic membrane nucleates the first calcification of the chamber wall, which proceeds by minor growth on the interior surface and more significant growth on the outer surface, including a growth layer of new calcite that covers all previously deposited chambers (Erez, 2003). The final layer of calcite produced in this way is termed gametogenic calcite as it immediately precedes the release of gametes and the demise of the adult foraminifera. The biomineralization process has been the subject of a recent detailed review (Dove *et al.*, 2003).

Coccolithophorid algae form calcite coccoliths intracellularly in a specialized vesicle, the coccolith vesicle, which does not contain advected seawater. Coccoliths are produced one at a time in the coccolith vesicle and subsequently extruded through the cell membrane to form an interlocking layer of plates covering the cell exterior (Anning *et al.*, 1996). The coccolith vesicle has a small volume and must be refilled many times with Ca in the course of production of a single coccolith (Langer *et al.*, 2006).

In this chapter, the concepts behind some of the most important indicators based on minor and trace element partitioning into these biogenic carbonates are reviewed. The emphasis is on the divalent cations which substitute readily for Ca in the calcite lattice such as Mg, Sr, Ba, Cd and anions which are envisioned to substitute for the CO_3^{2-} ion in the calcite lattice such as $B(OH)_3^-$ and $UO_2(CO_3)_2^{-2}$. The three most common approaches for employing ion partitioning in studies of past ocean conditions are described: (1) to infer element/Ca ratios of seawater; (2) to infer the temperature of the ocean; and (3) to infer the *pH* or carbonate ion chemistry of the ocean. Finally, some uniquely biological inferences that can be made from ion partitioning in coccoliths, and some solutions to interpret multiple influences on ion partitioning in foraminiferal carbonates are discussed.

Empirical data on ion partitioning in calcite and models of partitioning are typically established for conditions where calcite is precipitated from a reservoir which is essentially infinite relative to the mass of calcite precipitated, and in which solution conditions are typically assumed to be at steady state. In the surface ocean, the mass of calcite particles is very small relative to seawater, so that if biological processes were insignificant, the ion partitioning might be expected to be analogous to the experimental conceptions. However, in the cases of partitioning in the biominerals, we are always referring to an 'apparent partitioning' relative to seawater as the fluid. In reality, biomineralization occurs from a discrete reservoir isolated by the organism, and this fluid composition may differ from seawater but in most cases cannot be characterized independently. Consequently, the apparent partitioning includes true partitioning effects between the biomineralizing fluid and the calcite, as well as biases that arise from differences between seawater and the biomineralizing fluid composition.

2. Recording element/Ca ratios of the seawater

The use of element/Ca ratio of the carbonate to infer the element/Ca ratio in the solution (*i.e.* the ocean) is one of the most widely used concepts of ion partitioning in marine carbonates. Strictly speaking, the ion-partitioning coefficient between the solution (seawater) and carbonate is not constant but rather a function of solution chemistry, as expressed in Roozeboom diagrams (see Prieto, 2010). However, for many ions of interest in the ocean the abundance of the minor component is orders of magnitude smaller than that of Ca. In the case of such dilute solid solutions, the variation in partitioning coefficient is expected to be more restricted. Consequently, most authors have adopted the concept of a single partitioning coefficient expected to apply over the range of seawater compositions investigated.

The residence time of Ca in the ocean is long, of the order of 2 million years, compared to the 1000 year timescale for mixing of the ocean. Consequently Ca concentrations in the modern ocean vary by $<2\%$, showing only a slight depletion in the surface ocean due to biogenic carbonate formation. Over geological timescales of millions of years, significant variations in Ca concentration could arise, but most studies focus on shorter intervals of time for which Ca concentrations can be assumed essentially constant. Commonly, the interest is in trace elements which have short residence times compared to ocean mixing and therefore strong spatial and temporal gradients in composition in the ocean. Two examples are the Cd concentration in deep seawater and the Ba concentration in surface seawater.

The Cd concentration of seawater varies significantly in the deep ocean because Cd is taken up in the production of biological particles in the surface ocean and gradually released back to seawater as the sinking biogenic particles remineralize in the deeper ocean and the concentration of their components accumulates along the flow pathway of deepwater in the ocean (Boyle, 1991). The correlation of Cd concentration with important biological nutrients like P is so strong that the Cd/Ca ratios in foraminiferal carbonate is proposed as an indicator of nutrient content in the deep ocean (Boyle, 1995). Boyle showed that generally the Cd/Ca ratio of bottom-dwelling (benthic) foraminifera throughout the deep ocean follows the enrichments in Cd and P in deep water along the deep circulation pathway.

In coastal regions there is a strong inverse correlation between the Ba concentration of seawater and the salinity of the surface seawater because riverine runoff is enriched in Ba relative to seawater (Fig. 1a). Consequently Ba/Ca ratios in foraminifera are

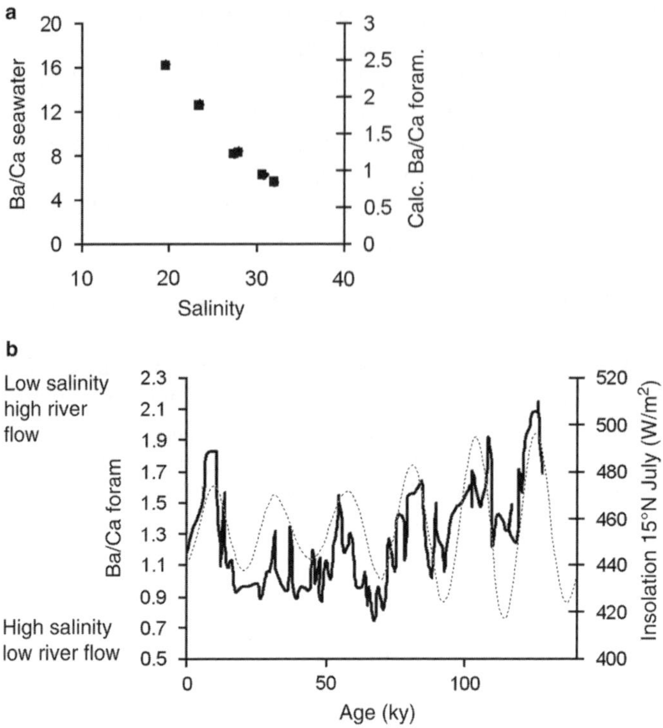

Fig. 1. (***a***) Ba/Ca ratio in seawater off the Congo delta correlated to sea surface salinity, and the predicted Ba/Ca ratio in planktic foraminifera shells (seawater data from Weldeab *et al.* 2007); calculation of equilibrium foraminiferal composition derived from Lea & Spero (1994). Seawater Ba/Ca units and foraminifera units are both μmol/mol. (***b***) Bold black curve: measured Ba/Ca ratio in planktic foraminifera from a sediment core off the Congo delta over the last 130,000 years, interpreted as salinity variations (data from Weldeab *et al.*, 2007). Dashed grey curve indicates the variation in summer (July) insolation at 15°N latitude which follows the Earth's 23,000 year precession of the orbit. High summer insolation increases the intensity of the monsoon rainfall, the total river flux, and causes minima in salinity in the delta region of the Congo.

proposed as an indicator of river runoff and past changes in the hydrological cycle. In laboratory cultures with seawater modified to have variable Ba/Ca ratios, foraminiferal Ba/Ca ratios covary with those of seawater (Lea & Spero, 1994) although the apparent partitioning coefficient of 0.147 is greater than expected for abiogenic calcite precipitation from seawater. Because the range of seawater Ba/Ca ratios is large, any influence of temperature or carbonate ion concentration on partitioning is minor and foraminifera appear to faithfully record seawater chemistry. This relationship was recently used to reconstruct past variations in river input from the tropical Congo river system draining into the Gulf of Guinea over the last 130,000 years (Fig. 1b) (Weldeab *et al.*, 2007). Ba/Ca data in foraminifera indicate that the annual river flow here is controlled by summer monsoon rainfall which closely follows changes in intensity of summer sunlight in the northern tropical regions.

Coccolith carbonate chemistry also reflects mean seawater element ratios. In coccolith calcite grown in culture, the Ba/Ca ratio of the calcite covaries with the Ba/Ca ratio in seawater (Langer *et al.*, 2009). The apparent partitioning coefficient of 0.1 is greater than that expected for abiogenic calcite precipitation from seawater. A similar result was shown for Sr/Ca ratio of coccolith calcite, with average partitioning coefficients of 0.35, also greater than abiogenic precipitates. In both cases the binding reaction with the organic template coccolith-associated polysaccharides is hypothesized to increase the apparent partitioning (Langer *et al.*, 2006). However, in the case of coccolith Sr/Ca, variations in seawater Sr/Ca ratios are small and biological partitioning effects are the dominant control over partitioning, as detailed below.

3. Recording temperature of the ocean

Temperature may be a strong influence on the ion partitioning of some elements in calcite. The most notable example is Mg. Thermodynamic effects drive a strong increase in Mg partitioning in calcite with increasing temperature. Thermodynamic data estimated from the heat of reaction of the Mg substitution in calcite suggest that the Mg/Ca ratio of a thermodynamically ideal calcite will increase exponentially by $\sim$3% increase in Mg/Ca per °C (Koziol & Newton, 1995). The experimental precipitation fits well to an exponential function with this slope (Oomori *et al.*, 1987).

The Mg/Ca ratio of foraminiferal calcite grown in laboratory cultures does exhibit a strong temperature dependence (Nurnberg *et al.*, 1996; Lea *et al.*, 1999). A strong correlation is also observed between the temperature of the surface ocean and Mg/Ca of surface-dwelling foraminifera recovered from recent (last few thousand year) ocean sediments (*e.g.* Fig. 2a). However, the absolute Mg/Ca ratios are at least an order of magnitude smaller than expected from equilibrium partitioning from seawater at comparable temperatures. This observation suggests that foraminifera reduce the Mg concentration of the calcifying fluid, either by pumping Mg out and/or complexation (de Nooijer *et al.*, 2009b). Reduction in Mg content would reduce the inhibition of Mg on calcite precipitation and therefore reduce the saturation state required to form the calcite test. Variability in the efficiency and mechanism of Mg removal in different species may be partly responsible for the range of Mg contents among different species of foraminifera. The consistency of the relationship between Mg/Ca ratios and temperature for a given species implies that the Mg concentration in the calcifying fluid is maintained constant at different temperatures. Alternatively, some authors have suggested that the Mg temperature dependence in foraminifera may not arise from direct Mg substitution into a homogeneous foraminiferal shell but rather by variation in the relative proportion of Mg-rich and Mg-poor calcite types which may covary with temperature (Erez, 2003).

Foraminiferal Mg/Ca ratios have been used widely as an indicator of temperature changes in the tropical oceans over glacial cycles (Fig. 2b) (Lea *et al.*, 2000; Visser *et al.*, 2003). As expected, Mg/Ca peaks during the warm interglacial periods, like the present day, and decreases during the cold glacial maxima, *e.g.* during the most

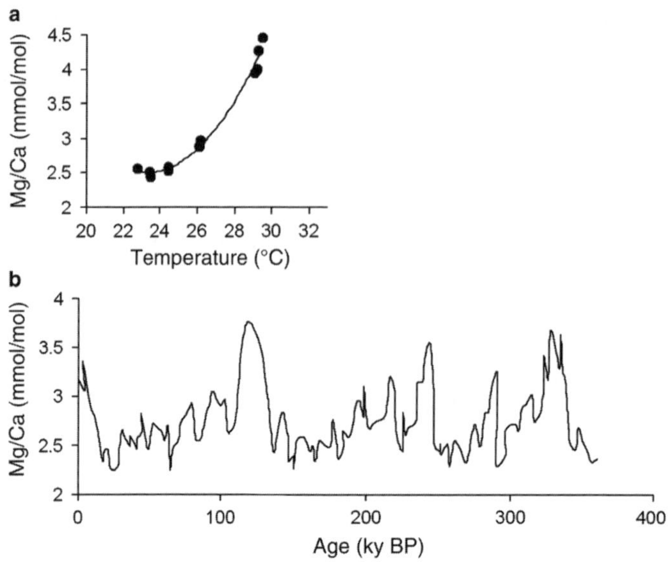

Fig. 2. (***a***) Representative relationship between Mg/Ca ratio and temperature in planktonic foraminifera, in this case *G. ruber* from surface sediment samples in the tropical Pacific region. (***b***) Measured Mg/Ca variations in *G. ruber* from the tropical Pacific (91°W) over the last several glacial cycles (all data from Lea *et al.*, 2000).

recent glacial maximum 21,000 years ago. According to the regional calibration shown in Figure 2a, these Mg/Ca changes indicate temperatures 2.8°C cooler than present during the last glacial maximum. The amplitude of temperature change across some previous glacial–interglacial cycles is as much as 5°C. These calculated temperature changes are considered much more realistic than previous estimates based on foraminiferal assemblages, which indicated negligible temperature change in the tropics during glacial cycles. The Mg/Ca data have been used to estimate the sensitivity of tropical temperature to atmospheric CO_2 oscillation from 180 to 280 ppm over glacial cycles.

Coccolith carbonate from a given strain produced in culture shows greater Mg/Ca ratios at higher temperatures, but absolute Mg/Ca concentrations vary by species and even strains of a given species (Stoll *et al.*, 2001). The absolute Mg/Ca ratios are an order of magnitude lower than those of foraminiferal carbonate. Because coccoliths must transport all cations used for calcification to the coccolith vesicle *via* active transport mechanisms, they appear to have developed strong selection against Mg uptake into the calcification space.

4. Recording *pH* and carbonate-ion concentration of the ocean

The *pH*-dependent speciation of some ions in the ocean could give rise to significant changes in partitioning of ions in calcite. Boron incorporation in calcite is strongly dependent on solution *pH*. Boron in seawater exists in two species, $B(OH)_3$ dominant at lower *pH* and $B(OH)_4^-$ dominant at higher *pH*, with a pK of 8.6 at 24°C. Because borate is the dominant species incorporated into the calcite lattice (Hemming *et al.*, 1995), B incorporation in calcite is greater for higher solution *pH*, as has been observed in abiogenic experiments in which the incorporation of B in the carbonate covaries with the proportion of B present as $B(OH)_4^-$ (Sanyal *et al.*, 2000). The inferred mechanism of B incorporation in carbonate

$$CaCO_3 + B(OH)_4^- \longleftrightarrow Ca(HBO_3) + HCO_3^- + H_2O \tag{1}$$

implies an exchange distribution coefficient of:

$$K_D = [B/Ca]_{calcite}/[B(OH)_4^-/HCO_3^-]_{aq} \tag{2}$$

In precipitation of abiogenic carbonate in Mg-free seawater, the partition coefficient ranged from 0.0004 to 0.0005 (Sanyal *et al.*, 2000).

Foraminiferal culture experiments with *O. universa* have shown apparent B partitioning coefficients which range by an order of magnitude, from 0.00033 to 0.0034 (Fig. 3) (Sanyal *et al.*, 1996, 2001). A comparable range was inferred for a number of species of foraminifera collected from surface sediments (Foster, 2008). In all cases the B partitioning is calculated assuming that the $B(OH)_4^-/HCO_3^-$ ratio during precipitation is identical to that of extracellular seawater. The apparent partitioning coefficients covary inversely with the carbonate ion concentration or *pH* of the seawater. One interpretation is that the carbonate ion concentration significantly affects the precipitation mechanisms or surface chemistry of the growing crystal and therefore affects partitioning. Alternatively, recent studies have suggested that foraminifera increase the *pH* of the calcification space to promote oversaturation (de Nooijer *et al.*, 2009b). Consequently, another possible interpretation is that the B incorporation reflects precipitation at the *pH* of the calcifying space which is greater than the seawater *pH* by progressively greater amounts as seawater *pH* decreases. In the culture experiments, the partitioning coefficient of B in *O. universa* at the highest seawater *pH*, the *pH*

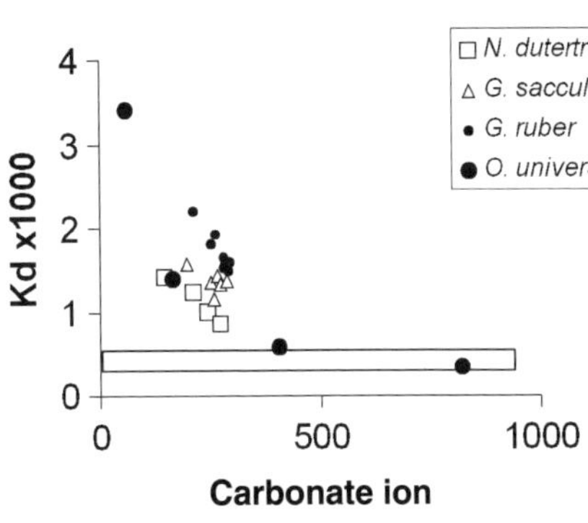

Fig. 3. B partitioning in calcite from three genera of foraminifera collected from recent (surface) sediments, compared to pre-industrial carbonate ion concentration in the seawater at their growth habitat; *N. dutertrei*, *G. sacculifer*, and *G. ruber*. Data from (Foster, 2008). B partitioning is also shown from foraminifera *O. universa* grown in culture (data from Sanyal *et al.*, 1996). In all cases the B partitioning is calculated assuming that the $B(OH)_4^-/HCO_3^-$ ratio during precipitation is identical to that of extracellular seawater. The rectangle indicates the range of B partitioning observed in abiogenic calcite precipitation (Sanyal *et al.*, 2000).

which is probably closest to that of the calcifying space, is 0.00034, close to the value in abiogenic experiments. It is likely that further work to understand biomineralization in foraminifera will be needed to reach robust interpretations of variations in B/Ca in fossil foraminifera. Coccolithophorid uptake of boron appears to be under a high degree of cellular control and is discussed in the subsequent section.

Uranium in seawater is almost completely complexed to carbonate groups. The U in the uranyl UO_2^{2+} ion may be surrounded by 1 to 3 carbonate groups depending on the seawater *pH* (Djogic *et al.*, 1986). At typical seawater *pH* range, most will have tri-carbonate complexation, but substitution in calcite may be dominated by the substitution of the uncharged monocarbonate complex or the doubly charged bicarbonate complex $UO_2(CO_3)_2^{2-}$ (Russell *et al.*, 1994). As the abundance of these two latter complexes depends on the carbonate ion concentration and *pH*, it may impart a *pH* dependence on total U partitioning. Foraminiferal culture experiments exhibit a strong inverse relationship between U/Ca ratios and *pH*. Offsets between species are consistent with elevated near-cell *pH* in symbiont-bearing species which has been inferred from a number of modelling and culture studies (Russell *et al.*, 2004).

5. Growth rate and carbonate ion effects on ion partitioning in foraminifera

In recent years, the partitioning of many other trace elements (Sr, Mg, Cd, Li) in foraminiferal carbonate has been shown to be sensitive to the seawater *pH* or carbonate ion concentration, even though these elements do not have extensive *pH*-dependent speciation in seawater as was the case for B or U. Culture studies of planktic foraminifera, *O. universa*, reveal seawater carbonate ion concentration to be positively correlated with Sr/Ca ratio and negatively correlated with Mg/Ca ratios (Fig. 4) (Russell *et al.*, 2004). Several other planktic species (*G. ruber*, *G. siphonifera*), however, showed no dependence of Sr partitioning on seawater carbonate ion concentration (Kisakurek *et al.*, 2008). Rather, diverse effects of salinity, carbonate ion, and temperature on Sr/Ca in planktic foraminifera *G. ruber* and *G. siphonifera* all define a single positive correlation with calcite production rate (Kisakurek *et al.*, 2008). Given that in several species of planktonic foraminifera, larger carbonate ion concentrations produce more heavily calcified foraminiferal shells, it is possible that some aspects of 'carbonate ion effect' are manifestations of a calcite production rate effect. For example, higher seawater carbonate ion concentration increases the shell wall thickness in *O. universa* and in other species produces a greater shell mass per shell size class (Russell *et al.*, 2004). Alternatively, the carbonate ion concentration in the calcifying space may not be identical to that of seawater, especially since foraminifera elevate the *pH* in the calcification space to enhance calcification (de Nooijer *et al.*, 2009b). Consequently, it is possible that a calcite production rate is in fact a manifestation of variation in the carbonate ion concentration in the calcification reservoir as a response to control of calcification *pH* by the organism.

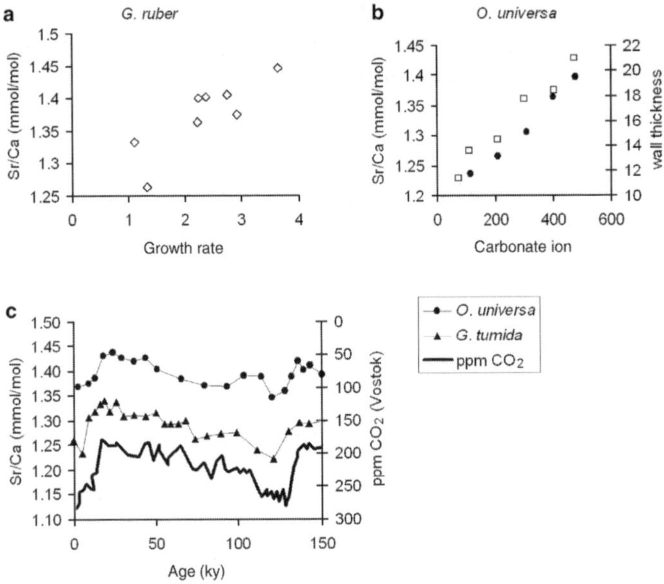

Fig. 4. (*a*) Apparent Sr partitioning as a function of growth rate in foraminifera *G. ruber* grown in a laboratory culture (data from Kisakurek *et al.*, 2008). (*b*) Apparent Sr partitioning as a function of growth rate in foraminifera *O. universa* grown in a laboratory culture (data from Russell *et al.*, 2004). (*c*) Variation in Sr/Ca ratio of foraminifera over the past 130,000 years, which may arise from carbonate ion and/or growth-rate effects (data from Stoll *et al.*, 1999). Also shown is the curve of atmospheric CO_2 from Vostok ice cores (Petit *et al.*, 1999). When atmospheric CO_2 was lower during glacial stages, surface ocean carbonate ion concentration would have been higher, which coincides with elevated Sr/Ca ratios.

Several mechanisms could be invoked to explain the dependence of apparent partitioning coefficients on calcite production rate or carbonate ion concentration. If calcite growth rate is the mechanism, it is typically attributed to a surface enrichment effect (Watson, 1996; Tang *et al.*, 2008). In the surface enrichment model, a different partitioning of trace metals on the surface of the crystal lattice compared to interior lattice sites gives rise to entrapment effects when the growth rate of the crystal is high relative to the diffusion or reorganization of coordination in the near-surface zone of the crystal. This model predicts that partitioning would vary with the linear crystal growth rate. In addition, the degree of surface enrichment of the trace element could vary depending on the *pH*, saturation state, and temperature of the solution. For example, the slope of the partitioning dependence on calcite production rate varies as a function of temperature in abiogenic experiments (Tang *et al.*, 2008).

Aside from the surface enrichment effect, partitioning could be influenced directly by carbonate ion or saturation state because these may change the growth mechanism or step velocity, or the Ca^{2+}/CO_3^{2-} ratio which affects step velocity and partitioning coefficients (Arvidson *et al.*, 2005; Davis *et al.*, 2005).

In benthic foraminifera, a dependence on carbonate ion concentration is inferred principally from the changes in effective partitioning observed in foraminifera living at greater depths in a water mass of homogeneous chemistry. For example Sr/Ca ratios typically decrease and Cd/Ca ratios increase with increasing water depth as carbonate ion concentration and saturation state decrease (Elderfield *et al.*, 1996). One interpretation is that these variations in apparent trace element partitioning could be consistent with variable degrees of depletion of a biomineralization reservoir with initial seawater composition, following Rayleigh Distillation (Elderfield *et al.*, 1996). With greater degrees of depletion, Sr/Ca ratios of the calcifying solution would increase whereas Cd/Ca ratios would decrease. Surprisingly, greater degrees of depletion of the biomineralisation reservoir are inferred for shallower depths which would have a higher saturation state and theoretically require lower degree of depletion to produce a given unit of carbonate. An alternative interpretation is that these two elements have different surface enrichment characteristics and therefore have opposite responses to growth-rate variations (Watson, 2004).

Outside of laboratory cultures, it may be difficult to infer growth rates from foraminifera present in marine sediments. Two approximations have been proposed, size-normalized shell weight, and shell size. Primary foraminifera shell weight is clearly influenced by carbonate saturation during growth in many species, though partial dissolution of shells during their sinking through the water column and repose in the sediments may impart an additional secondary signal on shell weights in many species (Barker *et al.*, 2004). Shell size may be indicative of average growth rates though much debate remains. Some workers suggest that foraminiferal growth is initially fast and then slows towards maturity, inferring that smaller individuals reflect those arrested in the fast-growth stage (Elderfield *et al.*, 2002). In contrast, others suggest that larger shells reflect higher average growth rates (Schmidt *et al.*, 2008). Finally, at least in some settings, greater shell size may reflect greater symbiont activity, reflecting in the ocean a shallower, high-light environment (Spero *et al.*, 2009).

To date, these carbonate-ion effects have been regarded principally as an unwanted secondary influence on ion partitioning which jeopardizes interpretations of temperatures from Mg/Ca or nutrient levels from Cd/Ca. However, with further calibration work they may be useful for deciphering past changes in the ocean's carbonate system. For example, it was suggested that small variations in Sr/Ca ratios over glacial cycles were more likely explained by changes in Sr partitioning due to carbonate ion or other vital effects than changes in seawater chemistry (Fig. 4c). This is because the long residence times of Sr and Ca in the ocean limit the degree of change in seawater Sr/Ca ratio over glacial timescales of 10^4 years (Stoll *et al.*, 1999). The dependence of Sr/Ca on carbonate ion revealed in culture studies suggested that the amplitude of Sr/Ca variation could be due to increased carbonate ion concentrations during glacial times (Russell *et al.*, 2004); increased carbonate ion concentrations would be consistent with the glacial–interglacial atmospheric CO_2 cycles. Recent culture data suggest that a factor of two increase in growth rates of foraminifera could account for the observed 3–5% increase in Sr/Ca ratios of planktic foraminifera (Kisakurek *et al.*, 2008).

6. Biological effects in coccoliths which affect their ion partitioning

6.1. Sr/Ca in coccoliths

Coccolithophorid calcification is fundamentally different from that of coral and foraminifera and therefore gives rise to some unique ion partitioning effects. Comparisons of the Sr/Ca ratio of coccoliths from surface sediments in the Equatorial Pacific showed a much greater variation in ion partitioning compared to those of planktonic foraminifera (Stoll & Schrag, 2000). Higher Sr/Ca ratios correlated with higher average productivity and community calcite production rates. Several culture studies of coccolithophorids reveal that cell growth rates or calcite production rates themselves are not responsible for strong variations in Sr partitioning. Varying light levels for algal cultures, at constant temperature and moderate nutrients, produced a large range in steady-state coccolithophorid cell division and calcification rates, but only modest (10%) increases in coccolith Sr/Ca (Stoll *et al.*, 2002a, 2002b). However, nutrient limitation of growth rates resulted in larger variations in Sr/Ca ratios (Fig. 5a) (Rickaby *et al.*, 2002; Stoll *et al.*, 2007a). These larger-amplitude responses in culture studies are consistent with the range in Sr/Ca ratios observed in coccoliths collected in sediment traps in the ocean spanning periods of peak bloom productivity and periods of low productivity (Fig. 5b) (Stoll *et al.*, 2007a). Culture experiments using variable nutrient concentrations without affecting growth rate do not induce large changes in Sr partitioning (Langer *et al.*, pers. comm.).

Several mechanisms have been proposed to explain the variations in Sr partitioning. Rickaby *et al.* (2002) suggest that there may be a rate-dependent discrimination between the biological transport of Sr^{2+} and Ca^{2+} associated with either passive transport through ion channels or active pumping via carrier proteins. If there were a selectivity against Sr in this symport which were reduced at higher pumping rates, the Sr/Ca of the calcifying fluid could increase. Given that coccolith Sr/Ca does not increase with growth and calcification rate under all culture conditions, the variations in pumping rates must respond to nutrient limitation and not growth rate per se. Alternatively, other workers suggest that the coccolithophorid cell behaves as a relatively closed system with respect to Ca for calcification, using nearly all the Ca and Sr taken up by the cell during calcite precipitation and refilling the calcification reservoir thousands of times during the production of a single coccolith. One hypothesis is that nutrient limitation in coccolithophores triggers carbon dumping of extracellular polysaccarides which preferentially bind extracellular Sr and reduce the Sr/Ca ratio taken into the calcifying vesicle, depressing coccolith Sr/Ca ratios (Langer *et al.*, 2006).

The dependency of coccolith Sr/Ca ratios on productivity in natural settings has been exploited to evaluate past productivity changes over climate cycles. For example, in the Late Quaternary, changes in summer sunlight in the subtropics have driven cycles in upwelling intensity in the Arabian Sea (Fig. 5c) (Stoll *et al.*, 2007b). These upwelling cycles are reflected by changes in productivity in coccolithophorid populations as indicated by Sr/Ca ratios in coccolithophorid *Calcidiscus leptoporus*.

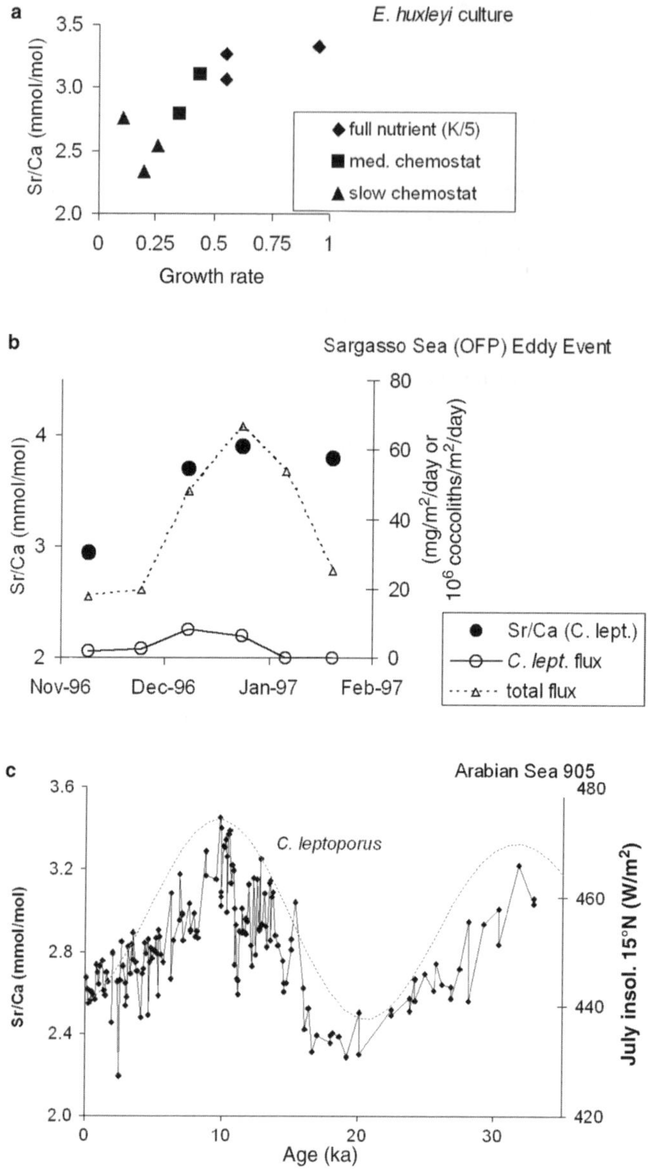

Fig. 5. Coccolith Sr/Ca ratios in (***a***) nutrient-limited continuous cultures of *Emiliania huxleyi*, in which the growth rate was controlled by nitrate availability (data from Stoll *et al.*, 2007). (***b***) Coccolith Sr/Ca ratios of *Calcidiscus leptoporus* collected in sediment traps in the Sargasso Sea before and during a productivity bloom event in the fall of 1996. Sr/Ca ratios increase significantly with the onset of high productivity during the bloom. (***c***) Sr/Ca ratios in *C. leptoporus* from a sediment core in the Arabian Sea spanning the last 30,000 years, interpreted as productivity cycles. Productivity is greatest during periods of high summer insolation, which drive a strong temperature gradient between land and ocean and hence stronger wind-driven upwelling to fuel productivity.

6.2. B/Ca in coccoliths

Because coccoliths are produced entirely intracellularly, the uptake of boron into the cell may differ significantly from that of organisms which precipitate calcite from seawater vacuoles. B uptake by the coccolithophorid cell is expected to be exclusively by boric acid diffusion across the lipid bilayer of the cell membrane (Stoll *et al.*, submitted). Alternative sources such as uptake of the charged borate species require special transporters which would entail a higher energetic cost to the cell. Once inside the cell, boron is likely to speciate to borate (in addition to boric acid) only within the alkaline coccolith vesicle, with the abundance of borate depending on the coccolith vesicle *pH*. There, as in other carbonates, the borate would partition into coccolith calcite. Boron behaviour in coccolithophorid biomineralization is thus a two-tiered system. The first tier is as boric acid uptake into the cell, in which the uptake would depend on the extracellular concentration of boric acid which decreases as seawater *pH* increases. The second tier, borate speciation in the coccolith vesicle, results in increased borate as the *pH* of the coccolith vesicle rises. Thus, if *pH* in the coccolith vesicle is constant, then coccolith B/Ca ratios decrease with increasing seawater *pH* (Fig. 6). Covariation

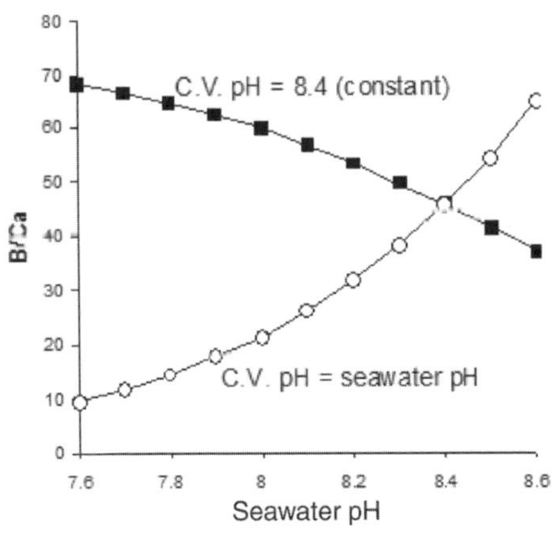

Fig. 6. Modelled B/Ca ratio in coccolith calcite (in μmol/mol) as a function of seawater *pH*, under two different assumptions – either constant *pH* (homeostasis) in the coccolith vesicle, or *pH* in the coccolith vesicle identical to that of seawater. The model assumes that boric acid is taken up into the cell and speciates to borate (along with boric acid) in the coccolith vesicle. The model assumes a constant K_D of 0.0005 in the calcite. Calculations derived from *Stoll et al.* (submitted).

of coccolith vesicle *pH* with seawater *pH* would result in increasing B/Ca with increasing seawater *pH*. Examples of both types of responses may be seen in cultures of coccolithophorids produced at different seawater *pH* (Stoll *et al.*, submitted). If seawater *pH* is independently constrained, the B/Ca ratio may give information on the *pH* of the calcification vesicle which may be correlated to calcification rate.

7. Solutions to multiple influences on ion partitioning

An ideal situation for palaeoceanographic studies would be a trace element, partitioning of which was regulated by a single variable (temperature or seawater composition).

A collection of trace elements, partitioning of each of which was controlled by single, independent variables, might give a complete picture of past changes in the ocean. The reality, however, is that trace-element partitioning is typically regulated by multiple factors such as temperature, salinity, carbonate ion concentration, or ocean chemistry. Over many timescales, these factors are closely coupled. For example, over Quaternary glacial cycles, ocean temperature fell along with atmospheric CO_2 which raised the carbonate ion concentration of the surface ocean, at the same time as growth of ice sheets raised the mean salinity of the ocean. Thus it is impractical to assume that one factor varied while others remained constant.

In this reality, the best solution is to combine multiple trace element systems, or trace element and isotopic systems, to fully constrain the multiple influences on partitioning. Over glacial cycles, the coupled use of Mg/Ca ratios and oxygen isotopic ratios in foraminifera is showing promise; both are affected by temperature and salinity, but to differing degrees.

Another elegant solution is the coupling of Mg/Ca and Li/Ca. Both are influenced by temperature and carbonate-ion concentrations but temperature dominates Mg/Ca and carbonate ion concentration dominates Li/Ca. In a global database of portioning in recent (surface sediment) foraminifera it has been possible to establish two calibration equations with two unknowns to apply the dual system in the palaeo-record (Lear & Rosenthal, 2006). This approach, together with careful selection of sites likely to experience minimal changes in carbonate ion concentration over time, has proven crucial to accurate diagnosis of the sequence of cooling and ice-sheet growth across the Eocene–Oligogene boundary. This important transition from 'greenhouse' to 'icehouse' climates was previously inferred to reflect mainly the growth of ice sheets both in Antartica and the Northern Hemisphere, because Mg/Ca ratios in benthic foraminifera remained stable across the transition and were interpreted as reflecting no cooling (Lear *et al.*, 2000). Subsequently it was suggested that the cooling effect expected in Mg/Ca (a decrease in Mg/Ca) was offset by an increase in the deep-sea carbonate ion concentration (increased Mg/Ca due to improved preservation of the more soluble, high Mg fraction of the shell). Work in shallow sites unaffected by the carbonate-ion concentration increase did indeed reveal the expected Mg/Ca decrease consistent with cooling (Lear *et al.*, 2008), and consistent with early glaciation only on the polar continent of Antarctica. Ongoing studies by Lear and others are coupling Li/Ca and Mg/Ca in open ocean sites to confirm the interpretation.

References

Anning, T., Nimer, N., Merrett, M.J. & Brownlee, C. (1996) Costs and benefits of calcification in coccolithophorids. *Journal of Marine Systems*, **9**, 45–56.

Arvidson, R.S., Davis, K.J. & Luttge, A. (2005) The relationship between cation/anion ratio, step velocities and bulk dissolution rate in calcite. *Geochimica et Cosmochimica Acta*, **69**, A779.

Barker, S., Kiefer, T. & Elderfield, H. (2004) Temporal changes in North Atlantic circulation constrained by planktonic foraminiferal shell weights. *Paleoceanography*, **19**, PA3008.

Boyle, E. (1995) Last-glacial-maximum North Atlantic deep-water – On, off or somewhere in-between. *Philosophical Transactions of the Royal Society of London Series B – Biological Sciences*, **348**, 243–253.

Boyle, E.A. (1991) Quaternary Ocean Paleochemistry. *Reviews of Geophysics*, **29**, 634–638.

Cohen, A.L. & Gaetani, G.A. (2010) Ion partitioning and the geochemistry of coral skeletons: solving the mystery of the vital effect. In: *Ion Partitioning in Ambient Temperature Aqueous Systems* (M. Prieto, editor). EMU Notes in Mineralogy, **10**, xx–xx. European Mineralogical Union and the Mineralogical Society of Great Britain & Ireland.

Davis, K.J., Arvidson, R.S. & Luttge, A. (2005) Resolving the role of $Ca^{2+}/CO_{2}{}_{3}^{2-}$ ratio in calcite dissolution and growth: consequences for biomineral formation. *Abstracts of Papers of the American Chemical Society*, **229**, 020–GEOC.

de Nooijer, L.J., Langer, G., Nehrke, G. & Bijma, J. (2009a) Physiological controls on seawater uptake and calcification in the benthic foraminifer Ammonia tepida. *Biogeosciences*, **6**, 2669–2675.

de Nooijer, L.J., Toyofuku, T. & Kitazato, H. (2009b) Foraminifera promote calcification by elevating their intracellular pH. *Proceedings of the National Academy of Sciences of the United States of America*, **106**, 15374–15378.

Djogic, R., Sipos, L. & Branica, M. (1986) Characterization of Uranium(Vi) in Seawater. *Limnology and Oceanography*, **31**, 1122–1131.

Dove, P.M., De Yoreo, J.J. & Weiner, S., editors (2003) *Biomineralization*. Reviews in Mineralogy and Geochemistry, **54**. Mineralogical Society of America, Chautilly, Virginia and The Geochemical Society, St-Louis, Missouri, USA.

Elderfield, H., Bertram, C.J. & Erez, J. (1996) Biomineralization model for the incorporation of trace elements into foraminiferal calcium carbonate. *Earth and Planetary Science Letters*, **142**, 409–423.

Elderfield, H., Vautravers, M. & Cooper, M. (2002) The relationship between shell size and Mg/Ca, Sr/Ca, delta O-18, and delta C-13 of species of planktonic foraminifera. *Geochemistry, Geophysics, Geosystems*, **3**, 1052.

Erez, J. (2003) The source of ions for biomineralization in foraminifera and their implications for paleoceanographic proxies. In: *Biomineralization* (P.M. Dove, S. Weiner & J.J. De Yoreo, editors), Mineralogical Society of America, Washington D.C., 115–149.

Foster, G.L. (2008) Seawater pH, PCO_2 and $[CO_3^{2-}]$ variations in the Caribbean Sea over the last 130 kyr: A boron isotope and B/Ca study of planktic forminifera. *Earth and Planetary Science Letters*, 271, 254–266.

Hemming, N.G., Reeder, R.J. & Hanson, G.N. (1995) Mineral-fluid partitioning and isotopic fractionation of boron in synthetic calcium-carbonate. *Geochimica et Cosmochimica Acta*, **59**, 371–379.

Kisakurek, B., Eisenhauer, A., Bohm, F., Garbe-Schonberg, D. & Erez, J. (2008) Controls on shell Mg/Ca and Sr/Ca in cultured planktonic foraminiferan, Globigerinoides ruber (white). *Earth and Planetary Science Letters*, **273**, 260–269.

Koziol, A.M. & Newton, R.C. (1995) Experimental determination of the reactions magnesite plus quartz equals enstatite plus CO_2 and magnesite equals periclase plus CO_2, and enthalpies of formation of enstatite and magnesite. *American Mineralogist*, **80**, 1252–1260.

Langer, G., Gussone, N. & 7 coauthors (2006) Coccolith strontium to calcium ratios in Emiliania huxleyi: The dependence on seawater strontium and calcium concentrations. *Limnology and Oceanography*, **51**, 310–320.

Langer, G., Nehrke, G., Thoms, S. & Stoll, H. (2009) Barium partitioning in coccoliths of Emilliania huxleyi. *Geochimica et Cosmochimica Acta*, **73**, 2899–2906.

Lea, D.W. & Spero, H.J. (1994) Assessing the reliability of paleochemical tracers – barium uptake in the shells of planktonic formainifera. *Paleoceanography*, **9**, 445–452.

Lea, D.W., Mashiotta, T.A. & Spero, H.J. (1999) Controls on magnesium and strontium uptake in planktonic foraminifera determined by live culturing. *Geochimica et Cosmochimica Acta*, **63**, 2369–2379.

Lea, D.W., Pak, D.K. & Spero, H.J. (2000) Climate impact of late quaternary equatorial Pacific sea surface temperature variations. *Science*, **289**, 1719–1724.

Lear, C.H. & Rosenthal, Y. (2006) Benthic foraminiferal Li/Ca: Insights into Cenozoic seawater carbonate saturation state. *Geology*, **34**, 985–988.

Lear, C.H., Elderfield, H. & Wilson, P.A. (2000) Cenozoic deep-sea temperatures and global ice volumes from Mg/Ca in benthic foraminiferal calcite. *Science*, **287**, 269–272.

Lear, C.H., Bailey, T.R., Pearson, P.N., Coxall, H.K. & Rosenthal, Y. (2008) Cooling and ice growth across the Eocene–Oligocene transition. *Geology*, **36**, 251–254.

Nurnberg, D., Bijma, J. & Hemleben, C. (1996) Assessing the reliability of magnesium in foraminiferal calcite as a proxy for water mass temperatures. *Geochimica et Cosmochimica Acta*, **60**, 803–814.

Oomori, T., Kaneshima, H., Maezato, Y. & Kitano, Y. (1987) Distribution coefficient of Mg^{2+} ions between calcite and solution at 10–50°C. *Marine Chemistry*, **20**, 327–336.

Petit, J.R. and 18 co-authors (1999) Climate and atmospheric history of the past 420,000 years from the Vostok ice core, Antarctica. *Nature*, **399**, 429–436.

Prieto, M. (2010) Thermodynamics of ion partitioning in solid solution-aqueous solution systems. In: *Ion Partitioning in Ambient Temperature Aqueous Systems* (M. Prieto, editor). EMU Notes in Mineralogy, **10**, xx–xx. European Mineralogical Union and the Mineralogical Society of Great Britain & Ireland.

Rickaby, R.E.M., Schrag, D.P., Zondervan, I. & Riebesell, U. (2002) Growth rate dependence of Sr incorporation during calcification of Emiliania huxleyi. *Global Biogeochemical Cycles*, **16**, 1006, doi: 10.1029/2001GB001408.

Russell, A.D., Emerson, S., Nelson, B.K., Erez, J. & Lea, D.W. (1994) Uranium in foraminiferal calcite as a recorder of seawater uranium concentrations. *Geochimica et Cosmochimica Acta*, **58**, 671–681.

Russell, A.D., Honisch, B., Spero, H.J. & Lea, D.W. (2004) Effects of seawater carbonate ion concentration and temperature on shell U, Mg, and Sr in cultured planktonic foraminifera. *Geochimica et Cosmochimica Acta*, **68**, 4347–4361.

Sanyal, A., Hemming, N.G., Broecker, W.S., Lea, D.W., Spero, H. & Hansen, G.N. (1996) Oceanic pH control on the boron isotopic composition of foraminifera: Evidence from culture experiments. *Paleoceanography*, **11**, 513–517.

Sanyal, A., Nugent, M., Reeder, R.J. & Buma, J. (2000) Seawater pH control on the boron isotopic composition of calcite: Evidence from inorganic calcite precipitation experiments. *Geochimica et Cosmochimica Acta*, **64**, 1551–1555.

Sanyal, A., Bijma, J., Spero, H. & Lea, D.W. (2001) Empirical relationship between pH and the boron isotopic composition of Globigerinoides sacculifer: Implications for the boron isotope paleo-pH proxy. *Paleoceanography*, **16**, 515–519.

Schmidt, D., Elliot, T. & Kasemann, S. (2008) The influences of growth rates on planktonic foraminifers as proxies for palaeostudies, a review. In: *Biogeochemical Controls on Palaeoceanographic Environmental Proxies* (W.E.N. Austin and R.H. James, editors). Special Publication, **303**. Geological Society of London, 73–85.

Spero, H.J., Schmidt, M.W., Lea, D.W. & Lavagnino, L. (2009) Evidence for a southern pattern of deglacial surface warming in the Eastern Equatorial Pacific. *Transactions of the American Geophysical Union*, **90**, Fall Meeting Supplement, Abstract PP14C–06.

Stoll, H.M. & Schrag, D.P. (2000) Coccolith Sr/Ca as a new indicator of coccolithophorid calcification and growth rate. *Geochemistry, Geophysics, Geosystems*, **1**, 1–24.

Stoll, H.M., Schrag, D.P. & Clemens, S.C. (1999) Are seawater Sr/Ca variations preserved in Quaternary foraminifera? *Geochimica et Cosmochimica Acta*, **63**, 3535–3547.

Stoll, H.M., Encinar, J.R., Alonso, J.I.G., Rosenthal, Y., Probert, I. & Klaas, C.M. (2001) A first look at paleotemperature prospects from Mg in coccolith carbonate: Cleaning techniques and culture measurements. *Geochemistry, Geophysics, Geosystems*, **2**, 1047, doi:10.1029/2000GC000144.

Stoll, H.M., Klaas, C.M., Probert, I., Encinar, J.R. & Alonso, J.I.G. (2002a) Calcification rate and temperature effects on Sr partitioning in coccoliths of multiple species of coccolithophorids in culture. *Global and Planetary Change*, **34**, 153–171.

Stoll, H.M., Rosenthal, Y. & Falkowski, P. (2002b) Climate proxies from Sr/Ca of coccolith calcite: Calibrations from continuous culture of Emiliania huxleyi. *Geochimica et Cosmochimica Acta*, **66**, 927–936.

Stoll, H.M., Ziveri, P., Shimizu, N., Conte, M.H. & Theroux, S. (2007a) Relationship between coccolith Sr/Ca ratios and coccolithophore production and export in the Arabian Sea and Sargasso Sea. *Deep Sea Research II*, doi: 10.1016/j.dsr2.2007.01.003.

Stoll, H.M., Shimizu, N., Archer, D. & Ziveri, P. (2007b) Coccolithophore productivity response to greenhouse event of the Paleocene–Eocene thermal maximum. *Earth and Planetary Science Letters*, **258**, 192–206.

Stoll, H.M., Langer, G., Shimizu, N. & Kanamaru, K. (submitted) B/Ca ratios in coccoliths and relationship to calcification vesicle pH and dissolved inorganic carbon. *Geochimica et Cosmochimica Acta.*

Tang, J.W., Kohler, S.J. & Dietzel, M. (2008) Sr^{2+}/Ca^{2+} and Ca-44/Ca-40 fractionation during inorganic calcite formation: I. Sr incorporation. *Geochimica et Cosmochimica Acta*, **72**, 3718–3732.

Visser, K., Thunnell, R. & Stott, L.D. (2003) Magnitude and timing of temperature change in the Indo-Pacific warm pool during deglaciation. *Nature*, **421**, 152–155.

Watson, E.B. (1996) Surface enrichment and trace-element uptake during crystal growth. *Geochimica et Cosmochimica Acta*, **60**, 5013–5020.

Watson, E.B. (2004) A conceptual model for near-surface kinetic controls on the trace-element and stable isotope composition of abiogenic calcite crystals. *Geochimica et Cosmochimica Acta*, **68**, 1473–1488.

Weldeab, S., Lea, D.W., Schneider, R.R. & Andersen, N. (2007) 155,000 years of West African monsoon and ocean thermal evolution. *Science*, **316**, 1303–1307.

EMU Notes in Mineralogy, Vol. 10 (2010), Index, 417–420

Subject and Author Index

Numbers refer to the page where a definition or explanation of and/or a *figure* or a **table** on a given subject is found. Author names are given with the number of the first page of the paper in which they are involved.

DOI: 10.1180/EMU-notes.10.index